Crystal-Field Engineering of Solid-State Laser Materials

This book is concerned with the underlying science and design of laser materials. It emphasizes the principles of crystal-field engineering and discusses the basic physical concepts that determine laser gain and nonlinear frequency conversion in optical crystals.

A concise review of the essential science is presented, including symmetry groups, the implications of covalency and the important theories of electronic structure, energy transfer, radiationless transitions, and the spectra of laser-active centres in crystals. The predictive capabilities of crystal-field engineering are developed to show how modification of the symmetry and composition of optical centres can improve laser performance. Applications of the principles of crystal-field engineering to a variety of optical crystals are also discussed in relation to the performances of laser devices.

This book will be of considerable interest to physical, chemical and materials scientists and to engineers involved in the science and technology of solid state lasers. It will be used by senior undergraduate and postgraduate students as well as by established scientists.

BRIAN HENDERSON was born on March 26, 1936. He was educated at the University of Birmingham, graduating with a B.Sc. and Ph.D. in Physical Metallurgy. He was appointed a Senior Scientific Officer at AERE Harwell (1962) where he researched the physical and spectroscopic properties of defects in n-irradiated insulators. In 1968 he was appointed Reader in Physics at the University of Keele. In 1974 he moved to Trinity College Dublin where, as Head of the Department of Physics, he established a research school in solid-state physics. He returned to the UK in 1984 where as Head of the Physics Department in the University of Strathclyde in Glasgow he established a world class facility for the growth of laser and optically nonlinear crystals. During this time he was also elected Dean of the Faculty of Science and Vice-Principal of the University. His researches have yielded more than 220 publications, which includes 5 books, 17 critical reviews, 150 refereed papers, 20 conference papers and 30 abstracts in conference digests. He supervised the thesis research of 30 Ph.D. students at the Universities of Keele and Strathclyde and Trinity College Dublin.

RALPH H. BARTRAM, born on 16 August 1929, earned his academic degrees at New York University: B.A. (1953), M.S. (1956) and Ph.D. (1960). He was employed in industrial research and development in microwave electronics at Sylvania Electric Products, Inc. and GTE Laboratories Inc. from 1953 to 1961. He joined the Faculty at the University of Connecticut in 1961, was appointed Professor in 1971, and served as department head from 1986 to 1992. He supervised 14 Ph.D. dissertations and retired in 1992. He is currently Professor Emeritus. Professor Bartram has more than 100 research publications and a comparable number of abstracts. He has four patents on microwave devices and is co-author of two books. His academic honorary memberships include Phi Beta Kappa, Sigma Xi, Phi Kappa Phi, Sigma Pi Sigma and Phi Eta Sigma. His biography is listed in American Men and Women of Science and in Who's Who in the East, Who's Who in America, Who's Who in the World and Who's Who in Science and Engineering.

CAMBRIDGE STUDIES IN MODERN OPTICS

Series Editors

P. L. KNIGHT
Department of Physics, Imperial College of Science, Technology and Medicine

A. MILLER
Department of Physics and Astronomy, University of St. Andrews

Crystal-Field Engineering of Solid-State Laser Materials

TITLES IN PRINT IN THIS SERIES

Holographic and Speckle Interferometry (2nd ed.)
R. Jones and C. Wykes

The Elements of Nonlinear Optics
P. N. Butcher and D. Cotter

Optical Solitons – Theory and Experiment
edited by J. R. Taylor

Particle Field Holography
C. S. Vikram

Ultrafast Fiber Switching Devices and Systems
M. N. Islam

Optical Effects of Ion Implantation
P. D. Townsend, P. J. Chandler, and L. Zhang

Diode–Laser Arrays
edited by D. Botez and D. R. Scifres

The Ray and Wave Theory of Lenses
A. Walther

Design Issues in Optical Processing
edited by J. N. Lee

Atom–Field Interactions and Dressed Atoms
G. Compagno, R. Passante, and F. Persico

Compact Sources of Ultrashort Pulses
edited by I. Duling

The Physics of Laser–Atom Interactions
D. Suter

Optical Holography – Principles, Techniques, and Applications (2nd ed.)
P. Hariharan

Theoretical Problems in Cavity Nonlinear Optics
P. Mandel

Measuring the Quantum State of Light
U. Leonhardt

Optical Properties of Semiconductor Nanocrystals
S. V. Gaponenko

Vertical-Cavity Surface-Emitting Lasers
edited by C. Wilmsen, H. Temkin, and L. Coldren

Crystal-Field Engineering of Solid-State Laser Materials

BRIAN HENDERSON
University of Cambridge

RALPH H. BARTRAM
University of Connecticut

CAMBRIDGE
UNIVERSITY PRESS

CAMBRIDGE UNIVERSITY PRESS
Cambridge, New York, Melbourne, Madrid, Cape Town, Singapore, São Paulo

Cambridge University Press
The Edinburgh Building, Cambridge CB2 2RU, UK

Published in the United States of America by Cambridge University Press, New York

www.cambridge.org
Information on this title: www.cambridge.org/9780521593496

First published 2000
This digitally printed first paperback version 2005

A catalogue record for this publication is available from the British Library

Library of Congress Cataloguing in Publication data

Henderson, B.
Crystal-field engineering of solid-state laser materials / Brian Henderson, Ralph H. Bartram
p. cm.–(Cambridge studies in modern optics)
ISBN 0 521 59349 2
1. Solid-state lasers–materials. 2. Optical materials. 3. Crystals. I. Bartram, Ralph H.
II. Title III. Cambridge studies in modern optics (Unnumbered)
QC374.H46 2000
621.36′61–dc21   99-054437

ISBN-13 978-0-521-59349-6 hardback
ISBN-10 0-521-59349-2 hardback

ISBN-13 978-0-521-01801-2 paperback
ISBN-10 0-521-01801-3 paperback

Contents

Preface

Almost four decades have passed since Maiman invented the ruby ($Cr^{3+}: Al_2O_3$) laser. The intervening years have witnessed many major achievements and much innovation, culminating in a plethora of devices that impinge in many walks of life. Fundamental studies and technology have been pursued with vigour, separately and in tandem, with resulting applications in materials and device processing, optical, communication and information sciences, and medical and paramedical sciences. Such applications have the potential to materially transform our everyday living at home, at work and at leisure. Optical devices that were once regarded as revolutionary but which are taken now for granted include the fibre laser, responsible for dramatically improving the quality and cost of intercontinental telephone calls, and the compact disc, now the ubiquitous storage medium for optical information. The scope of optical materials discussed in this monograph is very broad, indicating the potential sweep of exciting and novel developments in the short and long term futures.

Ruby lasers of the kind invented by Maiman, and discussed in this monograph, operate on the Cr^{3+} R-line emission near 694 nm: they are pumped by the ruby's absorption of some of the visible light from a xenon flashlamp. The He–Ne laser was reported soon thereafter, since when laser action has been observed on many thousands of transitions in the gas phase, including 200 or so on neutral Ne alone. Pulsed N_2 (337 nm) and CW He–Ne lasers (633 nm, 1.15 μm and 3.39 μm), the Ar^+ (488 nm, 514 nm) and Kr^+ ion (647 nm) lasers and the CO (10 μm) laser are among the numerous commercially developed gas lasers. The many gas, liquid and solid state lasers currently available provide coherent light outputs that can be single frequency, tunable, continuous wave (CW) or pulsed. Irrespective of the gain medium or output all lasers involve the processes of stimulated absorption, stimulated emission and spontaneous emission that were identified by Einstein in 1917. Tremendous advances have been made in the development of solid state lasers since the invention of the ruby laser. Complicit in these developments have been diverse theoretical and experimental scientists, technologists and engineers. A major attraction of this field has been its multi-disciplinarity, which requires a mastery of quantum mechanics and electromagnetic theory, of spectroscopy and optics, of materials science, crystallography and crystal growth as well as diverse elements of engineering. Research and development funding from industry, government agencies, research councils and foundations provides a dynamism that continues to attract many inventive researchers.

This text does not deal with laser engineering. Rather does it discuss the basic physical concepts that determine laser gain and nonlinear frequency conversion in optical crystals and their applications in solid state lasers. As a precursor to the main themes, Chapter 1 presents a purview of the operating principles of lasers, comparing three-level with four-level pumping, CW with pulsed operation and low power with high power lasers. Some of the discussion and theoretical results from subsequent chapters are anticipated and used to this end. Chapter 2 introduces the reader to formal group theory, describing the actions of symmetry operators that facilitate the codification of the atomic structures of solids into crystal classes, Bravais lattices, point-group and space-group symmetries. This chapter also develops representation theory and the relation between group theory and quantum mechanics. Such abstract concepts are a cornerstone of crystal field engineering; they are used in Chapter 3 to describe the crystal structures of natural and synthetic minerals and of some gemstones. The structures, compositions and symmetries of important optical crystals are then reviewed ahead of detailed descriptions of experimental techniques for growing optical crystals from melts and from fluxes. Some specialized techniques of more limited application, such as hydrothermal growth and laser heated pedestal growth are described more briefly.

Subsequently, the predictive capabilities of crystal field engineering are developed to show how modifications to the symmetry and composition of optical centres can improve laser performance. Chapter 4 reviews the theory of the electronic structures of transition-metal and rare-earth metal ions and of colour centres in ionic crystals and the contributions to their energy eigenvalues by free-ion terms and spin–orbit, electron–electron and crystal–field interactions. The influence of ionic coordination on the symmetry of the Hamiltonian and the classification of eigenvalues and eigenfunctions of electrons at impurities and point defects in solids is emphasized. The theoretical apparatus needed to understand optical transitions on laser-active centres in inorganic solids is presented in Chapter 5 and used there to determine the selection rules and line shapes of crystal field spectra of transition-metal and rare-earth metal ions in the presence of electron–phonon coupling. Chapter 6 describes the nature of radiationless transitions, with examples from the optical spectra of impurity ions and colour centres, including luminescence quenching of laser-active centres. The essential features of excitation transfer, excited state absorption and upconversion luminescence are described in Chapter 7. The significance of covalency to the electronic structure of transition-metal ions and its importance in determining the spectroscopic properties of solids, which was recognized by Van Vleck (1935) soon after the development of quantum mechanics, cannot be ignored in laser physics. Molecular orbital theories that include ion-size effects through the covalent overlap of wavefunctions on neighbouring ions are described in Chapter 8 and applied to energy level and spectroscopic line-shape calculations of impurity ions and point defects. The extensions of crystal-field and molecular-orbital theories using computer-based techniques to simulate the long-range extension of optical centres into their ionic environment are described and illustrated by application to laser crystals activated by Ti^{3+}, Cr^{3+} and Cr^{4+} ions.

The principles of crystal-field engineering, applied to the development of materials for application in solid-state lasers, are surveyed in Chapter 9. The broadening of optical transitions by static and dynamic distortions of the environment are discussed in relation to the wavelength tunability of solid-state lasers, using the theoretical bases developed in Chapters 4–6. Laser efficiency and threshold are discussed prior to consideration of empirical rules for designing practical laser systems. Odd-parity distortions are shown to enhance radiative transition rates and to induce the optical nonlinearities associated with frequency conversion processes. The consequent roles of structure–property relationships in optically nonlinear crystals such as the niobates, phosphates and borates are also developed. Finally, the applications of the concepts of crystal-field engineering to optical crystals are discussed in Chapter 10, consideration being given to a wide range of material systems, some successful and others not so successful, in their application in laser devices.

The present text has been written not only for researchers but also for senior undergraduate and postgraduate students with some understanding of optics, quantum physics, spectroscopy and condensed matter sciences, subjects which the authors have taught for more than three decades. It is a pleasure to record our indebtedness to friends and colleagues whose stimulating comradeship over these many years has helped us to develop the subject matter of this monograph, especially those who have permitted us to use their original material. In this monograph the theoretical bases of crystal-field engineering are illustrated with many examples of their applications in solid-state lasers, which nevertheless constitute a small subset of topics in the literature. The selections of topics are personal ones and the authors are cognizant of much excellent work to which reference has not been made. The authors are grateful to Professor Willy Firth, at whose suggestion the task of writing this book was undertaken. One of us (B.H.) owes particular debts of gratitude to Professor Michael Pepper and the Semiconductor Physics Group at the Cavendish Laboratory, Cambridge University, where he worked on the manuscript as an Academic Visitor, and to Roy Taylor (Imperial College) and Mitsuo Yamaga (Gifu University) for much helpful advice on the contents of Chapters 9 and 10. Since patience is indeed a virtue, we thank also Drs Philip Meyler and Simon Capelin at Cambridge University Press: who had enough to keep us up to the task long after others would have torn up the contract! And to our wives, who had patience even more abundantly, for their forbearance, support, care and love, we offer heartfelt thanks.

Brian Henderson
Cambridge

Ralph Bartram
Connecticut

1

An introduction to lasers

1.1 Historical notes

Laser is an acronym for Light Amplification by Stimulated Emission of Radiation. The operating principles of the laser were originally elucidated for devices operating at microwave frequencies (masers). The maser was invented by Gordon, Zeiger and Townes (1955), who used an inverted population between the excited vibrational levels of ammonia. Extension of the wavelength range of the maser into the optical regime was proposed by Schawlow and Townes (1958), whose various laser schemes all comprised a gain medium, an excitation source to pump atoms or ions in the gain medium into higher energy levels and a mirror feedback system to enable one or multiple passes of the emitted radiation through the laser medium. The special qualities that distinguish laser light from other optical sources include extreme brightness, monochromaticity, coherence and directionality. Also the laser output is linearly polarized and very frequency stable. After development over forty years modern lasers operate over wavelength ranges from the mid-infrared, through the visible and beyond into the ultraviolet and vacuum ultraviolet ranges.

The first operational laser used synthetic rubies, corundum crystals containing ~0.1 wt.% of Cr_2O_3, as the gain medium, pumped with white light from a helical flashlamp to oscillate on the sharp R-line at a wavelength of 693.4 nm [Maiman (1960)]. Soon afterwards the He–Ne laser was operated on the $3s \rightarrow 2p$ (632.8 nm), $2s \rightarrow 2p$ (1.15 μm) and $3s \rightarrow 3p$ (3.39 μm) transitions of atomic Ne. Since that time 203 laser lines have been identified with neutral Ne involving transitions in the energy level spectrum up to $n = 7$ as well as more than 1300 laser lines from atoms and ions of 50 or so elements in the gas phase. These include the familiar wavelengths featured in commercial laser devices such as the Ar^+ laser (479 nm, 488 nm and 514 nm) and Kr^+ ion laser (647 nm) and the He–Cd laser (325 nm, 354 nm and 442 nm). Gas lasers operate on sharp line transitions: they are not wavelength tunable. This is also the case for hundreds of thousands of laser emissions based on vibrational–rotational transitions of molecules in the gas phase. The CO_2 laser operating at 10.6 μm is one of the most efficient (> 30%) and powerful lasers currently available, yielding CW powers of more than 100 kW and pulsed energies exceeding 10 kJ. The second solid state system to be successfully operated as a laser was $Sm^{2+}:CaF_2$, followed shortly after by $Tm^{2+}:CaF_2$ and $Dy^{2+}:CaF_2$. The ease of crystal growth in these cases allied to the ability to control the divalent state of the rare-earth ion was important in their early development. Since then almost all the trivalent rare-earth ions have served as the optical centre in lasers involving a rapidly increasing diversity of host crystals.

Fixed wavelength operation is the characteristic of the many trivalent rare-earth ion (RE^{3+}) lasers, such as the Nd^{3+} : glass and Nd^{3+} : $Y_3Al_5O_{12}$ (YAG) lasers, that began to appear in the early 1960s. The Nd^{3+}-doped solid state lasers operate primarily at *ca* 1.06 μm, and less efficiently at 946 nm, 1.358 μm and 1.833 μm. Other RE^{3+}-activated lasers, including Pr^{3+}, Tb^{3+}, Ho^{3+}, Er^{3+} and Tm^{3+}, also operate at multiple, fixed wavelengths in the visible and near-infrared wavelength ranges, in a variety of suitable solid state environments. Apart from YAG, $YLiF_4$ (YLF) is one of the most useful crystals, but there are many others, including $CaWO_4$, $LiNbO_3$, YVO_4, LaF_3, $YAl_3(BO_3)_4$, $YAlO_3$, and the fluoroapatites and fluorovanadates of Ca^{2+} and Sr^{2+}. A number have been commercially developed in bulk and niche markets. The most successful have been Nd^{3+}- and Er^{3+}-doped materials, including glasses of numerous compositions and crystals such as YAG, $YLiF_4$ (or YLF), and yttrium orthovanadate (YVO_4). Although these rare-earth ion doped crystals may support laser action on several transitions, they are all fixed wavelength laser systems. This follows from their rich energy level structure in which the $4f^n$ electrons occupy compact orbitals and are shielded from the neighbouring ionic environment by the less energetic $5s$ and $5p$ outer shells of electrons. In consequence, the spectra of the $4f^n$ ions of the rare-earths are not strongly affected by the static or dynamic positions of neighbouring ligands. In a first approximation the energy levels of the RE^{3+} ions are characterized by free ion L, S and J values: the weak effects of the static crystal-field is to split these levels approximately about the centre of gravity of each crystal-field multiplet. Hence, there is very great similarity between the spectra of a rare-earth ion over many different host crystals.

Soon after the announcements of the single frequency ruby [Maiman (1960)] and Nd-YAG [Geusic *et al.* (1964)] lasers, Johnson *et al.* (1962, 1963) reported laser action at cryogenic temperatures in the broad emission bands of Ni^{2+}- and Co^{2+}-doped crystals of MgF_2. The commercial development of these wavelength tunable lasers was suppressed for more than two decades by the success of ruby, Nd : YAG and Nd : glass lasers as well as the evolution of many organic dye lasers. The gain medium in a dye laser consists of an organic dye molecule (e.g. rhodamine 6G) dissolved in a suitable solvent (water, alcohol, etc.). A large number of dye molecules are used to realize lasers covering a wavelength spectrum from 320–1500 nm, with each dye providing a gain bandwidth of *ca* 30–80 nm. For example, the Rh6G dye laser is continuously tunable over wavelengths from *ca* 560 nm to 640 nm. Combining a diffraction grating or prism-tuning element with such a broad emission bandwidth allows the laser output to be continuously tuned over the entire emission spectrum with a laser linewidth less than 10 GHz (or *ca* 10^{-3} nm). Usually dye lasers are pumped with flashlamps, Ar^+ or Kr^+ lasers, or even frequency-doubled or frequency-tripled Nd-YAG lasers.

The renaissance in the development of solid state lasers began with the F_A centre laser [Mollenauer and Olson (1974)], and was dramatically advanced by the announcement of the alexandrite laser [Walling *et al.* (1979)]. Prior to that the only commercial solid state lasers were the fixed wavelength ruby and Nd : YAG lasers. Efficient, broadband tuning was the domain of Ar (or Kr) ion pumped dye lasers. Although the potential of phonon-terminated transitions in four-level lasers was recognized soon after the invention of the ruby laser, they had been little developed. The Cr^{3+} : $BeAl_2O_4$ laser operates up to 400 K without loss in efficiency so that cryogenic cooling and precautions against optical and/or thermal bleaching are unnecessary. At the band peak (730 nm) the slope efficiency is close to being theoretically optimal. It benefits from the close proximity of ^2E and $^4T_2$ levels, the former being a population storage level that feeds the emission in the $^4T_2 \rightarrow {}^4A_2$ vibronic band. However, excited state absorption (ESA) transitions reduce the laser efficiency outside the band peak [Walling *et al.* (1985)].

That alexandrite, a much admired gemstone, should be used as a tunable laser gain medium was hardly surprising, given the earlier developments of the ruby and YAG lasers. Indeed, the optical properties of ruby ($Cr^{3+}:Al_2O_3$) were studied for more than two centuries before Maiman (1960) invented the ruby laser, and in consequence its quantum electronic properties were better understood than those of any other crystal. Furthermore, good quality synthetic crystals were available at modest cost. Nevertheless, the three-level pumping cycle (§1.3) would not have supported laser action had the luminescence quantum efficiency of $Cr^{3+}:Al_2O_3$ been much less than 100%. Diamond was also studied extensively by spectroscopic methods prior to much later reports of laser action in the UV based on the *H*3 centre [Rand and DeShazer (1985)]. Of course, diamonds, rubies, garnets and sapphires have rarity value as gemstones and research into their optical properties was natural. Indeed, many natural minerals have characteristics required of laser gain media, including good thermomechanical and optomechanical properties. Nevertheless, in the nineteen-fifties and sixties, apart from consulting the literature [Burns (1993), Sugano *et al.* (1970)], there was little to guide the designers of commercial laser systems. This is no longer the case, given many experimental and theoretical innovations over the past thirty years. The development of techniques for growing high quality single crystals has been especially important.

Since the first commercial exploitations of tunable solid state lasers with the appearance of colour centre and alexandrite ($Cr^{3+}:BeAl_2O_4$) lasers, many novel solid state lasers have been developed, some tunable, others single frequency. Only a few have found their way into the market place. Where tunability is deemed essential the Ti-sapphire laser has been most successful, being tunable from *ca* 710 nm to 1.1 μm. The $Cr^{3+}:LiSrAlF_6$ laser also appears to have all the ingredients for commercial success, especially when pumped by laser diodes. The most successful single frequency lasers have been Nd^{3+}-doped YAG, YLF and glasses. The development of glass fibre and semiconductor diode lasers has been dramatic, fuelling the needs of the communications and information technology industries.

Developments that parallel the applications of solid state lasers across the wavelength range from *ca* 300–1200 nm have also proceeded with novel optically nonlinear crystals for harmonic frequency generation, optical parametric oscillation and amplification. Such crystals are capable of efficient extension of the wavelength ranges of many solid state lasers. Franken *et al.* (1961) first reported optical nonlinear interactions in α-quartz, excited by a Nd:YAG laser at 1.064 μm to generate a second harmonic beam with wavelength of 532 nm. This was soon followed by the theoretical formulation of nonlinear optics by Bloembergen (1965) and his colleagues [Armstrong *et al.* (1962), Bloembergen and Shen (1964)]. Since that time a considerable body of theoretical and experimental research into the phenomenology and materials of nonlinear optics has made this a mature branch of the optical sciences [Shen (1984), Butcher and Cotter (1990) and Chen (1993)].

Nonlinear optical crystals play important roles in solid state lasers by extending their operating wavelength ranges into the ultraviolet and infrared ranges inaccessible to conventional lasers. Theoretical models of the mechanisms leading to enhanced non-linear susceptibilities led to families of compounds being researched for potential uses as harmonic generators, optical parametric oscillators and amplifiers, sum and difference frequency mixing and optical rectification. Particularly noteworthy for their applicability in practical devices have been the ferroelectric perovskites ($LiNbO_3$, $KNbO_3$), the dihydrogen phosphates and arsenates (KH_2PO_4, $NH_4H_2PO_4$), the titanyl phosphates and arsenates ($KTiOPO_4$, $CsTiOAsO_4$) and the crystalline borates (β-Ba_2BO_4, LiB_3O_5). That this is but a small subset of the many nonlinear crystals studied in research laboratories is a vivid reflection of the fact that few such crystals are deployed in commercial lasers.

No one crystal has the requisite optical and nonlinear optical properties to be the material of choice for the entire wavelength range from 150–3000 nm. Indeed, particular wavelengths require compromises to be made in respect of material properties. Such matters are discussed briefly in subsequent chapters, in terms of the structural origins of laser gain and nonlinear frequency conversion in the odd-parity distortions of atomic building blocks of common optical materials. In this chapter a synopsis of the essential phenomenology of solid state lasers is preceded by a simplified discussion of the underlying physical concepts which are rigorously developed in subsequent chapters that detail theoretical and experimental manifestations of optical interactions in condensed matter.

1.2 Principles of lasers

1.2.1 Stimulated and spontaneous radiative transitions

In his epoch making paper Einstein (1917) identified three fundamental interactions between a radiation field and an atomic system: stimulated absorption, spontaneous emission and stimulated emission. A parallel beam of linearly polarized and monochromatic radiation of intensity $I_0(\omega, \hat{\varepsilon})$ is attenuated exponentially with distance, x, in an absorbing medium according to the Beer–Lambert law (§5.1.4),

$$I(\omega, \hat{\varepsilon}) = I_0(\omega, \hat{\varepsilon}) \exp[-\alpha(\omega, \hat{\varepsilon})x], \tag{1.1}$$

where the coefficient of stimulated absorption is given by

$$\alpha(\omega, \hat{\varepsilon}) = A(\omega, \hat{\varepsilon}) \left[N_a \left(\frac{g_b}{g_a} \right) - N_b \right] \frac{V}{v}, \tag{1.2}$$

and $A(\omega, \hat{\varepsilon})$ is the spontaneous transition probability for emission into the volume V of a photon of mode $(\omega, \hat{\varepsilon})$, v is the velocity of light in the medium, the N are population densities in the ground (a) and excited states (b) and the g are the statistical weights of the states. In high symmetry sites in crystals the *absorption coefficient* should be independent of polarization and is given as

$$\alpha(\omega) = \frac{\pi^2 c^2}{\pi^2 \omega^2} \left(N_a \frac{g_b}{g_a} - N_b \right) A_{ba}\, g(\omega), \tag{1.3}$$

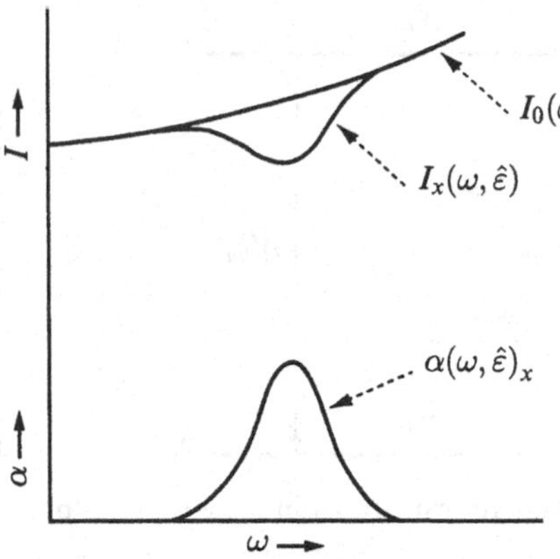

Figure 1.1. The intensity of a beam of polarized radiation $I_0(\omega, \hat{\varepsilon})$ is reduced on passing through a solid sample of length x [adapted from Henderson and Imbusch (1989)].

where $g(\omega)$ is a shape function which recognizes that the absorption covers a band of frequencies. The *absorption strength* is defined by

$$\int \alpha(\omega)\, d\omega = \frac{\lambda^2}{4n^2}\left(\frac{g_b}{g_a}N_a - N_b\right)A_{ba}$$
$$\cong \frac{N_a g_b \lambda^2}{4 g_a n^2} A_{ba}, \tag{1.4}$$

since, if the light beam is not very intense, $N_a \gg N_b$. The absorption coefficient $\alpha(\nu)$, defined in terms of the frequency of radiation (ν) instead of the angular frequency (ω) is written as

$$\int \alpha(\nu)\, d\nu = \frac{1}{2\pi}\int \alpha(\omega)\, d\omega \cong \frac{N_a g_b \lambda^2}{8\pi g_a n^2} A_{ba}. \tag{1.5}$$

Experimentally, $I_0(\nu, \hat{\varepsilon})$ represents the transmission of the system in the absence of the absorbing medium. In practice $I_0(\nu, \hat{\varepsilon})$ and $I(\nu, \hat{\varepsilon})$ are compared electronically in a spectrophotometer and the absorption coefficient $\alpha(\nu, \hat{\varepsilon})$ is found from

$$\alpha(\nu, \hat{\varepsilon})x = \frac{1}{2.303}\log_{10}\frac{I_0(\nu, \hat{\varepsilon})}{I(\nu, \hat{\varepsilon})}, \tag{1.6}$$

The variations of $I_0(\nu, \hat{\varepsilon}), I(\nu, \hat{\varepsilon})$ and $\alpha(\nu, \hat{\varepsilon})$ with wavelength are shown schematically in Fig. 1.1. Finally, the absorption strength is related to the absorption cross section, σ_{ab}, through

$$\int \alpha(\nu)\, d\nu = N\left(\frac{g_b}{g_a}\right)\left(\frac{\lambda_0^2}{8\pi n^2}\right)A_{ba}N\sigma_{ab}, \tag{1.7}$$

where λ_0 is the vacuum wavelength at the centre of the absorption band.

Three radiative transitions illustrated in Fig. 1.2 connect ground (a) and excited (b) states, assuming radiation at the Bohr frequency condition appropriate to atomically

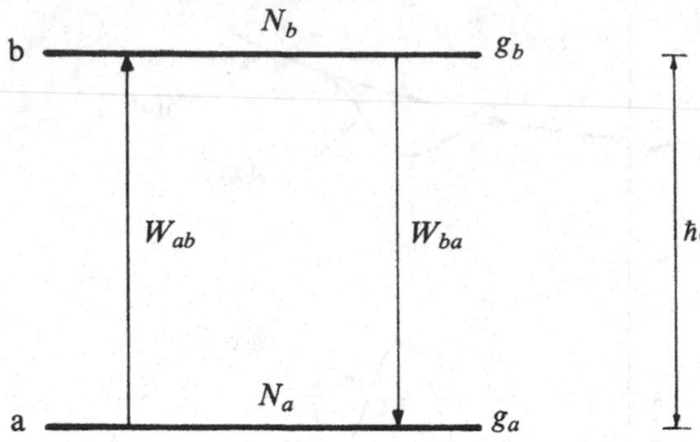

Figure 1.2. Radiative transitions in a two-level atomic system.

sharp resonance lines, Eqs. (5.61) and (5.85),

$$E_b - E_a = \hbar\omega. \tag{1.8}$$

Einstein (1917) had specified that the probability of a transition stimulated by the electromagnetic field was proportional to the energy density of the radiation field at the transition frequency, $u(\omega)$. The probability of exciting absorption transitions between states a and b is then given by

$$W_{ab} = B_{ab}u(\omega), \tag{1.9}$$

where the constant of proportionality, B_{ab}, labels the stimulated absorption coefficient. The analogous expression for the emission probability between states b and a is

$$W_{ba} = B_{ba}u(\omega) + A_{ba}. \tag{1.10}$$

The first term in Eq. (1.10) represents the emission stimulated by the radiations, to which is added the probability of the spontaneous emission (coefficient A_{ba}) which would occur in the absence of the introduced radiation. The rate equations that govern the population densities N_a and N_b of levels a and b are then

$$\frac{dN_a}{dt} = -N_aB_{ab}u(\omega) + N_b(B_{ba}u(\omega) + A_{ba}), \tag{1.11}$$

and

$$-\frac{dN_a}{dt} = \frac{dN_b}{dt}. \tag{1.12}$$

Since the stimulated absorption (g_aB_{ab}) and emission (g_bB_{ba}) rates are equal, and $dN_a/dt = 0$ at equilibrium, Eq. (1.11) leads to

$$\frac{A_{ba}}{B_{ba}} = \left(\frac{g_bN_a}{g_aN_b} - 1\right)u(\omega). \tag{1.13}$$

Assuming that the energy density in the radiation field, $u(\omega)$, is given by Planck's blackbody radiation formula (§5.1.5)

$$u(\omega) = \frac{\hbar\omega^3/\pi^2 v^3}{\exp(\hbar\omega/kT) - 1}, \tag{1.14}$$

it is evident from a comparison of $u(\omega)$ in Eqs. (1.13) and (1.14) that

$$\frac{N_a}{N_b} = \frac{g_a}{g_b} \exp\left(\frac{\hbar\omega}{kT}\right), \tag{1.15}$$

and that

$$\frac{A_{ba}}{B_{ba}} = \frac{\hbar\omega^3}{\pi^2 v^3}. \tag{1.16}$$

The total emission transition probability in Eq. (1.10) is given by

$$W_{ba} = A_{ba}\left[1 + \frac{1}{\exp(\hbar\omega/kT) - 1}\right] = A_{ba}[1 + n_\omega(T)]. \tag{1.17}$$

Similarly, the transition probability due to stimulated emission is

$$W_{ab} = B_{ab}u(\omega) = \left(\frac{g_b}{g_a}\right) A_{ba} n_\omega(T), \tag{1.18}$$

where $n_\omega(T)$ is the photon occupancy in the electromagnetic mode at angular velocity ω [Loudon (1983)]. This is just the number of photons at angular velocity ω at equilibrium in a blackbody cavity radiator at temperature T. The relationship between the Einstein A_{ba} coefficient and the radiative life time follows from Eq. (1.12), with $u(\omega) = 0$, i.e.

$$\frac{dN_b}{dt} = -N_b A_{ba},$$

which on integrating over time gives

$$N_b(t) = N_b(0)\exp(-A_{ba}t).$$

Since the radiant energy per second is $I(\omega)_t = A_{ba}N_b(t)\hbar\omega$ it follows that at time $t > 0$ the intensity decays exponentially according to

$$I(\omega)_t = I(\omega)_0 \exp(-t/\tau_R), \tag{1.19}$$

where the radiative life time (τ_R) is equal to the inverse of the Einstein A_{ba} constant. Equation (1.19) implies that, after switching off an excitation source at time $t = 0$, the emitted radiation intensity $I(\omega)$ will decay exponentially over time, with time constant equal to τ_R. Generally this is only the case for samples containing low concentrations of optical centres and measurements made at low temperature, where *energy transfer* and *nonradiative decay* are less important. As discussed in Chapters 6 and 7, luminescence decay patterns are more complex for crystals co-doped with several different impurities. For example, $Nd^{3+}:Cr^{3+}$:YAG samples excited by flashlamp at 77 K clearly show that the Cr^{3+} R-line emission decays radiatively with time constant τ_R, see Fig. 7.16. However, the Nd^{3+} emission shows contributions from the fast radiative decay of isolated Nd^{3+} and others excited by Cr^{3+}–Nd^{3+} energy transfer. Such excitation transfer enhances the efficiency of many rare-earth ion laser systems.

1.2.2 *Light amplification by stimulated emission*

Equation (1.3) shows that a beam of light passing through an absorbing sample is always attenuated. This is expressed in frequency terms by the optical absorption coefficient $\alpha(\nu)$ given by

$$\alpha(\nu) = \frac{\lambda^2}{8\pi n^2 \tau_R}\left[\left(\frac{g_b}{g_a}\right)N_a - N_b\right]g(\nu). \tag{1.3a}$$

In thermal equilibrium the inequality $N_a \gg N_b$ is satisfied and $\alpha(\nu)$ is positive. However, if $\alpha(\nu)$ is negative then the intensity in the beam will increase. This is achieved when the excited state population exceeds the ground state population, i.e. $N_b > (g_b/g_a)N_a$: the excess population in the excited state is

$$\Delta N = N_b - \frac{g_b}{g_a}N_a > 0, \tag{1.20}$$

showing that there is a *population inversion* in the excited atomic system, and accordingly the intensity $I(\nu)$ increases such that

$$I(\nu) = I_0(\nu)\exp[\gamma(\nu)x], \tag{1.21}$$

where the *gain coefficient per pass*, $\gamma(\nu)$, is given by

$$\gamma(\nu) = \frac{\lambda^2 \Delta N}{8\pi n^2 \tau_R}g(\nu). \tag{1.22}$$

In practice the spontaneous emission makes little contribution to the gain; since it radiates isotropically, it depopulates the upper state without adding to the amplification in the direction of the beam. To amplify the beam requires that feedback is introduced in the optical system. This is most simply achieved by adding highly reflecting parallel mirrors to the sample. Emitted photons are then reflected back and forth along the axis of the sample. This simple arrangement of mirror–sample–mirror is referred to as a *laser cavity*. In practice the end mirrors may, or may not, be attached to the laser rod. The spontaneously emitted photons are lost through the cavity walls, except for those travelling along the cavity axis which may *stimulate* emission from other atoms in their excited state. Since these photons are created by the stimulated emission they travel in the direction of the stimulating photon. In consequence, for each passage of the beam through the active medium there is a gain of intensity given by Eq. (1.21). The single pass gain of the medium, G_M, is defined by

$$G_M \geq I(x)/I_0 = \exp[\gamma(\nu)x], \tag{1.23}$$

where x is the length of the medium. If the loss per pass is $\exp(-\Delta)$, then the modified single pass gain is given by

$$G_M \geq I(x)/I_0 = \exp[\gamma(\nu)x - \Delta], \tag{1.24}$$

implying that amplification of the beam occurs when $G_M \geq 1$; this defines the laser threshold condition as

$$(\Delta/x) \leq \gamma(\nu).$$

Accordingly, the gain afforded by stimulated emission per single pass of the beam through the medium must exceed the losses in the cavity. This *threshold condition* is re-defined as

$$\frac{\lambda^2 c \Delta N \tau_c}{8\pi n^3 \tau_R} g(\nu) \geq 1, \tag{1.25}$$

where τ_c is the cavity decay time defined in terms of the time taken for the beam to decrease to e^{-1} of its original intensity. Since in a time t the beam makes ct/nl passes through the medium, the intensity after this time, in the absence of gain, may be written as

$$I(t) = I_0 \exp(-\Delta)^{ct/nl} = I_0 \exp(-t/\tau_c), \tag{1.26}$$

where $\tau_c = nl/c\Delta$. Obviously the processes contributing to the loss Δ must be minimized in order to reduce the pump intensity at threshold.

1.2.3 Major loss mechanisms

The most obvious loss from the amplified beam of stimulated emission occurs at the mirrors. Usually one mirror is made to be 100% reflective at the laser wavelength. The useful laser radiation is usually taken out of the cavity at the second mirror, which is usually less than 100% reflective. The reflection loss after two passes through the medium is just $R_1R_2 \simeq \exp[-(1-R_1R_2)]$. There are other losses by diffraction and by scattering and absorption processes: these we treat together as a loss per pass of $\exp(-\beta)$. The total loss per pass is therefore

$$\Delta = \left[\frac{1}{2}(1 - R_1R_2) + \beta\right]. \tag{1.27}$$

There are several different absorption losses. The sample may contain other optical centres that have absorption and/or emission bands that overlap those of the laser active centre. There was a problem of infrared absorption in early Ti-sapphire laser crystals which overlapped the emission band, thus reducing the efficiency and increasing the threshold power required to initiate laser operation. *Excited state absorption* can limit laser performance. Since the emitting state of the laser-active centre may be quite long-lived ($\tau_R > 10^{-6}$), the excited centre can absorb another pump or emission photon and be excited into an even higher eigenstate of the system. These excited state absorptions can be sufficiently strong that laser action may be precluded or at best seriously affected. Excited state absorption may also lead to upconverted luminescence.

1.3 Types of solid state laser

1.3.1 Three- and four-level lasers

Although the discussion of stimulated and spontaneous transitions only considered a two-level system, population inversion cannot be achieved by resonantly pumping two such levels. A small number of lasers (e.g. ruby) are said to be three-level lasers. Almost all others, and the most useful ones, are four-level lasers. Before considering specific

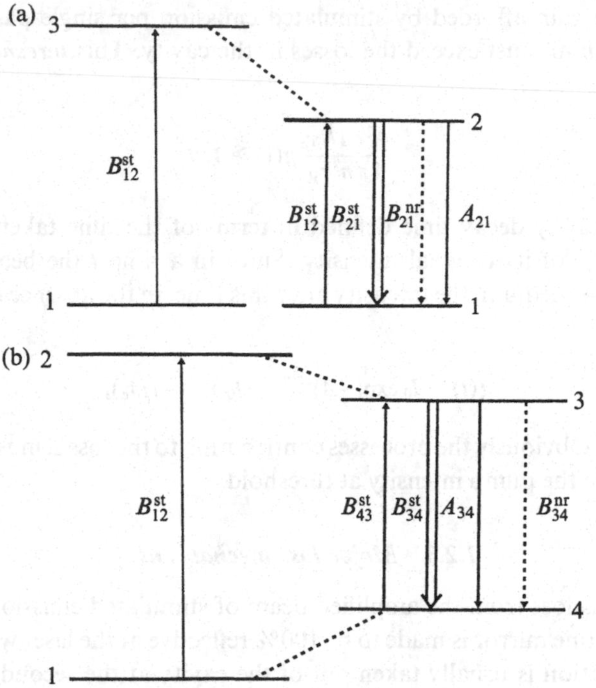

Figure 1.3. Energy levels in (a) three-level and (b) four-level laser systems.

examples, the characteristic features of three-level (Fig. 1.3a) and four-level (Fig. 1.3b) lasers are outlined.

The three-level laser in Fig. 1.3a is pumped via a transition from level 1 to level 3, which subsequently decays very rapidly by nonradiative processes to level 2. A population inversion $N_2/N_1 > 1$ is achieved between level 2 and level 1 if decay from level 2 to level 1 occurs rather slowly. At least 50 % of all the atoms must be maintained in level 2 for such a system to exceed the threshold condition. This condition is easily derived using a rate equation analysis. First the population N_3 is effectively zero in the presence of very fast nonradiative decay ($W_{32}^{nr} \sim 10^{13}\,s^{-1}$) from level 3 to level 2. In consequence the pump rate W_p effectively excites directly into level 2. Level 2 then decays radiatively by spontaneous (A_{21}) and stimulated processes (W_{21}^{st}): in addition there can be direct re-absorption of the emitted radiation (W_{12}^{st}). The rate of change of population of level 2 is then given by

$$\frac{dN_2}{dt} = -(W_{21}^{st} + A_{21})N_2 + (W_p + W_{12}^{st})N_1. \tag{1.28}$$

Assuming that the statistical weights $g_2 = g_1$ so that $W_{21} = W_{12}$, and that under steady state conditions $dN_2/dt = 0$, the population difference, ΔN, is given by

$$\Delta N = N_2 - N_1 = \frac{W_p - A_{21}}{W_p + A_{21} + 2W_{12}^{st}} N_0, \tag{1.29}$$

in which $N_0 = N_1 + N_2$. Amplification occurs when $W_p > A_{21}$, so that $\Delta N > 0$.

The ruby laser is the archetypal three-level laser system in which the population terminus is the ground state. A typical ruby laser rod contains *ca* 0.05 wt.% Cr_2O_3 (i.e. 1.58×10^{19} Cr^{3+} cm^{-3}) which substitute for Al^{3+} in the Al_2O_3 structure. In his original laser Maiman (1960) used a ruby rod having highly polished and parallel end faces placed at the centre of a spiral Xe flashlamp. The blue and green wavelengths from the flashlamp are absorbed in the broad $^4A_2 \rightarrow {}^4T_2, {}^4T_1$ absorption bands with peak cross sections $\sigma_p = 0.72 \times 10^{-19}$ cm^2 at 543 nm and 3.81×10^{-19} cm^2 at 398 nm, respectively, in π-polarization (E $\|$ c-axis) and $\sigma_p = 2.28 \times 10^{-19}$ cm^2 at 558 nm and 2.25×10^{-19} cm^2 at 412 nm, respectively, in σ-polarization (E $\perp$ c-axis) at 300 K. The excited ions relax rapidly by phonon cascade into ^2E storage levels with almost 100% quantum efficiency, from which they emit into the R_1-line at 694.3 nm with a bandwidth of 1.2 nm at 300 K. At this temperature the stimulated emission cross section of the R-line in σ-polarization is 2.5×10^{-20} cm^2 and the luminescence decay time is 3.1 ms. The ruby laser operates as a three-level device only because the unit quantum efficiency of the R-line emission which, in concert with the long-lived ^2E storage level, enables more than 50% of the Cr^{3+} ions to be cycled from the ground state via the pump bands into the ^2E level, given a suitably intense white light source. Whether pulsed or continuous wave (CW), the output of the ruby laser is a series of sharp, relaxation oscillations, which occur because the rate of stimulated emission increases rapidly above threshold and soon exceeds the rate at which the lamp provides pump radiation. Thus the population inversion is depleted until it falls below threshold. In ruby crystals containing larger Cr^{3+} concentrations (>0.1 wt.% Cr_2O_3) laser operation can also be stimulated on the Cr^{3+}–Cr^{3+} pair lines at 700.9 nm and 704.1 nm [Schawlow and Devlin (1961)].

The ruby laser is an example of a fixed frequency laser; others include rare-earth doped lasers such as Nd^{3+} : YAG and gas lasers (He–Ne, Kr^+, Ar^+). However, these are more usually four-level lasers, an energy level scheme for which is shown in Fig. 1.3b. The optical pumping cycle involves excitation from the ground level 1 into a highly excited level 2, which de-excites very rapidly (10^{-13} s) by nonradiative transitions into the emitting level 3. The lasing transition from level 3 to level 4 has a life time long ($\sim 10^{-6}$ s) compared with the nonradiative life times of level 2 and level 4. The system returns to the ground level 1 from level 4 by nonradiative processes. Because the nonradiative rates W_{23}^{nr} and W_{41}^{nr} are very large (*ca* 10^{13} s^{-1}), the population of levels 2 and 4 are always small ($N_2, N_4 \sim 0$). In consequence, it is not necessary to pump very hard into the $1 \rightarrow 2$ transition to invert the populations between levels 3 and 4, unlike the case of the three-level laser.

Below threshold, where the rate of stimulated emission W_{34}^{st} is small, the population difference

$$\Delta N \cong N_3 \cong W_p \frac{N_1}{A_{34}} \tag{1.30}$$

increases linearly up to the threshold value of ΔN_t, at which the population inversion N_3/N_4 saturates because W_{34}^{st} starts to increase very rapidly. The population difference above threshold ΔN_t is

$$\Delta N_t = \left(\frac{W_p N_1}{A_{34} + W_{34}^{st}} \right). \tag{1.31}$$

Since for a directed beam the intensity density is given by $I(v) = u(v)v$, where $u(v)$ is the energy density in the beam, v is the velocity of the radiation and $W_{34}^{st} = B_{34}u(\omega) = B_{34}u(\nu)/2\pi$, we can solve Eq. (1.28) to give

$$I(v) = \frac{2\pi\, vA_{34}}{B_{34}}\left(\frac{W_p N_1}{A_{34}\Delta N_1} - 1\right), \tag{1.32}$$

showing that the laser beam intensity increases linearly with pump power once W_p exceeds the threshold pump power W_p^t. Below W_p^t the intensity is zero because the stimulated emission rate never exceeds the losses in the laser system.

1.3.2 Fixed wavelength lasers

The ruby laser is a fixed wavelength laser, operating by the amplification of radiation emitted by Cr^{3+} dopant ions in the narrow R-line transition at 694.3 nm. In contrast, Nd^{3+} lasers can oscillate at numerous fixed wavelengths associated with the many spectral lines portrayed in the partial luminescence spectrum of Nd^{3+} : YAG in Fig. 1.4. However, Nd^{3+}-YAG lasers differ from the ruby laser by operating on a 'four-level' optical cycle. The four-level character of the Nd-YAG laser is not obvious from the many crystal-field levels of the ^4I_J multiplets illustrated in Fig. 1.4 and implied in the Dieke (1968) diagram (Fig. 4.9). Since the optical pumping cycles of Nd^{3+} and other RE^{3+} ions reflect their multi-level electronic structure, it is appropriate to discuss four-level pumping in some detail, and to anticipate the spectroscopy of $4f^n$ ions presented in Chapters 4, 5 and 10.

For the $4f^q$-configuration ions electron–electron interactions in the spherically symmetric environment of the nucleus and fully occupied core orbitals yield the LS-terms of the free ions. On the Dieke diagram, Fig. 4.9, the multiple energy levels of Nd^{3+} (and other RE^{3+} ions) are represented by horizontal lines at the approximate energy. Crystal field splittings, near degeneracies and overlaps of some levels are indicated by the thicker horizontal lines. For example, the free ion terms of Nd^{3+} ($4f^3$) illustrated in Fig. 4.9 are ^4I, ^4F, ^2H, ^4S, ^2G, ^4G, ^2D, ^2P and ^4D, in order of increasing energy. These terms are split into their J-fine structure levels by spin–orbit coupling, which brings some individual levels from different terms into near degeneracy. This effect is compounded by the splitting of the J-levels by interaction with the crystal-field, as discussed in §2.5.5 and §4.5. The number of crystal-field levels into which each J-value splits for all 32 crystallographic point groups is given in Tables 2.19 and 2.20 for integral and half-integral J-values, respectively. The lowest lying term, ^4I_J, of free Nd^{3+} ions splits into $^4I_{9/2}$, $^4I_{11/2}$, $^4I_{13/2}$ and $^4I_{15/2}$ levels, which are split into five, six, seven and eight levels *et seq.* by the additional crystal-field of D_2 symmetry at the Y^{3+} site in YAG. The crystal-field splittings of other multiplets follow from these tables. In consequence, there is an overlap of five levels from $^2H_{9/2}$ with three from $^4F_{5/2}$ yielding a cluster of absorption lines from the $^4I_{9/2}$ ground levels which provide for efficient pumping centred at *ca* 800 nm. Similar overlap and small LS-splitting of ^4G_J provides strong pump bands near 565 nm ($^4G_{7/2}$, $^4G_{5/2}$) and 535 nm ($^4G_{9/2}$).

The thermally-populated crystal-field levels of the $^4I_{9/2}$ ground state comprise level 1 of the four-level system in Fig. 1.3b. Pump radiation from a white light source is absorbed in crystal-field transitions to manifolds near 800 nm, near 750 nm, near 690 nm, near 625 nm, near 570 nm, near 525 nm, near 510 nm and near 465 nm etc., etc.

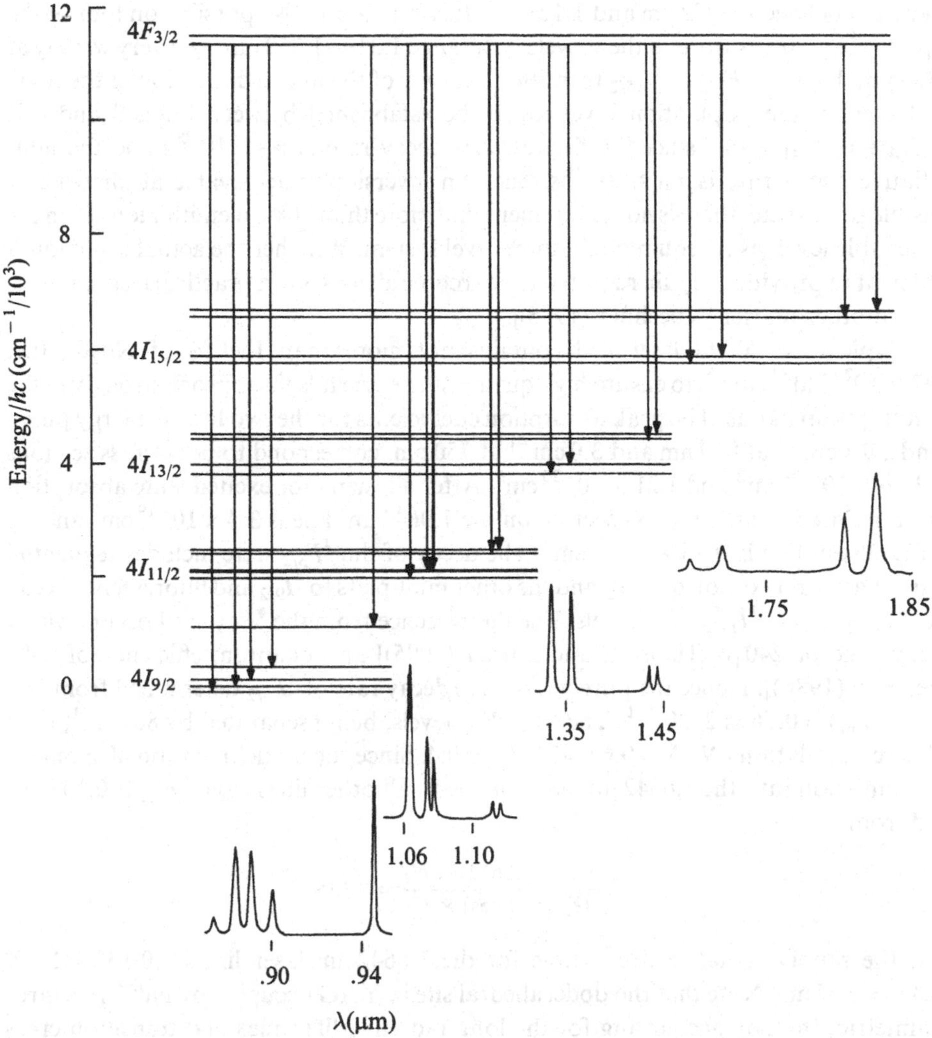

Figure 1.4. Luminescence transitions between the low-lying energy levels of Nd^{3+} in YAG measured at 77 K [after Henderson and Imbusch (1989)].

Such overlaps and homogeneous broadening of the spectra by thermal dephasing at 300 K creates broad, but structured, pump bands that are ideal for flashlamp pumping. Generally, these excited states decay nonradiatively ($\tau_{NR} \sim 10^{-10}$ s) into the metastable $^4F_{3/2}$ state. This affords great simplification of the multi-level model since all these excited levels are treated collectively as level 2. The two crystal-field levels in Fig. 1.4 derived from $^4F_{3/2}$, and between which population is shared at 300 K, are treated as level 3. The terminus of the laser transition, level 4, can be any crystal-field level of the $^4I_{9/2}$, $^4I_{11/2}$, $^4I_{13/2}$ and $^4I_{15/2}$ levels. Laser operation at four wavelengths between 0.890 and 0.9462 nm is observed for the $^4F_{3/2} \rightarrow {}^4I_{9/2}$ transitions. These are rather weak laser lines because this is a quasi-three-level system. In Fig. 1.4 the downward arrows involving the $^4F_{3/2} \rightarrow {}^4I_{11/2}$ transitions, and which connect the crystal-field levels associated with the usual Nd^{3+} : YAG laser wavelengths near 1.064 μm, yield eight laser lines between 1.052 μm and 1.123 μm. Laser action is also possible on four of the $^4F_{13/2} \rightarrow {}^4I_{13/2}$ transitions in the wavelength range 1.319–1.358 μm, and very weakly at 1.833 μm due to a $^4F_{3/2} \rightarrow {}^4I_{15/2}$ transition. In view of the above discussion of the four-level system, the population inversion to be established between levels 3 and 4 is $N_3/N_4 \cong \tau_{34}/\tau_{41} \approx 10^5$, since the fluorescence decay rate is $\tau_{32} \cong 10^{-5}$ s and the non-radiative decay rate is $\tau_{41} \cong 10^{-10}$ s. Such an inversion is achievable at almost any sensible pump rate: there is no requirement that more than 50% of emitter ions be in the metastable level, as is required of a three-level system. Whether the actual inversion is sufficient to provide the gain required to overcome all losses in a practical laser cavity is quite another matter [Koechner (1976)].

A typical Nd-YAG laser rod contains not more than 1 wt.% of Nd_2O_3 (i.e. 2.07×10^{20} Nd^{3+} cm^{-3}) to ensure high quality single crystals that absorb strongly in the various pump bands. The peak absorption coefficients for the two lowest energy pump bands, 9.1 cm^{-1} at 809 nm and 3.0 cm^{-1} at 750 nm, correspond to peak cross sections of 3.66×10^{-20} cm^2 and 1.21×10^{-20} cm^2. After allowing for excited state absorption the stimulated emission cross section on the 1.0642 μm line is 3.4×10^{-19} cm^2 and on the 1.338 μm line is 0.95×10^{-19} cm^2. The decay of the $^4F_{3/2}$ state includes sequential nonradiative relaxation to $^4I_{15/2}$ and the other multiplets to $^4I_{9/2}$ and fluorescence decay into $^4I_{15/2}$, $^4I_{13/2}$, $^4I_{11/2}$, $^4I_{9/2}$ levels. The fluorescence from the $^4F_{3/2}$ level occurs with a decay time of 240 μs [Hansom and Poirier (1995)] and quantum efficiency of 56% [Siegman (1986)]. Hence the purely radiative decay rate of $^4F_{3/2}$ determined from $W_r/(W_r + W_{nr}) = 0.56$ is 2330 s^{-1}. The two $^4F_{3/2}$ levels, being separated by 80 cm^{-1}, have relative populations $N_1/N = 0.6$ and $N_2/N = 0.4$. Since the branching ratio of spontaneous emission into the 1.0642 μm line relative to all other lines from $^4F_{3/2}$ is 0.135, we find from

$$\frac{W_r \text{ (at 1.0642)} \times N_2}{W_r \text{ (all lines)} \times N} = 0.135$$

that the purely radiative decay time for the 1.0642 μm laser line is $(0.135 \times 2330/0.4)^{-1} = 127$ ms. Note that the dodecahedral site in YAG occupied by Nd^{3+} is centrosymmetric, thereby accounting for the long radiative life times and transition cross sections. These quantities, including the branching ratios, can be quite different in other hosts offering RE^{3+} substitutional sites that lack inversion symmetry.

In addition to Nd^{3+} ($4f^3$) ions as laser active centres in different crystals, there have been many studies of four-level lasers activated by Pr^{3+} ($4f^2$), Ho^{3+} ($4f^{10}$), Er^{3+} ($4f^{11}$) and Tm^{3+} ($4f^{12}$) ions. The Er^{3+} ion in various hosts is laser active in the green (near 550 nm), in the red (near 680 nm) and in the infrared at *ca* 1.50 μm. Pr^{3+} is even more proficient in having more than twenty laser active lines in the visible and near-infrared regions. Ho^{3+} and Tm^{3+}, although individually laser active, are of much interest when doped together into crystals, since easy energy transfer paths between these ions facilitate the efficient pumping of laser lines in the eye-safe regions beyond 2.0 μm.

1.3.3 Wavelength tunable solid state lasers

It has been emphasized so far that rare-earth-activated lasers are in general fixed wavelength systems, although they have potential for operating at other wavelengths by various optically nonlinear frequency shifting techniques (harmonic generation,

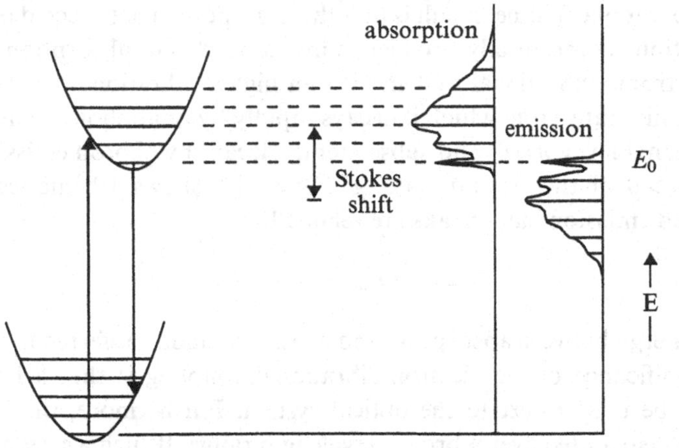

Figure 1.5. Representing schematically the broadening of optical transitions by the electron–phonon interaction.

Raman effect and optical parametric oscillation). However, when doped into glassy matrices instead of crystals there is also some potential for tunability. In crystals at 300 K the luminescence line width of Nd^{3+} at 1.064 is typically 1 nm. As a contrast in a typical SiO_2-based glass fibre the FWHM of this same resonance line is typically 30–50 nm. Such tunability has not normally been exploited in Nd^{3+}-glass laser systems, possibly because in CW operation using tunable, narrow linewidth techniques there are inherent thermal effects which preclude good performance. More usually, tunable solid state lasers operate on vibronically-broadened transitions of transition-metal ions or colour centres, although the Tl^0 centre in alkali halides and Ce^{3+} and Yb^{3+} rare-earth ions are interesting exceptions to such a general classification.

In most rare-earth ion lasers each of the four levels is a different electronic level, almost unaffected by the vibrations of other ions of the crystal. The *configurational coordinate* model of a vibronically-broadened solid state laser is shown in Fig. 1.5: it involves only two electronic levels but each is coupled to the vibrations of the crystal, assumed harmonic [§5.2.4, Henderson and Imbusch (1989)]. In this simple model the ground state energy E_G includes purely electronic and vibrational parts. Assuming the vibrational energy to be quantized as for a harmonic oscillator, then

$$E_G = E_g + \left(m + \tfrac{1}{2}\right)\hbar\omega. \tag{1.33}$$

In similar fashion the excited state energy is written as

$$E_E = E_e + \left(n + \tfrac{1}{2}\right)\hbar\omega, \tag{1.34}$$

where the same frequency of lattice phonons is assumed in the ground and excited states. However, since the electronic charge distributions are different in the two states, the vibrational centres are slightly different.

An absorption transition at low temperature starts from the $m = 0$ vibrational state, but from any configuration position within the ground state parabola. Vertical transitions from this state to the excited state sample a range of positions along the excited state parabola, although the most probable transition occurs when the ions are at the centre of the parabola. The net result is that the absorption line expected of a two-level atomic transition is vibronically broadened into a band. The absorption of a photon takes the electronic-vibrational system into an higher vibrational level $n > 0$ of the excited electronic state, from which it decays rapidly by multiphonon emission to the lowest vibrational level, $n = 0$, and subsequently decays by photon emission to vibrational level $m > 0$ of the ground state, g. Figure 1.5 shows schematically that the absorption and emission band peaks are related by

$$E_{\text{abs}} = E_{\text{em}} + (m + n)\hbar\omega. \tag{1.35}$$

This shift in energy between absorption and emission bands is referred to as the *Stokes shift*. The significance of the electron-vibrational coupling is that a broadband of energies may be used to excite the optical system. Furthermore, the luminescence output is also distributed over a broad wavelength range. If such a system can sustain laser oscillations, then the laser output can be tuned over the entire width of the luminescence band.

The first vibronically tuned solid state laser used $Ni^{2+}:MgF_2$ crystals as the gain medium; subsequently the vibronic emission band of Co^{2+}, Ni^{2+} and V^{2+} ions in other fluoride crystals was also reported as the basis for tunable laser action at 300 K [Johnson *et al.* (1962), (1963), (1964), (1966a,b)]. Anion vacancy colour centres also provide for tunable laser action in numerous ionic crystals, and such near-infrared lasers were commercially exploited from *ca* 1974 onwards [Mollenauer (1987) (1992)]. There are many structural variations of these laser-active colour centres, some more stable than others. Colour centre lasers usually require cryogenic temperatures for efficient and stable operation. An example of CW operation of an F_2^+-like centre laser in NaCl is shown in Fig. 1.6. More generally, lasers based on so-called F_A and F_2^+ type centres emit in the range 1.0–3.0 μm depending upon the host crystal. The $Tl^0(1)$ centre laser in alkali halides is a first cousin of the *F*-like centre lasers: a neutral $Tl^0(1)$ atom is trapped in the nearest neighbour site of a halide ion vacancy. Its principal absorption band near 1.05 μm permits convenient pumping by a Nd : YAG laser at 1.06 μm of the tunable laser output centred at *ca* 1.5 μm.

Perhaps the most important development in tunable solid state lasers following the ruby laser was the announcement in the late 1970s of the alexandrite ($Cr^{3+}:BeAl_2O_4$) laser, continuously tunable from 695–750 nm, being operational well above 300 K [Walling *et al.* (1979) (1987)]. This discovery stimulated an immense research effort into other tunable transition-metal ion lasers, but especially including host crystals containing Cr^{3+} ions occupying weak field sites where the broadband emission is observed rather than the sharp line system typified by $Cr^{3+}:Al_2O_3$ and Cr^{3+} : YAG. Broadband laser action from a number of Cr^{3+}-doped hosts has been reported, including $Cr^{3+}:KZnF_3$ [Dürr and Brauch (1986)], $Cr^{3+}:Gd_3Sc_2Ga_3O_{12}$ and other oxide garnets [Struve and Huber (1985)]. Most recently, there has been much research on Cr^{3+}-doped colquiriites ($LiCaAlF_6$, $LiSrAlF_6$ and $LiSrGaF_6$). The Cr^{3+} : LiSAF laser is exciting considerable interest in a diode-pumped scheme, with diodes operating at *ca* 670 nm and broadband tunability being evident from 725–1180 nm [Payne *et al.* (1988a,b)].

Co^{2+} in MgF_2 has enjoyed modest commercial success as a broadband near-infrared laser, continuously tunable in the range 1.5–2.4 μm [Moulton (1986a)]. Other transition-metal ions that support laser action are Ni^{2+} in MgO, $KMgF_3$, MgF_2

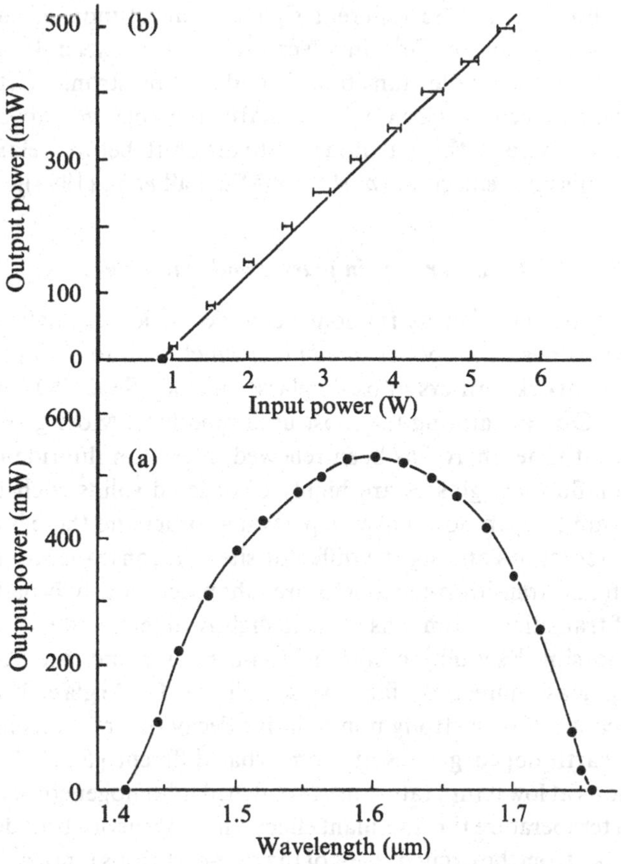

Figure 1.6. CW laser performance of $F_{2A}^{+}(O^{2-})$ centres in a NaCl laser. In (a) the output power pumped by 6 W of 1.064 μm power from a TEM Nd : YAG laser is plotted as a function of laser wavelength, and in (b) the output power is plotted as a function the power input by the pump laser. The power absorbed by the crystal was *ca* 40% of the input power and a 10% output coupler was used for both plots, after Pinto *et al.* (1985), who originally identified the defect as the F_{2A}^{+} centre [see Pinto *et al.* (1986) and Sennaroglu and Pollock (1991)].

and $MgAl_2O_4$ [Moulton (1985)] as well as Cr^{4+} in $MgSiO_4$ and Cr^{4+}-doped YAG [Petricevic *et al.* (1988)]. The collective tuning range of Ni^{2+} lasers extend from *ca* 1.15 μm to 1.8 μm. However, the efficiencies of such lasers are limited by nonradiative decay and excited state absorption. The laser performance of Ti^{3+} : Al_2O_3 is almost that to be expected from an ideal four-level laser [Moulton (1982a,b), (1986a,b)]. Broad blue-green absorption bands provide for pumping over wide wavelength ranges (400–600 nm). The laser output is continuously tunable from *ca* 700–1100 nm using two or three different mirror sets. Many types of Ti-sapphire laser are now available commercially, including ultrashort pulse and single frequency laser systems. The Cr^{4+} ion also emits in the near-infrared from *ca* 1.15–1.55 μm. In the two most successful laser hosts, $MgSiO_4$ (forsterite) [Petricevic *et al.* (1988)] and Ca^{2+} : YAG, there are several different crystallographic sites and valence states for the Cr^{4+} ion [Angert *et al.* (1988)] which increase the threshold pump power and reduce the laser efficiency. There is currently much interest in Ce^{3+} $(4f^1)$ and Yb^{3+} $(4f^{13})$, which have rather similar electronic structures after allowing for the different sign and magnitude of the spin–orbit interaction. The laser action of Ce^{3+} involves excitation between 4*f* and 5*d* levels, which are allowed electric dipole transitions broadened by strong electron–phonon interactions of the 5*d*-electron. For $Ce^{3+}:LiCaAlF_6$ the polarized absorption bands occur near 260 nm: there is then a modest Stokes shift before relaxation into a broadband laser emission centred at *ca* 340 nm [Marshall *et al.* (1994)].

1.3.4 Laser ions in glasses and waveguides

Glassy solids may be regarded as random networks of ionic polyhedra (*formers*) bound together by other ionic species (*modifiers*) which tend to disrupt the network. The most useful network formers in oxide glasses include SiO_4, PO_4 and BO_3 polyhedra: Na_2O and CaO are among the most used modifiers. Most glasses are oxide-based, but in recent times there has been renewed interest in fluoride-based glasses. Whether oxide or fluoride, glasses are highly disordered solids such that the ionic arrangements around dopant ions have no particular order and there is a distribution of crystal-field interactions among the different sites. In consequence, the spectra of typical rare-earth and transition-metal ions are inhomogeneously broadened because of the overlap of transitions from ions at their slightly different sites. This has special significance for possible laser action in transition-metal ion and rare-earth ion doped glasses including glass fibres. So far, the search for Cr^{3+}: glass lasers has been unsuccessful, because of very strong nonradiative decay of Cr^{3+} ions in their excited states. The rare-earth doped glasses are somewhat different §5.5.7. Line spectra of Nd^{3+} and Er^{3+} ions at low temperature are broadened inhomogeneously by a factor of 10–100: at room temperature the dominant effect is homogeneous broadening (§5.5.7). Nonradiative decay from the excited levels of the rare-earth ions is not as effective as for transition-metal ions, and they support laser action over a wider spread of frequencies than their crystalline counterparts.

Fibre lasers are important applications of rare-earth doped glasses. Nd^{3+}-doped fibre lasers are particularly important in optical communication systems, with first generation systems operating at 1.06 μm and subsequent long distance communication being carried on 1.32 μm laser radiation. The Er^{3+} fibre amplifier has also been an important development in this context. The inhomogeneous broadening of rare-earth transitions permits more effective overlap of the levels involved in excited state absorptions, leading to upconverted luminescence laser action. The first upconverted luminescence laser at room temperature used a Ho^{3+}-doped fibre pumped using a Kr^+ laser operating at 647.1 nm. In this case the sequential absorption of two 647.1 nm photons led to laser action at 550 nm. A doubly-doped fibre laser activated by Pr^{3+} and Yb^{3+} has been reported which provides laser action in blue, green and red spectral regions.

The success of glass fibre lasers has stimulated work on active crystalline fibres and on crystalline waveguides. Crystal fibres are of much reduced quality relative to glass fibres. Although many activated crystal fibres have been reported, few have been shown to support laser action. Perhaps not surprisingly, the successful ones include $Cr^{3+}:Al_2O_3$ and Nd^{3+}: YAG fibres. Currently, there is considerable interest in crystalline waveguide lasers. Some of the earliest waveguide lasers were produced by ion implantation techniques [Townsend *et al.* (1994)]. The energy of the implant ion is used to control its penetration depth in the target sample. Any difference between the refractive indices of materials in the bulk and implanted regions causes the confinement of the laser beam in the waveguide. Various rare-earth doped crystals including Nd^{3+} : YAG have been used in the waveguide configuration. More recently there has been much interest in solid state slab lasers for diode-pumping, with excellent results using Nd : glass slab lasers [Faulstich *et al.* (1996)] and Nd : YAG planar waveguide lasers. In the latter case a 200 μm thick active layer of 1% Nd-doped YAG is diffusion bonded between 400 μm cladding layers of undoped YAG, and provides weak guiding in the active layer, the multimode guide supporting 48 *TE* and *TM* modes [Pelaez-Millas *et al.* (1997)]. Much thinner single mode slab waveguides have been grown by liquid phase epitaxy techniques [Hanna *et al.* (1995)].

1.4 Pumping solid state lasers

A number of laser-pumping schemes have been used with solid state lasers. As noted above, the ruby laser was first pumped using a helical Xe flashlamp in a diffusely reflecting pump enclosure. The ruby rod extracts most of the radiation at blue and green wavelengths in exciting the Cr^{3+} ions to their excited states. Once in the excited levels there is a rapid relaxation to the emitting levels lying some 14 400 cm^{-1} or 694 nm above the ground level. The quantum efficiency of this pumping process is almost 100 %. In almost all-solid-state lasers subsequently developed the helical Xe flashlamp was replaced by one or more straight flashlamps or arc-lamps placed parallel to the laser rod on the axis of an highly reflecting, elliptical pump cavity.

1.4.1 Gas lasers

Gas discharge lasers are common pump sources for solid state lasers. The first such discharge laser was the He–Ne laser. A low pressure (3–5 torr) He–Ne mixture enclosed in a quartz plasma tube is excited via a DC discharge operating typically at 1–2 kV yielding a DC current of *ca* 10 mA. The discharge excites He atoms into the 2^3S and 2^1S levels some 19.81 eV and 24.6 eV, respectively, above the ground state. This energy is transferred from excited He atoms to the $n = 2$ and $n = 3$ levels of Ne. The $3s \rightarrow 3p$ transitions of Ne at 3.39 μm, $2s \rightarrow 2p$ transitions at 1.15 μm and $3s \rightarrow 2p$ transitions at 633 nm all sustain laser action. The He–Ne laser is a low power system yielding only (say) 0.5 to 20 mW of typical power output. As such it is not used to pump other laser systems, although it is a useful spectroscopic source for use with red absorbers. Other inert gas lasers such as the Ar^+ and Kr^+ lasers are used to pump solid state gain media. These inert gas lasers are also discharge excited to give from hundreds of milliwatts to tens of watts of CW oscillation at various wavelengths. The Kr^+ laser produces a red laser excitation at 647.1 nm, whereas Ar^+ lasers produce lines in the green, blue and near UV regions. Both Ar^+ and Kr^+ are quite useful pump sources for pumping various solid state laser systems, although the requirement that the particular resonance line must fall within the absorption spectrum of the centre under investigation may be quite restricting of choice.

1.4.2 Dye lasers

In contrast to gas lasers, organic dyes provide continuous ranges of tunable laser output from allowed singlet–singlet transitions of organic molecules: these transitions are homogeneously broadened by vibronic interaction to cover bandwidths of 30–40 nm. A profusion of dye-solvent mixtures enables efficient laser outputs covering a range from *ca* 350 nm to 1000 nm. Such organic dyes are ideal gain media for wavelength tunable lasers operating on a four-level pumping cycle. Although the luminescence may cover 30–40 nm, a dispersing prism inside the laser cavity can reduce the dye laser output to *ca* 2 nm and still permit the laser to maintain up to 50% of the output power. Then by slight adjustment (rotation) of the prism or (more easily) the output coupler the dye laser may be tuned over almost the entire luminescence bandwidth of the dye. In consequence, the tuned laser output has the potential for both narrow bandwidth and high power.

1.4.3 Semiconductor diode laser pumping

Combinations of gas and dye lasers for pumping solid state lasers are rather bulky and much of the present day technology focuses on miniaturization of the entire system. In such miniature solid state lasers the pump source is inevitably a semiconductor diode laser. These laser diodes offer electrical to optical conversion efficiency of *ca* 30–50 %. Although the emissions from diode lasers are quite narrow (1–2 nm) they may be temperature tuned to the peaks of appropriately overlapping absorption bands. The packaging of the laser diode, a Peltier cooler for heat extraction and temperature tuning electronics, although making for compact, efficient and reliable devices, also causes them to be quite expensive at present. Individual emitters are limited to output powers of about 50 mW. However, arrays of 500 or so single diodes emitting from large areas of around 500 μm × 1 μm can give output powers of about 4 W at 800 nm [Botez and Scifres (1994)]. Several multi-element arrays can also be mounted on a 1 cm bar to give even higher power, up to 20 W. The earliest diode-pumped laser systems took advantage of the fortuitous overlap between the GaAs/GaAlAs quantum well laser output near 800 nm and the overlapping $^4I_{9/2} \rightarrow {}^4F_{5/2,3/2}$ absorption lines of $Nd^{3+}:Y_3Al_5O_{12}$ (YAG). Present developments are aimed at both shorter and longer wavelength diodes using other III–V compounds and their alloys. Although there are still major problems before commercial exploitation is possible, the II–VI compound semiconductor alloys based on Zn(Cd)S(Se) have the capability for laser output over the blue–green range of wavelengths. Furthermore, there is very great excitement and research effort into the growth and optical properties of the III-N compounds AlN–GaN–InN, which have the capability for diode devices covering the entire wavelength range from 300–650 nm [Nakamura *et al.* (1995)].

1.5 Laser output

1.5.1 Output coupling and tuning

§1.2 examined the conditions under which useful output may be extracted from a laser oscillator, consisting of gain medium and partially-reflecting, carefully aligned end mirrors. It was shown that if the round trip gain minus losses (including spurious absorptions, reflection, transmissions, etc.) is less than unity then the radiation will decrease in intensity on each pass and eventually die away after a few passes. However, if the round trip gain $\gamma(\nu)l - \Delta$, including losses, exceeds unity then the amplitude of radiation will build up exponentially on each pass, growing into a coherent self-sustained oscillation between the end mirrors that define the laser cavity. A part of the cavity loss is the transmission of the output coupler through which the laser beam is extracted. Depending on the efficiency of the laser process, the transmission of output coupler may vary from 2–3% up to 10–20%.

The original ruby cavity, involving dielectric mirror coatings on parallel end-faces of the rod, has long since been replaced by mirrors displaced along the axis from the gain medium. These mirrors may be curved or planar, one being fully reflecting at the laser wavelength and the other partially transmitting. The laser output is an intense beam (mW up to W), of polarized, coherent radiation which is highly monochromatic and frequency stable. Polarization of the output beam usually follows from the use of Brewster's angle mirrors.

The Ar^+ laser is often used to pump vibronically tuned lasers, including dye lasers and solid state lasers. This laser system may operate simultaneously on all UV and visible region lines. However, by introducing a dispersing element into the cavity ahead of the output coupling mirror, it is possible to tune through the multi-line output, selecting each line in turn. In the Ar^+ laser the dispersing element is usually a small prism, which is fixed in position in the cavity. The tuning operation involves mechanically tilting the output coupler through a few degrees to select the desired wavelength. This same tuning mechanism is used in solid state and dye lasers also, although the prism tuning element is usually replaced by a diffraction grating or a multiple element birefringent plate. Such dispersive elements may be used to reduce the output bandwidth to only $\sim$0.1 nm, compared with that achieved with a prism of *ca* 2 nm. For applications in very high resolution spectroscopy an external Fabry–Perot etalon may be used to reduce the bandpass still further, to only 5–10 MHz. A ring dye laser controlled by an interferometer may produce a stable output of several hundred milliwatts over a spectral bandwidth of 100–200 kHz.

1.5.2 Threshold and slope efficiency

As Eq. (1.32) shows, a plot of laser intensity $I(\nu)$ against pump power W_p increases only very little below threshold as the population inversion grows (Eq. (1.31)). Above threshold the population inversion ΔN_t does not grow further: it is 'clamped' at ΔN_t because of the rapidly increasing rate of *stimulated emission* that builds up in the cavity as the pump rate is increased above threshold. Cavity losses including the transmission (T) of the output coupler increase the threshold power above which the intensity of stimulated emission increases. The threshold power, P_{th}, is

$$P_{th} = \frac{h\nu_p(A_p + A_l)}{4\sigma_e \tau \eta_p}(L + T), \tag{1.36}$$

where L is the passive cavity loss, $h\nu_p$ is the energy of the pump photons, the A are the cross sections of the pump (p) and laser (l) beams, σ_e is the cross section for stimulated emission, τ is the luminescence life time and η_p is the pump efficiency. Since in practice σ_e is wavelength dependent, so too is P_{th}, which is also directly proportional to the total cavity losses $(L + T)$. For transition-metal and rare-earth ion lasers high reflectivity mirrors are used, with T being typically 1–3%. However, the output transmission in dye lasers and colour centre lasers may be as high as 10–20%.

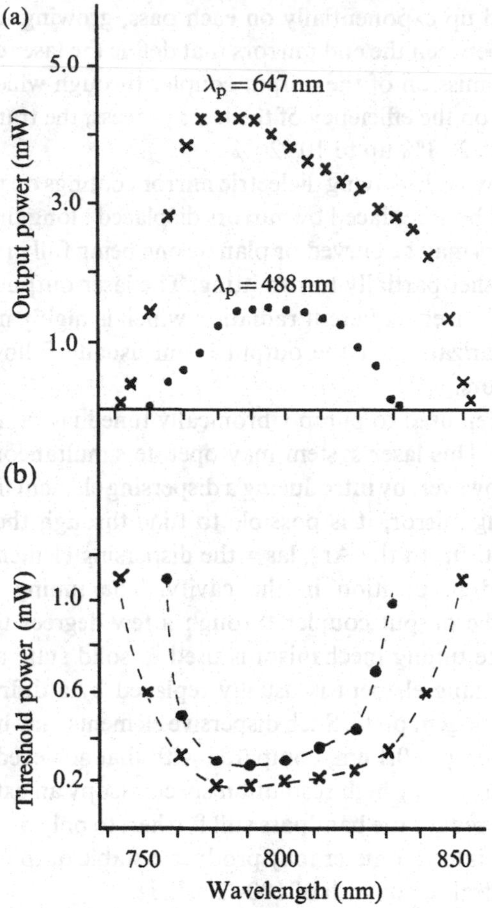

Figure 1.7. (a) The output tuning curve for a Cr^{3+} : GSGG laser pumped with 1.7 W of CW Ar^+ laser output at 488 nm (•) and 1.1 W of CW Kr^+ laser output at 647 nm (+). (b) The threshold tuning curve at the same pump wavelengths [after Struve and Huber (1985)].

Typical input–output characteristics for a NaCl : $F_2^+(O^{2-})$ centre laser with a 10 % output coupler in Fig. 1.6 show the expected linear behaviour of the output above a threshold power of *ca* 0.4 W. Also shown is the output tuning curve, which is related to the wavelength variation of the stimulated emission cross section. Figure 1.7 shows the output and threshold power tuning curves for a Cr^{3+} : GSGG laser pumped at wavelengths of 488 nm and 647 nm. The reduced output power and increased threshold under 488 nm pumping reflects the ESA at this wavelength. Fitting the data to Eq. (1.36) yields a peak emission cross section of 7×10^{-20} cm^2 [Struve and Huber (1985)].

To minimize the cavity losses as the radiation bounces back and forth between the cavity mirrors the multiply-reflected beams must interfere constructively, such that

$$2nl = p\lambda \quad \text{or} \quad \nu_p = p\frac{c}{nl}, \tag{1.37}$$

where n is the refractive index of medium, l is the cavity length and p is the mode number, normally very large. Below threshold all modes within the gain envelope are populated. Only those axial modes that fall within the gain envelope of the spectral line are amplified above threshold. However, as population inversion gradually increases, one mode will satisfy the threshold condition before the others. All the excited atoms are forced to emit into this mode by the stimulated emission process. It is this single mode which is amplified. In a typical ruby laser the length of the rod may be $l = 0.1$ m and $n = 1.7$ with $\lambda = 700$ nm. The mode number p derived from Eq. (1.33) is *ca* 5×10^5, and the mode separation is $\Delta\nu_p \cong 10^9\ \mathrm{s}^{-1}$ or 10^{-3} nm. The halfwidth of a typical atomic resonance line is only *ca* 0.1 nm, so that only a few modes will fall within the gain envelope, even for the very broad (50 nm) gain profile of a tunable solid state laser. The action of manipulating the dispersive-element in the cavity is then to shift that mode within the band pass of the gain medium through the gain envelope. This apparently simple process is not quite so straightforward as it appears, especially in high resolution ring lasers, in which mode-hopping is a problem that must be resolved.

There are several important characteristics of the laser threshold that are noteworthy. First, as the pump power is increased above threshold there is a sudden enhancement of the output power which then increases linearly with further increases in pump power. At threshold the upper state population is 'clamped' by the stimulated emission process: in consequence so is the intensity of the randomly emitted spontaneous emission. The output beam suddenly narrows spectrally as the signal beam changes from the broadband spontaneous emission to the few longitudinal modes of the cavity that fall within the linewidth of the optical transition. There is a concomitant spatial narrowing of the beam associated with the few transverse cavity modes excited: this process defines also the collimation and spatial coherence of the beam.

The experimental slope efficiency for an $(F_{2A}^+)(O^{2-})$ centre laser determined from Fig. 1.6 is about 15 %. This measurement is not necessarily equal to the intrinsic slope efficiency, η_i, obtained from the gradient of Eq. (1.32), which is related to the pumping efficiency, η_p, by

$$\eta_p = \left(\frac{\lambda_p}{\lambda_l}\right)\left(\frac{2\pi\nu}{B_{34}}\right)\left(\frac{N_1}{\Delta N_t}\right) = \left(\frac{\lambda_p}{\lambda_t}\right)\eta_i. \tag{1.38}$$

The ratio of pump wavelength to laser wavelength is sometimes referred to as the *quantum defect*. Both η_i and the losses L can be derived from measurements of the experimental slope efficiency η at different values of the transmission of the output coupler, since

$$\eta = \eta_p(T/(T+L)). \tag{1.39}$$

A plot of $1/\eta$ versus $1/T$ should be linear with slope equal to L/η_p and intercept $1/\eta_p$ [Caird *et al.* (1988)]. From Eqs. (1.36) and (1.38) it is evident that P_{th} and η increase as the transmission of the output coupler is increased.

1.5.3 Continuous wave (CW) and pulsed laser output

The spectral characteristics of the output are determined by the interplay of the gain medium and the resonant cavity. Since only a few axial cavity modes lie within the bandwidth of the optical transition, only that axial mode closest to line centre will be excited. The intensity in this mode increases rapidly above threshold; other modes, however, never cross threshold and are not excited. In such an idealized laser system oscillating on a single mode the beam is expected to be spectrally very pure. Unfortunately, hole burning in the optical line shape leads to line broadening, and to obtain very pure spectral output, i.e. single frequency, hole burning must be overcome. The use of a unidirectional ring laser oscillator completely eliminates spatial hole burning and permits high average power, single frequency operation. Ring dye lasers and traditional solid state ring oscillators have been operated with linewidths down to 0.1–1 MHz. However, recently developed semiconductor diode-pumped microchip lasers can produce single frequency operations with linewidths down to 5 kHz and output powers up to 20–30 mW [Zayhowski (1996, 1999)].

There are many applications for pulsed outputs from lasers. It is possible to produce quite short pulses (10^{-7} s) at very high powers of up to 10^9 W. Such lasers are referred to as *giant-pulse lasers* or *Q-switched lasers* because the lasers incorporate an optical cavity with a controlled Q-factor. If the transmission of the output coupler is transiently reduced to zero the population inversion will build up quickly to many times the threshold inversion. If the output coupler is now suddenly reopened the population inversion rapidly returns to the threshold value with the output of a very intense burst of laser radiation. This oscillation burst is sufficiently powerful that it rapidly depletes the inverted population to well below the new cavity loss level, after which the oscillation signal in the cavity dies out almost as quickly as it emerged. The necessary modulation of the transmission of the output coupler may be achieved with a variety of techniques. The commonest Q-switching devices are electro-optic and acousto-optic switches, which may have switching times of duration 10 ns or less.

Passive Q-switching is also fairly common. An easy to saturate absorbing element is inserted in the cavity. The laser inversion builds up during pumping until the gain in the cavity exceeds the introduced absorption, following which laser oscillation begins. Even at a relatively low laser oscillation level the amplified radiation saturates the absorption and opens up the output coupler again, resulting in a short, intense laser pulse. Organic dyes and transition-metal doped ionic crystals are frequently used as saturable absorbers [Siegman (1986)].

Mode-locking of lasers is another important technique for generating ultrashort light pulses. There have been particularly exciting developments in recent years of mode-locked solid state lasers. As discussed already, the axial mode spacing frequency is given by $\nu_p = pc/2nl$ and is typically $\sim 10^9$ Hz. Active mode-locking involves the insertion of a modulator in the cavity which is driven to provide amplitude or phase modulation at the axial mode spacing frequency, thereby locking the modes together. In consequence, the laser output takes the form of a train of pulses with repetition rate equal to the modulation frequency. In active mode-locking the pulse duration τ_p is proportional to the reciprocal square root of the product of the modulation frequency (ν_m) and the gain bandwidth ($\Delta\nu$). As we have seen $\nu_m \sim 10^9$ Hz, although higher frequency modulation will result in shorter pulses. However, it is the $(\Delta\nu)^{-1/2}$ dependency of the pulse duration which is most important since it represents the effect of the gain medium. For example, this factor results in shorter pulses generated in a mode-locked Nd-YLF laser than in Nd-YAG because the 1.06 μm laser line is a factor of two broader in the fluoride host. By matching the cavity length to the modulation frequency the laser mode-locks. However, the shortest mode-locked pulses produced in such a simple Nd^{3+}-doped laser system have been of the order of 2 ps, longer by far than the sub-picosecond duration available from bandwidth limited pulses. To make full use of the gain bandwidth requires more sophisticated mode-locking procedures to be used, including self mode-locking that relies on optical nonlinearities of the gain medium at the very high light intensities available in short pulse lasers. Using nonlinear mode-locking techniques, pulse durations of only 20 fs have been reported for $Ti^{3+}:Al_2O_3$ lasers. Comparably short pulses have been generated in $Cr^{3+}:LiSrAlF_6$ lasers, which have the additionally attractive feature that they absorb at wavelengths near 670 nm and may be pumped using the red laser diodes that are currently becoming available.

1.5.4 Nonlinear frequency conversion

The operational wavelength ranges of many lasers can be extended using nonlinear optical techniques. Harmonic generation was the first nonlinear optical effect to have been used for the purpose of frequency conversion. The developments of new types of nonlinear crystal and resonant cavity enhancement techniques have led to very efficient harmonic generation systems. Indeed, using such crystals as LiB_3O_5 it is now quite practical to produce the second (532 nm), third (355 nm) and fourth (262 nm) harmonics of the Nd : YAG laser. Figure 1.8 shows a schematic diagram of a frequency doubled Nd : YLF laser using an enhancement cavity. Two laser diodes are used as the pump laser, mode-locked at 1.047 μm, to give the high peak powers required for harmonic conversion. Frequency doubling is achieved using an LBO operating at *ca* 410 K, non-critically phase-matched at the pump wavelength. Some 6 W of pump power yields 1.8 W incident on the cavity. Almost 1 W of usable power was obtained at 523 nm, representing one of the most efficient techniques of producing short pulses in the visible region [Malcolm and Ferguson (1991a,b)].

Improvements in materials and laser performances have revitalized the development of optical parametric oscillators (OPOs) to cover all of the UV, visible and near-infrared regions of the electromagnetic spectrum. The OPO is one of the most effective methods of producing tunable coherent radiation. The pump beam generates signal and idler beams in a three-wave mixing process according to the energy matching condition

$$\hbar\omega_p = \hbar\omega_s + \hbar\omega_i, \qquad (1.40)$$

where the subscripts p, s and i identify the frequencies (ω) of pump, signal and idler beams. In an OPO the frequencies ω_s and ω_i are determined by the need to phase-match the interaction. An OPO is tuned by varying the phase-matching condition by changing the orientation or temperature of the nonlinear optical crystal.

Many inorganic and organic crystals have been designed for application in nonlinear optics. The most commercially successful have been lithium niobate ($LiNbO_3$), potassium niobate ($KNbO_3$), potassium titanyl phosphate (KTP) and arsenate (KTA), β-barium borate (BaB_2O_4, BBO) and lithium triborate (LiB_3O_5, LBO). All have been used as harmonic generating crystals and in optical parametric oscillators (OPO) and amplifiers (OPA). Figure 1.9 shows the temperature tuning characteristics of an LBO OPO pumped using a diode-pumped Nd : YLF laser [Ferguson (1994)]. At

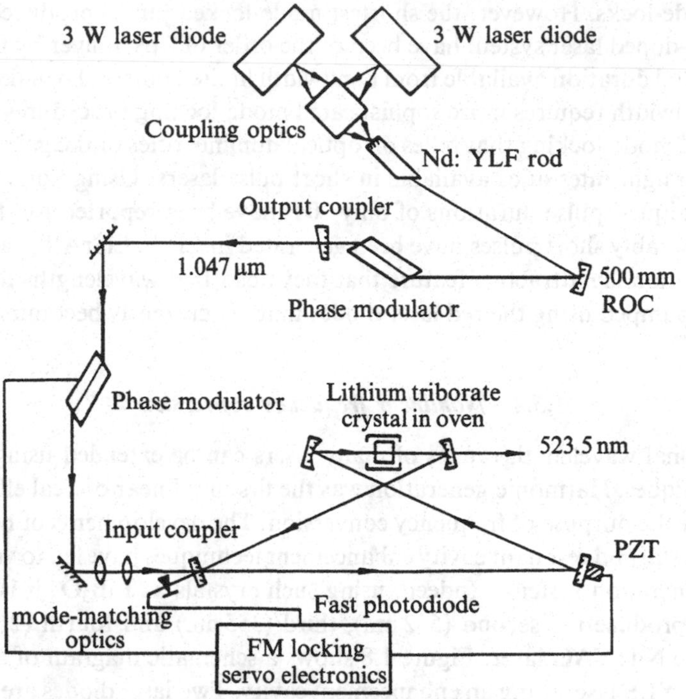

Figure 1.8. A resonant enhancement cavity for frequency doubling a Nd:YLF laser [after Ferguson (1994)].

temperatures in the range 400–440 K the OPO device will operate on four different wavelengths. However, at an operational temperature of 460–470 K the OPO works on a single frequency. The output is very stable and the threshold is only 70 mW. Varying the crystal temperature in the range 390 K to 470 K permits continuous tuning from 650 nm to 2900 nm.

1.6 Motivation, scope and organization of the book

This is not a treatise on all aspects of lasers, solid state or otherwise: there are many excellent texts on operational characteristics of all classes of gain media. Rather does this monograph, after the initial brief survey, deal with the physics of solid state gain media and of nonlinear optical crystals. There have been many exciting developments in these areas. First was the operation of broadband tunable lasers, including a variety of colour centre lasers operating at low temperatures (77–150 K) and the alexandrite ($Cr^{3+}:BeAl_2O_4$) laser that operated up to *ca* 400 K. The latter in particular stimulated a remarkable series of studies of tunable Cr^{3+}-doped gain media, over fifty of which support laser action at room temperature. Soon after the invention of the alexandrite laser, Moulton reported that $Ti^{3+}:Al_2O_3$ lased over an even wider tuning range, from *ca* 700 nm to 1100 nm. This was followed by development of the colquiriite family of fluoride crystals, which when doped with Cr^{3+} cover a similarly broad tuning range at room temperature. Both $Ti:Al_2O_3$ and $Cr^{3+}:LiSAF$ lasers are extremely efficient since nonradiative decay and excited state absorption are weak. Other recently discovered

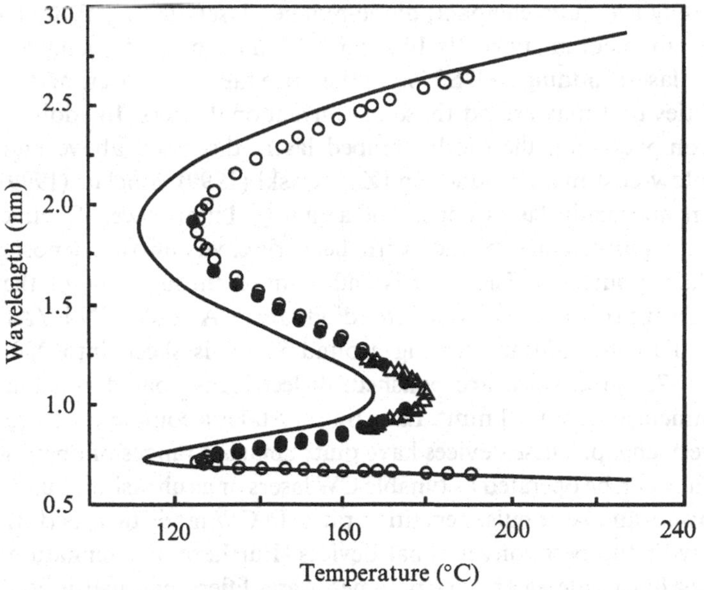

Figure 1.9. The tuning range for non-critical phase matching of an LBO optical parametric oscillator [after Ferguson (1994)].

tunable solid state lasers include Cr^{4+} lasers for application in the near-infrared (1.2 μm–1.7 μm) and Ce^{3+} lasers operating in the ultraviolet (290–330 nm). The developments with tunable lasers have not left untouched research on single frequency lasers using rare-earth doped ionic crystals as gain media. The 1970s witnessed the commercial development of Nd : YAG and Nd : YLF lasers. There have since been reported hundreds of rare-earth ion: host crystal combinations that support laser action. The research has followed a parallel path to that on tunable solid state lasers, in that more complex and lower symmetry crystals have featured strongly in which line broadening and enhanced absorption and emission cross sections are advantageous.

The second major breakthrough in the recent past has been the emergence of the laser diode to pump miniature solid state laser systems. Laser diodes (LDs) convert electrical energy into optical energy with great efficiency (~ 30%), and currently available LDs are capable of delivering up to 20 W of CW energy at selected wavelengths. Since they may (within reason) be tuned to peaks in the absorption bands of various gain media, they are ideal pump sources for many solid state lasers. The early LDs based on GaAs/AlGaAs quantum wells emitted at *ca* 800–810 nm, and can be temperature-tuned to the strongest absorption peak among the overlapping split components of the $^4I_{9/2} \rightarrow {}^4F_{5/2}$, $^4H_{9/2}$ transitions of Nd-YAG (Nd-YLF). The resulting diode-pumped Nd-YAG lasers are compact, efficient and reliable, operating at 1.06 μm or 1.32 μm with conversion efficiency from pump diode to solid state laser as high as 50%. However, the efficiency of diode-pumping the 946 nm line on the $^4F_{3/2} \rightarrow {}^4I_{9/2}$ transition of Nd-YAG is much less efficient. Although much pioneering work used Nd-YAG, recent emphasis has been on diode-pumping Nd : YLF and Nd : YVO_4, especially for ultrashort lasers. Diode-pumped Nd : YAG, Nd : YVO_4 and Nd : YLF lasers are ideal pump sources for harmonic generation and parametric oscillation schemes involving monolithic, resonant enhancement cavities [Ferguson (1994)].

Although they are quite compact, diode-pumped lasers use a substantial piece of crystal as the gain medium, typically $10 \times 2.5 \times 2.5\,mm^3$ for diode-pumped Nd-doped YAG or YLF lasers, adding *ca* US$500 to the price tag of the laser. *Microchip lasers* have capabilities that may exceed those of conventional lasers. In addition, they are even more compact than the diode-pumped lasers discussed above and have the potential for low-cost mass production [Zayhowski (1999), Sinclair (1999)]. In their simplest form microchip lasers consist of a thin (< 1 mm) slice of active medium, polished flat and parallel on two sides, with dielectric cavity mirrors deposited directly onto the polished surfaces. The laser is end-pumped through one of the dielectric mirrors with an appropriate semiconductor diode array. A boule of Nd-YAG of length 200 mm and diameter 100 mm costing around $5000 is sliced into 200 wafers of thickness 500–750 μm, which are polished, dielectrically coated and then cut into pieces with dimensions $\approx 1 \times 1\,mm^2$. Each tiny crystal is a complete laser cavity and is potentially very cheap. These devices have quite amazing ranges of operational characteristics. They can be operated as tunable CW lasers or as ultrashort pulse lasers with high peak powers and large pulse repetition rates. In CW mode their performances are comparable with the best conventional devices. Furthermore, miniature nonlinear optical devices; harmonic generators, parametric amplifiers, parametric oscillators and Raman amplifiers can convert the output wavelengths of these diode-pumped lasers across the entire spectrum from 5000–190 nm.

The rapid development of the technology for producing high power GaAs/AlGaAs LDs originally stimulated many devices pumped near 800 nm. There is now much more attention to diodes operating at other wavelengths. There is considerable emphasis on III–V semiconductor LDs to pump broadband Cr^{3+} emitters at 670 nm, and for pumping Cr^{4+}- and Er^{3+}-lasers in the near-infrared. These developments were made possible because molecular beam epitaxy (MBE) and metallorganic chemical vapour deposition (MOCVD) techniques facilitate an ability to engineer the band structure of a semiconductor low dimensional structure, thereby changing the output to meet particular demands. *Band structure engineering* refers to the application of appropriate growth techniques to change the composition and dimensions of quantum well structures. The very rapid developments of GaAs/AlGaAs laser diodes operating at 800 nm and 1.3 μm were driven by the large market for such devices in the information and communication industries. However, there is now emphasis on devices operating over much wider spectral ranges. Ongoing successes with devices in the range 300–750 nm using the III–V nitrides will play a dominant role over the next decade. Already light emitting diodes (LEDs) and LDs operating over the whole of this range of wavelengths have been developed and marketed in Japan. Although the earliest devices had short operating life times, improvements in fabrication techniques have led to device life times exceeding 50 k hours.

In this same spirit *crystal field engineering* is being used to vary the efficiencies and wavelength ranges of solid state lasers based on colour centres, transition-metal ions and rare-earth ions in crystals. In this case the environment of an isolated optical centre in a crystal is engineered to give improved performance. The engineering is accomplished through the crystal growth process in which the structure and composition of the unit cell is modified in a controlled and known manner. As a simple example, the family of Cr^{3+}-doped garnets may be grown to a state of considerable perfection using the Czochralski technique. In $Y_3Al_5O_{12}$ (YAG) and $Y_3Ga_5O_{12}$ (YGG) the Cr^{3+} ion at

300 K emits through the sharp $^2E \rightarrow {}^4A_2$ transition, and laser action in these cases mimics that of the ruby laser, supporting single frequency operation near 695 nm in both crystals. In contrast, the Gd-garnets $Gd_3Sc_2Al_3O_{12}$ (GSAG) and $Gd_3Sc_2Ga_3O_{12}$ (GSGG) with their larger unit cells provide a weaker crystal-field environment for the Cr^{3+} ion, so that single frequency or broadband, tunable laser operation is possible with good efficiency. Changing the composition and unit cell dimensions further to Lu_3La_2 Ga_3O_{12} (LLGG) leads to broadband laser tuning, but at longer wavelengths than either Cr^{3+} : GSAG or Cr^{3+} : GSGG. However, crystal-field engineering affects other facets of the operational characteristics of solid state lasers. Not only are the unit cell dimensions modified, so too are the symmetry characteristics of the unit cell. The symmetry elements determine the width of optical transitions through the effect of dynamic even parity distortions of the environment (even parity phonons), as well as the transition rates and polarizations of transitions by static and dynamic odd-parity distortions. Crystal-field engineering is based on the crystal-field theory of Bethe (1929), which treated the electrostatic field of the host as a static perturbation on the energy levels of the free (point) ions, then classifying the new levels as representations of the symmetry group of the structure. In consequence, crystal-field theory combines elements of quantum mechanics and group representation theory within a single framework. The applications of these ideas to transition-metal and rare-earth ion spectroscopy were developed by Sugano, Tanabe and Kamimura (1970) and for rare-earth ions by Judd (1962, 1963) and Ofelt (1962). There have been many recent expositions of the subject, notably in the context of experimental results by Henderson and Imbusch (1989). All were restricted to crystal-field theory within the point ion lattice approximation, despite the fact that the covalency present in all ionic crystals requires the specific recognition of finite ion sizes [Van Vleck (1935)].

2

Symmetry considerations

2.1 Introduction

The distinguishing feature of crystalline solids is their symmetry, manifest microscopically in their X-ray diffraction patterns and macroscopically in crystal morphology. An ideal crystal is an infinite regular repetition in space of identical structural units. The symmetry of a particular ideal crystal is specified by the set of symmetry elements, comprising rotations, reflections and translations, which leave it invariant. Real crystals are not only of finite extent, but also contain a variety of imperfections such as inclusions of minority phases, grain boundaries, dislocations, impurities and point defects; the latter two are especially relevant in the present context. An isolated impurity or point defect in an otherwise ideal crystal obviously removes translational symmetry. It may also reduce the residual point symmetry in some circumstances, exemplified by ions with degenerate electronic states (the Jahn–Teller effect), impurities in the form of small molecules, small substitutional cations which move off-centre, vacancy pairs, bipolarons, etc. A principal theme of the present monograph is exploration of the ways in which the properties of a laser-active centre are controlled or affected by its crystalline surroundings, including their residual point symmetry. We shall find that it is often useful to distinguish a *dominant site symmetry*, determined by coordination alone, which is somewhat higher than the actual point symmetry.

The formal description of symmetry exploits a branch of mathematics called 'Group Theory', which is not a physical theory in the sense of 'Quantum Theory', but is rather a collection of principles deduced from a chosen set of axioms. Group theory is useful not only for the description and labeling of symmetry properties, but also for the classification of energy levels and wave functions and the determination of optical selection rules. Principles of group theory are considered first, beginning with abstract groups and proceeding to symmetry groups and their matrix representations. Operator groups serve to establish the relevance of group theory to quantum mechanics. Crystallographic point groups and space groups are introduced and applied to a representative crystal structure. Finally, Lie groups and their representations are considered. New terms are italicized at their initial occurrence. Several useful theorems are stated without proof; although the proofs are straightforward, and are contained in many standard references [Weyl (1931); Wigner (1959); Hamermesh (1962); Tinkham (1964); Lax (1974); Cornwell (1984); Hahn 1995)], they are omitted here for the sake of brevity.

Symmetry principles developed in the present chapter find extensive application in subsequent chapters. They will be seen to be essential to an understanding of the optical properties of laser materials.

2.2 Principles of group theory

2.2.1 *Abstract groups*

The theory of abstract groups proceeds from a set of four simple axioms as follows. An *abstract group* is a set of distinct elements, together with a law of composition (group multiplication), such that:

(1) The set is closed under group multiplication ($ab = c$).
(2) Group multiplication is associative [$(ab)c = a(bc)$].
(3) The set contains an identity element ($ae = ea = a$).
(4) Every element has an inverse in the set ($aa^{-1} = a^{-1}a = e$).

Note that there is only one law of composition, group multiplication, which should not be confused with ordinary multiplication. In general, group multiplication does not commute. If it does for all elements, the group is *Abelian*. The *order*, g, of a finite group G is the number of its elements.

If a subset of elements of a group, G, satisfies group axioms with the same law of composition, it comprises a *subgroup*, H, of order h. Trivial subgroups are the entire group, G, and the identity element, e. A group can be decomposed into a union of a subgroup and its *left cosets*,

$$G = H + aH + bH + \cdots + kH, \tag{2.1}$$

where a is an element of G not in H, b is an element of G not in H or aH, etc. Left cosets are disjoint and their elements are distinct; hence Lagrange's theorem: $g = mh$, where the integer m is the *index* of H in G. The elements $a, b, \ldots, k$ are *left-coset generators*. Note that any element of a left coset can serve as a left-coset generator. *Right coset* decomposition can be performed analogously.

Conjugation provides an alternative criterion for classification of group elements. Element a is *conjugate* to element b if there is an element u in G such that $b = uau^{-1}$. Conjugation is an equivalence relation since it is reflexive ($a = eae^{-1}$), reciprocal [$b = uau^{-1} \rightarrow a = u^{-1}b(u^{-1})^{-1}$] and transitive [$b = uau^{-1}$ and $c = vbv^{-1} \rightarrow c = (vu)a(vu)^{-1}$]. Consequently, group elements can be uniquely sorted into *classes* of mutually conjugate elements. The *conjugate subgroup* aHa^{-1} has the same algebraic structure as subgroup H. If $aHa^{-1} = H$ for all a in G, then H is an *invariant* subgroup. Invariance implies that right and left cosets are identical ($aH = Ha$), and that H contains elements only in complete classes.

An invariant subgroup, H, and its cosets are closed under *coset multiplication*, since, with repeated product elements counted only once,

$$(aH)(bH) = abHH = abH. \tag{2.2}$$

The invariant subgroup and its cosets satisfy group axioms under coset multiplication, with H as the identity and $a^{-1}H$ as the inverse of aH, since the associative law for cosets is assured by the corresponding law for group elements. The resulting group is called the *factor group*, and is designated by the symbol G/H. A many-to-one correspondence between elements of two groups G and G' which preserves algebraic structure ($ab = c \rightarrow a'b' = c'$) is a *homomorphic mapping*. The elements of G which map onto the identity element e' of G' form an invariant subgroup called the kernel, K, of the mapping, and cosets of K map onto the remaining elements of G'. The *homomorphic image* G' is *isomorphic* (in one-to-one correspondence) with the factor group G/K.

A group G is the *direct product* of some of its invariant subgroups, called *direct factors*, provided elements of different subgroups commute and every element of G is uniquely expressible as a product of elements, one from each subgroup. The notation is $G = H_1 \otimes H_2 \otimes \cdots \otimes H_n$.

A finite group is specified by its multiplication table, whose entries are products of a left-hand factor from an edge column and a right-hand factor from an edge row. The multiplication table must have the following properties:

(1) The edge row and column are replicated in the table.
(2) Each element occurs once in each row and column.
(3) Rectangles with the identity e in the upper left-hand corner have the form:

$$\begin{matrix} e & \cdots & y \\ \vdots & & \vdots \\ x & \cdots & xy \end{matrix}$$

All of the properties of the elements of an abstract group are defined by its multiplication table; the elements have no other meaning, in contrast with elements of symmetry groups, matrix groups and operator groups considered below.

2.2.2 *Symmetry groups*

Symmetry operations are rotations, reflections and translations which leave an object invariant; i.e., by virtue of its symmetry, the object appears to be the same after reorientation. The set of symmetry operations satisfies group axioms with successive application as the law of composition, and is thus a *symmetry group*. The identity operation corresponds to no net reorientation, e.g., rotation through a multiple of 2π about any axis or double reflection in any mirror plane, and the inverse of any operation corresponds to reversing its sense. Multiplication is obviously closed and associative. Symmetry groups are always isomorphic with abstract groups.

As an example of a finite symmetry group, consider the group D_3, of order $g = 6$, of all proper rotations which leave an equilateral triangle invariant. Elements include the identity (E), rotations through 120° (C_3) and 240° (C_3^2) about an axis perpendicular to the plane of the triangle which passes through its centre, and rotations through 180° (C_a, C_b and C_c) about three axes in the plane of the triangle which pass through its centre and its vertices. The sense of rotation is related to the positive direction of an axis by a right-hand rule. The directions of the two-fold axes a, b and c are related to that of the three-fold axis as illustrated in Fig. 2.1. One can construct the multiplication table of D_3 by successive application of symmetry operations, observing the following conventions: The operation corresponding to the right-hand element in a product is applied first, and successive rotations are performed about axes fixed in space rather than in the body. The multiplication table displayed in Table 2.1, with rows and columns labelled by group elements, is obtained in this manner.

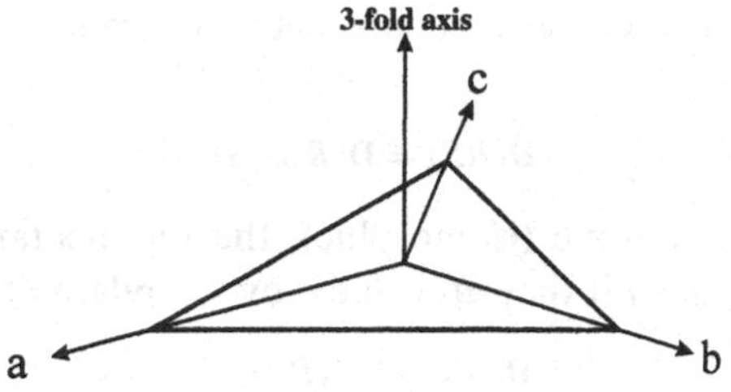

Figure 2.1. Symmetry axes of an equilateral triangle.

Table 2.1. *Multiplication table for group D_3*

D_3	E	C_3	C_3^2	C_a	C_b	C_c
E	E	C_3	C_3^2	C_a	C_b	C_c
C_3	C_3	C_3^2	E	C_c	C_a	C_b
C_3^2	C_3^2	E	C_3	C_b	C_c	C_a
C_a	C_a	C_b	C_c	E	C_3	C_3^2
C_b	C_b	C_c	C_a	C_3^2	E	C_3
C_c	C_c	C_a	C_b	C_3	C_3^2	E

Table 2.2. *Multiplication table for factor group D_3/H*

D_3/H	H	C_aH
H	H	C_aH
C_aH	C_aH	H

One can readily verify that the multiplication table for D_3 has the required properties.

The elements of D_3 fall into three conjugate classes: $\{E\}$, $\{C_3, C_3^2\}$ and $\{C_a, C_b, C_c\}$. Symmetry groups provide a geometrical interpretation of conjugate classes: Elements in the same class correspond to rotations through the same angle about axes which are related to one another by other elements of the group. This concept can be generalized to improper rotations as well.

In addition, four non-trivial subgroups can be identified: one invariant subgroup $H=\{E, C_3, C_3^2\}$, of index 2, with coset $C_aH=\{C_a, C_b, C_c\}$, and the subgroups $\{E, C_a\}$, $\{E, C_b\}$ and $\{E, C_c\}$ of index 3. Furthermore, it is evident that the multiplication table of the factor group D_3/H is as shown in Table 2.2. Finally, elements C_3 and C_a are group generators, since all elements can be expressed as products of these two.

2.2.3 *Matrix representations*

A *matrix group* is a set of square, non-singular matrices which satisfies group axioms with matrix multiplication as the law of composition. A matrix group is a *matrix representation* of an abstract group if it is a homomorphic image of the abstract group;

i.e., there is a many-to-one correspondence between group elements and matrices such that

$$\mathbf{D}(RS) = \mathbf{D}(R)\mathbf{D}(S). \tag{2.3}$$

If the correspondence is unique (isomorphic), the representation is *faithful*. Two representations are *equivalent* if they are related by a similarity transformation:

$$\mathbf{D}'(R) = \mathbf{U}\mathbf{D}(R)\mathbf{U}^{-1}, \tag{2.4}$$

where the square, non-singular matrix $\mathbf{U}$ is not necessarily an element of the matrix group. Every representation is equivalent to a *unitary representation*, in which every element is represented by a unitary matrix. Only unitary representations are considered henceforth.

If all matrices in a representation are partitioned in the same way,

$$\mathbf{D}(R) = \begin{pmatrix} \mathbf{D}'(R) & 0 \\ 0 & \mathbf{D}''(R) \end{pmatrix} = \mathbf{D}'(R) \oplus \mathbf{D}''(R), \tag{2.5}$$

then $\mathbf{D}'(R)$ and $\mathbf{D}''(R)$ are also representations. If a given representation is equivalent to a partitioned representation, it is *reducible*; otherwise, it is *irreducible*. *Schur's lemma* states that any matrix which commutes with all matrices of an irreducible representation is a constant matrix. Pre-eminent among the theorems which follow from Schur's lemma is the *grand orthogonality theorem*,

$$\sum_R D_{ij}^{(\mu)*}(R) D_{kl}^{(\nu)}(R) = \frac{g}{d_\mu} \delta_{\mu\nu}\, \delta_{ik}\, \delta_{jl}, \tag{2.6}$$

where the sum is over elements of a group of order g, and μ and ν label inequivalent irreducible representations of dimension d_μ and d_ν, respectively. It is evident from Eq. (2.6) that a finite group has a finite number of inequivalent irreducible representations of finite dimension. In fact, the number of inequivalent, irreducible representations is just equal to the number of classes, and their dimensions satisfy the relation

$$\sum_\mu d_\mu^2 = g. \tag{2.7}$$

The character of a representation is the set of traces, $\chi(R) \equiv \sum_i D_{ii}(R)$, for all R.

By virtue of trace invariance under similarity transformation, the characters of equivalent representations are identical and the traces are the same for all elements in a class. Consequently, one can display all of the characters of a given finite group, G, in a unique *character table* with rows and columns labeled by irreducible representations Γ_μ and classes C_j, respectively, as shown in Table 2.3.

Table 2.3. *Character table of a finite group*

G	C_1	C_2	$\cdots$	C_k
Γ_1	$\chi_1^{(1)}$	$\chi_2^{(1)}$	$\cdots$	$\chi_k^{(1)}$
Γ_2	$\chi_1^{(2)}$	$\chi_2^{(2)}$		
$\cdots$	$\cdots$		$\cdots$	
Γ_k	$\chi_1^{(k)}$			$\chi_k^{(k)}$

Table 2.4. *Irreducible representations of group D_3*

D_3	E	C_3	C_3^2	C_a	C_b	C_c
A_1 (Γ_1)	1	1	1	1	1	1
A_2 (Γ_2)	1	1	1	−1	−1	−1
E (Γ_3)	$\begin{pmatrix}1 & 0\\ 0 & 1\end{pmatrix}$	$\begin{pmatrix}\varepsilon & 0\\ 0 & \varepsilon^2\end{pmatrix}$	$\begin{pmatrix}\varepsilon^2 & 0\\ 0 & \varepsilon\end{pmatrix}$	$\begin{pmatrix}0 & 1\\ 1 & 0\end{pmatrix}$	$\begin{pmatrix}0 & \varepsilon^2\\ \varepsilon & 0\end{pmatrix}$	$\begin{pmatrix}0 & \varepsilon\\ \varepsilon^2 & 0\end{pmatrix}$

Table 2.5. *Character table of crystallographic point group D_3*

D_3	E	$2C_3$	$3C_2$
A_1 (Γ_1)	1	1	1
A_2 (Γ_2)	1	1	−1
E (Γ_3)	2	−1	0

Row orthogonality of the character table follows trivially from the grand orthogonality theorem, Eq. (2.6),

$$\sum_i \frac{g_i}{g}\chi_i^{(\mu)^*}\chi_i^{(\nu)} = \delta_{\mu\nu}, \tag{2.8}$$

where g_i is the number of elements in the ith class. This theorem enables one to determine how many times each inequivalent irreducible representation is contained in a given reducible representation, $\mathbf{D}(R)$. By definition, there exists a unitary matrix $\mathbf{U}$ such that

$$\mathbf{U}\mathbf{D}(R)\mathbf{U}^+ = \sum_\nu a_\nu \mathbf{D}^{(\nu)}(R), \tag{2.9}$$

where the summation sign denotes the direct sum. It follows from Eqs. (2.8) and (2.9) that the number of times, a_μ, that the μth irreducible representation is contained in $\mathbf{D}(R)$ is given by

$$a_\mu = \frac{1}{g}\sum_i g_i \chi_i^{(\mu)^*}\chi_i. \tag{2.10}$$

This relation finds extensive application in crystal-field theory.

As an example, consider the irreducible representations of group D_3, displayed in Table 2.4 where $\varepsilon = \exp(2\pi i/3)$. Representation A_1 is the *identity representation.* Note that representation A_2 is isomorphic with the factor group D_3/H, and that representation E is faithful. The corresponding character table is shown in Table 2.5. Classes are denoted by the number of elements, g_i, followed by a typical element.

2.2.4 *Symmetry and quantum mechanics*

Operator groups provide another type of representation which serves to establish the connection between symmetry and quantum mechanics. We begin by considering a group of transformations of coordinates:

$$\mathbf{r}' = R\mathbf{r}, \tag{2.11}$$

where R is a real, proper or improper orthogonal transformation. There is a corresponding operator, O_R, which rotates a field:

$$O_R\psi(\mathbf{r}') = \psi(\mathbf{r}) \to O_R\psi(\mathbf{r}) = \psi(R^{-1}\mathbf{r}). \tag{2.12}$$

The rotated function, $O_R\psi$, has the same value at the rotated point, $\mathbf{r}'$, as the original function, ψ, at the original point, $\mathbf{r}$. The operator group is a homomorphic image of the group of transformations, since

$$O_S O_R\psi(\mathbf{r}) = O_R\psi(S^{-1}\mathbf{r}) = \psi(R^{-1}S^{-1}\mathbf{r}) = \psi[(SR)^{-1}\mathbf{r}] = O_{SR}\psi(\mathbf{r}); \tag{2.13}$$

therefore, algebraic structure is preserved.

The transformation, R, in Eq. (2.11) can be viewed alternatively as an inverse rotation of the coordinate axes which alters the Cartesian components of a fixed vector, $\mathbf{r}$. A corresponding interpretation then applies for the operators O_R, and even for the symmetry operations which leave the appearance of an object invariant. Labeling of symmetry elements by the rotations of the coordinate axes rather than those of the object itself alters the appearance of the multiplication table. The difference is only semantic, however, since isomorphism ensures preservation of irreducible representations and characters.

Consider a set of d independent functions, $\psi_i(\mathbf{r})$, with $d \leq g$, which transform among themselves under operations of the group:

$$O_R\psi_i(\mathbf{r}) = \sum_j \psi_j(\mathbf{r})D_{ji}(R). \tag{2.14}$$

It can be shown readily that the matrices of coefficients, $\mathbf{D}(R)$, form a d-dimensional matrix representation of the group of transformations. The functions $\psi_i(\mathbf{r})$ are called *basis functions* for the representation. If the matrices of coefficients form an irreducible representation, $\mathbf{D}^{(\mu)}(R)$, then the basis function $\psi_i^{(\mu)}(\mathbf{r})$ *belongs* to the ith row of the μth irreducible representation. Orthogonality of basis functions for irreducible representations follows from the grand orthogonality theorem:

$$(\psi_i^{(\mu)}, \phi_j^{(\nu)}) = \delta_{ij}\delta_{\mu\nu}\frac{1}{d_\mu}\sum_k (\psi_k^{(\mu)}, \phi_k^{(\mu)}); \tag{2.15}$$

i.e., basis functions belonging to different rows or to different irreducible representations are orthogonal, and the scalar product is independent of row. Note that ψ and ϕ may belong to two entirely different sets of basis functions. It will become apparent that this extremely powerful theorem has many useful consequences.

A symmetry operator, O_R, transforms another operator, Ω, according to the recipe

$$\Omega' = O_R\Omega O_R^{-1}, \tag{2.16}$$

which preserves the algebraic structure of equations. Suppose that the Hamiltonian, H, is *invariant* under a group of operations, G, called the *group of the Hamiltonian*; i.e., H is transformed into itself for all R in G,

$$O_R H O_R^{-1} = H. \tag{2.17}$$

Application of O_R to the time independent Schrödinger equation then yields

$$O_R H \psi_i^{(\nu)}(\mathbf{r}) = H O_R \psi_i^{(\nu)}(\mathbf{r}) = E^{(\nu)} O_R \psi_i^{(\nu)}(\mathbf{r}). \tag{2.18}$$

Since $O_R\psi_i^{(\nu)}(\mathbf{r})$ is an eigenfunction of H with eigenvalue $E^{(\nu)}$, it may be expanded in the set of d_ν degenerate eigenfunctions with eigenvalue $E^{(\nu)}$,

$$O_R \psi_i^{(\nu)}(\mathbf{r}) = \sum_{j=1}^{d_\nu} \psi_j^{(\nu)}(\mathbf{r}) D_{ji}^{(\nu)}(R). \tag{2.19}$$

It follows that the square matrices $\mathbf{D}^{(\nu)}(R)$, with elements $D_{ij}^{(\nu)}(R)$, form a matrix representation of G, for which the degenerate eigenfunctions form a basis. If the representation is irreducible, the degeneracy is *symmetry-induced*; otherwise, it is *accidental*. In the latter case, which reflects fortuitous values of parameters, it is still possible to choose linear combinations of degenerate eigenfunctions which partition the representation into irreducible blocks. It can be shown from Shur's lemma and the orthogonality of basis functions that the degeneracy of eigenfunctions is a necessary as well as a sufficient condition that they transform as bases for irreducible representations of G. Thus, eigenvalues can be labeled by irreducible representations of the group G of the Hamiltonian, and eigenfunctions by the rows of these representations.

2.2.5 *Coupled systems*

A recurring problem in quantum mechanics is that of two nearly independent systems which are coupled by a weak interaction. The symmetry classification of the eigenvalues and eigenfunctions of the coupled system is greatly facilitated by the following considerations.

Given basis functions $\psi_i^{(\mu)}(\mathbf{r}_1)$ and $\phi_j^{(\nu)}(\mathbf{r}_2)$ for irreducible representations $\mathbf{D}^{(\mu)}(R)$ and $\mathbf{D}^{(\nu)}(R)$ of group G, respectively, then products $\psi_i^{(\mu)}(\mathbf{r}_1)\phi_j^{(\nu)}(\mathbf{r}_2)$ form a basis for the *Kronecker product representation*

$$\mathbf{D}^{(\mu\times\nu)}(R) = \mathbf{D}^{(\mu)}(R) \otimes \mathbf{D}^{(\nu)}(R). \tag{2.20}$$

In a concise notation, in which $\boldsymbol{\psi\phi}$ denotes a row vector with $d_\mu d_\nu$ elements $\psi_i^{(\mu)}(\mathbf{r}_1)\phi_j^{(\nu)}(\mathbf{r}_2)$, the transformation properties can be expressed as

$$O_R \boldsymbol{\psi\phi} = \boldsymbol{\psi\phi}\, \mathbf{D}^{(\mu\times\nu)}(R). \tag{2.21}$$

The Kronecker product representation is reducible in general, and can be partitioned into irreducible representations by a unitary transformation,

$$\mathbf{U}\mathbf{D}^{(\mu\times\nu)}\mathbf{U}^{+} = \sum_\sigma a_\sigma \mathbf{D}^{(\sigma)}(R), \tag{2.22a}$$

$$\mathbf{U}\mathbf{U}^{+} = \mathbf{U}^{+}\mathbf{U} = \mathbf{I}, \tag{2.22b}$$

$$a_\sigma = \frac{1}{g}\sum\nolimits_i g_i \chi_i^{(\mu)} \chi_i^{(\nu)} \chi_i^{(\sigma)*}. \tag{2.22c}$$

It follows that elements of the row vector $\psi\phi\mathbf{U}^+$ are bases for irreducible representations of G,

$$O_R\psi\phi\mathbf{U}^+ = \psi\phi\mathbf{U}^+\mathbf{U}\mathbf{D}^{(\mu\times\nu)}(R)\mathbf{U}^+ = \psi\phi\mathbf{U}^+\sum_\sigma a_\sigma \mathbf{D}^{(\sigma)}(R). \tag{2.23}$$

Consequently, these elements may be written in the form

$$\Psi_k^{(\lambda\tau_\lambda)}(\mathbf{r}_1, \mathbf{r}_2) = \sum_{i,j} \psi_i^{(\mu)}(\mathbf{r}_1)\phi_j^{(\nu)}(\mathbf{r}_2)\langle\mu\nu ij|\lambda\tau_\lambda k\rangle, \tag{2.24}$$

where the *coupling coefficients* $\langle\mu\nu ij|\lambda\tau_\lambda k\rangle$ are elements of the matrix $\mathbf{U}^+$. The index τ_λ distinguishes multiple occurrences of irreducible representation λ in the reduction of the Kronecker product representation; it may be dispensed with in the special case of a *simply reducible* group, for which no irreducible representation occurs more than once in the reduction of a Kronecker product representation.

Consider two subsystems with Hamiltonians H_1 and H_2 which are coupled by a weak interaction H'. The Hamiltonian for the combined system is then

$$H = H_0 + H', \tag{2.25a}$$

$$H_0 \equiv H_1 + H_2, \tag{2.25b}$$

and the group of the Hamiltonian, H_0, of the uncoupled system is

$$G_0 = G_1 \otimes G_2, \tag{2.26}$$

since H_0 is invariant under independent operations R_1 and R_2 of G_1 and G_2, respectively. Suppose further that G_1 and G_2 are isomorphic, and that the effect of the interaction, H', is to reduce the symmetry so that the Hamiltonian, H, of the coupled system is invariant only under simultaneous operations $R_1 = R_2 \equiv R$, and not under independent operations. The group G of H is then isomorphic with G_1 and G_2, and is the subgroup of G_0 for which $R_1 = R_2 \equiv R$. Since irreducible representations of G_0 are of the form $\mathbf{D}^{(\mu)}(R_1) \otimes \mathbf{D}^{(\nu)}(R_2)$, the subset of matrices representing the subgroup G, the *subduced* representation, is just the Kronecker product representation defined by Eq. (2.20).

Eigenvalues $E_0^{(\mu\nu)}$ of H_0 are labeled by irreducible representations of G_0. The effect of the perturbation H' is to split each level of H_0 into a set of levels labeled by the irreducible representations of G contained in the reduction of the Kronecker product representation, Eq. (2.22a). The number of levels derived from $E_0^{(\mu\nu)}$ is $\sum_\sigma a_\sigma$, with residual degeneracy d_σ of each level, such that

$$d_\mu d_\nu = \sum_\sigma a_\sigma d_\sigma. \tag{2.27}$$

2.3 Crystal symmetry

2.3.1 Translation groups

An ideal crystal is composed of atoms or ions such that the atomic arrangement looks the same from points $\mathbf{r}$ and $\mathbf{r}'$, where

$$\mathbf{r}' = \mathbf{r} + l_1\mathbf{d}_1 + l_2\mathbf{d}_2 + l_3\mathbf{d}_3, \tag{2.28}$$

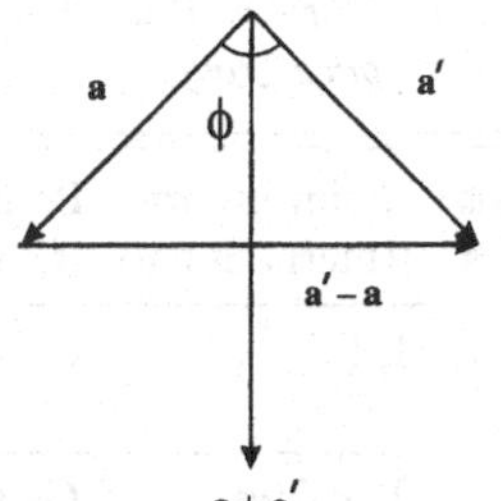

Figure 2.2. Restrictions of the rotation angle in a crystal lattice. All of the indicated vectors must be lattice vectors no shorter than **a**, and $\phi = 2\pi/n$, where n is a positive integer.

the l_i are integers, and the $\mathbf{d}_i$ are fundamental translation vectors. A translation operation $\{(\varepsilon|\mathbf{R}_l\}$ consists of a rigid translation of the entire crystal by a *lattice vector* $\mathbf{R}_l$, defined by

$$\mathbf{R}_l \equiv \mathbf{r}' - \mathbf{r} = l_1\mathbf{d}_1 + l_2\mathbf{d}_2 + l_3\mathbf{d}_3. \tag{2.29}$$

The entire set of translations which leave the ideal crystal invariant constitutes a *translation group*.

A three-dimensional *lattice* is an array of points in space with translational symmetry in the three non-coplanar directions $\hat{\mathbf{d}}_1$, $\hat{\mathbf{d}}_2$ and $\hat{\mathbf{d}}_3$. A *basis* is a unit assembly of atoms or ions associated with each lattice point. Translational symmetry is preserved in the addition of a basis to a lattice, but not necessarily all of the symmetry.

2.3.2 Crystallographic point groups

A *point group* is a group of symmetry operations applied about a point which leave an object invariant; its elements include rotations, reflections and inversions. *Crystallographic point groups* are severely restricted by the translational symmetry of an ideal crystal. Assume that **a** is one of the shortest non-vanishing lattice vectors, and that a rotation through an angle ϕ about a particular symmetry axis leaves the crystal invariant and takes **a** into **a**′. Then $\mathbf{a} - \mathbf{a}'$ and $\mathbf{a} + \mathbf{a}'$ are also lattice vectors, and, by hypothesis, the following inequalities are satisfied:

$$|\mathbf{a}| \leq |\mathbf{a}' - \mathbf{a}|, |\mathbf{a} + \mathbf{a}'|. \tag{2.30}$$

It is apparent from Fig. 2.2 that the angle of rotation is restricted to the following values:

$$\phi = 2\pi/n, \quad (n = 1, 2, 3, 4, 6). \tag{2.31}$$

Crystallographic point groups with only a single n-fold axis, $n = 1, 2, 3, 4, 6$, include Abelian, *cyclic* groups C_n, of order n, with elements C_n^p such that $C_n^n = E$. (Unfortunately, the names of group elements in Schoenflies notation are easily confused with the names of the groups themselves.) Dihedral groups D_n, $n = 2, 3, 4, 6$ of order $2n$, are formed by adjoining n perpendicular 2-fold axes to a single n-fold axis; of these, only D_2 is Abelian. Polyhedral groups include the octahedral group O, of order 24, and the

Table 2.6. *The seven symmetry systems, fourteen Bravais lattices, thirty-two crystallographic point groups and two hundred and thirty space groups*

System	Restrictions	Bravais lattices	Point groups (International)	Point groups (Schoenflies)	Space group numbers
Triclinic	$a \neq b \neq c$, $\alpha \neq \beta \neq \gamma$	P	$1, \bar{1}$	C_1, C_i	1, 2
Monoclinic	$a \neq b \neq c$, $\alpha = \gamma = 90° \neq \beta$	P, C	$2, m, 2/m$	C_2, C_s, C_{2h}	3–5, 6–9, 10–15
Orthorhombic	$a \neq b \neq c$, $\alpha = \beta = \gamma = 90°$	P, C, I, F	$222, mm2,$ mmm	$D_2, C_{2v},$ D_{2h}	16–24, 25–46, 47–74
Tetragonal	$a = b \neq c$, $\alpha = \beta = \gamma = 90°$	P, I	$4, \bar{4}, 4/m,$ $422, 4mm,$ $\bar{4}2m, 4/mmm$	$C_4, S_4, C_{4h},$ $D_4, C_{4v},$ D_{2d}, D_{4h}	75–80, 81–82, 83–88, 89–98, 99–110, 11–112, 123–142
Trigonal	$a = b = c$, $\alpha = \beta = \gamma \neq 90°$	R	$3, \bar{3},$ $32,$ $3m, \bar{3}m$	$C_3, S_6,$ $D_3,$ C_{3v}, D_{3d}	142–146, 147–148, 149–155, 156–161, 162–167
Hexagonal	$a = b \neq c$, $\alpha = \beta = 90°$, $\gamma = 120°$	P	$6, \bar{6}, 6/m,$ $622, 6mm,$ $\bar{6}2m, 6/mmm$	$C_6, C_{3h}, C_{6h},$ $D_6, C_{6v},$ D_{3h}, D_{6h}	168–173, 174, 175–176, 177–182, 183–186, 187–190, 191–194
Cubic	$a = b = c$, $\alpha = \beta = \gamma = 90°$	P, I, F	$23, m\bar{3},$ $432, \bar{4}3m,$ $m\bar{3}m$	$T, T_h,$ $O, T_d,$ O_h	195–199, 200–206, 207–224, 215–220, 221–230

tetrahedral group T, of order 12. The octahedral group O, which leaves a cube or octahedron invariant, has three perpendicular four-fold axes, four three-fold axes and six two-fold axes. The tetrahedral group T, which leaves a regular tetrahedron invariant, is a subgroup of O with four three-fold axes and three two-fold axes.

Additional crystallographic point groups are obtained by adjoining *rotation-reflections* S_n, consisting of rotation through an angle $2\pi/n$, followed by reflection in a plane perpendicular to the axis of rotation. Special cases include reflection ($\sigma = S_1$) and inversion ($I = S_2$). New groups generated from groups C_n include C_s, of order 2, with elements (E, σ); C_i, of order 2, with elements (E, I); S_4, of order 4, with elements (E, C_2, S_4, S_4^3); and S_6, of order 6, with elements $(E, C_3, C_3^2, S_6, S_6^3, S_6^5)$. They also include groups C_{nh} and C_{nv}, obtained, respectively, by adjoining *horizontal* and *vertical* mirror planes to C_n; i.e., planes which, respectively, are perpendicular to and include the n-fold axis. Additional dihedral groups D_{nh} and D_{nd} are generated from D_n by adjoining horizontal and diagonal mirror planes, respectively, where a diagonal mirror plane is a vertical mirror plane which bisects the angle between two-fold axes. Finally, additional polyhedral groups T_h, T_d and O_h are generated from T and O by adjoining horizontal and diagonal mirror planes. The thirty-two crystallographic point groups listed in Table 2.6 exhaust the possibilities. They are isomorphic with just eighteen distinct abstract groups, and several can be represented as direct products of invariant subgroups which are themselves crystallographic point groups.

Table 2.7. *Character table of crystallographic point group C_i*

C_i	E	I
$A_g(\Gamma_1^+)$	1	1
$A_u(\Gamma_1^-)$	1	−1

Table 2.8. *Character table of crystallographic point group D_{3d}*

D_{3d}	E	$2C_3$	$3C_2$	I	$2C_3I$	$3C_2I$
A_{1g} (Γ_1^+)	1	1	1	1	1	1
A_{2g} (Γ_2^+)	1	1	−1	1	1	−1
E_g (Γ_3^+)	2	−1	0	2	−1	0
A_{1u} (Γ_1^-)	1	1	1	−1	−1	−1
A_{2u} (Γ_2^-)	1	1	−1	−1	−1	1
E_u (Γ_3^-)	2	−1	0	−2	1	0

As an example, the group D_{3d} of symmetry operations which leave a regular rhombohedron invariant is generated by adjoining the inversion I to the group D_3 of proper rotations which leave an equilateral triangle invariant. It can be represented as the direct product $D_{3d}=D_3 \otimes C_i$. The character table of the Abelian group C_i, which has only one-dimensional representations, is shown in Table 2.7. The subscripts g and u stand for *gerade* and *ungerade*, German for even and odd, respectively. The character table for D_{3d}, shown in Table 2.8, is then obtained trivially from that for D_3 and C_i by keeping in mind that the inversion is represented by +1 in gerade representations and by −1 in ungerade representations.

2.3.3 *Space groups*

A space group element $\{\alpha|\mathbf{t}\}$ is a symmetry operation which leaves an ideal crystal invariant. It effects a proper or improper orthogonal transformation α, followed by a translation through $\mathbf{t}$, not necessarily a lattice vector,

$$\mathbf{r}' = \alpha\mathbf{r} + \mathbf{t}. \tag{2.32}$$

The orthogonal transformation α must be an element of a crystallographic point group G which contains an identity element ε, and the set of elements $\{\varepsilon|\mathbf{R}_l\}$ is an invariant subgroup of pure translations. It should be noted that $\{\alpha|\mathbf{0}\}$ is not necessarily an element of the space group. If it is, for all α in G, then these elements form a subgroup and the space group is *symmorphic*. Non-symmorphic space groups contain elements of the form $\{\alpha|\mathbf{v}(\alpha)\}$, where $\mathbf{v}(\alpha)$ is not a lattice vector. The corresponding symmetry operations include a *screw operation*, in which α is a rotation through $2\pi/n$ and $\mathbf{v}(\alpha)$ is an advance through $(1/n)$th of the repeat distance parallel to the rotation axis, and a *glide operation* in which α is a reflection in a mirror plane and $\mathbf{v}(\alpha)$ is an advance through half the repeat distance parallel to the mirror plane. There are 230 distinct space groups in three dimensions, of which 73 are symmorphic.

Given a particular crystallographic point group of orthogonal transformations α, the requirement that $\alpha\mathbf{R}_l$ be a lattice vector imposes restrictions on the fundamental

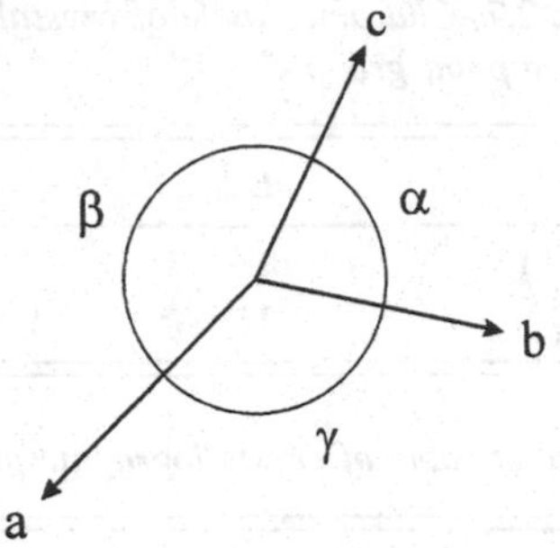

Figure 2.3. Linear combinations of fundamental translation vectors on which restrictions defining the seven symmetry systems in Table 2.6 are imposed.

translation vectors $\mathbf{d}_i$. Consequently, the thirty-two crystallographic point groups can be classified in seven *crystal systems*, defined by restrictions on the lengths and angles of linear combinations **a**, **b** and **c** of the $\mathbf{d}_i$, as shown in Fig. 2.3 and Table 2.6. The parallelepiped defined by the fundamental translation vectors $\mathbf{d}_i$ is a *primitive unit cell.* These cells may be stacked to fill all space by displacing them through all lattice vectors $\mathbf{R}_l$. The vectors **a**, **b** and **c** define a *conventional unit cell* which may contain more than one lattice point, and which may be stacked to fill all space by displacement through a subset of lattice vectors. In general, conventional unit cells have a more obvious relation to point symmetry than do primitive unit cells. There are fourteen distinct *Bravais lattices*, corresponding to different ways of associating the fundamental translation vectors $\mathbf{d}_i$ with vectors **a**, **b** and **c**. Within each crystal system, the Bravais lattices are classified by the arrangement of lattice points within the conventional unit cell as P (primitive), I (body centred), F (face centred) and C (base centred), as illustrated in Fig. 2.4. The fourteen Bravais lattices listed in Table 2.6 also include a primitive rhombohedral unit cell, conventionally designated R. In the hexagonal system, the primitive cell is always hexagonal, whereas in the trigonal system, the primitive cell is sometimes rhombohedral and sometimes hexagonal, depending on the space group. Since either system can be referred to either type of axes, the trigonal system in contemporary usage is combined with the hexagonal system in a single *hexagonal crystal family* [Hahn (1995)]. Point groups may be associated with Bravais lattices in a variety of ways within each crystal system. In Schoenflies notation, different space groups derived from the same point group are distinguished by numerical superscripts in an arbitrary sequence. The *international* (*Hermann–Mauguin*) notation for space groups is rather more informative, since both the point symmetry and the Bravais lattice are identified, with additional symbols denoting screw and glide operations. Both notations, as well as space-group numbers listed in Table 2.6, are employed in subsequent chapters.

As an example of a non-symmorphic space group, consider the group D_{3d}^6 ($R\bar{3}c$), associated with the trigonal R Bravais lattice with rhombohedral unit cell. The element $\{I|\mathbf{v}\}$ is adjoined to subgroups $\{\alpha_{D_3}|\mathbf{0}\}$ and $\{\varepsilon|\mathbf{R}_l\}$, where $\mathbf{v} = \frac{1}{2}(\mathbf{d}_1 + \mathbf{d}_2 + \mathbf{d}_3)$, to generate elements $\{\alpha_{D_3}|\mathbf{R}_l\}$ and $\{\alpha_{D_3}I|\mathbf{R}_l + \mathbf{v}\}$. The symmetry elements of this group are illustrated in Fig. 2.5a. It is evident from this figure that points of inversion symmetry do not coincide with points of D_3 symmetry, and no point has the full D_{3d} symmetry.

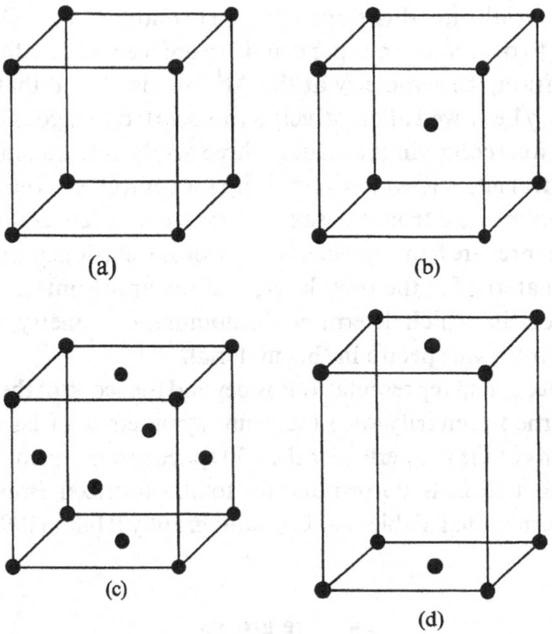

Figure 2.4. Examples of unit cells for Bravais lattices: (a) primitive cubic (P), (b) body centred cubic (I), (c) face centred cubic (F) and (d) base centred tetragonal (C).

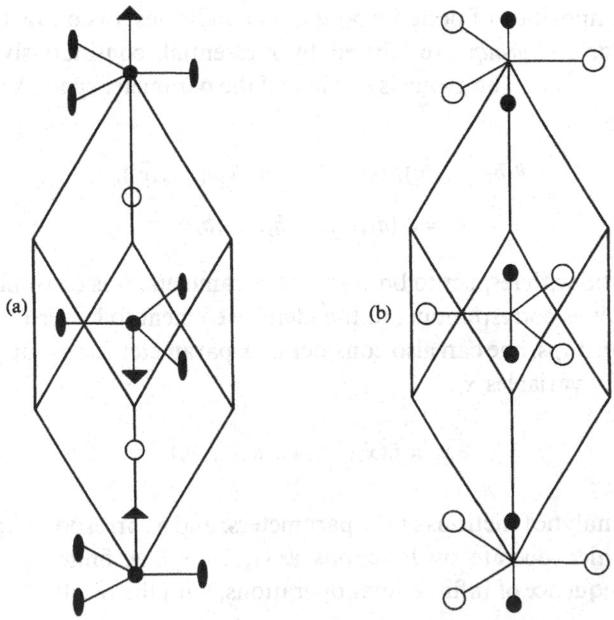

Figure 2.5. (a) Symmetry elements in the rhombohedral R primitive unit cell of space group D_{3d}^6. Open circles denote centres of inversion (point group C_i) and filled circles, points of D_3 symmetry. Ellipses and triangles label two-fold and three-fold axes, respectively. (b) Rhombohedral R primitive unit cell of corundum (α-Al_2O_3). Filled and open circles label Al^{3+} and O^{2-} ions, respectively. The basis contains two formula units. Not drawn to scale.

An example of a crystal with this symmetry is corundum (α-Al_2O_3), illustrated in Fig. 2.5b. There are two formula units per primitive unit cell, and neither ion occupies a high symmetry position; the symmetry at the Al^{3+} site is C_3 and that at the O^{2-} site is C_2. Corundum may be viewed alternatively as a distorted hexagonal close packing of O^{2-} ions with Al^{3+} ions occupying two out of three octahedral interstices, and with the unoccupied octahedral interstices coinciding with the centres of inversion. Three of the coordinating oxygen ions are from the same primitive unit cell as the aluminum ion, while the remaining three are from adjacent cells. Corundum doped with substitutional Cr^{3+} is the active material for the ruby laser, and the approximate octahedral coordination of the cation site, which determines its dominant symmetry, is essential to the understanding of chromium spectra in this material.

The theory of space group representations is beyond the scope of the present volume, since we are concerned primarily with the point symmetries of laser-active centres. Detailed descriptions of the properties of the 230 space groups, including illustrations of the conventional unit cells corresponding to the fourteen Bravais lattices, are available in the International Tables for Crystallography [Hahn (1995)].

2.4 Lie groups

2.4.1 Lie algebras

Although space groups are, in principle, of infinite order by virtue of the translation symmetry of ideal crystals, they have discrete elements and can be converted readily to finite groups by imposition of periodic boundary conditions. In contrast, the elements of an r-parameter *Lie group* are labeled by r essential, continuously varying real parameters $a_1, a_2, \ldots, a_r$. The group is *compact* if the parameter space is finite. The law of composition is

$$R(b_1, \ldots, b_r)R(a_1, \ldots, a_r) = R(c_1, \ldots, c_r), \tag{2.33a}$$

$$c_i = \phi_i(a_1, \ldots, a_r, b_1, \ldots, b_r), \tag{2.33b}$$

where ϕ_i is analytic with respect to both sets of parameters. It is convenient to choose the parameter values corresponding to the identity element to be zero.

As with finite groups, one can also consider an r-parameter Lie group of transformations of a set of n variables x_i:

$$x_i' = f_i(x_1, \ldots, x_n; a_1, \ldots, a_r), \tag{2.34}$$

where the f_i are analytic functions of the parameters, and a corresponding Lie group of operators O_R which operate on functions $\psi(x_1, \ldots, x_n)$. A finite operation can be generated by a sequence of infinitesimal operations, with the result

$$O_R = \exp\left(-\sum_{\mu=1}^{r} a_\mu X_\mu\right), \tag{2.35a}$$

$$X_\mu \equiv \sum_{i=1}^{n} \frac{\partial f_i(x_1, \ldots, x_n, a_1, \ldots, a_r)}{\partial a_\mu}\bigg|_{a=0} \frac{\partial}{\partial x_i}. \tag{2.35b}$$

The *infinitesimal operators* X_μ, which serve as group generators, are elements of a non-associative *Lie algebra* with the product defined as the commutator:

$$[X_\sigma, X_\tau] = \sum_\kappa c^\kappa_{\tau\sigma} X_\kappa. \tag{2.36}$$

The *structure constants* $c^\kappa_{\tau\sigma}$ define the Lie algebra of a particular Lie group.

Matrix representations of the Lie group can be constructed from matrix representations of the Lie algebra:

$$\mathbf{D}[R(a_1, \dots, a_r)] = \exp\left[-\sum_{\mu=1}^{r} a_\mu \mathbf{D}(X_\mu)\right]. \tag{2.37}$$

Note that if $\mathbf{D}(R)$ is unitary, then $D(X)$ is anti-hermitian. Basis functions for the Lie algebra are also basis functions for the Lie group. It can be shown that Casimir's operator C, defined by

$$C \equiv \sum_\rho \sum_\sigma g^{\rho\sigma} X_\rho X_\sigma, \tag{2.38a}$$

$$\sum_\rho g^{\mu\rho} g_{\rho\nu} \equiv \delta_{\mu\nu}, \tag{2.38b}$$

$$g_{\mu\nu} \equiv \sum_\alpha \sum_\beta c^\beta_{\mu\alpha} c^\alpha_{\nu\beta}, \tag{2.38c}$$

commutes with all of the infinitesimal operators X_μ. Consequently, C commutes with all elements of the Lie group, and, by virtue of Schur's lemma, it is represented by a constant matrix in any irreducible representation, which can then be labeled by one of its eigenvalues.

2.4.2 *The rotation group*

Crystallographic point groups suffice to describe the site symmetry of laser-active centres in crystals. However, as will become apparent in subsequent chapters, the effects of the crystalline environment are generally treated as perturbations on the electronic structure of a free ion with spherical symmetry, whose description requires a group of continuously infinite order, the *full rotation group* $O(3)$. This group is a direct product of its subgroups,

$$O(3) = O^+(3) \otimes C_i, \tag{2.39}$$

where the *rotation group* $O^+(3)$ is the compact, three-parameter Lie group of proper orthogonal transformations in three dimensions.

The parameters for a rotation through an angle θ about an axis $\hat{\mathbf{n}}$ can be chosen as θn_x, θn_y and θn_z, with corresponding infinitesimal operators X_x, X_y and X_z. The Lie algebra of the rotation group is specified by

$$[X_\alpha, X_\beta] = \varepsilon_{\alpha\beta\gamma} X_\gamma, \tag{2.40}$$

where $\varepsilon_{\alpha\beta\gamma}$ is 1 for a cyclic permutation of its indices, -1 for an anticyclic permutation and 0 otherwise. It is convenient to define hermitian operators J_μ by

$$J_\mu \equiv i X_\mu, \tag{2.41}$$

which then satisfy commutation relations

$$[J_\alpha, J_\beta] = i\varepsilon_{\alpha\beta\gamma} J_\gamma, \tag{2.42}$$

where $i = \sqrt{-1}$. Then a general rotation operator can be expressed in the form

$$O_R = \exp(-\theta\, \hat{\mathbf{n}} \cdot \mathbf{X}) = \exp(i\theta\, \hat{\mathbf{n}} \cdot \mathbf{J}), \tag{2.43}$$

where **X** and **J** are defined by

$$\mathbf{X} \equiv \hat{\mathbf{i}} X_x + \hat{\mathbf{j}} X_y + \hat{\mathbf{k}} X_z, \tag{2.44a}$$

$$\mathbf{J} \equiv \hat{\mathbf{i}} J_x + \hat{\mathbf{j}} J_y + \hat{\mathbf{k}} J_z. \tag{2.44b}$$

Casimir's operator is given by

$$C = -\tfrac{1}{2}\mathbf{X}^2 = \tfrac{1}{2}\mathbf{J}^2, \tag{2.45}$$

with the consequence

$$[\mathbf{J}^2, J_\mu] = 0. \tag{2.46}$$

It follows that irreducible representations of the rotation group are labeled by eigenvalues of $\mathbf{J}^2$.

It is evident from Eq. (2.43) that the Lie algebra of $O^+(3)$ is identical with the algebra of angular momentum operators in quantum mechanics. In a representation in which both $\mathbf{J}^2$ and J_z are diagonal, these operators have simultaneous eigenkets $|jm\rangle$, where j is a positive integer or half integer and m, which labels the rows and columns of each matrix in a $2j+1$-dimensional representation, has one of the values $-j, -j+1, \ldots, j$. Matrix elements are derived from the following relations:

$$\mathbf{J}^2 |jm\rangle = j(j+1)|jm\rangle, \tag{2.47a}$$

$$J_z|jm\rangle = m|jm\rangle, \tag{2.47b}$$

$$J_\pm|jm\rangle = \sqrt{(j \mp m)(j \pm m + 1)}\,|j, m \pm 1\rangle, \tag{2.47c}$$

$$J_\pm \equiv J_x \pm i J_y. \tag{2.47d}$$

A $2j+1$-dimensional irreducible representation of the rotation group is then constructed from the relation

$$\mathbf{D}^{(j)}(O_R) = \exp(i\theta\, \hat{\mathbf{n}} \cdot \mathbf{D}^{(j)}(\mathbf{J})), \tag{2.48}$$

$$\mathbf{D}^{(j)}(\mathbf{J}) \equiv \hat{\mathbf{i}}\mathbf{D}^{(j)}(J_x) + \hat{\mathbf{j}}\mathbf{D}^{(j)}(J_y) + \hat{\mathbf{k}}\mathbf{D}^{(j)}(J_z). \tag{2.49}$$

For proper orthogonal transformations of three real variables x, y and z, the infinitesimal operators are components of $\mathbf{X} = -\mathbf{r} \times \nabla$; then $\mathbf{J} = \mathbf{L}$, the orbital angular momentum operator, and irreducible representations of the rotation group $O^+(3)$ are labeled by integral quantum numbers $j = l$. The spherical harmonics $Y_l^m(\theta, \phi)$, defined by

$$Y_l^m(\theta,\phi) \equiv \frac{(-1)^{(m+|m|)/2}}{\sqrt{4\pi}} \sqrt{\frac{(2l+1)(l-|m|)!}{(l+|m|)!}}\, P_l^{|m|}(\cos\theta)\exp(im\phi), \tag{2.50a}$$

$$P_l^m(x) \equiv \frac{1}{2^l l!}(1-x^2)^{m/2}\frac{d^{l+m}(x^2-1)^l}{dx^{l+m}}, \tag{2.50b}$$

which satisfy the normalization condition

$$\int_0^{2\pi}\int_0^{\pi} Y_{l'}^{m'}(\theta,\phi)^* Y_l^m(\theta,\phi)\sin\theta\, d\theta\, d\phi = \delta_{ll'}\delta_{mm'}, \tag{2.51}$$

are simultaneous eigenfunctions of $\mathbf{L}^2$ and L_z with eigenvalues $l(l+1)$ and m, respectively, where l is a positive integer or 0 and m has one of the values $-l, -l+1, \ldots, l$. The spherical harmonics $Y_l^m(\theta, \phi)$ are then basis functions for irreducible representations of the rotation group. They are also basis functions for irreducible representations of the full rotation group, with parity $(-1)^l$, but that is an incidental feature of spherical harmonics which does not apply to rotation-group basis functions in general.

2.4.3 *SU*(2)

The group of unitary, unimodular transformations of two complex variables,

$$\begin{pmatrix}\xi'\\ \eta'\end{pmatrix} = \begin{pmatrix} a & b\\ -b^* & a^*\end{pmatrix}\begin{pmatrix}\xi\\ \eta\end{pmatrix}, \tag{2.52a}$$

$$|a|^2 + |b|^2 = 1, \tag{2.52b}$$

is a three-parameter, compact Lie group called $SU(2)$, which has the same Lie algebra as the rotation group $O^+(3)$. By virtue of their unitary and unimodular properties, the elements of this group effect real, orthogonal transformations in three dimensions by the recipe

$$\mathbf{h}' = \mathbf{u}\mathbf{h}\mathbf{u}^{-1}, \tag{2.53a}$$

$$\mathbf{h} \equiv \begin{pmatrix} -z & x+iy\\ x-iy & z\end{pmatrix}, \tag{2.53b}$$

$$\mathbf{u} \equiv \begin{pmatrix} a & b\\ -b^* & a^*\end{pmatrix}. \tag{2.53c}$$

Since $\mathbf{u}$ and $-\mathbf{u}$ effect the same transformation, there is a two-to-one correspondence between elements of the two groups such that $O^+(3)$ is a homomorphic image of $SU(2)$.

Irreducible representations of $SU(2)$ with integral quantum numbers are identical with the irreducible representations of $O^+(3)$ considered previously. Irreducible representations of $SU(2)$ with half-integral quantum numbers, which arise in the description of the intrinsic angular momentum of the electron and other fermions, are *double-valued representations* of $O^+(3)$, in that each element of the rotation group is represented by two distinct matrices.

An explicit expression which can be derived for the matrices of irreducible representations of $SU(2)$ as a function of Euler angles [Hamermesh (1962)] will be omitted since it is not needed in the present context. Higher $SU(n)$ groups, which are useful in the description of complex lanthanide spectra [Judd (1963)] as well as in the theory of elementary particles, are beyond the scope of the present volume.

2.4.4 *Application to coupled systems*

The general problem of coupled systems was considered in §2.2.5. In the special case in which the Hamiltonian H_0 of the uncoupled system is invariant under independent operations of the rotation group applied to each subsystem, the group G_0 of the uncoupled system is

$$G_0 = O^+(3) \otimes O^+(3). \tag{2.54}$$

It follows from Eq. (2.43) that elements of G_0 can then be expressed in the form

$$O_{R_1R_2} = \exp[i(\theta_1 \hat{\mathbf{n}}_1 \cdot \mathbf{J}_1 + \theta_2 \hat{\mathbf{n}}_2 \cdot \mathbf{J}_2)]. \tag{2.55}$$

The Hamiltonian H of the coupled system is invariant only under simultaneous rotations of the two subsystems through the same angle θ about the same axis $\hat{\mathbf{n}}$,

$$O_R = \exp[i\theta \hat{\mathbf{n}} \cdot (\mathbf{J}_1 + \mathbf{J}_2)] \equiv \exp(i\theta \hat{\mathbf{n}} \cdot \mathbf{J}). \tag{2.56}$$

Since $\mathbf{J}_1$ and $\mathbf{J}_2$ commute, $\mathbf{J} = \mathbf{J}_1 + \mathbf{J}_2$ satisfies the commutation relations (2.42) and generates the group G of H.

The rotation group is simply reducible; i.e., each irreducible representation occurs at most once in the reduction of the Kronecker product representation,

$$\mathbf{U}\mathbf{D}^{(j_1 \times j_2)}(R)\mathbf{U}^+ = \sum_{j=|j_1-j_2|}^{j_1+j_2} \mathbf{D}^{(j)}(R). \tag{2.57}$$

Eigenkets of $\mathbf{J}^2$ and J_z in the coupled representation can be expressed in terms of those in the uncoupled representation in the form

$$|jm\rangle = \sum_{m_1=-j_1}^{j_1} \sum_{m_2=-j_2}^{j_2} |j_1m_1\rangle |j_2m_2\rangle \langle j_1 j_2 m_1 m_2 | jm\rangle. \tag{2.58}$$

Selected values of the coupling coefficients $\langle j_1 j_2 m_1 m_2 | jm\rangle$, called *Clebsch–Gordon coefficients*, are tabulated in Table 2.9; more extensive tabulations are readily available [Rotenberg *et al.* (1959), Weissbluth (1978)]. They are related to 3-j symbols by

$$\begin{pmatrix} j_1 & j_2 & j \\ m_1 & m_2 & -m \end{pmatrix} \equiv \frac{(-1)^{j_1-j_2+m}}{\sqrt{2j+1}} \langle j_1 j_2 m_1 m_2 | jm\rangle. \tag{2.59a}$$

Clebsch–Gordon coefficients vanish unless the following conditions are satisfied:

$$\mathbf{j}_1 + \mathbf{j}_2 + \mathbf{j} = 0, \tag{2.59b}$$

$$m_1 + m_2 = m. \tag{2.59c}$$

Table 2.9. *Clebsch–Gordon coefficients $\langle lsm_l m_s | jm_j \rangle$ for $l=1$ and $s=1/2$*

$l=1$	$s=1/2$	$j=3/2$				$j=1/2$	
m_l	m_s	$m_j=3/2$	$m_j=1/2$	$m_j=-1/2$	$m_j=-3/2$	$m_j=1/2$	$m_j=-1/2$
1	1/2	1					
1	−1/2		$\sqrt{1/3}$			$\sqrt{2/3}$	
0	1/2		$\sqrt{2/3}$			$-\sqrt{1/3}$	
0	−1/2			$\sqrt{2/3}$			$\sqrt{1/3}$
−1	1/2			$\sqrt{1/3}$			$-\sqrt{2/3}$
−1	−1/2				1		

Eq. (2.59b), the *triangle condition*, is represented concisely by $\Delta(j_1 j_2 j)$. A closed-form expression for the Clebsch–Gordon coefficients [Racah (1942)] is omitted here.

2.4.5 *Wigner–Eckart theorem*

Spherical tensor operators T_{LM} are defined by the relations [Racah (1942)]

$$[J_x \pm iJ_y, T_{LM}] = \sqrt{(L \mp M)(L \pm M + 1)}\, T_{LM\pm 1}, \tag{2.60a}$$

$$[J_z, T_{LM}] = MT_{LM}. \tag{2.60b}$$

An equivalent definition [Rose (1957)] is

$$O_R T_{LM} O_R^{-1} = \sum_{M'=-L}^{L} T_{LM'} D^L_{M'M}(R), \tag{2.61}$$

where R is an element of $O^+(3)$; i.e., operators T_{LM}, of *rank* L, transform as a basis for a $2L+1$-dimensional irreducible representation of the rotation group. Spherical tensor operators are multiplied according to the rule

$$T_{LM} = \sum_{M_1=-L_1}^{L_1} T_{L_1 M_1} T_{L_2 M-M_1} \langle L_1 L_2 M_1 M - M_1 | LM \rangle. \tag{2.62}$$

The *Wigner–Eckart theorem* [Wigner (1931), Eckart (1930)] states that matrix elements of spherical tensor operators between angular momentum eigenkets $|jm\rangle$ are expressible in the form

$$\langle j'm' | T_{LM} | jm \rangle = \frac{1}{\sqrt{2j'+1}} \langle j' \| T_L \| j \rangle \langle jLmM | j'm' \rangle, \tag{2.63}$$

where the reduced matrix element $\langle j' \| T_L \| j \rangle$ is independent of the quantum numbers m, M and m'. An obvious corollary of the Wigner–Eckart theorem is the *replacement theorem*, which states that matrix elements of two spherical tensor operators of the same rank, T_{LM} and U_{LM}, are proportional within a manifold of constant j, L and j'. This corollary permits substitution of one spherical tensor operator for another, within a constant of proportionality equal to the ratio of their reduced matrix elements.

2.4.6 *Racah coefficients*

Coupling of three angular momenta can be accomplished by first coupling two, and then coupling the third to their resultant. Since the coupled representation is not simply reducible in this case, the resultant is obviously not unique, but depends on the coupling sequence. Eigenkets in the coupled representation, obtained by two different coupling sequences, are related by a unitary transformation:

$$|(j')jm\rangle = \sum_{j''} |(j'')jm\rangle R_{j''j'}(j), \tag{2.64a}$$

$$\mathbf{J}' = \mathbf{J}_1 + \mathbf{J}_2, \tag{2.64b}$$

$$\mathbf{J}'' = \mathbf{J}_2 + \mathbf{J}_3, \tag{2.64c}$$

$$R_{j''j'}(j) = \sum_{m_1}\sum_{m_2} \langle j_1 j_2 m_1 m_2 | j' m_1 + m_2\rangle \langle j_2 j_3 m_2 m_3 | j'' m_2 + m_3\rangle \\ \times \langle j' j_3 m_1 + m_2 m_3 | jm\rangle \langle j_1 j'' m_1 m_2 + m_3 | jm\rangle, \tag{2.64d}$$

where the right-hand side of Eq. (2.64d) is independent of $m = m_1 + m_2 + m_3$. *Racah coefficients* W are defined by the relation

$$R_{j''j'}(j) \equiv [(2j''+1)(2j'+1)]^{1/2} W(j_1 j_2 j j_3; j' j''), \tag{2.65}$$

and the related 6-j symbols are defined by

$$\begin{Bmatrix} j_1 & j_2 & j' \\ j_3 & j & j'' \end{Bmatrix} \equiv (-1)^{j_1+j_2+j_3+j} W(j_1 j_2 j j_3; j' j''). \tag{2.66}$$

An extensive tabulation is available in Rotenberg *et al.* (1959). A closed-form expression [Racah (1942)] is omitted here.

Racah coefficients find application in the evaluation of reduced matrix elements associated with application of the Wigner–Eckart theorem to coupled systems. For example, the interaction Hamiltonian $H'(1,2)$ which couples two subsystems is often expressible as the *contraction* of two spherical tensors of rank L,

$$H'(1,2) = \sum_{M=-L}^{L} (-1)^M T_{LM}(1) U_{L,-M}(2) \equiv T_L(1) \cdot U_L(2). \tag{2.67}$$

Reduced matrix elements of $H'(1,2)$ in the coupled representation can then be expressed in terms of reduced matrix elements of $T_{LM}(1)$ and $U_{LM}(2)$ in the uncoupled representation by the relation

$$\langle j_1' j_2' j \| T_L(1) \cdot U_L(2) \| j_1 j_2 j\rangle \\ = \langle j_1' \| T_L(1) \| j_1\rangle \langle j_2' \| U_L(2) \| j_2\rangle (-1)^{j_1'+j_2-j} [(2j_1'+1)(2j_2'+1)]^{1/2} W(j_1 j_2 j_1' j_2'; jL). \tag{2.68}$$

As a second example, consider a tensor operator $T_{LM}(1)$ which operates only on part of a coupled system with coordinates 1 and 2. The reduced matrix element in the coupled system can then be expressed in terms of that in the uncoupled system by the *generalized projection theorem,*

$$\langle j_1' j_2' j' \| T_L(1) \| j_1 j_2 j \rangle$$
$$= \delta_{j_2' j_2} (-1)^{j_2+L-j_1'-j'} [(2j_1'+1)(2j+1)]^{1/2} W(j_1 j j_1' j'; j_2 L) \langle j_1' \| T_L(1) \| j_1 \rangle. \tag{2.69}$$

Finally, reduced matrix elements of the product of spherical tensor operators defined by Eq. (2.62) are given by

$$\langle \gamma' j' \| T_L \| \gamma j \rangle$$
$$= \sqrt{2L+1} \sum_{j''\gamma''} (-1)^{L_2+j'+j''} W(L_2 L j'' j'; L_1 j) \langle \gamma' j' \| T_{L_1} \| \gamma'' j'' \rangle \langle \gamma'' j'' \| T_{L_2} \| \gamma j \rangle, \tag{2.70}$$

where γ denotes all of the parameters which distinguish different functions with the same value of j. Equations (2.68)–(2.70) find immediate application in the calculation of atomic energy levels in Chapter 4 and transition probabilities in Chapter 5.

2.5 Some additional applications of group theory

2.5.1 Evaluation of matrix elements

The Wigner–Eckart theorem facilitates the evaluation of matrix elements of spherical tensor operators. The point-group analogue of the Wigner–Eckart theorem [Koster (1958)] is

$$\langle \lambda k | T_i^{(\mu)} | \nu j \rangle = d_\lambda^{-1/2} \sum_{\tau_\lambda} \langle \lambda \tau_\lambda \| T^{(\mu)} \| \nu \rangle \langle \mu \nu i j | \lambda \tau_\lambda k \rangle^*, \tag{2.71}$$

where the tensor operator $T_i^{(\mu)}$ is defined by

$$O_R T_i^{(\mu)} O_R^{-1} = \sum_{j=1}^{d_\mu} T_j^{(\mu)} D_{ji}^{(\mu)}(R), \tag{2.72}$$

and the coupling coefficients $\langle \mu \nu i j | \lambda \tau_\lambda k \rangle$ are elements of the unitary matrix $\mathbf{U}^+$, defined in §2.2.5, which effects the reduction of the Kronecker product representation $D^{(\mu\times\nu)}(R)$. Values of coupling coefficients for crystallographic point groups are tabulated by Koster *et al.* (1963). Equations (2.71) and (2.72) find application in the derivation of selection rules and the calculation of transition probabilities for optical transitions between crystal-field levels of impurities and colour centres in solids.

2.5.2 Projection operators

A common feature of the application of the variational principle to atomic and molecular structure calculations is the specification of trial functions as linear combinations of symmetry-adapted basis functions. The spherical harmonics, Eqs. (2.50), provide the required basis functions in the case of spherical symmetry. An effective method for generating analogous symmetry-adapted functions for the crystallographic point groups employs operators defined by

$$P_i^{(\mu)} \equiv \frac{d_\mu}{g} \sum_R D_{ii}^{(\mu)}(R)^* O_R, \tag{2.73}$$

Table 2.10. *Basis functions for group D_3 projected from spherical harmonics*

Representation	Row	Basis function
A_1	a_1	$Y_l^m(\theta,\phi)+(-1)^l Y_l^{-m}(\theta,\phi), m=3n$
A_2	a_2	$Y_l^m(\theta,\phi)-(-1)^l Y_l^{-m}(\theta,\phi), m=3n$
E	x	$Y_l^m(\theta,\phi), m=3n-1$
E	y	$(-1)^l Y_l^{-m}(\theta,\phi), m=3n-1$

where R is a point group element and $\mathbf{D}^{(\mu)}(R)$ is an irreducible matrix representation. It follows from the grand orthogonality theorem, Eq. (2.6), that these operators are idempotent and exclusive, and that they resolve the identity operator,

$$P_i^{(\mu)} P_j^{(\nu)} = \delta_{\mu\nu}\,\delta_{ij}\,P_i^{(\mu)}, \tag{2.74a}$$

$$\sum_\mu \sum_i P_i^{(\mu)} = O_E. \tag{2.74b}$$

Consequently, $P_i^{(\mu)}$ is a projection operator; its application to a specified function $\psi(\mathbf{r})$ then yields either a basis function for the ith row of the μth irreducible representation, or nothing,

$$P_i^{(\mu)}\psi(\mathbf{r}) = \begin{cases} \psi_i^{(\mu)}(\mathbf{r}), \\ 0. \end{cases} \tag{2.75}$$

Partner functions $\psi_{ji}^{(\mu)}(\mathbf{r})$ for other rows of the μth irreducible representation, which transform together with $\psi_i^{\mu}(\mathbf{r})$, are generated by the auxiliary operator

$$Q_{ji}^{(\mu)} \equiv \frac{d_\mu}{g}\sum_R D_{ji}^{(\mu)}(R)^* O_R. \tag{2.76}$$

As an example, basis functions for group D_3, projected from spherical harmonics, are listed together with their partner functions in Table 2.10.

2.5.3 *Subduced representations*

Crystallographic point groups are finite subgroups of the full rotation group $O(3)$, Eq. (2.39), and the corresponding subduced representations are in general reducible. In a $(2J+1)$-dimensional irreducible representation of the rotation group, $O^+(3)$, a rotation through angle θ about the z-axis is represented by a diagonal matrix with elements

$$\langle JM|\exp(i\theta J_z)|JM'\rangle = \delta_{MM'}\exp(i\theta M), \tag{2.77}$$

and trace

$$\chi^{(J)}(\theta) = \frac{\sin(J+\frac{1}{2})\theta}{\sin\frac{1}{2}\theta}. \tag{2.78}$$

The trace depends only on the angle of rotation and not on the axis, since all rotations through the same angle are in the same class. The traces for the full rotation group are

Table 2.11. *Character table of isomorphic crystallographic point groups O and T_d*

O	E	$8C_3$	$3C_2$	$6C_2'$	$6C_4$
T_d	E	$8C_3$	$3C_2$	$6\sigma_d$	$6S_4$
A_1 (Γ_1)	1	1	1	1	1
A_2 (Γ_2)	1	1	1	−1	−1
E (Γ_3)	2	−1	2	0	0
T_1 (Γ_4)	3	0	−1	−1	1
T_2 (Γ_5)	3	0	−1	1	−1

Table 2.12. *Rotation group compatibility table for octahedral point group O*

J	Representations
0	A_1
1	T_1
2	$E+T_2$
3	$A_2+T_1+T_2$
4	$A_1+E+T_1+T_2$
5	$E+2T_1+T_2$
6	$A_1+A_2+E+T_1+2T_2$

obtained by multiplying $\chi^{(J)}(\theta)$ by ±1 for even (g) and odd (u) representations, respectively. The number of times each point-group irreducible representation Γ_μ is contained in each subduced representation is then readily obtained by application of Eq. (2.10),

$$a_\mu = \frac{1}{g}\sum_i g_i \chi_i^{(\mu)*} \chi^{(J)}(\theta_i). \tag{2.79}$$

As examples, consider the crystallographic point groups O_h and T_d, which reflect the dominant symmetry in octahedral or cubic coordination and in tetrahedral coordination, respectively. For the octahedral point group, $O_h = O \otimes C_i$, it suffices to determine the irreducible representations of point group O contained in the subduced representation of the group of proper rotations, $O^+(3)$, since inversion symmetry is preserved. The character table of the isomorphic groups O and T_d appears in Table 2.11. The resulting *rotation-group compatibility table* for point group O is displayed in Table 2.12, and the *full-rotation-group compatibility table* for point group T_d is displayed in Table 2.13.

Reduction of the subduced representation is accomplished by a unitary transformation; accordingly, basis functions for irreducible representations of crystallographic point groups can be expressed as linear combinations of basis functions for the full rotation group [Griffith (1961)]. These symmetry-adapted linear combinations can be generated by means of projection operators, as discussed in Section 2.5.2. They find application in both the crystal potential and the wave functions for crystal-field states.

Table 2.13. *Full rotation group compatibility table for tetrahedral point group T_d*

J (g)	Representations	J (u)	Representations
0	A_1	0	A_2
1	T_1	1	T_2
2	$E+T_2$	2	$E+T_1$
3	$A_2+T_1+T_2$	3	$A_1+T_1+T_2$
4	$A_1+E+T_1+T_2$	4	$A_2+E+T_1+T_2$
5	$E+2T_1+T_2$	5	$E+T_1+2T_2$
6	$A_1+A_2+E+T_1+2T_2$	6	$A_1+A_2+E+2T_1+T_2$

Table 2.14. *Characters of additional representations of isomorphic double point groups O^* and T_d^**

O^*	E	R	$8C_3$	$8C_3R$	$3C_2$ $3C_2R$	$6C_4$	$6C_4R$	$6C_2'$ $6C_2'R$
T_d^*	E	R	$8C_3$	$8C_3R$	$3C_2$ $3C_2R$	$6S_4$	$6S_4R$	$6\sigma_d$ $6\sigma_dR$
E' (Γ_6)	2	−2	1	−1	0	$\sqrt{2}$	$-\sqrt{2}$	0
E'' (Γ_7)	2	−2	1	−1	0	$-\sqrt{2}$	$\sqrt{2}$	0
U' (Γ_8)	4	−4	−1	1	0	0	0	0

Table 2.15. *Rotation group compatibility table for additional representations of double point group O^**

J	Representations
1/2	E'
3/2	U'
5/2	$E''+U'$
7/2	$E'+E''+U'$
9/2	$E'+2U'$
11/2	$E'+E''+2U'$
13/2	$E'+2E''+2U'$

Subduced double-valued representations of $SU(2)\otimes C_i$ corresponding to crystallographic point groups are called *double-group representations*, since they may be regarded as single-valued representations of *double point groups*, distinguished by an asterisk, in which a rotation through 2π is regarded as an element R distinct from the identity element E. Irreducible representations of a crystallographic point group are also irreducible representations of the corresponding double group, but the latter has *additional representations* as well. The characters of the additional representations of the isomorphic double point groups O^* and T_d^* are listed in Table 2.14; the rotation group compatibility table for the additional representations of O^* is presented in Table 2.15; and the full-rotation group compatibility table for the additional representations of T_d^* is presented in Table 2.16.

Table 2.16. *Full rotation group compatibility table for additional representations of double point group T_d^**

J (g)	Representations	J (u)	Representations
1/2	E'	1/2	E''
3/2	U'	3/2	U'
5/2	$E''+U'$	5/2	$E'+U'$
7/2	$E'+E''+U'$	7/2	$E'+E''+U'$
9/2	$E'+2U'$	9/2	$E''+2U'$
11/2	$E'+E''+2U'$	11/2	$E'+E''+2U'$
13/2	$E'+2E''+2U'$	13/2	$2E'+E''+2U'$

2.5.4 *Time-reversal invariance*

In addition to the spatial symmetry of its Hamiltonian operator, the Schrödinger equation possesses a symmetry called *time-reversal invariance* which imposes additional degeneracies. The time-reversal operator T is related to the complex-conjugate operator K since the complex conjugate of the Schrödinger equation is

$$H\psi^* = i\hbar\frac{\partial\psi^*}{\partial(-t)}; \tag{2.80}$$

thus $\psi^*(-t)$ is a solution of the Schrödinger equation in negative time. For a system of q electrons, the complete time-reversal operator is given by

$$T = UK, \tag{2.81}$$

where U is a unitary operator which operates on the spin coordinates. A general result is

$$T^2\psi = \pm\psi, \tag{2.82}$$

where the upper sign applies for an even number of electrons, and the lower sign for an odd number. In the latter case, one can show that $T\psi$ is orthogonal to ψ, which leads to *Kramers' theorem*: All states are at least doubly degenerate for an odd number of electrons [Kramers (1930)]. This *Kramers degeneracy* can only be removed by an external magnetic field.

The relation of time-reversal invariance to spatial symmetry arises from the fact that, if ψ transforms as a basis function for irreducible representation $\mathbf{D}(R)$ of the group of the Hamiltonian, then $T\psi$ transforms as a basis function for $\mathbf{D}^*(R)$. Three cases can be identified [Wigner (1931), Tinkham (1964)]:

(a) $\mathbf{D}$ and $\mathbf{D}^*$ are equivalent to the same real irreducible representation.
(b) $\mathbf{D}$ and $\mathbf{D}^*$ are inequivalent.
(c) $\mathbf{D}$ and $\mathbf{D}^*$ are equivalent but cannot be made real.

These cases can be distinguished by the *Frobenius–Schur test* [Frobenius and Schur (1906)] based on the sum of traces of the squares of elements, as indicated in Table 2.17. The implications for degeneracy in each case depend on whether the spin is integral or half-integral; i.e., on whether the system has an even or odd number of electrons.

Table 2.17. *Effect of time-reversal invariance on degeneracy*

Case	$\sum_R \chi(R^2)$	Integral spin	Half-integral spin
a	g	No extra degeneracy	Doubled degeneracy
b	0	Doubled degeneracy	Doubled degeneracy
c	$-g$	Doubled degeneracy	No extra degeneracy

Table 2.18. *Character table of crystallographic point group* C_3

C_3	E	C_3	C_3^2
A (Γ_1)	1	1	1
E (Γ_2)	1	ω^2	$-\omega$
E (Γ_3)	1	$-\omega$	ω^2

$\omega = \exp(i\pi/3)$

As an example of the effect of time-reversal invariance on degeneracy, consider the character table of crystallographic point group C_3, shown in Table 2.18.

Point group C_3 is Abelian, with only one-dimensional irreducible representations. Nevertheless, since representations Γ_2 and Γ_3 clearly belong to case *b*, they label degenerate energy levels and so are conventionally combined in a two-dimensional representation *E*. It should be emphasized that the degeneracy in this case is doubled for both even and odd numbers of electrons; thus Kramers' degeneracy is not the only effect of time-reversal invariance.

2.5.5 *Crystal-field levels*

Subduced representations of the full rotation group find application in crystal-field theory, as described in Chapter 4, since the perturbing crystal field reduces the symmetry of the free-ion Hamiltonian and splits each free-ion energy level into a set of crystal-field levels. The numbers and degeneracies of these crystal-field levels are determined both by the numbers and dimensions of the point-group irreducible representations contained in the subduced representations and by additional degeneracies arising from time-reversal invariance.

The numbers of crystal-field levels for subgroups of the rotation group are listed in Tables 2.19 and 2.20. The subgroups in Table 2.19 include all of the crystallographic point groups which contain only proper rotations, listed in the first row. The table also serves for their direct products with point group C_i, listed in the second row, since these are subgroups of the full rotation group which preserve inversion symmetry. The remaining crystallographic point groups are listed in Table 2.20. The irreducible representations of these groups are different for even and odd free-ion levels, but the numbers of crystal-field levels are the same in each case. Time-reversal invariance imposes Kramers' degeneracy, evident in the tables, as well as additional degeneracies for point groups T, T_h, C_6, C_{6h}, C_4, C_{4h}, C_3, C_{3h}, S_6, and S_4, which are also reflected in the tables.

Table 2.19. *Number of crystal-field levels into which a free-ion level J splits for crystal fields of various symmetries. Point groups in the first row contain only proper rotations and those in the second row are their direct products with point group C_i*

J	Degeneracy $(2J+1)$	O O_h	T T_h	D_6 D_{6h}	D_4 D_{4h}	D_3 D_{3d}	C_6 C_{6h}	C_4 C_{4h}	C_3 S_6	D_2 D_{2h}	C_2 C_2	C_1 C_i
0	1	1	1	1	1	1	1	1	1	1	1	1
1	3	1	1	2	2	2	2	2	2	3	3	3
2	5	2	2	3	4	3	3	4	3	5	5	5
3	7	3	3	5	5	5	5	5	5	7	7	7
4	9	4	4	6	7	6	6	7	6	9	9	9
5	11	4	4	7	8	7	7	8	7	11	11	11
6	13	6	6	9	10	9	9	10	9	13	13	13
7	15	6	6	10	11	10	10	11	10	15	15	15
8	17	7	7	11	13	11	11	13	11	17	17	17
1/2	2	1	1	1	1	1	1	1	1	1	1	1
3/2	4	1	1	2	2	2	2	2	2	2	2	2
5/2	6	2	2	3	3	3	3	3	3	3	3	3
7/2	8	3	3	4	4	4	4	4	4	4	4	4
9/2	10	3	3	5	5	5	5	5	5	5	5	5
11/2	12	4	4	6	6	6	6	6	6	6	6	6
13/2	14	5	5	7	7	7	7	7	7	7	7	7
15/2	16	5	5	8	8	8	8	8	8	8	8	8
17/2	18	6	6	9	9	9	9	9	9	9	9	9

Table 2.20. *Number of crystal-field levels into which a free-ion level J splits for crystal fields of various additional symmetries. The irreducible representations of groups listed in this table are different for even and odd free-ion levels, but the numbers of crystal-field levels are the same in each case*

J	Degeneracy $(2J+1)$	T_d	D_{3h}	C_{6v}	C_{3v}	C_{3h}	D_{2d}	C_{4v}	S_4	C_{2v}	C_s
0	1	1	1	1	1	1	1	1	1	1	1
1	3	1	2	2	2	2	2	2	2	3	3
2	5	2	3	3	3	3	4	4	4	5	5
3	7	3	5	5	5	5	5	5	5	7	7
4	9	4	6	6	6	6	7	7	7	9	9
5	11	4	7	7	7	7	8	8	8	11	11
6	13	6	9	9	9	9	10	10	10	13	13
7	15	6	10	10	10	10	11	11	11	15	15
8	17	7	11	11	11	11	13	13	13	17	17
1/2	2	1	1	1	1	1	1	1	1	1	1
3/2	4	1	2	2	2	2	2	2	2	2	2
5/2	6	2	3	3	3	3	3	3	3	3	3
7/2	8	3	4	4	4	4	4	4	4	4	4
9/2	10	3	5	5	5	5	5	5	5	5	5
11/2	12	4	6	6	6	6	6	6	6	6	6
13/2	14	5	7	7	7	7	7	7	7	7	7
15/2	16	5	8	8	8	8	8	8	8	8	8
17/2	18	6	9	9	9	9	9	9	9	9	9

2.5.6 *Pauli principle*

In addition to spatial symmetry and time-reversal invariance, the Hamiltonian for a system of N electrons is invariant under permutations of their space and spin coordinates. The set of these permutations is a group of order $N!$ called the *symmetric group* S_N of *degree* N [Wigner (1959)]. Eigenfunctions of the N-electron Hamiltonian must then transform as basis functions for irreducible representations of S_N. These representations always include two one-dimensional representations for $N \geq 2$: the identity representation in which every element is represented by $+1$, and a fully antisymmetrical representation in which transpositions are represented by -1. The *Pauli principle* states that the only solutions of the Schrödinger equation for a system of N identical particles which are physically meaningful are those which transform as the identity representation of the symmetric group S_N for particles of integral spin, and those which transform as the fully antisymmetrical representation for particles of half-integral spin. The wave functions for a system of N electrons, with spin 1/2, must then be antisymmetric under exchange of the space and spin coordinates of any electron pair.

2.5.7 *Selected tables for symmetry groups*

Several additional tables are included here for reference in subsequent chapters.

Table 2.21. *Irreducible representations contained in the Kronecker product representations for point-groups O and T_d*

A_1	A_2	E	T_1	T_2	O, T_d
A_1	A_2	E	T_1	T_2	A_1
	A_1	E	T_2	T_1	A_2
		A_1+A_2+E	T_1+T_2	T_1+T_2	E
			$A_1+E+T_1+T_2$	$A_2+E+T_1+T_2$	T_1
				$A_1+E+T_1+T_2$	T_2

Table 2.22. *Character table for crystallographic point groups D_4, C_{4v} and D_{2d}*

D_4	E	C_2	$2C_4$	$2C_2'$	$2C_2''$
C_{4v}	E	C_4^2	$2C_4$	$2\sigma_v$	$2\sigma_d$
D_{2d}	E	C_2	$2S_4$	$2C_2'$	$2\sigma_d$
A_1 (Γ_1)	1	1	1	1	1
A_2 (Γ_2)	1	1	1	-1	-1
B_1 (Γ_3)	1	1	-1	1	-1
B_2 (Γ_4)	1	1	-1	-1	1
E (Γ_5)	2	-2	0	0	0

Table 2.23. *Compatibility table for crystallographic point group O_h with point groups C_{4v} and C_{2v}*

O_h	A_{1g}	A_{2g}	E_g	T_{1g}	T_{2g}	A_{1u}	A_{2u}	E_u	T_{1u}	T_{2u}
C_{4v}	A_1	B_1	A_1+B_1	A_2+E	B_2+E	A_2	B_2	A_2+B_2	A_1+E	B_1+E
C_{2v}	A_1	B_1	A_1+B_1	B_1+A_2 $+B_2$	A_1+A_2 $+B_2$	A_2	B_2	A_2+B_2	A_1+B_1 $+B_2$	A_1+B_1 $+A_2$

Table 2.24. *Character table for crystallographic point group C_{2v}*

C_{2v}	E	C_2	σ_v	σ_v'
A_1 (Γ_1)	1	1	1	1
B_1 (Γ_2)	1	−1	1	−1
A_2 (Γ_3)	1	1	−1	−1
B_2 (Γ_4)	1	−1	−1	1

Table 2.25. *Character table for Lie group $D_{\infty h}$*

$D_{\infty h}$	E	$C(\phi)$	σ_v	I	$IC(\phi)$	$I\sigma_v$
Σ_g^+	1	1	1	1	1	1
Σ_u^+	1	1	1	−1	−1	−1
Σ_g^-	1	1	−1	1	1	−1
Σ_u^-	1	1	−1	−1	−1	1
Π_g	2	$2\cos\phi$	0	2	$2\cos\phi$	0
Π_u	2	$2\cos\phi$	0	−2	$-2\cos\phi$	0
Δ_g	2	$2\cos 2\phi$	0	2	$2\cos 2\phi$	0
Δ_u	2	$2\cos 2\phi$	0	−2	$-2\cos 2\phi$	0
...	...	...	...	...	...	...

Table 2.26. *Character table of crystallographic point group D_2*

D_2	E	C_z	C_y	C_x
A_1 (Γ_1)	1	1	1	1
B_1 (Γ_3)	1	1	−1	−1
B_2 (Γ_2)	1	−1	1	−1
B_3 (Γ_4)	1	−1	−1	1

Table 2.27. *Character table for additional representations of double point group C_{4v}^**

C_{4v}	E	R	C_4 C_4^3R	C_4^3 C_4R	C_4^2 C_4^2R	$2\sigma_v$ $2\sigma_vR$	$2\sigma_d$ $2\sigma_dR$
$E_{1/2}$ (Γ_6)	2	−2	$\sqrt{2}$	$-\sqrt{2}$	0	0	0
$E_{3/2}$ (Γ_7)	2	−2	$-\sqrt{2}$	$\sqrt{2}$	0	0	0

Table 2.28. *Full rotation group compatibility table for additional representations of double point group C_{4v}^**

$J(g)$	Representations	$J(u)$	Representations
1/2	$E_{1/2}$	1/2	$E_{1/2}$
3/2	$E_{1/2}+E_{3/2}$	3/2	$E_{1/2}+E_{3/2}$

Table 2.29. *Compatibility table for crystallographic point group T_d with point group D_{2d}*

T_d	A_1	A_2	E	T_1	T_2
D_{2d}	A_1	B_1	A_1+B_1	A_2+E	B_2+E

Table 2.30. *Irreducible representations contained in the Kronecker product representation for point groups D_4, C_{4v} and D_{2d}*

A_1	A_2	B_1	B_2	E	D_4, C_{4v}, D_{2d}
A_1	A_2	B_1	B_2	E	A_1
	A_1	B_2	B_1	E	A_2
		A_1	A_2	E	B_1
			A_1	E	B_2
				$A_1+A_2+B_1+B_2$	E

Table 2.31. *Character table for additional representations of double point group C_3^**

C_3^*		E	R	C_3	C_3R	C_3^2	C_3^2R
$E_{1/2}$	Γ_4	1	-1	ω	$-\omega$	ω^2	$-\omega^2$
	Γ_5	1	-1	$-\omega^2$	ω^2	$-\omega$	ω
$B_{3/2}$ (Γ_6)		1	-1	-1	1	1	-1

$\omega=\exp(i\pi/3)$

Table 2.32. *Character table for additional representations of double point group D_3^**

D_3^*		E	R	C_3, C_3^2R	C_3^2, C_3R	$3C_2'$	$3C_2'R$
$E_{3/2}$	Γ_4	1	-1	-1	1	i	$-i$
	Γ_5	1	-1	-1	1	$-i$	i
$E_{1/2}$ (Γ_6)		2	-2	1	-1	0	0

Table 2.33. *Compatibility table for crystallographic point group O and double point group O^* with point groups C_3 and D_3 and double point groups C_3^* and D_3^**

O, O^*	A_1	A_2	E	T_1	T_2	E'	E''	U'
D_3, D_3^*	A_1	A_2	E	A_2+E	A_1+E	$E_{1/2}$	$E_{1/2}$	$E_{1/2}+E_{3/2}$
C_3, C_3^*	A	A	E	$A+E$	$A+E$	$E_{1/2}$	$E_{1/2}$	$E_{1/2}+2B_{3/2}$

3

Optical crystals: their structures, colours and growth

3.1 Natural minerals and gemstones

There are several thousand naturally occurring minerals, inorganic compounds of fixed chemical composition and regular sub-microscopic structure. Not more than 100 or so of these are recognized as gemstones although the number may vary as fashions change and new sources are found. Gemstones are hard and durable to withstand regular use without damage. But most particularly, they are beautiful, in their colour and lustre, especially when cut, faceted or polished for personal adornment. Beauty notwithstanding, the most important quality of a gemstone is scarcity. To be rare is to be greatly valued.

Minerals that are the basis of gems are found in rocks which form over aeons in time: where they are found reflects the process of continuous formation. There are three main classes of gem-bearing rocks. *Sedimentary rocks* are formed by the accumulation of eroded rock fragments, which settle over time, are compressed and again harden into rock. They are set down in layers which eventually emerge from below the earth's surface. Gypsum is a typical sedimentary rock with important variations in alabaster and selenite. Opal and tourmaline occur as veins in such sedimentary rocks as shale. *Igneous rocks* solidify from molten rock deep beneath the earth's surface, sometimes escaping in lava flows from erupting volcanoes. The slower the rate at which the molten rocks cool, the larger are the gems that grow within them. In consequence, gems grow at high temperature, under huge hydrostatic pressures. *Metamorphic rocks* are changed by high temperature and pressure from sedimentary or igneous rocks into new minerals: marble formed from limestone under intense pressure at elevated temperature may contain rubies.

Hall (1994) provides a concise guide to the identification of some 130 gemstones, giving their gemmological names, chemical formulae and important physical properties. Almost all are oxides or silicates: just a few are phosphates, carbonates and sulphates. Fluorite is the only representative of the halides, although LiCAF is the mineral colquiriite ($LiCaAlF_6$) discovered at Colquiri in Bolivia [Fleischer *et al.* (1981)]. Most gemstones are single crystals, although a few (e.g., obsidian, coral, opal, lapis lazuli) are not. Common gems, such as quartz and garnet, are distributed worldwide, but most rare gems are found in unusual geological locations. Wherever the appropriate minerals are found, they contain only a tiny proportion of gems. Gem production facilities are sited only where the quantities of gem-quality materials are sufficient to make extraction economically viable. Diamonds are mined mainly in southern Africa, Russia, Brazil and Australia. Rubies are extracted from mineral deposits in Afghanistan, East Africa, Thailand and especially Burma, and sapphires come mainly from Australia, Sri Lanka, Thailand and the United States. The twelve most popular gems are worn as 'birthstones'; particular local selections are determined by availability, custom and fashion.

Gemmology is the complete study of each stone from its natural state, embedded in a host rock, to the cleaved, faceted, carved and polished article. Earth scientists measure the hardness, cleavage, fracture strength and specific gravity of gems and relate them to crystal structure and chemical bonding. The tenpoint scale of hardness, devised by the mineralogist Friederich Moh, measures the ability of a stone to resist scratching. Each mineral is compared with ten 'testers': talc (1), gypsum (2), calcite (3), fluorite (4), apatite (5), orthoclase (6), quartz (7), topaz (8), corundum (9) and diamond (10). A tester will scratch the mineral below it in the list but not the one above it. The Knoop test quantifies this scale by measuring the size of an indent made by a diamond point in the surface of a mineral subjected to some common force. The harder the material the smaller the indentation and the larger the number on the Knoop scale. Ten stages are identified on the Knoop scale to correspond with Moh's scale. In terms of diamond indentation the Moh scale is distinctly nonlinear.

The link between the strength of a solid and its hardness reflects the internal structure. This linkage is complex, as is the mechanism by which solids break [Cottrell (1964)]. Gems that break along one or more atomic planes in the crystal are said to cleave. Gems that undergo *perfect* cleavage have perfectly smooth surfaces; they include diamond, fluorite, calcite, topaz and baryte. *Distinct* cleavages, although not perfectly smooth, also have clearly visible cleavage surfaces. In aquamarine the cleavage surface is *indistinct*: it breaks along a surface not related to its internal structure and its fracture surfaces are uneven and irregular. The fracture surface in obsidian is shell-like or conchoidal. A fine-grained mineral such as dumortierite will fracture along an uneven surface, whereas an interlocking texture, such as nephrite, will give a splintery fracture.

Physical scientists seek to understand how the bonding between atoms determines their arrangement on a regular, symmetrical crystal lattice. *Cubic* crystals are the most symmetrical: they may be cleaved or cut into cubic, octahedral or pentagonal dodecahedral prisms. At least four three-fold axes are required to reproduce the cubic unit cell. Diamond (10), garnets (7.5), pyrites (6), spinel (8) and fluorite (4) are all cubic; hardnesses are given in parentheses. *Hexagonal* crystals have six-fold symmetry. The best known examples are emerald (7.5), aquamarine (7.5) and apatite (5). Scheelite (5), rutile (6) and zircon (7.5) are tetragonal with one four-fold axis and quartz (7), onyx (7), ruby (9), sapphire (9), tourmaline (7.5) and calcite (3) are trigonal. Finally, topaz (8) and chrysoberyl (8.5) are orthorhombic (three two-fold axes), diopside (5.5) and meerschaum (2.5) are monoclinic (one two-fold axis) and turquoise (6) is triclinic with no axes of symmetry.

Although some gemstones such as rock crystal, diamond and scheelite are usually colourless, and fluorite, zircon, sapphire and apatite sometimes coloured, others are brilliantly coloured always. There are seven categories of colour: colourless, white or silver, pink to red, yellow to brown, green, blue or violet and black. Each category has three sub-divisions; always, usually and sometimes coloured. However, these divisions are imprecise and physical scientists prefer to work with the optical absorption spectra, or other quantitative fingerprints, of individual stones. Self-coloured or *idiochromatic* gems are coloured by those constituent elements essential to their chemical composition. Peridot is coloured green by the Fe^{2+} present in magnesium iron silicate, $(MgFe_2)SiO_4$. By contrast, *allochromatic* gems are coloured by trace impurities. Amorphous and cubic stones are single coloured. Pure corundum (Al_2O_3) is colourless and trace amounts of Cr_2O_3 turns corundum to pink to dark red depending upon concentration. Small amounts of iron and/or titanium create the blue, green and yellow colourations of sapphires. Some gems have two, three or more unevenly distributed colours resulting from the zoning of different impurities during growth. The complex borosilicate tourmaline can exhibit as many as ten shades of colour within a single stone. Such other crystals as alexandrite ($Cr^{3+} : BeAl_2O_4$) and aquamarine ($Fe^{3+} : Be_3Al_2(SiO_3)_6$) are *pleochroic*, displaying different colours when viewed from different directions. Hexagonal and tetragonal gems display two colours and are *dichroic*. Orthorhombic, monoclinic or triclinic gems are *trichroic*. Lower symmetry crystals are also *birefringent*: light rays are divided into two as they pass through the crystal to give two images. The difference in the refractive index, measured for the two rays, gives the magnitude of the birefringence.

Lustre refers to the overall appearance of a gemstone, in respect of the intensity of light reflected by the stone. *Splendent*, *resinous*, *earthy* or *dull* are descriptions given to indicate the intensity of the reflected radiation. Splendent indicates mirror-like powers of reflection. Earthy or dull suggests that little light is reflected. When a stone's lustre is comparable to diamond it is described as *adamantine*. These are the most sought-after stones. Some species of gemstone (e.g. garnets) vary in their lustre: the hessonite ($Ca_3Fe_2(SiO_4)_3$) garnets are adamantine. Other gems owe their colouration to interference (opals, moonstone, hematite), variously being referred to as opalescent, iridescent and adularescent. The lustre is brought to best effect by fashioning a number of *facets* on the surface to give the stone its final shape or *cut*. Because gemstones have regular, internal structures, the positions of the facets necessary to show off a stone to best effect can be calculated with mathematical precision to make the stone bright and sparkling with flashes of colour or fire. Some spectacular properties of gems are derived from internal microstructures. Star sapphires contain needlelike rutile (TiO_2) crystals precipitated along three directions of the crystal perpendicular to the three-fold axis of the crystal. These give rise to characteristic six-rayed stars. The 'catseye' effect in chrysoberyl is also caused by rutile inclusions.

3.2 Synthetic and imitation gemstones

Synthetic gemstones are chemically and physically identical with natural gemstones except that they are man-made. The production of synthetic gems is a significant, modern industry grossing over $\$10^9$ per annum in world-wide sales. Such manufacturing facilities as there are have evolved through developments over thousands of years, timescales still short compared to the geological production cycle of natural gems. In earlier times it was comparatively easy to grow from aqueous solutions large single crystals of the alums, copper sulphate and nickel sulphate. But the hardness, colour, lustre and fire of gems are a consequence of their chemical make-up, strongly bonded components requiring elevated temperatures (and sometimes high pressure) to produce single crystals from their chemical components. The necessary high temperature technology has emerged only over the past one hundred or so years.

The first growth of gem-quality synthetic crystals in the laboratory was reported by the French chemist Edmond Frémy. He developed a *flux growth* technique for emeralds (Cr^{3+}: $Be_3Al_2(SiO_3)_6$). The intimately mixed ingredients were dissolved in a flux at high temperature and crystals measuring several cubic millimetres in volume nucleated spontaneously from the high temperature solution on cooling. The whole heating, melting and cooling cycle lasted several months. Later, Frémy also grew tiny crystals of ruby by melting Al_2O_3 and Cr_2O_3 in a crucible. A mass of tiny ruby crystals were recovered from the crucible after solidification. The next important development was the invention of the *flame fusion* process for the growth of rubies. A powdered mixture of ingredients is dropped into a gas furnace where they melt in the very high temperature flame (> 2500 K). The liquid droplets fall onto a pedestal where they solidify into single crystals. As the pedestal and growing crystal are slowly withdrawn they form a long, cylindrical boule, the dimensions of which are determined by the temperature profile of the flame and the rotation/withdrawal rate of the pedestal.

Imitation gemstones are not usually fashioned from the same material as natural gems: they have the appearance of the stones that they seek to imitate but they are easily distinguishable by their different chemical and physical properties. The most precious gemstones are diamond, ruby, sapphire and emerald. All have been imitated by growth into single crystals that may support laser action. They are not the only inorganic crystals which, when doped with the right activator, may be used in solid state lasers, but they and their surrogates are among the most important. The French manufacturers Gilson produce successful imitations of lapis lazuli, turquoise, coral and opal. Glasses used for centuries to imitate rubies, sapphires and emeralds were easily distinguished from natural gems by the ease with which they scratch and wear and by their warmth to the touch. The several different single crystals grown as imitation diamonds have characteristic faults: strontium titanate ($SrTiO_3$) is softer, zircon is heavier, as is YAG which also lacks fire.

Doped spinel ($MgAl_2O_4$) crystals were originally grown by the Verneuil technique to imitate ruby, sapphire, blue zircon and chrysoberyl. The quality of synthetic red spinel, often used as imitation ruby, is better than that of synthetic ruby grown by flame-fusion because of the lower incidence of growth faults. Generally, synthetic gems have different growth faults from the natural stones that they replicate because characteristic faults are often particular to the growth process. The dominant defects in Verneuil-grown rubies are gas bubbles with well-defined crystallographic habits. Man-made boules that are too large in diameter may crack as a consequence of internal stresses that develop on cooling from high temperatures. Flux-grown emeralds tend to contain flux inclusions, distributed in characteristic 'veil' and 'feather' patterns. Many other minerals are grown into single crystals for scientific study or technological application. These applications have stimulated the global development of crystal growth techniques for metallic, semiconducting and insulating materials.

3.3 Laser and other optical materials

Although almost all crystalline and glassy nonmetals have potential uses in optical technology, glasses are the dominant optical material providing *ca* 90% of all bulk optical components. They are easily made from inexpensive components to have a high degree of compositional homogeneity, and subsequently can be moulded and/ or machined to shape at low cost. Glasses are the materials of choice for bulk applications given these reasons, but also because of their high transmittance at near-ultraviolet, visible and near-infrared wavelengths. They are made from hundreds of different mixtures of inorganic compounds with a vast range of physical properties. Optical glasses have their compositions altered to vary their transmittance, indices of refraction, dispersion, thermo-optic and thermomechanical properties.

Inorganic crystals are used mainly in specialized applications. Some crystals (e.g. heavy metal halides) transmit out to longer wavelengths than common glasses and others (e.g. fluorides) to shorter wavelengths, and are used where low scattering, high thermal conductivity, hardness and strength are required. Other applications take advantage of higher quantum efficiency in luminescence, and directionally-dependent optical and nonlinear optical characteristics. Uniformity of the refractive index in a component is one reason for choosing single crystals for applications in lasers, coherent optics, harmonic generation and acousto-optics. The scattering of light by high quality single crystals is only *ca* 5–10 per cent of that common in glasses. Purity of starting materials is required to avoid optical nonuniformities, impurity absorption as well as void-, bubble- and crack-formation.

Since there is no long-range order in glasses they are *isotropic*. Crystals are differentiated from glasses by their long-range crystallographic order. The ordered arrangements of atoms or ions in crystals lead to periodic structures, and the possibility of directionally-dependent, i.e. *anisotropic*, properties. A quantitative description of the properties of crystals requires a knowledge of their structure and symmetry, since this dictates the number of directionally-dependent terms required for their full specification in tensor form [Nye (1985)]. The crystal class is described by its *point group* in Schönfliess or International notation §2.3.2: both are quoted in the text when each crystal is first mentioned. The point group of the crystal and the rank of the property tensor define the number of tensor components. Scalars such as energy, temperature and heat capacity are represented by tensors of zero rank. Glassy or amorphous solids, being isotropic, require the least number of terms. Table 3.1 lists the number of components of tensors of rank 1 to 3 for some optical crystals. The symmetry permutation condition reduces the number of independent coefficients required. For example, the second order optical susceptibility $\chi^{(2)}_{ijk}$ is a third rank tensor with eighteen independent coefficients. However, potassium niobate ($KNbO_3$), an orthorhombic crystal with space group C^{14}_{2v} (Amm2, #38), has only five independent non-zero coefficients, and trigonal $LiNbO_3$, space group C^{6}_{3v} (R3c, #161) has eight non-zero coefficients, four of which are independent. The *principal values* of a property are measured along crystal axes as defined in Table 2.6. Just as the symmetry of a crystal is described by the space group, so are particular sites in the crystal described by their point group. Beryl ($Be_3Al_2(SiO_3)_6$) is hexagonal with space group D^{2}_{6h} (P6/mcc, #192): the individual site symmetries are D_2 for Be^{2+} and Si^{4+}, D_3 for Al^{3+}, C_S and C_1 for two different O^{2-} sites. The colour of emerald is due to Cr^{3+} substitution at the Al^{3+} sites, which are six-fold coordinated to nearest neighbour O^{2-} anions.

Table 3.1. *Tensor characteristics of some optical crystals*

Crystal	Class	Point group	Tensor coefficients*		
			Rank 1	Rank 2	Rank 3
rocksalt, NaCl	Cubic	O_h	0	3(1)	0
periclase, MgO	Cubic	O_h	0	3(1)	0
fluorite, CaF_2	Cubic	O_h	0	3(1)	0
YAG, $Y_3Al_5O_{12}$	Cubic	O_h	0	3(1)	0
diamond, C	Cubic	O_h	0	3(1)	0
gallium arsenide, GaAs	Cubic	T_d	0	3(1)	3(1)
emerald, $Be_3Al_2(SiO_3)_6$	Hexagonal	D_{6h}	0	3(2)	0
lithium iodate, $LiIO_3$	Hexagonal	C_6	1(1)	3(2)	7(4)
BBO, β-BaB_2O_4	Trigonal	C_{3v}	1(1)	3(2)	8(4)
LiCAF, $LiCaAlF_6$	Trigonal	C_{3v}	1(1)	3(2)	8(4)
lithium niobate, $LiNbO_3$	Trigonal	C_{3v}	1(1)	3(2)	8(4)
α-quartz, SiO_2	Trigonal	D_3	0	3(2)	5(3)
sapphire, Al_2O_3	Trigonal	C_{3i}	0	3(2)	5(2)
fluorite, MgF_2	Tetragonal	D_{4h}	0	3(2)	0
zircon, $ZrSiO_4$	Tetragonal	D_{4h}	0	3(2)	0
YLF, $LiYF_4$	Tetragonal	C_{4h}	0	3(2)	0
KDP, KH_2PO_4	Tetragonal	D_{2d}	0	3(2)	3(2)
ADP, $(NH_4)H_2PO_4$	Tetragonal	D_{2d}	0	3(2)	3(2)
alexandrite, $BeAl_2O_4$	Orthorhombic	D_{2h}	0	3(3)	0
YAP, $YAlO_3$	Orthorhombic	D_{2h}	0	3(3)	0
LBO, LiB_3O_5	Orthorhombic	C_{2v}	1(1)	3(3)	5(5)
$KNbO_3$	Orthorhombic	C_{2v}	1(1)	3(3)	5(5)
KTP, $KTiOPO_4$	Orthorhombic	C_{2v}	1(1)	3(3)	5(5)
CBO, CsB_3O_5	Orthorhombic	D_2	0	3(3)	3(3)

* The numbers of independent tensor coefficients are given in parentheses. Important optical tensors are electric field, **E** and electric polarization, **P** (rank 1), dielectric constant, ε_r, and susceptibility, χ (rank 2), and second order susceptibility χ_{ijk} (rank 3).

3.3.1 *Octahedral and distorted octahedral structures*

In perfect octahedral symmetry (O_h) there are six anions equidistant from the central cation along the $\pm x$, $\pm y$ and $\pm z$-axes. Figure 3.1a illustrates the three C_2-rotations and six C_4-rotations about the $\langle 100 \rangle$ axes, six C_2'-rotations around the $\langle 110 \rangle$ axes and eight C_3-rotations about the $\langle 111 \rangle$ axes of the cube, characteristic of the face-centred cubic (O_h^5, Fm$\bar{3}$m, #225) structure of the alkali halides (LiF, NaF, NaCl, KCl) and alkaline earth oxides (MgO, CaO, SrO and BaO). The alkali halides have octahedral, O_h, point symmetry at both cation and anion sites. They are the host crystals of colour centre (F_A and F_2^+ type) and $Tl^0(1)$ centre lasers. Many inorganic crystals contain distorted six-fold coordinated MX_6 building blocks. The symmetry-adapted distortions of the octahedron in Fig. 3.2 are classified as the irreducible representations of the equilibrium symmetry of the centre and its environment. *Even-parity distortions*, A_{1g}, E_g and T_{2g}, retain the inversion centre of the perfect octahedron and *odd-parity distortions*, T_{1u} and T_{2u}, destroy the inversion symmetry. The displacements may be static or dynamic, depending on whether they are determined by the time-averaged position of the ions or their vibrations, respectively.

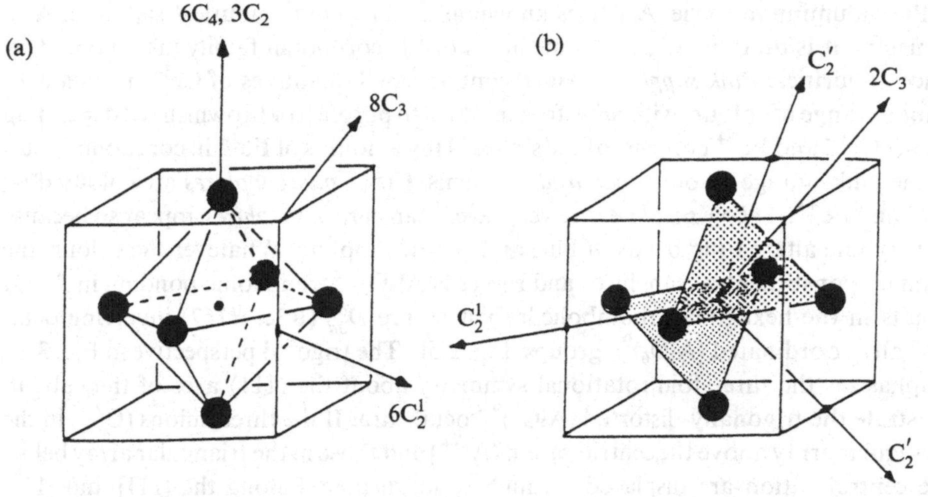

Figure 3.1. Two views of the octahedral arrangement of ions showing six equally charged anions at face-centres and a cation at the body-centre of the cube. Rotational invariance is maintained under the operation of the C_2, C_2', C_3, and C_4 rotations shown. In (a) the octahedron is outlined by the lines that link adjacent anions: (b) emphasizes the three-fold symmetry about the ⟨111⟩ axes. A T_{2g} distortion parallel to a ⟨111⟩ axis reduces the O_h symmetry to D_3, and reduces the number of rotation operators relative to the perfect octahedron.

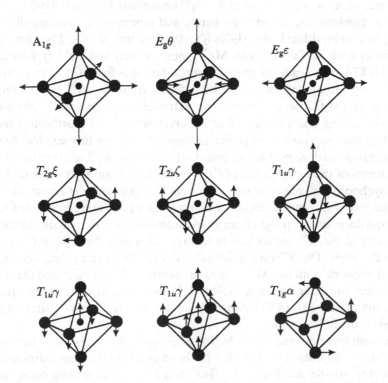

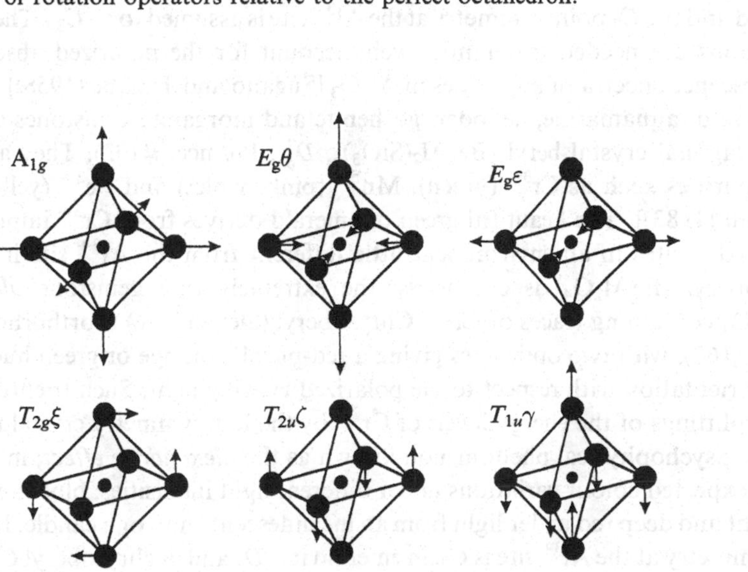

Figure 3.2. The normal modes of distortion of an octahedral complex.

Pure aluminium oxide, Al_2O_3, is known as *corundum* or colourless sapphire. As a gemstone it is quite rare. Usually members of the corundum family are coloured by trace impurities. *Pink sapphires*, which contain small quantities of Cr^{3+}, form a continuous range of colours with *ruby*, from pinkish to purplish or brownish red depending on the Cr^{3+} and Fe^{3+} contents of the stones. Tiny amounts of Fe^{3+} in corundum result in the pink-orange coloured *paparadscha* gems. Clear, *blue sapphires* are coloured by Ti^{3+} and Fe^{3+}. *Yellow sapphires* are very rare. Many *green sapphires* appear so because of very fine alternating bands of blue and yellow sapphire. Whatever the colour, the primary constituent of sapphires and rubies is Al_2O_3. Strong ionic bonding in Al_2O_3 results in the hexagonal (rhombohedral) structure D_{3d}^6 (R$\bar{3}$c, #167) involving octahedrally-coordinated $(AlO_6)^{9-}$ groups, Fig. 2.5b. The trigonal perspective in Fig. 3.1b emphasizes the three-fold rotational symmetry about the $\langle 111 \rangle$ axes of the cube to illustrate the trigonally-distorted $(AlO_6)^{9-}$ octahedra. If the three anions (O^{2-}) in the triangular array above the central cation (Al^{3+}) and those in the triangular array below the central cation are displaced through equal distances along the [111] and $[\bar{1}\bar{1}\bar{1}]$ directions, respectively, the symmetry is reduced from O_h to D_{3d} by an even-parity T_{2g} distortion. This leaves the distorted octahedron invariant under $3C_2$ and $2C_3$ rotations, only. Odd-parity T_{1u} distortions move the central cation (Al^{3+}) away from the position midway between the two triangles of O^{2-} ions and reduce the size of one of the triangles of O^{2-} ions relative to that of the other. T_{2u} distortions counter-rotate one of these triangles through an angle $\phi = 4.3°$ relative to the other. Sometimes the T_{2u} distortion is ignored and the C_3 point symmetry at the Al^{3+} site is assumed to be C_{3v}. The odd-parity distortions are needed to quantitatively account for the polarized absorption and luminescence spectra of impurities in Al_2O_3 [Sugano and Tanabe (1958)].

Emerald, aquamarine, helioder, goshenite and morganite gemstones all comprise the hexagonal crystal beryl ($Be_3Al_2(SiO_3)_6$, D_{6h}^2, P6/mcc, #192). They are coloured by impurities such as Cr^{3+} (green), Mn^{5+} (pink–violet) and Fe^{3+} (yellow or blue) [Nassau (1983)]. The beautiful green of emerald derives from Cr^{3+} impurities occupying Al^{3+} sites in an environment little different from the Al^{3+} site in ruby. Pure chrysoberyl ($BeAl_2O_4$) is colourless: the extremely rare gemstone *alexandrite* is $BeAl_2O_4$ containing traces of Cr^{3+}. Chrysoberyl (alexandrite) is orthorhombic, (D_{2h}^{16}, Pnma, #62), with two optic axes giving a red-purple, orange or green hue depending upon orientation with respect to the polarized viewing light. Such trichroism follows from splittings of the energy levels of Cr^{3+} by the low symmetry crystal field. This is *not* the psychophysical phenomenon known as the *alexandrite effect*, in which there are unexpected colour variations under different light intensities: blue-green in bright sunlight and deep red under light from an incandescent lamp or a candle. In corundum the symmetry at the Al^{3+} site is C_3; in emerald it is D_3 and in chrysoberyl C_S on one site and C_i on the other. The O^{2-} ions in chrysoberyl, Fig. 3.3, form an almost close-packed hexagonal network, with the Al^{3+} ions at the octahedral interstices and the Be^{2+} ions on tetrahedral sites. The two inequivalent Al^{3+} sites in $BeAl_2O_4$ occur in equal numbers. Approximately 80% of Cr^{3+} impurity ions occupy the C_S mirror sites, which play an exclusive role in laser action.

In the isostructural compounds MgF_2, ZnF_2 and MnF_2, the M^{2+} cation is surrounded by six equidistant F^- ions, but the bond angles in the plane perpendicular to the symmetry axis deviate from 90°. These rutile (TiO_2)-structured compounds are tetragonal with space group D_{4h}^{14} (P4$_2$/mnm, #136) and D_{2h} symmetry at the cation site.

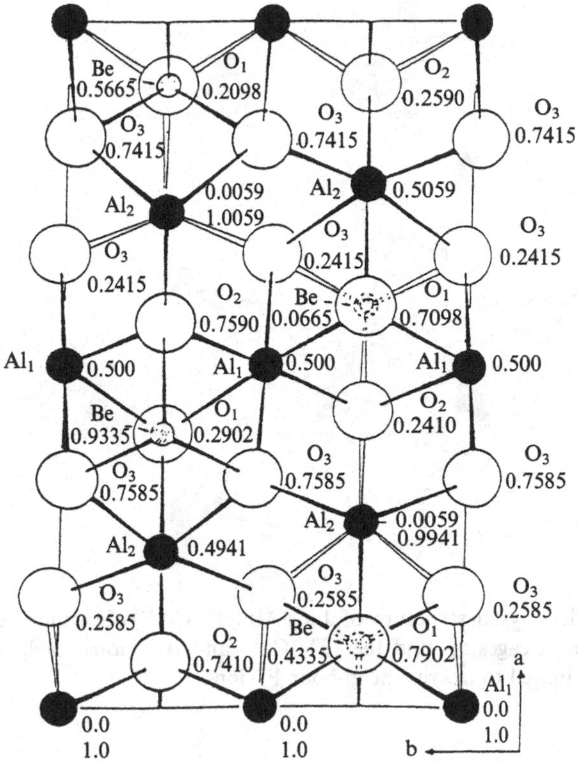

Figure 3.3. A projection of the chrysoberyl crystal structure in the *a*–*b* plane [after Farrell *et al.* (1963)].

The dominant component of the crystal field is octahedral with weaker orthorhombic perturbations. Impurities substituted at the Mg^{2+} or Zn^{2+} sites, such as Co^{2+} ($3d^7$) and Ni^{2+} ($3d^8$), are of interest for applications as near-infrared lasers. The fluoride perovskites (e.g. $KMgF_3$ and $KZnF_3$) are also cubic, O_h^1(Pm3m, #221), with octahedral O_h point symmetry at the K^+ and Mg^{2+} (Zr^{2+}) cation sites and D_{4h} symmetry at the F^- sites. The Cr^{3+}, Co^{2+} and Ni^{2+}-doped perovskites were among the first reported examples of tunable laser gain media. In fact, Cr^{3+} impurities are almost always located in six-fold coordinated ligand environments in ionic crystals, because both the ionic radius and octahedral crystal field stabilization energy favour six-fold coordination [Caird (1986)]. The elpasolites, A_2BMX_6, in which A and B are monovalent cations, M is a trivalent cation (Al^{3+}, Sc^{3+}, Ga^{3+}, Y^{3+}) and X is a monovalent anion (F^-, Cl^-), are face-centred cubic with space group O_h^5, (Fm3m #225). This structure is derived from perovskite by replacing alternate divalent cations (Mg^{2+}, Zr^{2+}) by monovalent and trivalent cations and by doubling the unit cell in each direction. Elpasolite crystals can accommodate substitutional trivalent impurities at rigorously octahedral M^{3+}-sites without charge compensation. The fluoride garnets $Na_3M_2Li_3F_{12}$, with $M = Al^{3+}$, Sc^{3+} etc, have a body-centred Bravais lattice belonging to the cubic crystal class O_h^{10} (Ia$\bar{3}$c, #230).

The trigonally distorted MF_6^{3-} octahedron is an ideal location for laser-active Cr^{3+} ions [Caird *et al.* (1988)] as in the case in the colquiriite-structured compounds,

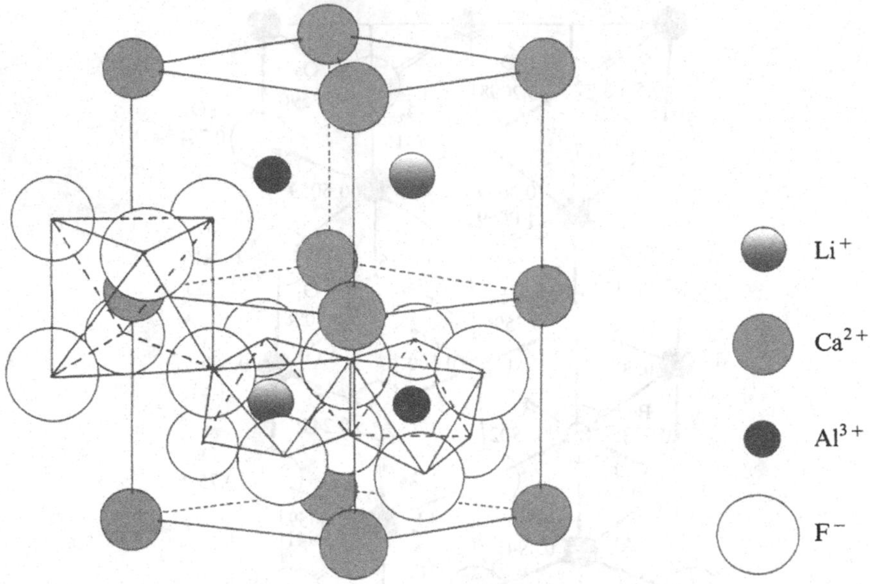

Figure 3.4. The crystal structure of $LiCaAlF_6$ (LiCAF) showing the distorted octahedral anion cages around the Li^+, Ca^{2+} and Al^{3+} ions, each of which is six-fold coordinated to nearest neighbour F^- ions.

$LiCaAlF_6$ (LiCAF), $LiSrAlF_6$ (LiSAF), and $LiSrGaF_6$ (LiSGaF). As Fig. 3.4 shows, in the LiCAF structure there are three connected MF_6 octahedra, $M = Li^+$, Ca^{2+} and Al^{3+}, which experience dominant T_{2g} and T_{2u} distortions [Viebahn (1971)]. Viewed in trigonal perspective, as Fig. 3.1b, the $(AlF_6)^{3-}$ octahedron has triangular arrays of F^- ions above and below the Al^{3+} ion that are moved slightly closer together than in the perfect octahedron. Although all Al–F distances remain equal, the F–Al–F bond angles are slightly changed from 90° so that the octahedron exhibits D_{3d} rather than O_h symmetry. The T_{2u} distortion destroys the inversion symmetry by a counter-rotational displacement of the upper triangle of F^- ions through $\phi = 4.3°$ relative to the lower one: the angle $\phi = 7.2°$ in LiSAF and 8.2° in LiSGaF. Colquiriites are trigonal, with space group D_{3d}^4 (P$\bar{3}$c1, #165) having two formula units per unit cell, in which Cr^{3+}: dopants substitute on M^{3+} (Al, Ga) site. In contrast, Ce^{3+} ions substitute on the divalent Ca^{2+} sub-lattice rather than the trivalent Al^{3+} sub-lattice as a consequence of size-factor considerations, charge compensation for the excess positive charge on Ce^{3+} relative to Ca^{2+} being effected by one Li^+ ion vacancy for each Ce^{3+} ion substituent.

3.3.2 Tetrahedral structures

Diamonds are face-centred cubic with point group O_h^7 (Fd$\bar{3}$m, #227). The local symmetry at each carbon atom is T_d, being determined by the four (sp^3)-covalent bonds directed along the equivalent [111], [11$\bar{1}$], [1$\bar{1}$1], [$\bar{1}$11] directions of the cubic cell. Type Ib diamonds are coloured yellow by nitrogen impurities, predominantly present as single substitutional atoms. These may be converted to *H*3 centres, two nitrogen atoms bridged by a carbon vacancy, by a complex annealing–electron irradiation–annealing process. The $H3$ centres sustain efficient, stable laser action in the spectral range 500–600 nm [Rand and De Shazer (1985)]. Unfortunately, Type Ib stones are rare, and those with a concentration of substitutional N-atoms sufficient for laser action, even rarer. The further development of lasers based on $H3$ centres in natural type Ib diamonds would be prohibitively expensive. Nevertheless, since diamonds have excellent optical, thermal and mechanical properties their device application is desirable. Rand (1986) pointed out that synthetic diamonds, available commercially at low cost, might be suitable hosts for laser-active centres. Lasers based on synthetic diamonds have yet to be reported.

Interest in four-fold coordinated structures was stimulated by the invention of the forsterite (Cr^{4+} : Mg_2SiO_4) laser [Petricevic *et al.* (1988)]. Forsterite is an orthorhombic crystal (D_{2h}^{16}, Pnma, #62) in which Cr^{4+} substitutes at the Si^{4+} site which has distorted tetrahedral (C_s) symmetry. Studies on similar materials have included the isostructural Ca_2GeO_4, Zn_2SiO_4, $Ca_2Al_2SiO_7$ and Y_2SiO_5. The most efficient Cr^{4+} laser has been reported as the Cr^{4+}-doped garnets, in which the Cr^{4+} ion is stabilized in tetrahedral sites (§3.3.3, §5.4.2, §9.3.2).

3.3.3 The garnet structure

The garnets are cubic with space group O_h^{10} (Ia$\bar{3}$d, #230). They can be represented *chemically* and *structurally* by $A_3B_2C_3O_{12}$ related to the cation–ligand polyhedra in Fig. 3.5 in which the A-sites are dodecahedral, the B-sites are octahedral and the C-sites are tetrahedral. These simplified descriptions ignore ionic displacements from ideal positions that reduce the symmetries to D_{2d} at A, C_{3i} at B and D_2 at C. Most naturally occurring garnets are silicates. The *grossular garnets*, composed of calcium aluminium silicate, $Ca_3Al_2Si_3O_{12}$, display many colours associated with trace impurities; gooseberry green from V^{3+} and Cr^{3+}, orange-brown due to Mn^{5+} and Fe^{3+} impurities and pink due to Fe^{3+}. Other allochromatic garnets include Fe^{3+} : Cr^{3+} : $Mg_3Al_2(SiO_4)_3$ or *pyrope*, which is bright red. Certain garnets are idiochromatic, including the fiery red *almandines* ($Fe_3Al_2(SiO_4)_3$) and bright orange *spessartines* ($Mn_3Al_2(SiO_4)_3$). Dark orange-to-red spessartines also contain traces of iron. The attractive, bright green colour of *uvarovite* garnet ($Ca_3Cr_2(SiO_4)_3$) is due to the intrinsic Cr^{3+} constituent of these crystals. Although these various silicate garnets can be cut into fine gemstones, they have not found technological application. This is the domain of the man-made rare-earth oxide garnets of which YAG (a laser host) and $Gd_3Ga_5O_{12}$ (GGG) (an optoelectronic substrate) are the lead players.

An alternative *chemical* formula can be used to represent the rare-earth garnets, $A_3B_xC_{5-x}O_{12}$ with $0 < x \leq 2$, in which the A-sites are reserved for Y^{3+}, Gd^{3+} or Lu^{3+} cations, Al^{3+}, Sc^{3+} or Ga^{3+} ions fill the B-positions and C-sites are occupied by Al^{3+} or Ga^{3+} ions [Winkler (1981)]. This formula symbolizes the binary oxides $Y_3Al_5O_{12}$ (YAG) or $Lu_3Ga_5O_{12}$ (LGG), tertiary oxides $Y_3Sc_2Al_3O_{12}$ (YSAG) or $Gd_3Sc_2Ga_3O_{12}$ (GSGG) and mixed or non-stoichiometric garnets such as $Gd_3Sc_{1.84}Ga_{3.16}O_{12}$, in which a small Ga^{3+} excess substitutes on the Sc^{3+} B-sites. For these $A_3B_xC_{5-x}O_{12}$ garnets the lattice parameter a_0 varies linearly with the *effective ionic radius*, r_{eff}, defined by

$$r_{\mathrm{eff}} = \left(\frac{x}{5}\right) r_{\mathrm{B}} + \left(\frac{5-x}{5}\right) r_{\mathrm{C}} \tag{3.1}$$

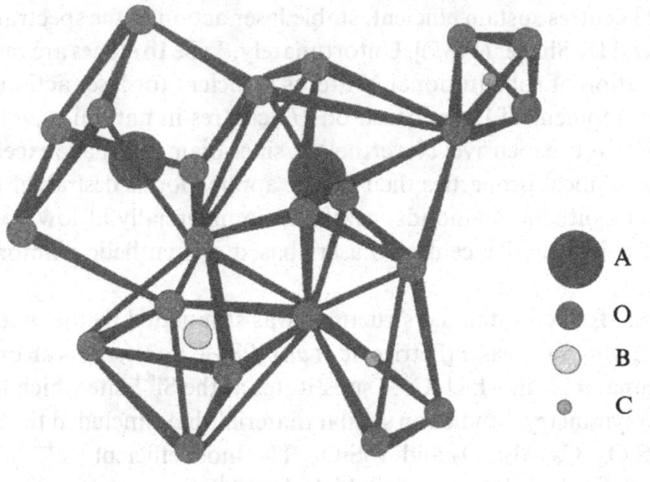

Figure 3.5. The metal–oxygen polyhedra in the $A_3B_2C_3O_{12}$ garnet structure.

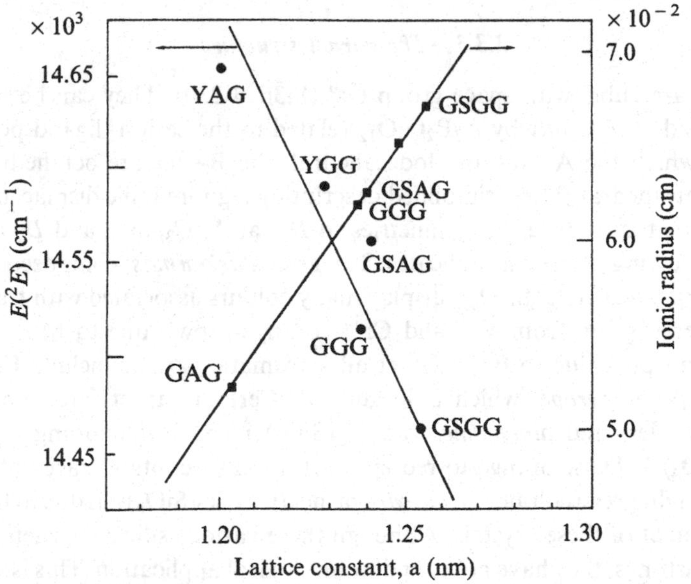

Figure 3.6. The linear dependence of the lattice constant, a_0, on the effective ionic radius for some Gd-garnets. Also shown are the energies of the ^2E state of Cr^{3+} ions in some Gd and Y-based garnets.

where r_B and r_C are the ionic radii of the components B and C, respectively. This concept of an effective ionic radius allows the properties of all the garnets represented by the formula $A_3B_xC_{5-x}O_{12}$ to be compared on a single plot. In Fig. 3.6, the plot of r_{eff} against the lattice constant, a_0, for Gd garnets having $x=0$ and $x=2$ is linear. In consequence, the two lowest-lying excited states of Cr^{3+} ions ^2E and $^4T_2$ have energies $E(^2E)$ and $E(^4T_2)$ which decrease linearly with a_0 over the range of a_0 values representative of the Y, La and Gd garnets. Data for the $E(^2E)$ state energy are also shown in Fig. 3.6. Impurity ions having optical transitions within the band gap are incorporated on A, B and C sub-lattices by adding suitable dopants to the melt before crystal growth. Trivalent rare-earth ions such as Ce^{3+}, Pr^{3+}, Nd^{3+}, Er^{3+} etc. substitute on the A-sites and trivalent transition-metal ions Ti^{3+}, Cr^{3+} and Fe^{3+} substitute on the octahedral B-sites. By co-doping with Cr and Ca^{2+} it is possible to introduce Cr^{4+} ions onto the tetrahedral C-sites of the oxide garnets. Of the numerous Cr^{4+}-activated laser gain media, Cr^{4+}-doped $La_3Al_5O_{12}$ (LAG) and $Y_3Al_5O_{12}$ (YAG) crystals are the most efficient.

3.3.4 *Apatites and related crystals*

The mineral apatite, $Ca_5(PO_4)_3F(Cl)$, is widely abundant across the globe; apatite gems are mined in Spain, Mexico, Sri Lanka and Burma. Gemstones are cut and faceted into hexagonal, rectangular or octagonal prisms, brightly coloured yellow, green, blue or violet. The apatites are chemically diverse within the formula $A_5(BO_4)_3X$, where $A = Ca^{2+}, Sr^{2+}$ or Ba^{2+}, $B = P^{5+}, V^{5+}$ or Si^{4+} and $X = F^-, Cl^-, OH^-$ or O^{2-} and tolerate a complex defect structure [McConnell (1973)]. The crystal structure of $Ca_5(PO_4)_3F(Cl)$ is hexagonal, belonging to the space group C_{6h}^2(P 6_3/m, #176). The A^{2+} ions occupy two sites in the unit cell, with 40% in the large A(I) site which has C_3 point symmetry and the remainder in lower symmetry (C_s) A(II) sites, which have reflection symmetry in the *ab*-plane and an irregular seven-fold coordination to six O^{2-} ions and one X^- ion. There are six parallel columns of A^{2+} ions around the central three-fold *c*-axis, linked by three shared O^{2-} ions to adjacent A^{2+} ions in the column. The laser-active RE^{3+} and Cr^{4+} ions substitute on the A(II)- and B-sites, respectively. The other $3d^2$ ions have been investigated in the fluoroapatites and fluorovanadates isomorphic with apatite ($Ca_5(PO_4)_3F(Cl)$) gemstones found in many parts of the world. In such crystals, the $3d^2$ ion occupies the P^{5+} site in the $(PO_4)^{3-}$ molecular unit.

3.3.5 *Nonlinear optical crystals*

At extreme intensities appropriate to laser radiation the macroscopic polarization, P, of a medium can be described in a Taylor series that contains terms nonlinear in the electric field, E:

$$\mathbf{P} = \varepsilon_0(\boldsymbol{\chi}^{(1)} \cdot \mathbf{E} + \boldsymbol{\chi}^{(2)} : \mathbf{EE} + \boldsymbol{\chi}^{(3)} : \mathbf{EEE} \ldots) \tag{3.2}$$

where ε_0 is the free space permittivity. The associated susceptibility χ and dielectric constant ε are also functions of the field;

$$\boldsymbol{\chi}(\mathbf{E}) = \boldsymbol{\chi}^{(1)} + \boldsymbol{\chi}^{(2)} \cdot \mathbf{E} + \boldsymbol{\chi}^{(3)} : \mathbf{EE} \ldots \tag{3.3}$$

and

$$\boldsymbol{\varepsilon}(\mathbf{E}) = \boldsymbol{\varepsilon}^{(1)} + \boldsymbol{\varepsilon}^{(2)} \cdot \mathbf{E} + \boldsymbol{\varepsilon}^{(3)} : \mathbf{EE} \ldots, \tag{3.4}$$

where $\boldsymbol{\varepsilon}^{(1)} = \varepsilon_0(1 + \boldsymbol{\chi}^{(1)})$ and $\boldsymbol{\varepsilon}^{(n)} = \varepsilon_0 \boldsymbol{\chi}^{(n)}$ for $n \geq 2$. The nonlinear terms in Eq. (3.2) become more important as the light intensity increases and lead to such nonlinear effects as harmonic generation, optical rectification, frequency mixing and parametric oscillation. Second harmonic generation (SHG) requires the second order optical susceptibility, $\boldsymbol{\chi}^{(2)}$, to be *ca* 1 pm/V. Generally, $\boldsymbol{\chi}^{(2)}$ is a tensor of rank 3, written in cartesian coordinates as

$$P_i^{(2)} = \varepsilon_0 \sum_{j,k} \chi_{ijk}^{(2)} E_j E_k. \quad (3.5)$$

The 27 components of $\chi_{ijk}^{(2)}$, reduced to 18 by the permutation condition $\chi_{ijk}^{(2)} = \chi_{ikj}^{(2)}$, are conveniently represented by a two-dimensional 3×6 tensor, the Kleinman d_{im}-tensor, in which notation the subscripts i and m on the d_{im} range from 1 to 3 and 1 to 6, respectively. The nonlinear optical coefficients $d_{im}^{(2)}$, related to the susceptibilities by $d^{(2)} = (1/2)\chi^{(2)}$, are used to assess materials for nonlinear optical applications. The independent components of the d-tensors for the 32 crystallographic point groups are given by Yariv (1975). They are non-zero only when there is asymmetry in the crystalline potential in some direction. All components of the susceptibility tensor are zero for centro-symmetric crystals. Since crystals are highly symmetrical, some d_{im} coefficients are zero and others may be equal, greatly reducing the labour involved in fully evaluating the d-tensor of a new nonlinear crystal. Only triclinic crystals with point group C_1 have 18 independent, non-zero Kleinmann d-coefficients. Other crystals listed in Table 3.1 have rather fewer, non-zero d_{im}-coefficients. Also listed in Table 3.1 are some laser crystals that have centres of inversion, including the cubic crystals YAG and MgO, tetragonal crystals such as MgF_2 and $LiYF_4$ and hexagonal crystals such as $Be_3Al_2(SiO_3)_6$. The second order susceptibilities of these and other centro-symmetric crystals are zero.

The first nonlinear optical crystal to be identified was α-quartz, a hexagonal crystal with space group D_3^6 ($P\bar{3}m$, #154). This crystal has five non-zero d-coefficients of which two are independent. Cubic (GaAs), trigonal ($LiNbO_3$) and tetragonal (KH_2PO_4) crystals, which lack an inversion centre, have finite nonlinear susceptibilities. Among the families of compounds that are optically nonlinear are the distorted perovskites ABO_3 (where A = Li, Na or K and B = Nb, Ta), the dihydrogen phosphates (AH_2PO_4, D_{2d}^{12}, ($I\bar{4}2d$, #122)) and arsenates (AH_2AsO_4) (where A = NH_4, K, Rb, Cs), and the titanyl phosphates such as $KTiOPO_4$ (C_{2v}^9, (Pna2, #33)). In these inorganic crystals the sources of the nonlinearities are the NbO_6 and PO_4 structural units. Representative $\chi^{(2)}$ values for members of these families are given in Table 3.2.

Rather few inorganic crystals are used in nonlinear optical devices, the most common being $LiNbO_3$ and $KNbO_3$, ammonium dihydrogen phosphate (ADP), potassium titanyl phosphate (KTP), and the crystalline borates LiB_3O_5 (LBO, C_{2v}^9 (Pna2, #33)) and β-BaB_2O_4 (BBO, C_3^4 (R3, #146)). However, the recent developments illustrate the fact that no ideal nonlinear optical material covering the entire range IR-to-visible-to-UV has yet been discovered. Reference to Table 3.2 shows that the $\chi^{(2)}$ values for $LiNbO_3$ and $KNbO_3$ are superior to the other compounds. Nevertheless, they cannot be used for SHG below 400 nm, at which wavelengths the borates and isomorphs of KDP have superior transmittance, and are more resistant to bulk laser damage. In consequence, BBO, LBO and KDP find applications for harmonic generation and parametric oscillation in the UV, where they have sufficient birefringence to offset material dispersion in phase-matching operations.

3.3.6 *Laser glasses*

There is an enormous range of glass compositions into which laser active ions may be doped. Apparently transition metal ions with the $3d^n$ configuration ($n = 1$ to 9) have

Table 3.2. *Some properties of nonlinear optical crystals* [*selected from Nikogosyan* (1997) *and Bass* (1995)]#

Property	Crystal					
	$LiNbO_3$	$KNbO_3$	KDP	KTP	LBO	BBO
Point group	C_{3v}	C_{2v}	D_{2d}	D_{2d}	C_{2v}	C_3
Transparency+ range (nm)	370–5500	400–5000	176–1400	350–4500	155–3200	189–3500
Birefringence (Δn)	0.0765	0.1478	0.0339	0.0938	0.0399	0.1126
Nonlinearity* [pm/v]	$d_{31}=-4.35$ $d_{33}=-27.2$ $d_{22}=-2.10$ $d_{21}=-2.1$ $d_{32}=-4.3$	$d_{31}=-11.3$ $d_{32}=11.9$ $d_{33}=-19.6$ $d_{24}=11.9$ $d_{15}=-11.3$	$d_{36}=0.39$ $=d_{14}$ $=d_{25}$ $d_{21}=d_{31}=$ $d_{15}=0$	$d_{31}=6.5$ $d_{32}=5.0$ $d_{33}=13.7$ $d_{24}=7.6$ $d_{15}=6.1$	$d_{31}=-0.67$ $d_{32}=+0.85$ $d_{33}=10.04$	$d_{31}=0.16$ $d_{33}=-0.10$ $d_{22}=0.22$
SHG cut-off (nm)	~1080	860	487	990	555	411
OPO tuning range (nm)	—	—	430–700	610–4200	415–2500	410–2500
Laser-induced damage threshold++ $\times 10^{-12}$ W/m^2	5	5	50	10	190	135

+ Measured at the 0 transmittance level.
* Measured at 1.064 μm; the sign of the tensor coefficient is not always determined.
++ For bulk crystals using 10 ps pulses at 1.064 μm radiation.
# Values for the d_{im} are not always fully determined and can vary widely between sources.

little potential for applications in laser glasses because of their sensitivity to the local environment. The absence of long-range order leads to a wide distribution of crystal field sites for the optical centres and their optical spectra are inhomogeneously broadened far more effectively than are the spectra of ions in crystals. The R-lines of Cr^{3+} ions in ruby and other oxides have low temperature halfwidths of *ca* 1 cm^{-1}, compared with the R-lines in glassy solids where the halfwidths may exceed 100 cm^{-1}. Furthermore, excited ions in the weaker field sites in glasses de-excite non-radiatively rather than radiatively so that the luminescence quantum efficiency is much reduced. Depending on the particular host glass, the quantum efficiency of Cr^{3+} emission is rarely greater than 1–5% at 300 K, militating against laser gain in these cases.

Rare-earth ions are less of a problem. At room temperature the lifetime broadening of emission lines is stronger than the disorder-induced inhomogeneous broadening, and the luminescence quantum efficiency is little different for rare-earth ions in glasses than in crystalline hosts. For this reason, in the early developments of laser systems the Nd-glass laser was almost as successful as the Nd-YAG crystal. Indeed, Nd-glass lasers have a number of advantages over Nd-YAG lasers. Much larger glass rods can be fabricated than YAG crystals, permitting the manufacture of very large Nd-glass amplifiers. In addition, the much broader gain bandwidth of the Nd-glass laser line makes possible shorter mode-locked pulses from Nd-glass lasers than from Nd-YAG. Finally, the stimulated emission cross section is lower for Nd^{3+} ions in glasses than in

YAG, commensurate with the broader gain bandwidth, which allows increased energy storage when used as an amplifier.

Most oxide glasses contain molecular ion structural units (e.g. SiO_4^{4-}, PO_4^{3-}, BO_4^{5-}). In the case of silicate glasses, the conversion of silica into glass leads to the randomization of the quartz network obtained by a dissymetric rotation of the SiO_4^{4-} tetrahedra relative to one another. The consequent distortion of these units follows mainly from the large spread in O–Si–O bond angles. The break up of the three-dimensional quartz network is effected by the addition of 10–20% Li_2O, Na_2O or K_2O as *network modifiers*. As the modifier content in the glass increases, so the long-range periodicity decreases, resulting in open structures with long distances between ligand ions and sites suitable for occupation by transition or rare-earth metal ions. Stabilizers such as Al_2O_3 and B_2O_3 are also added to prevent crystallization. The same general principles apply in the formation of silicate, germanate, borate and phosphate glasses, in that an intimate mix of glass former, modifier and stabilizer is made, melted in a Pt crucible and then cast into a mould. However, there are some general principles. Common oxide glasses for bulk applications transmit radiation in the range 0.35–2.5 μm: this range can be extended to shorter wavelengths by increasing the SiO_2, Al_2O_3 or MgO content and to longer wavelengths by adding GeO_2 or B_2O_3 [Nikogosyan (1997)]. The addition of BaO to an oxide glass results in glasses with high refractive index and lower than normal dispersion, whereas B_2O_3 glasses offer low indices with very low dispersion. Alternatively, heavy or light metal fluoride glasses, respectively, can be used for IR or UV applications. Normally the precise glass composition and thermal history is proprietorial to the manufacturer.

The vigorous interest in RE^{3+}-doped glasses in recent times has resulted from the many potential applications in telecommunications, information technology, metrology, medicine and sensing. Although optimization of fibre characteristics may be different in each area of application, the general principles of production and operation are the same. Basic fibre production starts with the manufacture of a large glass block. Although this *pre-form* has dimensions many, many times larger than the required fibre diameter, it has the same composition profile as the end product. This is achieved by melting and casting as for bulk glasses, or more usually by chemical vapour deposition (CVD) for layer-by-layer deposition of the pre-form to control the composition to give the appropriate refractive index profile. The standard dopants for silica fibres, other than the RE^{3+} ions, are GeO_2 and P_2O_5 to increase the refractive index of the core and B_2O_3 or SiF_4 to decrease the refractive index in the cladding. Usually two pre-forms are fabricated in separate crucibles and combined in the process of drawing down the fibre. This eliminates the need for the extreme size of pre-form required in producing long lengths of fibre.

A recent triumph of glass technology has been the realization of extremely long lengths (~30 km) of fibres between repeater links. This followed developments of extremely low loss fibres, first around the 800–900 nm window in the fibre transmission spectrum overlapping the output spectrum near 800–900 nm of GaAs/AlGaAs LDs, and now at the 1.3 μm (for low dispersion) and 1.55 μm (for minimum loss) windows to match the development of InGaAs/GaAsInP multi-quantum well lasers for these spectral regions. The fabrication of single-mode fibre for 1.55 μm transmission with low loss (< 0.1 db/km) and the availability of Er^{3+}-doped fibre amplifiers has confirmed the preferred operating wavelength as 1.55 μm for high speed communications.

Trivalent RE^{3+} ions other than Er^{3+} have been doped into optical fibres, resulting in a laser active gain medium, despite the absence of a well-defined process for optimization of the RE^{3+} profile in fibre lasers and amplifiers. Of particular interest has been the application of Pr^{3+}-doped and variously co-doped fibres for up-conversion lasers in the blue, green, yellow and red regions of the optical spectrum.

An important refinement in optical fibre fabrication has been the development of polarization-preserving optical fibre by 'poling' the fibre by application of an electric field at temperatures above the glass softening point, thereby aligning GeO_4^{4-}/SiO_4^{4-} molecule-ions along the fibre axis. This alignment is akin to long-range order in crystals, resulting in a nonlinear refractive index for the fibre at high laser intensities. The refractive index is then

$$n = n^{(1)} + n^{(2)} I, \tag{3.6}$$

in which the second order index $n^{(2)}$ is of order $2\text{–}3 \times 10^{-16}\,\text{cm}^2/\text{W}$. This is important for the shapes of optical pulses that propagate in a fibre. Under appropriate conditions the optical pulses propagate without change in shape: i.e. as zeroth-order *solitons*. Higher order solitons display periodic evolutions of shape during propagation down the fibre [Mollenauer *et al.* (1980)]. This observation is of considerable importance in the development of ultra-short pulse lasers [Mollenauer and Stolen (1984)] and in proposed soliton communication systems [Mollenauer *et al.* (1983a), Mollenauer (1985b), Hasegawa (1989)].

3.4 Growth of optical crystals

The symmetry, rarity and aesthetic beauty of many natural gemstones, combined with unique physical properties, stimulated early attempts to duplicate them in the laboratory. The growth of emerald, sapphire and ruby in the laboratory quickly led to developments by Czochralski (1917), Bridgman (1925) and Kyropoulos (1926) in the crystallization of metals from the melt. Their basic techniques have been much modified and improved upon. However, the demands of the semiconductor industry since the late 1940s, and of the optics and optoelectronics industries since the advent of the laser led to many further advances in crystal growth, which is now a multidisciplinary science built on foundations in solid state physics and chemistry, preparative and analytical chemistry, thermodynamics and statistical mechanics, electrochemistry and surface science, and embracing an array of laboratory practices. Both specialist and non-specialist can appreciate the present status of the field from the *Handbook of Crystal Growth* by Hurle (Volumes I–VI, 1993–97).

The growths of crystals from the melt by the Bridgman–Stockbarger technique (§3.4.1) and the Czochralski technique (§3.4.2) are particularly important; ninety per cent of commercially available bulk crystals have been grown by one variant or another of these processes. Nevertheless, several important optical materials cannot be grown from the melt, including LiB_3O_5, β-BaB_2O_4 and $YAl_3(BO_3)_4$, which are usually grown by high temperature solution growth (§3.4.5), also referred to as flux-growth. Materials which decompose, or have high vapour pressures at their melting temperature, and which are grown hydrothermally (§3.4.4), include α-quartz, β-alumina and certain rare-earth fluorides. The advantages of reduced dimensionality, so evident in semiconductor quantum well devices and glass fibre lasers, are also apparent in laser crystals. Recent growth techniques in this area include laser-heated pedestal growth and liquid-phase epitaxy.

Problems inherent in the growth of optical crystals derive both from the source materials and from the defect structures that result from high temperature processes. It is simple to specify the source materials to be used and their impurity levels: it is another matter to maintain high purity to close tolerances during preparation and growth. The purity of some oxides and fluorides can exceed 5Ns: such fine powders are easily contaminated during pre-growth preparation. Water absorption is a particular problem and handling/storage under dry, clean conditions is essential. Since the powders that make up the charge occupy large volumes relative to the final crystals, they must undergo densification by cold pressing, sintering or continuous charging at crucible temperatures above the melting point, all sources of contamination. In mixed systems the preferential loss of one component by evaporation may cause problems, and pre-sintering to prepare the final compound prior to melting can improve the stability of the melt. An example is the growth of $LiNbO_3$, for which the melt is formed from Li_2CO_3 and Nb_2O_5 because Li_2O is not stable. Failure to allow for such problems during preparation for growth can lead to compositional inhomogeneities and defect production in the grown crystal.

Point defects can absorb pump and emitted photons, and defect aggregates can cause light scattering when present in particulate form. This latter is especially the case in fluorides which, being sensitive to hydrolysis by even tiny amounts of water, may contain a distribution of hydroxy-fluoride precipitates. Scattering in fluorides can be overcome by hydrofluorination of the starting materials at elevated temperature (600–700 K) or, for mixed fluorides such as LiYF or LiCAF [Cockayne *et al.* (1981)], by converting the mixture of component fluorides into polycrystalline bars of the required compound by zone-refining. In LiYF and LiCAF the loss by evaporation of LiF can be ameliorated by operating the furnace under ambient gas pressures of 1.5–2.0 atmospheres. Pastor *et al.* (1975) contained scatter and improved melt congruency for fluoride growth by using HF gas as a reactive ambient atmosphere in the growth chamber.

Apart from quality control problems associated with the purity of the starting components, there are others associated with the defect structure of the grown crystals. Defects which lead to non-uniform optical properties degrade the crystal quality such that they are not usable in laser devices. Defects include voids, inclusions, precipitates, cellular structures, dislocation boundaries, growth striations, facets and colour centres. The extent to which they occur depends upon both the material and the growth technique, being determined by chemical factors or stress factors alone or in combination. In view of their dependence on method of crystal growth and on material, further discussion of the defect control is postponed until the interrelationship between crystal growth and high temperature phase equilibria has been outlined. However, control of microstructure is essential not only for reasons of laser efficiency, but also because there is an economic penalty for poor crystal quality.

The growth of single crystals of a material requires an ability to manipulate the phase relationships that exist between components; an essential in the crystal grower's armoury is a mastery of the thermodynamics of phase transitions and their consequences for crystal growth. A simple and elegant account of the construction of phase diagrams from considerations of the temperature and composition dependence of the

Gibbs free energy is given by Cottrell (1953). The application of phase diagrams to crystal growth is discussed by Neilsen and Monchamp (1970) and by Rosenberger (1979), and with specific application to optical crystals by Nassau (1971).

3.4.1 Growth by directional solidification

The growth of single crystals by slow directional solidification was pioneered by Bridgman (1925) for low melting point metals. He used a vertically mounted DC electric furnace to melt the charge and lowered an ampoule containing the charge through the heated zone. Later, Stockbarger (1963) used a similarly vertical furnace system to grow LiF. Closed ampoules are only necessary when the charge has a high vapour pressure at the melting temperature of the material being grown. Otherwise open crucibles are used in furnaces with a controlled growth atmosphere. The most common crucibles are graphite for growing fluorides, and platinum or iridium for growth of oxides. The crucibles were tapered near the bottom to promote self-seeding. A seed crystal of known orientation can be inserted through the point in the tapered section to provide the reference orientation for growth of new material on to the seed. Such directional growth is referred to as *Bridgman–Stockbarger* or *vertical Bridgman growth*. Horizontal directional growth was introduced by Chalmers in the 1950s also to study the crystallization of low melting point metals. This is the *horizontal Bridgman technique*. Both vertical and horizontal Bridgman techniques, in which the whole charge is melted, are *normal freezing techniques*. An alternative technology uses *gradient freeze* or *dynamic gradient freeze growth* in which the temperature profile of the furnace is moved along the charge by ramping the electrical power supplied to a single zone furnace or to a linear array of such furnaces. In *zone melting* a short section of the charge is melted and caused to traverse along the charge with the aim of purifying the resulting crystal [Pfann (1966)].

The ease of construction and application of the various Bridgman–Stockbarger furnaces led to growth of many metal, semiconductor and insulating crystals. However, their designs vary from material to material depending upon the melting temperature, the chemical reactivity of the melt and its vapour, vapour pressure and crystal orientation requirements. In Bridgman–Stockbarger growth the crystal-melt interface is moved through the length of the charge by translation of the crucible or the furnace. A vertical Bridgman–Stockbarger furnace is shown in Fig. 3.7 with an idealized temperature distribution applied along its axis. The power levels for the high temperature and low temperature zones are independently adjustable for ease of control of the axial temperature profile, and set so that the melting point of the crystal is at the mid-point of the adiabatic loss zone between the high and low temperature sections. A seed crystal, positioned just below this central point, is allowed to come to thermal equilibrium with the melted region just above it, before crucible and charge are mechanically lowered through the furnace into the low temperature zone. The power to the furnace windings is then gradually ramped down so that furnace, crucible and crystal are slowly cooled to the ambient temperature. Ideally the crystal should be chemically inert with respect to the crucible and have a larger coefficient of thermal expansion than the crucible, so that it contracts away from the crucible wall on cooling, making it easier to detach from the crucible. The multi-zone furnace shown in Fig. 3.7 may be applied in the gradient freeze technique, in which case the entire thermal

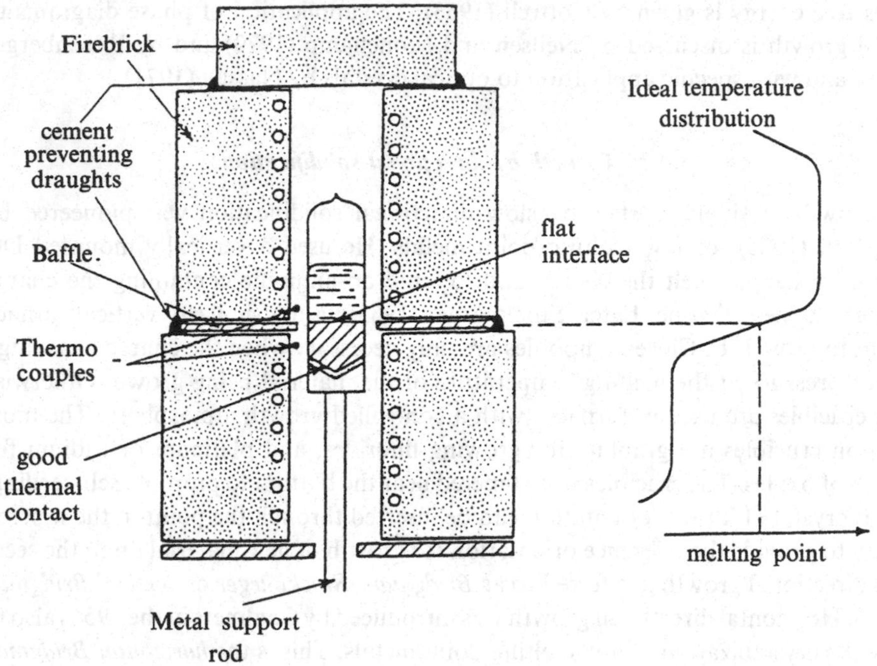

Figure 3.7. Sketch of a vertical Bridgman–Stockbarger furnace for single crystal growth from the melt [after Monberg (1994)].

profile is shifted along the furnace axis. With the advent of digital, multi-channel, integrated control systems it is possible to set power levels independently to as many as 50 heating zones each measuring one centimetre or more in length. The temperature gradient is traversed along the furnace by sequential changes in the power used to heat appropriately adjacent control zones. The smoothness of the moving temperature gradient is determined by the length of each independent furnace zone relative to the diameter of the furnace tube.

The vertical configuration for Bridgman–Stockbarger growth is used much more often than the horizontal configuration. $LiYF_4$ is an important host for laser active ions, especially the RE^{3+} dopants such as Ce^{3+}, Pr^{3+}, Nd^{3+}, Ho^{3+} and Er^{3+}. Such crystals are prone to the formation of precipitates that cause passive scattering losses. This problem can be partially ameliorated by holding the newly-grown crystal in the cold zone of the furnace within *ca* 100°C of the melting temperature of the crystal for periods of 24 hr or more. The same technique works quite well for pure LiCAF, Cr^{3+} : LiCAF and Ce^{3+} : LiCAF crystals. De Yoreo *et al.* (1991) at the Lawrence Livermore National Laboratory commissioned a commercial vertical Bridgman–Stockbarger furnace for growing large slab-shaped single crystals. The furnace was inductively heated at 25 kHz using a solid, cylindrical susceptor containing a rectangular channel along the axis to accommodate the tapered ampoule. The furnace specification called for available temperatures up to 2300 K stable to better than 0.1, for application to the growth of oxide and fluoride crystals. In growing Cr^{3+} : LiCAF a graphite susceptor was used in the hot and cold zones consisting of 10 individual discs each of 40 mm thickness. The thickness of the adiabatic zone was adjustable between

10–50 mm. Rectangular-shaped boules were grown measuring approximately 10 × 100 × 300 mm^3. Axial temperature gradients of between 6–16 K cm^{-1} and growth rates of 0.5–1.5 mm/hr were used. The as-grown crystals contained micron-sized scattering defects, which did not absorb visible or near-infrared radiation. Scattering losses of order 2–3 × 10^{-2} cm^{-1} were measured. The lower scattering losses resulted in crystals grown at the lower rates in smaller temperature gradients. De Yoreo *et al.* (1991) eliminated the scattering centres by annealing the Cr^{3+} : LiCAF crystals in flowing argon atmospheres within 50 K of the melting temperature for periods up to 14 days, reducing the passive scattering losses in the crystals by an order of magnitude. At these high temperatures the precipitates dissolve into the matrix and then diffuse out of the crystal with a diffusion constant of order 10^{-7} cm^2 s^{-1}. The particular defects observed in Cr^{3+} : LiCAF single crystals were not determined. Apparently different types occur for each material and perhaps for each growth technique. Different treatments have to be devised to eliminate them, wholly or partially, during the growth or post-growth processing. Fahey *et al.* (1986) report an elegant multilayer heat shield arrangement to provide the necessary temperature gradient for growth of Ti-sapphire crystals by the gradient freeze techniques with reduced particle scattering and near-infrared absorption, as compared with Ti-sapphire grown by Czochralski pulling.

3.4.2 Crystal pulling techniques

Crystal pulling from the melt was pioneered by Czochralski (1917), for growing single crystal fibres of low melting point metals. Subsequently the technique was further developed by Teal and Little (1950) for growing large, highly perfect single crystals of Ge. Crystal pulling from the melt, now universally referred to as Czochralski growth, was later extended to the growth of Si [Teal and Buehler (1952)], a range of III–V, II–VI and IV–VI compound semiconductors, pure metals and alloys, oxides and halides [Hurle and Cockayne (1994)]. Czochralski growth, as with Bridgman–Stockbarger growth, is particularly useful for growth from melts comprising single component compounds (such as Al_2O_3) and congruently (or near-congruently) melting intermediate phase compounds such as YAG, GSGG, YLF and LiCAF.

The principle features of a Czochralski puller are shown in Fig. 3.8: heating is applied to the charge via a radio frequency induction coil that couples RF power into the susceptor, usually graphite, platinum or iridium depending on material being melted. Usually the susceptor is also the crucible. The RF coil and crucible are electrically isolated from one another by an insulator made from silica, fused quartz or alumina. A pull-rod fitted with chuck containing the seed crystal and mounted co-axially above the melt in the crucible is lowered towards the melt surface until the seed is suspended just above the surface where it is allowed to come to thermal equilibrium just below the melt temperature. Once equilibrated, the seed is lowered until its end just dips below the melt surface. The melt temperature is then carefully adjusted until a meniscus is established between the melt and the seed crystal. Once a steady thermal state has been reached between seed crystal, meniscus and melt surface, the pull rod is lifted and slowly rotated: crystallization then occurs preferentially onto the seed crystal. At the start of pulling, the melt temperature and pull-rate are adjusted so that a thin neck is grown onto the seed to minimize defects propagating into the growing crystal from the seed. The diameter of the growing crystal is then increased out to that

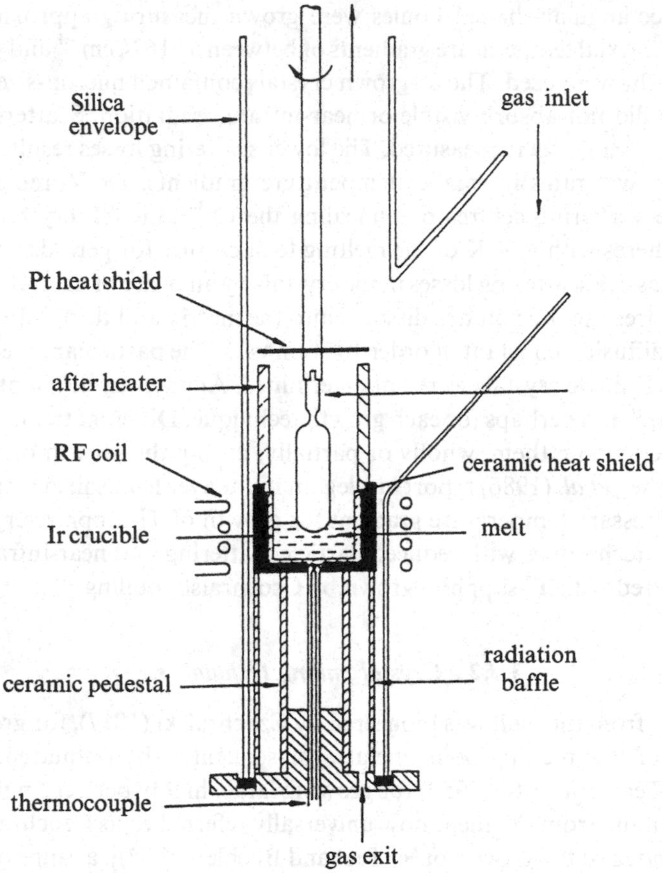

Figure 3.8. A schematic diagram of a Czochralski puller for growing single crystals of $CaWO_4$ and higher melting point oxides [after Cockayne *et al.* (1964)].

required by reducing the heater power (and consequently the melt temperature). Further minor adjustments of the melt temperature and pull/rotation rates are made to enable further growth at the specified boule diameter. Although adjustments to the pull rate, crystal rotation rate and heater power can be performed manually, they are now routinely carried out under computer control. There have been various approaches used to the pulling of single crystals using automatic servo-control systems. In growing insulating materials automatic diameter control is best achieved by servo-controllers that sense the change in weight of the crystal or the crucible plus melt. Although originally developed for growing oxides, automatic diameter control by weight sensing is now the established technique for all types of crystal because of its versatility, reliability and sensitivity [Hurle (1977)]. The rate of growth of the crystal, given the cylindrical symmetry of the puller, is equal to the pulling speed plus the rate at which the melt descends in the crucible, a condition achieved by careful adjustment of the heat flow between melt and crystal. Crystal rotation is essential to maintain radial heat flow at the melt-crystal interface, as this ensures radially uniform composition of the crystal.

The development of crystal growth apparatus for optical crystals was given great stimulus by the invention of the ruby laser [Maiman (1960)], soon to be followed by the

Nd : $CaWO_4$ [Johnson *et al.* (1962)] and Nd-YAG [Geusic *et al.* (1964)] lasers. Nassau and Broyer (1964) first applied Czochralski pulling to the growth of Nd : $CaWO_4$ crystals. This and refinements to the Czochralski furnace by Cockayne *et al.* (1964), Fig. 3.8, established the need for refractory support of the crucible concentric with the RF coil, heat shields and after-heater that controlled the thermal environment into which the crystal grows. A completely enclosed growth chamber enables the control of the furnace atmosphere by the introduction of neutral, oxidising or reducing gases. Many similar growth systems to that illustrated in Fig. 3.8 have been used for growing an ever widening array of oxides for applications in optics and electronics. Although other growth techniques have been used, the Czochralski technique is the most adaptable and generally applicable to growth from solid solutions and near-congruently melting compounds.

The applicable temperature range of Czochralski growth extends from the melting point of lead germanate (738°C) to magnesium aluminate spinel (2150 C) [Nassau (1971), Cockayne (1977)]. Among the many crystals amenable to pulling from the melt are various of the $A_3B_2C_3O_{12}$ garnets, and alexandrite, $BeAl_2O_4$. The furnace in Fig. 3.8 shows an iridium crucible, containing the melt supported and surrounded by quartz, firebrick and zirconia refractories. Passive after-heating of the crystal immediately above the growth region is provided by thick alumina felt supported between concentric alumina tubes. Such features are co-axial with the pull-rod and attached seed crystal, which accesses the melt through a hole in the centre of an iridium lid on the crucible. This system provides for low-thermal gradients in the post-growth region of the furnace, thereby minimising cracking of the garnet crystals. Typical growth rates of $< 1\,\mathrm{mm\,hr^{-1}}$ are successful for Nd^{3+} : YAG. Rather faster pull rates of $3-5\,\mathrm{mm\,hr^{-1}}$ are used for Cr^{3+} and Nd^{3+}-doped GSGG. When some 75–80% of the melt has crystallized, the crystals are pulled quickly from the melt into the region of the after-heater, where they eventually cool at the rate of the surrounding refractories.

The binary phase diagram of the Y_2O_3–Al_2O_3 system shows that both $Y_3Al_5O_{12}$ (YAG) and the $YAlO_3$ (YAP) melt at their stoichiometric compositions: both are excellent hosts for laser-active rare-earth ions such as Nd^{3+}, Er^{3+} and Pr^{3+}. Re^{3+} : YAG and YAP melts may be used for Czochralski growth without serious problem, although crystal quality may be compromised when the RE_2O_3 content is greater than about 1–1.5 mol.%. The growth of GSGG is rather different, with melts at the stoichiometric composition $Gd_3Sc_2Ga_3O_{12}$ producing variations in crystal homogeneity along the length of the boule. In fact, only melts with the congruent composition $\{Gd_{2.957}Sc_{0.043}\}[Sc_{1.862}Ga_{0.138}](Ga_3)O_{12}$, where the brackets { }, [ ] and ( ) indicate substitution at dodecahedral, octahedral and tetrahedral sites, *et seq.*, produce boules of uniform composition. Evidently some Sc^{3+} ions and Ga^{3+} ions, respectively, occupy dodecahedral and octahedral positions in addition to their normal sites. This follows from the non-unity values of the axial segregation coefficients, k_0, for host cations in GSGG and from the finite solid solution range at the GSGG intermediate phase [Brandle and Barns (1973)].

Usually k_0 is derived, assuming the normal freeze condition of perfect mixing, from the expression

$$C_s = C_0 k_0 (1 - f)^{(k_0 - 1)} \tag{3.7}$$

where C_0 is the initial concentration of the dopant in the melt, C_s is the concentration in the solid for the melt fraction, f, that has been crystallized [Scheil (1942)]: Eq. (3.7) is used to determine k_0 from experimental axial dopant profiles. When $k_0 = 1$ the concentration of dopant in the crystal is uniform along its length and equal to the initial melt concentration. Component distribution is not a problem when growing at the congruent point. However, allowance must be made when growing crystals at off-congruent compositions, such as stoichiometric GSGG. The random distribution of Sc^{3+} on Gd^{3+} sites and Ga^{3+} on Sc^{3+} sites in GSGG crystals grown from the melt at the congruent composition is a source of intrinsic disorder which has been probed spectroscopically. In most garnets k_0 is close to unity for Cr^{3+}, resulting in the uniform green colouration of Cr^{3+}-doped GSAG crystals. The sharp line spectra of Nd^{3+} dopants in the garnets are also sensitive to intrinsic disorder. In GSGG the value of k_0 is fairly high (≈ 0.7) and a uniform axial distribution of Nd^{3+} may be obtained in boules grown at fast pull rates. However, in Nd^{3+} : YAG, where $k_0 \cong 0.18$, constitutional supercooling occurs when the growth rate is too rapid and the resultant crystal quality is rather poor, especially at Nd^{3+} concentrations above *ca* 1.0%. The practical consequence is that Nd^{3+} : GSGG crystals may be pulled some 8–10 times faster than Nd^{3+} : YAG while maintaining excellent structural and optical quality.

Most garnets grown by the Czochralski technique have been rare-earth aluminium and gallium based, in which the rare-earth ions Y^{3+} or Gd^{3+} can be partially substituted by $Re^{3+} = Tb^{3+}, Dy^{3+}, Ho^{3+}, Er^{3+}, Tm^{3+}$ and Yb^{3+} on the A-sites. In addition GSGG, $Gd_3Sc_2Al_3O_{12}$ (GSAG), $Y_3Sc_2Al_3O_{12}$ (YSAG) and $Y_3Sc_2Ga_3O_{12}$ (YSGG) are extensively soluble in one another. Single crystals grown from the quaternary system, GSGG : GSAG : YSAG : YSGG are compositionally disordered, and display inhomogeneously broadened optical spectra when doped with Cr^{3+} ions or Nd^{3+} ions. In the latter case the disorder appears to play some role in the pulse duration of mode-locked lasers [Ober *et al.* (1992)]. The co-substitution of Mg (or Ca) with Zr on the B- and C-sites, respectively, in GGG single crystals has been used to produce lattice parameter-matched substrates for application in waveguide devices. The garnets are also important hosts for the Cr^{4+} impurity which preferentially occupies the tetrahedral C-sites. The Cr^{4+} oxidation state is stabilized by co-doping YAG with Ca^{2+} and Cr^{4+}, dopants which have k_0-values of 0.2 and 3.2 respectively. In consequence, the two impurity ions tend to segregate at opposite ends of the boule. The production of a fairly uniform distribution of Cr^{4+} in the boule requires a significant excess of Ca^{2+} in the melt relative to Cr-dopant. Indeed, the optimum results are obtained when the Ca/Cr ratio in the melt is 3–4 : 1, coupled with an oxidising anneal at 1600 K for 20–40 hours. The Cr^{4+} content so produced is extremely stable against optical bleaching and $Ca^{2+} : Cr^{4+}$: YAG is an excellent gain medium for an infrared laser system at 300 K, given the excellent optothermal and thermomechanical properties of the YAG host. Various other Cr^{4+}-doped laser crystals, including $Cr^{4+} : Mg_2SiO_4$ (forsterite), $Cr^{4+} : Ca_2GeO_4$ and a number of other Cr^{4+} doped garnets, are also grown by the Czochralski technique.

Sapphire (Al_2O_3) can also be pulled from the melt [Cockayne *et al.* (1967)]. The passive after-heater in Fig. 3.8 is replaced by an active after-heater, usually an Ir liner, which totally encloses the space into which the crystal is pulled. This furnace genre has been up-sized for industrial production and Al_2O_3 and ruby boules with dimensions 120–150 mm ϕ by 600–700 mm long are in production. Sapphire does not pose problems of non-congruent melting, and growth rates can be fast from 5Ns-pure melts. Ruby ($Cr^{3+}:Al_2O_3$), grown for many years by flame fusion, is now grown by the Czochralski method. The outstanding tunable laser material, Ti-sapphire ($Ti^{3+}:Al_2O_3$), is also pulled from the melt. The low distribution coefficient of Ti^{3+}, $k_0 = 0.14$, is determined by the reaction $4TiO_2 \rightarrow 2Ti_2O_3 + O_2$ at the melt surface. Control of the furnace atmosphere is essential to maximize the Ti^{3+} content of the crystals, a slightly reducing atmosphere enhancing the Ti^{3+} ions content at the expense of Ti^{4+} ions. However, there is an ubiquitous weak infrared absorption band with peak near 900 nm, from Ti^{4+}–Ti^{3+} pairs which underlies the Ti^{3+} luminescence peak. This absorption is substantially reduced by post-growth annealing in an Ar : 5% N_2 atmosphere at 1700–1900°C for extended periods (20–40 hr) [Kokta (1985, 1986), Fahey *et al.* (1986)]. Improved crystal quality has been reported for crystals grown by the gradient freeze technique, in which the omni-presence of scattering centres (small bubbles and inclusions) is greatly reduced [Fahey *et al.* (1986)]. Nevertheless, the concentration gradient of Ti^{3+} in as-grown boules, determined by the small value of k_0, presents a challenge to the crystal grower.

The growth of $LiNbO_3$ from the melt is complicated by the congruent composition being deficient in Li_2O (48.6 mol %). Good quality single crystals are grown at the congruent composition (48.6% Li_2O : 51.4% Nb_2O_5) [Byer *et al.* (1970)]. The as-grown crystals, which contain multiple ferroelectric domains, have excellent resistance to laser damage (10–20 GW cm^{-2}). After poling in an electric field just below the ferroelectric temperature (*ca* 1100°C) the crystals are single domain; there is a concomitant reduction in the laser damage threshold. However, the threshold against photorefractive damage can be improved by 10–100 times for single domain crystals by doping with 5–8 mol.% MgO. The MgO-doped $LiNbO_3$ crystals always show compositional striations perpendicular to the growth direction. Such macro-segregation can limit the length of crystal that can be phase-matched for second-harmonic generation. Phase-matching of the isomorphous $LiTaO_3$ cannot be achieved using bulk crystals. In an elegant refinement of the Czochralski technique, Wang *et al.* (1986) produced a periodic domain structure which permits quasi-phase matching for $LiTaO_3$. Periodically-poled lithium niobate (or PPNB), can also be produced by post-growth processing.

The ease of control of crystal orientation has led to some single crystal fluorides being grown by Czochralski technique rather than by the Bridgman–Stockbarger technique. A major difference between the Czochralski growth of oxides and fluorides is that the induction heating required for the elevated melting temperature of oxides can be replaced by resistance heating. The Czochralski method is more flexible than Bridgman–Stockbarger in enabling large boules to be grown for commercial application. Divalent metal fluorides, (MgF_2, ZnF_2) doped with transition metal ions as laser gain media were first grown at Bell Laboratories [Guggenheim (1961), Nassau (1961)]. Crystals with dimensions up to 15 mm ϕ by 100 mm in length were grown at pulling rates of 5–50 mm hr^{-1} using rotation rates of 20–40 rpm. In growing mixed fluorides such as $LiRF_4$, where R = Y, Ho, Er, Tm and Lu, the gas pressure in the growth chamber was raised to 1.5–2.0 atmospheres to suppress the loss by evaporation of the LiF component. Small particle scattering in such compounds can be eliminated by zone refining the starting mixture, followed by growth in a reactive ambient atmosphere containing HF and/or *in-situ* annealing of the grown crystal followed by rapid cooling. Crystal growth and rotation rates of 1–5 mm h^{-1} and 10–20 rpm are appropriate for $LiRF_4$ crystals. Similar techniques can be applied to the colquiriites LiCAF and LiSAF doped with Cr^{3+} or Ce^{3+}; particulate scattering of laser light can be substantially reduced by growth in a dry N_2 atmosphere while bubbling dry N_2 gas through the melt.

The Czochralski growth of apatite crystals was developed early as their potential for infrared lasers based on the trivalent rare-earth ion substituents was realized. Indeed, the earliest Nd^{3+} : CFAP crystals were superior in flashlamp-pumped laser performance at 1.06 μm to Nd : YAG [Mazelsky *et al.* (1968)]. Such studies were soon extended to other dopants (Er^{3+}, Ho^{3+}, Yb^{3+}) and to the Sr^{2+} fluoroapatites and silicate-oxyapatites [Steinbrugge *et al.* (1972)]. Unfortunately, laser-induced damage precluded their exploitation in commercial solid state lasers. A resurgence of interest in the apatites and related hosts was prompted by the recognition that Nd^{3+}- and Yb^{3+}-doped crystals were ideally suited for diode-pumping with $Al_xGa_{1-x}As$ and $In_xGa_{1-x}As$ laser diodes emitting at 810 nm and 980 nm, respectively [Payne *et al.* (1994), Zhang *et al.* (1994), DeLoach (1994)]. Such crystals feature high absorption and emission cross sections associated with the rare-earth ion substitution on the low symmetry A(II) site in the $A_5(BO_4)_3X$ lattice. Related crystals for which some commercial aspirations have recently been realized as diode-pumped gain media include the Nd^{3+}-doped orthovanadate YVO_4 and $GdVO_4$ [DeShazer (1994), Tucker *et al.* (1977)]. The Nd^{3+}-doped fluoroapatites $Ca_5(PO_4)_3F$ and $Sr_5(PO_4)_3F$ promise efficient laser action at 1.06 μm [Ohlmann *et al.* (1968), Mazelsky *et al.* (1968)], but especially in diode-pumped operation at 1.34 μm because of the superior branching to the $^4F_{3/2} \rightarrow {}^4I_{13/2}$ transition of the Nd^{3+} ion [Scott *et al.* (1994, 1995, 1997)]. Large single crystals up to 20 mm ϕ by 100 mm long have been grown by the Czochralski technique, although achieving the crystal quality required for commercial exploitation is not straightforward. For example, Nd^{3+} : $Sr_5(PO_4)_3F$ growth is plagued by the formation of $V^{3+}(3d^2)$ ions in tetrahedral sites, where the normal valence state should be V^{5+}. The effect was minimized by growing and/or post-growth annealing at 1400°C in a mildly oxidising atmosphere to promote the $V^{3+} \rightarrow V^{5+}$ oxidation reaction. Even so the Nd^{3+}-doped $Ca_5(PO_4)_3F$, and oxysilicate apatite $CaY_4(SiO_4)O$ all experience concentration quenching of the Nd^{3+} fluorescence due to the multiplicity of crystal field sites occupied by the Nd^{3+} dopant [Peale *et al.* (1995)]. This problem was eliminated by careful attention to the melt composition. Among the many other microscopic faults permitted by the rich defect structure of the apatites are elongated and dendrific solute inclusions, as well as decorated dislocation lines, control of which will require further research into the growth dynamics. Laser-quality crystals have to be carefully chosen from defect-free regions of the boule. Some studies have been made of the apatites as hosts for the isoelectronic $3d^2$ ions (V^{3+}, Cr^{4+}, Mn^{5+}, Fe^{6+}) which have potential as gain media in tunable near infrared lasers.

3.4.3 Other melt growth techniques

Several other growth techniques involve the melt and growing crystal having near identical compositions. Crucible-free *floating zone growth* involves a vertical rod of material, clamped at both ends, being heated over a very small molten zone, which is suspended between the two parts of the solid rod. The zone is moved through the length of the rod by the upward motion of the heater. As with zone refining, float zone melting can result in a redistribution of components or impurities according to their distribution coefficients [Pfann (1966)]. Repeated passes of the molten zone through the crystal can produce purification of a substantial portion of the final crystal. The starting rod is usually cast or melted in suitably shaped moulds. The length of stable molten zone, limited by the meniscus-like shape of the liquid, is primarily determined by the balance between the hydrostatic pressure in the melt and its surface tension. For rod diameters up to *ca* 20 mm the maximum length of stable zone increases linearly with rod diameter, changing gradually to a regime of constant length of zone at rod diameters between 20 and 30 mm, depending on the material being melted. When the diameter exceeds *ca* 30 mm, the maximum length of stable molten zone is of order 20 mm. The heating method is material dependent. RF induction heating is much used for Si, Ge and other semiconductors. Non-conductors require other heating techniques, such as resistance heating with perforated iridium strip-heaters, electric arc heating, electron beam heating and thermal or optical radiant heating, the latter being derived from a lamp or laser source. Float zone melting has been used to grow single crystals of sapphire [Cockayne *et al.* (1967)], lithium triborate (LBO) [Yangyang *et al.* (1991)], β-barium borate (BBO) [Tang and Route (1988)], magnesium aluminate spinel ($MgAl_2O_4$) [Gasson and Cockayne (1970)], $SrTiO_3$ [Tiller and Yen (1991)], gadolinium gallium garnet (GGG), YAG, YIG [Balbashov and Egorov (1981)], YVO_4 [Muto and Awazu (1969)] and many other optical materials.

The technique of float zone melting is conceptually similar to the growth of crystalline fibres from a small molten pool formed on the end of a charge rod and held in place by surface tension (pedestal growth). The use of CO_2 lasers to heat a controlled and confined volume of the source rod, referred to as laser heated pedestal growth (LHPG), was pioneered by Feigelson (1986) and Fejer *et al.* (1984). Growth is initiated by dipping a seed crystal into the molten zone, which is then slowly withdrawn. Simultaneously and slowly, the source rod is pushed into the molten zone so that its volume is kept constant. The pull rate of the growing fibre crystal and push rate of the source rod are carefully adjusted to maintain a stable molten zone. Elwell *et al.* (1985) reported a laser interferometric system of automatic diameter control for the fibre that is accurate to 0.01%. By repeated application the fibre diameter can be reduced from *ca* 1 mm to 50–100 μm over lengths of 100–150 mm. The Stanford group grew crystalline fibres of over fifty materials including halides, oxides, borides and silicates, and $Cr^{3+}:Al_2O_3$ and Nd^{3+} : YAG fibres so grown have been used in lasers. Periodically poled lithium niobate fibres have been grown with domain dimensions varying from 50–100 μm in width along the length of the fibre. Although clearly capable of yielding device quality material, the LHPG technique is best suited to the first growth of novel compounds for physical and optical property assessment. In this context Qi *et al.* (1996a,b, 1997)] studied a family of mixed niobates $RETiNbO_6$, with RE = Pr, Nd and Er, as potential diode-pumped laser gain media containing exceptionally large number densities of active ions coupled with enhanced absorption and emission cross sections.

3.4.4 Hydrothermal growth

Geological minerals, including gemstones, which form in the presence of water at high temperature and pressure are said to be of *hydrothermal origin.* The largest natural hydrothermal gemstone was a crystal of beryl weighing 1 kg. Crystals of α-quartz grown hydrothermally in the laboratory weigh typically 0.2–0.3 kg. Hydrothermal growth is important in producing larger and more perfect single crystals than can be achieved with other techniques. Growing crystals under hydrothermal conditions requires a pressure vessel to contain a corrosive solution at high temperature coupled to a means of producing, controlling and recording temperature and pressure. Designing and constructing a suitable hydrothermal apparatus, an *autoclave*, is difficult and material specific. Usually the autoclave consists of an external can with an internal liner: hydrothermal synthesis and growth using aqueous solvents or mineralizers to dissolve relatively insoluble materials under high temperature and pressure is carried out within the liner.

Liners made from low carbon steel, which is corrosion resistant in systems containing silica and NaOH, can be used for the growth of α-quartz [Laudise (1991)]. The supersaturated solution of nutrient, usually small particles of quartz, silica glass, high quality silica sand and silica gel, are dissolved in an alkaline solvent (NaOH, Na_2CO_3, KOH or K_2CO_3) in the hot zone of the autoclave liner at 355–370 °C. Crystallization takes place in the cooler (*ca* 350 °C) growth zone of the autoclave, nucleated on appropriately positioned seed crystals. The seed crystals are polished to a very fine finish to prevent the formation of defects in the crystal. Growth rate is strongly dependent on the orientation of the seed crystal. Growth on these seed crystals is fed by transport of material along the convection currents induced by the temperature gradient between solution and growth zones. A baffle is placed at the top of the hot zone to control transport of material at a pre-determined rate. Autoclave assemblies must withstand typical pressures of 0.1–10 kbar at temperatures of *ca* 700–800 °C. Usually the space between autoclave wall and liner is filled to the same extent as the liner to provide a pressure balance so that the liner supports virtually no pressure. When very high purity, device quality quartz is to be grown a silver liner is used in the autoclave. More than 200 different materials have been grown into single crystals using hydrothermal techniques [Byrappa (1994)].

The growth of α-quartz has been central to the development of hydrothermal technology. However, the application of α-quartz in electronic devices is currently being replaced by berlinite ($AlPO_4$), also grown hydrothermally from nutrient mixtures containing Al_2O_3, H_3PO_4, NH_4Cl and NaCl or HCOOH. In this case teflon liners are used: the temperature of solvent is slowly raised to 290–300 °C, uninterrupted growth periods of 2–3 weeks being required to obtain crystals 4–5 mm on edge. Borates are grown using copper linings for the autoclave, although noble metal liners are preferred for alkaline growth media. Potassium titanyl phosphate, $KTiOPO_4$ (KTP), like LBO and BBO used for applications in nonlinear optics, is also grown hydrothermally. Laudise *et al.* (1990) used a Pt-lined autoclave with internal dimensions 25 mm ϕ by 150 mm long in which TiO_2 was dissolved at 425 °C in a $K_2HPO_4 + HPO_3$ mineralizer in the nutrient zone. The temperature of the growth zone was 375 °C at pressures of 2 kbar or so. A growth rate of 0.14 mm/day was achieved on seed crystals with (011) faces. Hydrothermal KTP crystals have much improved tolerances for laser-induced damage than crystals grown from high temperature solutions. Among gems grown by hydrothermal methods are coloured quartz, emerald, ruby and sapphire. YAG, YGG and YIG have also been grown hydrothermally, as have rare-earth vanadates and phosphates such as $GdVO_4$, YVO_4, NdP_5O_{14} and TmP_5O_{14}.

3.4.5 *Growth from high temperature solutions (HTSG)*

The growth of crystals from high temperature solutions (HTSG), sometimes called flux growth, can be used when the melting temperature of a compound is too high for melt growth by available technologies or when the vapour pressure of a component in the melt is too high [Laudise (1970)]. The desired material is dissolved in a solvent at elevated temperature, which is then solidified by cooling. The crystals are extracted from the supersaturated solution as they are formed, or allowed to grow in the solution until the whole solidifies. HTSG can also be used for layer growth on single crystal substrates, normally called *liquid phase epitaxy*. The main advantages of the technique are that solvents can be found for almost any material and that the crystals grow at temperatures well below the melting temperature of the crystal. The major disadvantages are the slow growth rates and the unavoidable presence of solvent impurity ions and flux inclusions in the grown crystal.

The solubility of the compound or its components in the solvent should be quite high (20–50 wt.%) and decreasing with temperature to the melting temperature of the solvent. Solvents should have low vapour pressures. Solutions must have low viscosities (< 10 centipoise), be chemically stable at the requisite temperature and inert with respect to the crucible. The single crystal material should be the only stable solid phase on the phase diagram under the growth conditions and easily separable from the residual melt. Common fluxes for oxide growth include Li_2O, Na_2O, K_2O, BaO, PbO, Bi_2O_3, B_2O_3, SiO_2, P_2O_5, V_2O_5 and MoO_3 and the fluorides PbF_2, BaF_2, KF, or some combination of two or three of them. Various crucible materials can be used, including Pt and Ir for oxides, graphite for fluorides, and such others as silica glass, nickel, molybdenum and tantalum with particular fluxes. Nevertheless, Pt finds the most widespread application because of its high melting point and excellent resistance to oxidation and corrosion by fluxes. Long crucible life times require thicker walls, typically 2–3 mm; since Pt is quite soft at elevated temperature, the crucible can be supported in a refractory envelope. Because HTSG occurs at lower temperatures than melt growth it is usual to use resistively heated furnaces, rectangular muffle furnaces or vertical tube furnaces. The latter are particularly useful in seeded growth, usually called top-seeded solution growth (TSSG). Since thermal stability in the environment of the growing crystal should be high, certainly better than ±0.1 K, it is necessary to control furnace temperature with a flexible, computer-controlled programmer.

Many garnets have been grown from PbO/PbF_2 or BaO/BaF_2 fluxes, including YIG, GIG, Nd : YAG and Nd : YGG [Tolksdorff (1994)], although crystals tend to be rather small. Optical crystals used in solid state lasers, or for applications in harmonic generation, are grown by the TSSG technique, especially in growth from solutions with low volatility. The seeding rod is introduced from above at the saturation temperature, and as with Czochralski growth, at the centre of the molten surface, which should be the coldest spot in the solution. The seed crystal is rotated slowly, *ca* 10–20 rpm, and may or may not be slowly retracted (with the growing crystal) from the surface. The temperature gradient should be small and the solution cooled very slowly (1–2 K hr^{-1}). The Czochralski furnace shown in Fig. 3.8 with simple modifications would be applicable to the growth of laser gain media and such nonlinear optical crystals as KTP, KDP and KTA, $KNbO_3$, $K_xLi_{1-x}NbO_3$ and $LiNbO_3$.

There is great interest in borate crystals LiB_3O_5 (LBO), β-BaB_2O_4 and $CsLiB_6O_{10}$ (CLBO) for harmonic generation and optical parametric oscillation. These crystals are also grown by TSSG. LBO can be grown from Li_2O–B_2O_3 solutions at *ca* 810 °C, sometimes with a modifier to improve the viscosity, whereas BBO is grown from a Na_2O–BaO–B_2O_3 solution at ~900 °C. The double borates $RX_3(BO_3)_4$, where $R = La^{3+}$, Y^{3+} or Gd^{3+} and $X = Al^{3+}$ or Sc^{3+}, have trigonal crystal structure and space group D_3^7 (R32, #155); they have excellent optical and thermal properties. Furthermore, their crystal structure provides substitutional cation sites for rare earth and transition ions in which ion–ion interactions are weak, so that large concentrations of laser-active ions can be introduced on either cation sub-lattice. Both Nd^{3+}: $YAl_3(BO_3)_4$ (Nd : YAB) and Er : YAB have been grown by TSSG and used as gain media in single frequency lasers [Luo *et al.* (1989a,b)]. Unfortunately, such crystals have quality control problems due to flux inclusions. In studies of the Cr^{3+}-doped yttrium and gadolinium alumino- and scandio-borates, Wang *et al.* (1996) sought improved solvents for TSSG, pulling Cr^{3+} : $RX_3(BO_3)_4$ from $K_2Mo_3O_{10}$-based fluxes in the temperature range 970–1325 K, where R = Y or Gd and X = Al or Sc. In growing Cr^{3+}-doped YAB and $GdAl_3(BO_3)_4$ (GAB) they determined the optimal solution composition to be 22 wt.% YAB or GAB, 75 wt.% $K_2Mo_3O_{10}$ and 3 wt.% B_2O_3. The YAB or GAB crystals contained up to 5 mol.% of Cr_2O_3 dopant. The addition of B_2O_3 to the flux had three primary effects: reducing the viscosity, saturation temperature and volatility of the solution. The flux modifier also reduced the evaporative loss of solvent by over 20% and eliminated flux inclusions from the resulting crystals. Typical crystal sizes obtained using a 50 mm $\phi \times$ 50 mm Pt crucible were *ca* $12 \times 16 \times 22\,mm^3$, these dimensions being limited by crucible diameter. The crystals were grown during crystallization from the solution over the temperature range 1060–950 °C with a cooling rate of 2–4 °C/day and rotation rate of 4–6 rpm. After 30–50 days with the growth period complete the crystal was pulled free of the remaining solution and cooled to room temperature at a rate of 50 °C hr^{-1}. The conditions for growth of Cr^{3+} : GAB were only slightly different from those for the growth of Cr^{3+} : YAB. Crystals of Cr^{3+}-doped YSB and GSB required slightly higher temperature ranges (1300–1100 °C) and crystal rotation rates (15 rpm) using a $K_2Mo_3O_{10}$ solvent with a lithium tetraborate flux modifier. Very similar growth conditions were used in the growth of Ti^{3+}-doped GAB and YAB.

3.5 General materials considerations

Many and varied are the uses of optical crystals in technology: laser gain media, nonlinear optical devices, lattice-matched (or otherwise) substrates, shock-proof windows, surface acoustic wave (SAW) devices, scintillators, sensors and radiation detectors are some of their many applications. Such applications require highly specialized properties that can only be realized by crystals exhibiting a high degree of structural and chemical perfection. Having committed substantial resources to the development of single crystal growth of a novel material, it is not unreasonable to ask whether the crystal is fitted for the design purpose? A starting point for the evaluation of novel crystals is such structural data as the crystal classes, unit cell dimensions and constitutions, molecular weights and densities. In principle, such properties need only be determined once. In practice some of these data will be determined for each crystal as reassurance to the end user. An X-ray diffraction determination will confirm unit cell dimensions, hence purity, the presence of second phases and inhomogeneous strain distributions. Such a test will also confirm or deny that crystals with the same composition but grown by different techniques have identical structures. Recently it was revealed that in hydrothermally grown $LiKYF_5$ crystals only a single site was occupied by Nd^{3+}, whereas two almost identical Nd^{3+} sites exist in Czochralski-grown crystals of the same composition.

The optical characteristics of the pure crystal must be measured accurately, including the transmittance spectrum as well as the indices of refraction, their birefringence, wavelength dispersion and temperature dependence. For doped crystals it is necessary to measure the optical absorption and luminescence cross sections and the temperature dependence of the emission lifetime. Furthermore, the branching of the excited state population between radiative decay, excited state absorptions and nonradiative decay must be determined. The polarizations of optical spectra are also important in determining the optimum pumping geometry for both laser and nonlinear optical devices. Nonlinear optical crystals are important components in many lasers: it is necessary to know the tensor components of the second order susceptibility $\chi^{(2)}$. Higher order nonlinearities should be measured when applications in two-photon absorption, third or fourth harmonic generation, phase conjugation and the optical Kerr effect are expected [Shen (1984)]. A complete evaluation involves the measurement of a large number of properties. Unfortunately, such property data are frequently incomplete, sometimes imprecise and occasionally inapplicable for the use for which the material is required. Furthermore, measurements reported in different references are often made under quite different experimental conditions. This is almost always the case for the laser-induced damage thresholds quoted for bulk and surface samples, which are variously reported for pulse duration from picoseconds to nanoseconds, for single or multiple pulses at different repetition rates and laser wavelengths. This notwithstanding, several compilations of data are in the literature, with rather fulsome bibliographies [Nikogosyan (1997), Tropf *et al.* (1995)].

Most optical crystals are grown at high temperatures; inevitably some defects are formed during growth on a scale varying from the sub-microscopic to the macroscopic. For example, it is quite difficult to realize a homogeneous distribution of Nd^{3+} dopant ions in fluoride and oxide crystals. Although it is desirable to use crystals of 1% Nd : YLF in some lasers, it follows from the small distribution coefficient for Nd^{3+} in YLF $k \cong 0.2$ that the Nd^{3+} concentration will vary along the length of a boule from *ca* 0.5% in the first solid to be pulled from the melt to *ca* 1.0% at the end of the boule. The composition gradient is reduced in Nd^{3+} : $LiKYF_5$, where the distribution coefficient of $k \cong 0.8$ guarantees an almost homogeneous distribution of Nd^{3+} throughout the crystal. A value of $k_0 = 0.21$ is reported for Ti^{3+} in Al_2O_3, and the specified Ti^{3+} concentration required in laser crystals of between 0.1–0.2 at.% can result in colour changes along the length of an as-grown boule that can be detected by eye. When crystals are pulled in air by the Czochralski process there is an ubiquitous broad absorption band at *ca* 900 nm that self-absorbs the $^2E \rightarrow {}^2T_2$ laser emission. This band, the source of early problems in quality control for Ti-sapphire laser rods, was due to Ti^{4+}–Ti^{3+} pairs in Al_2O_3. Their impact as sources of laser loss was minimized either by crystal growth in a neutral or slightly reducing atmosphere or by post-grown annealing in a reducing atmosphere.

The distribution coefficients of rare-earth ions in crystals are determined by size factor effects rather than by crystal field effects. In the rare-earth-doped garnets Brandle and Barns (1975) expressed this geometrical influence in an empirical equation:

$$k_0(\mathrm{Nd}) = 1.029a - 1.2164, \tag{3.8}$$

where a is the lattice parameter. A homogeneous distribution of Nd^{3+} in garnet crystals results when $k_0(\mathrm{Nd}) > 0.6–0.7$ is combined with relatively high pull rates for crystals with $a > 1.240$ nm. This applies to the RE gallium garnets, where $k_{\mathrm{Nd}} = 0.62$ in GGG and 0.75 in GSGG. In contrast, the k_{Cr} distribution coefficient in the octahedral site in garnets may be anomalously high, with $k_{\mathrm{Cr}} = 2.4$ in Cr^{3+} : YAG rising to $k_{\mathrm{Cr}} = 3.3$ in Cr^{3+} : GGG. Such high values of k also result in inhomogeneities. Molecular orbital theory of the octahedral $(MO_6)^{9-}$ complexes show that $(AlO_6)^{9-}$ and $(GaO_6)^{9-}$ are less stable than $(ScO_6)^{9-}$ and $(CrO_6)^{9-}$. The consequence is that $k_{\mathrm{Cr}} \cong 1$ for Cr^{3+} in GSGG and YSGG, and very homogeneous crystals result from Czochralski growth at high pull rates (2–3 mm per hour). In these crystals the octahedral sites are sufficiently large (~ 0.42 nm) that Cr^{3+} ions occupy weak field sites, resulting in broadband luminescence at 300 K. These crystals also provide larger dodecahedral sites, resulting in higher distribution coefficients for Nd^{3+} and other RE^{3+} ion dopants.

A wide variation in macroscopic defects may result from high temperature growth. In Mg^{2+} : $LiNbO_3$ visible striations occur perpendicular to the growth direction: in these optically nonlinear materials the Mg^{2+} dopant is added to suppress optical damage at laser wavelengths above 700 nm. The resulting optical quality is poor and it is surprising that device manufacturers have continued to use such material when superior alternatives exist. Scattering centres are frequently observed in crystals grown from the melt or from a flux. These may be voids caused by melt instability during constitutional supercooling or particulate distributions that result from undesirable impurity. In garnet crystals grown from the melt contained in Ir crucibles there is an ubiquitous tendency for Ir platelets to form close to the crystal axis. The consequence is that regions of the boule close to the core, where most platelets are formed, are severely strained. Laser rods are then core-drilled well away from the central axis of the boule where the platelet-related stress patterns are much reduced. Finally, most laser rods after cutting from an as-grown crystal are mechanically polished using diamond or alumina paste on a revolving polishing disk. The apparent high polish nevertheless contains many microcracks of varying lengths and depths. These microcracks seriously reduce the resistance of laser rods to failure by thermal fracture. These microcracks may be reduced in size or even eliminated by chemical polishing subsequent to mechanical polishing [Marion (1985)]. Normally the defects introduced during growth are material and process specific, and particular recipes have to be formulated for each crystal.

4

Energy levels of ions in crystals

4.1 The Hamiltonian

4.1.1 Assumptions of crystal-field theory

The present chapter is concerned with the energy levels of impurity ions and colour centres in ionic crystals at the level of *crystal-field theory*, proposed by Bethe (1929) and elaborated by subsequent investigators [see McClure (1959)]. Iron-group transition-metals and lanthanides are emphasized. Effects of covalency are deferred to Chapter 8. Since the subject matter of this chapter is sufficiently mature that many of the standard references are no longer in print, it is useful to include concise summaries of the theory of atomic structure and of crystal-field theory in order to provide background information essential for crystal-field engineering.

Since the open electronic shells of transition-metal and rare-earth impurity ions are relatively compact, and are partially shielded from their crystalline environments by filled shells, a tight-binding model is appropriate as a first approximation. This model can be characterized by a non-relativistic Hamiltonian for the impurity ion which incorporates the interactions of its electrons with its nucleus, their spin–orbit interactions, and their interactions with the external crystal-field and with one another,

$$H = \sum_{i=1}^{N}\left[-\frac{\hbar^2}{2m}\nabla_i^2 - \frac{Ze^2}{(4\pi\varepsilon_0)r_i} + \xi(r_i)\mathbf{l}_i\cdot\mathbf{s}_i + V_c(\mathbf{r}_i)\right] + \frac{1}{2}\sum_{i\neq j=1}^{N}\frac{e^2}{(4\pi\varepsilon_0)r_{ij}}, \tag{4.1}$$

where the sums are over the N electrons of the ion. The crystal potential $V_c(\mathbf{r}_i)$ arises from the inert charge distribution of surrounding ions, and reflects the symmetry of that distribution. The spin–orbit interaction is a relativistic effect derived from the Dirac equation by expanding the Dirac Hamiltonian in powers of v/c; its origin is discussed more fully in Chapter 8. The operators $\mathbf{l}_i$ and $\mathbf{s}_i$ are, respectively, the orbital and spin angular momentum operators, in units of $\hbar$, of the ith electron, and $\xi(r_i)$ is given by

$$\xi(r) = \frac{\hbar^2}{2m^2c^2}\frac{1}{r}\frac{\partial U}{\partial r}, \tag{4.2}$$

where $U(r)$ is an approximate central potential which incorporates both the nuclear potential and the electrostatic interaction with the average charge distribution of the other electrons. Much weaker terms, including spin–other-orbit, orbit–orbit, Zeeman, hyperfine and quadrupole interactions, are omitted from Eq. (4.1).

4.1.2 *Hierarchy of perturbations*

The determination of impurity-ion energy levels proceeds from the *central-field approximation*,

$$H_{\text{cf}} = \sum_{i=1}^{N} h_i, \tag{4.3}$$

$$h = -\frac{\hbar^2}{2m}\nabla^2 + U(r). \tag{4.4}$$

The remaining terms in the Hamiltonian, $H - H_{\text{cf}}$, are then treated sequentially by perturbation theory. For free ions, the dominant perturbation is the difference between the exact mutual electrostatic interaction of the electrons and its central-potential equivalent, $\frac{1}{2}\sum_{i\neq j=1}^{N}[e^2/(4\pi\varepsilon_0)r_{ij}] - \sum_{i=1}^{N}[U(r_i) + Ze^2/(4\pi\varepsilon_0)r_i]$. For relatively light ions such as the iron-group transition metals, the remaining term in the free-ion Hamiltonian, the spin–orbit interaction, may be treated as a small perturbation on the electrostatic interaction (*Russell–Saunders coupling*). Since the strength of the spin–orbit interaction increases rapidly with atomic number, the two perturbations are more nearly comparable for lanthanide elements.

The position of the crystal-potential term in the hierarchy of perturbations in Eq. (4.1) is strongly dependent on the type of ion. In the case of iron-group transition-metal impurities in ionic crystals, the crystal potential is comparable with the mutual electrostatic interaction of the electrons and greatly exceeds the spin–orbit interaction. The relative positions of spin–orbit interaction and crystal potential are reversed for the lanthanides, by virtue of their higher atomic numbers and their more compact and more thoroughly shielded open-shell orbitals; for them, the crystal potential is a small perturbation on the spin–orbit interaction.

The crystal-field concept can be extended, somewhat more crudely, to heavy-metal ions such as thallium, for which all three perturbations are comparable. It can also be extended successfully to electron-excess colour centres such as the F centre. For the F centre and related centres, the crystal potential replaces the central potential and is thus no longer a perturbation, while the spin–orbit interaction is weak in the absence of a vacancy-centred nuclear potential. The mutual electrostatic interaction is of intermediate strength for colour centres with more than one electron, such as F centres in alkaline-earth chalcogenides and F' and F-aggregate centres in alkali halides.

4.2 Free-ion electronic structure

4.2.1 *Central-field approximation*

At the level of the central-field approximation, the Hamiltonian, H_{cf}, of Eq. (4.3) is invariant under independent rotations of the space and spin coordinates of the individual electrons. The Schrödinger equation is then separable, and can be solved for the atomic spin orbitals, $\phi_{nlm_lm_s}(\mathbf{r}, s)$,

$$h\phi_{nlm_lm_s}(\mathbf{r}, s) = \varepsilon_{nl}\phi_{nlm_lm_s}(\mathbf{r}, s), \tag{4.5}$$

where h is the one-electron Hamiltonian defined by Eq. (4.4), n is the principal quantum number, l is the orbital angular momentum quantum number, and quantum numbers

m_l and m_s are the projections of the orbital and spin angular momenta, respectively. It is convenient to employ *Hartree atomic units* in which the unit of length is the *Bohr radius*, $a_0 = (4\pi\varepsilon_0)\hbar^2/me^2$, and the unit of energy is the *Hartree*, equal to twice the Rydberg constant, $1\text{Hartree} = 2\text{Ry} = me^4/(4\pi\varepsilon_0)^2\hbar^2$. Equation (4.5) may be written in atomic units as

$$\left(-\frac{1}{2}\nabla^2 + U(r)\right)\phi_{nlm_lm_s}(\mathbf{r}, s) = \varepsilon_{nl}\phi_{nlm_lm_s}(\mathbf{r}, s), \tag{4.6}$$

which is separable in polar coordinates,

$$\phi_{nlm_lm_s}(\mathbf{r}, s) = \frac{1}{r}P_{nl}(r)Y_l^{m_l}(\theta, \phi)\chi_{m_s}(s), \tag{4.7}$$

where $Y_l^{m_l}(\theta, \phi)$ is a spherical harmonic, $\chi_{m_s}(s)$ is an eigenfunction of $\mathbf{s}^2$ and s_z, and the radial function $P_{nl}(r)$ satisfies the radial equation

$$\left\{\frac{d^2}{dr^2} + 2[\varepsilon_{nl} - U(r)] - \frac{l(l+1)}{r^2}\right\}P_{nl}(r) = 0, \tag{4.8}$$

which may be solved numerically when $U(r)$ is known.

An additional restriction is imposed by the *Pauli principle*, discussed in §2.5.6, which requires that the many-electron wavefunction be antisymmetrical under exchange of space and spin coordinates of any two electrons. The many-electron wavefunction ψ_{cf} in the central-field approximation is then the antisymmetrized product of occupied spin orbitals (*Slater determinant*) [Slater (1929)],

$$\psi_{\text{cf}} = \mathscr{A}\prod_{i=1}^{N}\phi_{k_i}(j) \equiv \frac{1}{\sqrt{N!}}\det|\phi_{k_i}(j)|, \tag{4.9a}$$

where $\mathscr{A}$ is the antisymmetrizer, defined by

$$\mathscr{A} = \frac{1}{\sqrt{N!}}\sum\nolimits_{\mathscr{P}}(-1)^p\mathscr{P}; \tag{4.9b}$$

$\mathscr{P}$ is an operator which permutes the electron space and spin coordinates $\mathbf{r}_j$, s_j, denoted concisely by j; p is the number of pair-wise permutations (transpositions) into which $\mathscr{P}$ may be decomposed; N is the number of occupied spin-orbitals; and k_i denotes the quantum numbers n_i, l_i, m_{li} and m_{si}. Each distinct energy level in the central-field approximation is associated with a degenerate *configuration*, labeled by the set of quantum numbers $\{n_i, l_i\}$,

$$E(\{n_i, l_i\}) = \sum_{i=1}^{N}\varepsilon_{n_il_i}. \tag{4.10}$$

Degenerate states within a configuration are distinguished by values of the quantum numbers $\{m_{li}, m_{si}\}$. In practice, the various self-consistent, iterative schemes for accomplishing the solution of the central-field model [Hartree (1957), Fischer (1977)] do not all rely necessarily on an explicit central potential $U(r)$, but the essential point is that well defined procedures exist for determining optimized radial functions $P_{nl}(r)$

with approximate energies which depend only on the quantum numbers n and l. These schemes will be considered further in Chapter 8.

4.2.2 *Electrostatic interaction*

The mutual electrostatic interaction of the electrons, given in atomic units by

$$G = \frac{1}{2}\sum_{i \neq j} g_{ij}, \tag{4.11a}$$

$$g_{ij} \equiv 1/r_{ij}, \tag{4.11b}$$

effects a sharp reduction in symmetry; the Hamiltonian including this term is invariant only under simultaneous rotations of the space and spin coordinates of all open-shell electrons. As a consequence, much of the degeneracy is removed, and the configuration splits into *terms* (or *multiplets*), labeled by quantum numbers L and S, corresponding, respectively, to the total orbital angular momentum operator $\mathbf{L} = \sum_i \mathbf{l}_i$ and the total spin angular momentum operator $\mathbf{S} = \sum_i \mathbf{s}_i$. Spin coordinates are still independent of space coordinates at this level of approximation, however. The coupling of the orbital angular momenta of the individual electrons by their mutual electrostatic interaction is effected by the torques which they exert on one another. The independent coupling of their spin angular momenta is a subtle manifestation of the *Pauli principle*; the *antisymmetrizer*, $\mathscr{A}$, defined by Eq. (4.9b), commutes with $\mathbf{S}$ but not with $\mathbf{s}_i$. It is *not* a consequence of the much weaker spin–spin interaction.

Electrons in closed shells make no contribution to the splitting of the configuration energy into terms, but rather shift all of the energy levels equally; furthermore, they make no net contribution to L and S. Accordingly, one can restrict consideration to electrons in open shells in the calculation of relative term energies. Even for open-shell electrons, the kinetic energy and nuclear potential energy, which are one-electron operators, do not contribute to the splitting. The combined contribution of kinetic energy and nuclear potential energy of the open-shell electrons is represented by a constant matrix within a single configuration. Antisymmetry is irrelevant in the evaluation of its diagonal matrix elements, which can be expressed as a sum of one-electron integrals, $\sum_i I(n_i l_i)$, the same for every term. The interactions of open-shell electrons with closed shells can also be absorbed into the constant matrix. Consequently, the splitting of the configuration into terms is effected solely by the mutual electrostatic interaction, G, of open-shell electrons, and, for the purpose of determining the relative term energies, one can regard G alone as the dominant perturbation. Accordingly, the additive contribution $\sum_i I(n_i l_i)$ will be omitted from subsequent formulas for term energies in the present chapter.

The relative term energies associated with the mutual electrostatic interaction of the open-shell electrons are obtained in first-order degenerate perturbation theory by diagonalizing the matrix of G within a single configuration. Diagonal matrix elements of G between Slater determinants are given by $\frac{1}{2}\sum_{i \neq j}(\langle ij \,|\, g \,|\, ij\rangle - \langle ij \,|\, g \,|\, ji\rangle)$, matrix elements between Slater determinants which differ in one spin orbital by $\sum_i(\langle ik' \,|\, g \,|\, ik\rangle - \langle ik' \,|\, g \,|\, ki\rangle)$, and matrix elements between Slater determinants which differ in two spin orbitals by $\langle k'l' \,|\, g \,|\, kl \rangle - \langle k'l' \,|\, g \,|\, lk \rangle$, where

$$\langle ij|g|rt\rangle = \delta(m_{si}m_{sr})\delta(m_{sj}m_{st})\delta(m_{li}+m_{lj}, m_{lr}+m_{lt}) \times \sum_{k=0}^{\infty} c^k(l_i m_{li}; l_r m_{lr}) c^k(l_t m_{lt}; l_j m_{lj}) R^k(ij; rt), \tag{4.12a}$$

$$c^k(l_1 m_1; l_2 m_2) = \sqrt{\frac{2l_2+1}{2l_1+1}} \langle k l_2 (m_1 - m_2) m_2 | l_1 m_1 \rangle \langle k l_2 0 0 | l_1 0 \rangle, \tag{4.12b}$$

$$R^k(ij; rt) = \int_0^{\infty}\int_0^{\infty} P_{n_i l_i}(r_1) P_{n_j l_j}(r_2) \frac{r_<^k}{r_>^{k+1}} P_{n_r l_r}(r_1) P_{n_t l_t}(r_2) dr_1 dr_2. \tag{4.12c}$$

Matrix diagonalization is greatly facilitated by transformation to the coupled representation spanned by linear combinations of Slater determinants which are eigenfunctions of $\mathbf{L}^2$, L_z, $\mathbf{S}^2$ and S_z, represented by kets $|\alpha LSM_L M_S\rangle$, since the orthogonality of basis functions precludes matrix elements of G connecting functions which differ in L, S, M_L or M_S. The parameter α distinguishes multiple occurrences of L, S which may occur in configurations with more than two open-shell electrons. The matrix is then partitioned into blocks labeled by these four quantum numbers, with matrix elements $\langle \alpha LSM_L M_S \,|\, G \,|\, \alpha' LSM_L M_S \rangle$, where blocks with the same values of L and S are identical. Eigenvalues of these blocks are then the term energies $E(\gamma, L, S)$ and their eigenvectors provide the coefficients of the correct linear combinations $|\gamma LSM_L M_S\rangle$ of unperturbed eigenfunctions $|\alpha LSM_L M_S\rangle$. The latter are simultaneous eigenstates of the commuting observables H, $\mathscr{A}$, $\mathbf{L}^2$, L_z, $\mathbf{S}^2$ and S_z [Dirac (1958)]. The term energies can be expressed finally as linear combinations of *Slater integrals* [Slater (1929)], defined by

$$F^k(n_i l_i; n_j l_j) \equiv R^k(ij, ij), \tag{4.13a}$$

$$G^k(n_i l_i; n_j l_j) \equiv R^k(ij, ji). \tag{4.13b}$$

The conventional notation for terms is ${}^{2S+1}L$, where $2S+1$ is the *multiplicity* and the numerical value of L is replaced by its letter designation: $0, 1, 2, 3, 4, 5, 6, 7, \ldots \rightarrow S, P, D, F, G, H, I, K, \ldots$ The first four letters are derived from features of spectroscopic lines (sharp, principal, diffuse and fundamental), and the remainder are alphabetical with the omission of J to avoid confusion. Since first-order perturbation theory is a marginal approximation for the mutual electrostatic interaction of the electrons, the calculated term energies can be improved significantly by including matrix elements of G between configurations. Alternatively, first-order theory is often employed with the Slater integrals of Eqs. (4.13) treated as adjustable parameters.

Of primary interest in crystal-field theory are the energy levels within a ground configuration with a single open shell; i.e., all of the open-shell electrons have the same values of quantum numbers n and l (*equivalent electrons*). The conventional notation for such configurations is $(nl)^q$, where q is the number of electrons in the open shell and numerical values of l are replaced by their letter designations: $0, 1, 2, 3, 4, 5, 6, 7, \ldots \rightarrow s, p, d, f, g, h, i, k, \ldots$. Thus iron-group transition-metal ions have $(3d)^q$ ground configurations and lanthanide ions have $(4f)^q$ ground configurations. The central problem in the calculation of free-ion term energies within the ground configuration is that of transformation to a coupled representation $|\alpha LSM_LM_S\rangle$. For $q=2$, an elementary method is available for generating the coupled representation of a two-electron configuration, which exploits the fact that Slater determinants are eigenfunctions of L_z and S_z. One begins by sorting the Slater determinants into bins labeled by M_L and M_S, and proceeds by application of angular-momentum ladder operators L_- and S_- to selected Slater determinants, together with orthogonality considerations, to generate functions $|LSM_LM_S\rangle$ as linear combinations of Slater determinants. The resulting antisymmetric functions $|LSM_LM_S\rangle$ are unique for the two-electron configuration, since the Kronecker product representations are simply reducible; accordingly, the parameter α is omitted. The *diagonal-sum rule* provides an even more direct method for obtaining term energies: The matrix of G is partitioned into blocks labeled by M_L and M_S in the uncoupled representation and is diagonal in the coupled representation. Trace invariance under unitary transformation leads to equations of the type

$$\sum_{L,S} E(L,S) = \sum_{\text{Slater det.}\,\mu} G_{\mu\mu} \tag{4.14}$$

within each M_L, M_S block. By calculating the diagonal matrix elements of G between Slater determinants for a sufficient number of blocks, one can solve the resulting system of linear equations for the term energies $E(L,S)$. These elementary methods fail for larger values of q whenever a given set of quantum numbers L, S occurs more than once (*duplicated terms*).

A more formal procedure for the two-electron configuration involves a reasonably straightforward application of vector coupling of angular momenta, Eq. (2.58),

$$|LM_L\rangle = \sum_{m_{l1}=-l_1}^{l_1} \sum_{m_{l2}=-l_2}^{l_2} |l_1m_{l1}\rangle|l_2m_{l2}\rangle\langle l_1l_2m_{l1}m_{l2}|LM_L\rangle, \tag{4.15a}$$

$$|SM_S\rangle = \sum_{m_{s1}=-1/2}^{1/2} \sum_{m_{s2}=-1/2}^{1/2} \left|\frac{1}{2}m_{s1}\right\rangle\left|\frac{1}{2}m_{s2}\right\rangle\left\langle\frac{1}{2}\frac{1}{2}m_{s1}m_{s2}\middle|SM_S\right\rangle. \tag{4.15b}$$

Antisymmetrization is then effected by applying either the antisymmetrizer, $\mathscr{A} = (1/\sqrt{2})(1-\mathscr{P}_{12})$, or the *symmetrizer*, $\mathscr{S} \equiv (1/\sqrt{2})(1+\mathscr{P}_{12})$, to the $|LM_L\rangle$, and then forming the products of the antisymmetric space functions with the symmetric *triplet* spin functions ($S=1$) and the products of the symmetric space functions with the antisymmetric *singlet* spin function ($S=0$). The functions $|LSM_LM_S\rangle$ can be expressed as a linear combination of Slater determinants by recombining terms. For inequivalent electrons, L ranges from $|l_1-l_2|$ to l_1+l_2 and, independently, $S=0$ or 1; for example, a $2p3p$ configuration has terms ${}^1S, {}^1P, {}^1D, {}^3S, {}^3P, {}^3D$. For equivalent electrons, however, some Slater determinants vanish and others become identical within a sign, so that some terms are eliminated and the wavefunctions for surviving terms must be renormalized. Only terms with even values of $L+S$ survive; for example, a $2p^2$ configuration has only terms ${}^1S, {}^3P, {}^1D$.

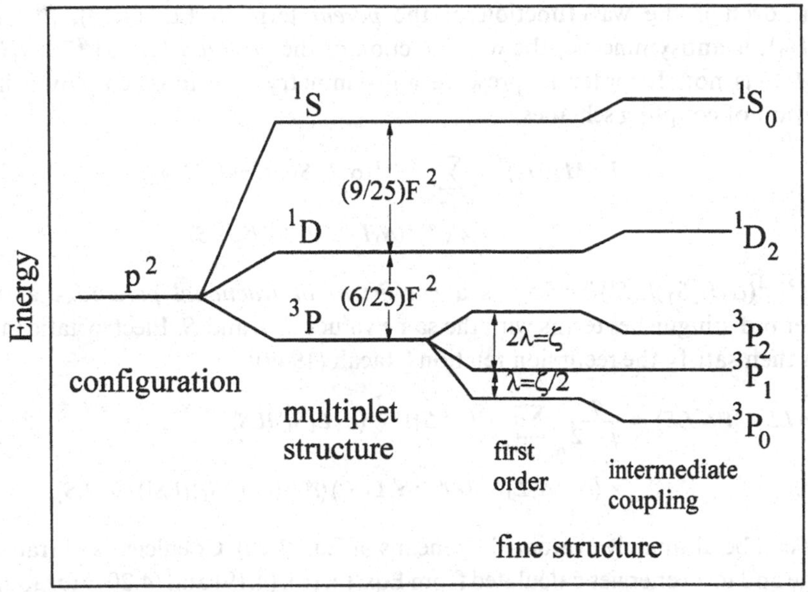

Figure 4.1. Multiplet structure and fine structure of a p^2 configuration.

Term energies of an $(nl)^2$ configuration are given by

$$E(l^2, L, S) = \langle l^2 LSM_LM_S|G|l^2 LSM_LM_S\rangle = \sum_{k=0}^{\infty} f_k F^k(nl, nl), \tag{4.16a}$$

$$f_k = (-1)^{2l-L}(2l+1)\langle lk00|l0\rangle^2 W(llll; Lk). \tag{4.16b}$$

Terms in the sum on the right-hand side of Eq. (4.16a) are non-vanishing only for even values of k between 0 and $2l$, inclusive. For example, term energies of an $(np)^2$ configuration are given by

$$E(^3P) = F^0(np, np) - \tfrac{5}{25}F^2(np, np), \tag{4.17a}$$

$$E(^1D) = F^0(np, np) + \tfrac{1}{25}F^2(np, np), \tag{4.17b}$$

$$E(^1S) = F^0(np, np) + \tfrac{10}{25}F^2(np, np). \tag{4.17c}$$

The integral $F^0(np, np)$ is common to all three terms and thus contributes nothing to the term separations. Thus the two term separations depend only on the single Slater integral, $F^2(np, np)$. The multiplet splitting of the $(np)^2$ configuration is illustrated in Fig. 4.1.

In a configuration with more than two equivalent electrons, one can generate symmetry-adapted wavefunctions by repeated vector coupling,

$$|l^{q-1}(\alpha_1 L_1 S_1)lLSM_LM_S\rangle$$

$$= \sum_{M_{L1}=-L_1}^{L_1} \sum_{m_l=-l}^{l} \sum_{M_{S1}=-S_1}^{S_1} \sum_{m_s=-1/2}^{1/2} |l^{q-1}\alpha_1 L_1 S_1 M_{L1} M_{S1}\rangle |nlm_lm_s\rangle$$

$$\times \langle L_1 l M_{L1} m_l | LM_L\rangle \langle S_1 \tfrac{1}{2} M_{S1} m_s | SM_S\rangle. \tag{4.18}$$

However, even if the wavefunction of the *parent* term in Eq. (4.18), $|l^{q-1}\alpha_1 L_1 S_1 M_{L1} M_{S1}\rangle$, is antisymmetric, the wavefunction of the *daughter* term, $|l^{q-1}(\alpha_1 L_1 S_1) lLSM_L M_S\rangle$, is not. In order to preserve antisymmetry, one must employ a linear combination of coupling schemes,

$$|l^q\alpha L\, SM_L M_S\rangle = \sum_{\alpha_1 L_1 S_1} |l^{q-1}(\alpha_1 L_1 S_1) lL\, SM_L M_S\rangle \times \langle l^{q-1}(\alpha_1 L_1 S_1) lLS|\}l^q\alpha LS\rangle, \tag{4.19}$$

where $\langle l^{q-1}(\alpha_1 L_1 S_1) lLS|\}l^q\alpha LS\rangle$ is a *coefficient of fractional parentage* and the parameter α distinguishes terms with the same values of L and S. Electrostatic matrix elements then satisfy the recursion relation [Racah (1949)]

$$\langle l^q\alpha LS|G|l^q\alpha' LS\rangle = \frac{q}{q-2}\sum_{\alpha_1\alpha_1' L_1 S_1} \langle l^q\alpha LS\{|l^{q-1}(\alpha_1 L_1 S_1) lLS\rangle \times \langle l^{q-1}\alpha_1 L_1 S_1|G|l^{q-1}\alpha_1' L_1 S_1\rangle \langle l^{q-1}(\alpha_1' L_1 S_1) lLS|\}l^q\alpha' LS\rangle, \tag{4.20}$$

and thus can be ultimately evaluated by means of Eq. (4.16). Coefficients of fractional parentage and term energies calculated from Eqs. (4.16), (4.19) and (4.20) are tabulated for p^q, d^q and f^q configurations [Nielson and Koster (1963)]. Alternatively, an explicit expression for these matrix elements is also available [Sobelman (1972, 1979), Weissbluth (1978), Condon and Odabasi (1980)]

$$\langle l^q\alpha LS|G|l^q\alpha' LS\rangle = \sum_{k=0}^{\infty}\frac{1}{2}(2l+1)\langle lk00|l0\rangle^2 F^k(nl,nl) \times \left[\frac{1}{2L+1}\sum_{\alpha'' L'' S''}(-1)^{L+L''}\langle l^q\alpha LS\|\mathbf{U}^{(k)}\|l^q\alpha'' L'' S''\rangle \times \langle l^q\alpha'' L'' S''\|\mathbf{U}^{(k)}\|l^q\alpha' LS\rangle - \frac{q}{2l+1}\delta_{\alpha\alpha'}\right], \tag{4.21}$$

where $\mathbf{U}^{(k)}$ is a sum of *unit tensor operators* $\mathbf{u}^{(k)}$, defined by

$$\mathbf{U}^{(k)} = \sum_i \mathbf{u}_i^{(k)}, \tag{4.22a}$$

$$\langle l\|\mathbf{u}^{(k)}\|l'\rangle \equiv \delta_{ll'}. \tag{4.22b}$$

Reduced matrix elements of $\mathbf{U}^{(k)}$, given by

$$\left\langle l^q\alpha LS\|\mathbf{U}^{(k)}\|l^q\alpha' L'S'\right\rangle = \delta_{SS'} q\sqrt{(2L+1)(2L'+1)}\sum_{\alpha L_1 S_1}(-1)^{L'+L_1+l+k} \times \langle l^q\alpha LS\{|l^{q-1}(\alpha_1 L_1 S_1) lLS\rangle\langle l^{q-1}(\alpha_1 L_1 S_1) lL'S'|\}l^q\alpha' L'S'\rangle W(LL'll;kL_1), \tag{4.23}$$

are tabulated for allowed values of k in p^q, d^q and f^q configurations [Nielson and Koster (1963)].

A single nl shell can accommodate a maximum of $2(2l+1)$ electrons, the degeneracy of the one-electron spin orbitals $\phi_{nlm_lm_s}(\mathbf{r},s)$ in the central-field approximation, since the Pauli principle requires that the electrons occupy distinct spin orbitals in order to ensure non-vanishing Slater determinants. *Conjugate* configurations $(nl)^q$ and

$(nl)^{2(2l+1)-q}$ have identical multiplet structure, since the electrostatic interaction of two holes is the same as that of two electrons. Hund's rules (1927a) dictate that the term of lowest energy in a configuration with a single open shell has the highest value of L consistent with the highest value of the multiplicity, $2S+1$.

4.2.3 *Spin–orbit interaction*

The spin–orbit interaction can be represented within each term as $\lambda(\gamma, L, S)\mathbf{L}\cdot\mathbf{S}$ by appealing to the Wigner–Eckart theorem; consequently, it couples the total orbital and total spin angular momenta, thus splitting the term energy into *fine-structure levels* labeled by the total angular momentum quantum number J, where $\mathbf{J}=\mathbf{L}+\mathbf{S}$. The eigenkets in the coupled representation are obtained by vector coupling of angular momenta,

$$|\gamma LSJM_J\rangle = \sum_{M_L=-L}^{L} \sum_{M_S=-S}^{S} |\gamma LSM_L M_S\rangle\langle LSM_L M_S|JM_J\rangle. \tag{4.24}$$

To first-order, the energies of the fine-structure levels relative to the unperturbed term energies are then given by

$$\langle\gamma LSJM_J|\lambda L\cdot S|\gamma LSJM_J\rangle = \tfrac{1}{2}\lambda[J(J+1) - L(L+1) - S(S+1)], \tag{4.25}$$

and their energy separation is given by the Landé interval rule,

$$E(J) - E(J-1) = \lambda J. \tag{4.26}$$

The foregoing approximation for fine-structure levels is adequate for iron-group transition-metals, but not for lanthanide elements; in that case, one must also take account of matrix elements between terms, which upset the Landé interval rule. This interval rule is also upset by spin–spin, spin–other-orbit and orbit–orbit interactions, much weaker perturbations which nonetheless preserve the symmetry of the spin–orbit interaction.

In an $(nl)^2$ configuration, the diagonal-sum rule provides a simple method for evaluating the spin–orbit constants $\lambda(L,S)$. When matrix elements of the spin–orbit interaction are calculated within manifolds of constant M_L and M_S in both an $|LSM_LM_S\rangle$ representation and a representation spanned by Slater determinants, trace invariance under unitary transformation leads to the following relations within each manifold:

$$M_L M_S \sum_{L,S} \lambda(L,S) = \sum_{\text{Slater det.}} \sum_i \zeta_{n_il_i} m_{li} m_{si}, \tag{4.27a}$$

$$\zeta_{nl} \equiv \int_0^\infty P_{nl}(r)^2 \xi(r)\, dr, \tag{4.27b}$$

since antisymmetry is irrelevant in calculating matrix elements of one-electron operators. By applying these relations to enough M_L, M_S blocks, one can solve for all of the spin–orbit constants $\lambda(L,S)$ in terms of the integrals $\zeta_{n_il_i}$. This simple method fails for configurations $(nl)^q$ with $q>2$ when there are duplicated terms.

The general formulas for the spin–orbit constant for a single term of an $(nl)^q$ configuration is

$$\lambda(l^q\gamma LS) = \zeta_{nl}\sqrt{\frac{(2l+1)(l+1)l}{L(2L+1)(L+1)S(2S+1)(S+1)}}\langle l^q\gamma LS\|\mathbf{V}^{(11)}\|l^q\gamma LS\rangle, \tag{4.28}$$

and an analogous formula for off-diagonal elements of the spin–orbit interaction between terms, important for intermediate coupling in lanthanide elements, is [Sobelman (1972)]

$$\begin{aligned}&\langle l^q\alpha LSJM_J|\sum_i \xi(r_i)\mathbf{l}_i\cdot\mathbf{s}_i|l^q\alpha' L'S'J'M'_J\rangle\\ &= \delta_{JJ'}\delta_{M_JM'_J}(-1)^{L'+S+J}\zeta_{nl}\sqrt{(2l+1)(l+1)l}\,W(LSL'S';J1)\langle l^q\alpha LS\|\mathbf{V}^{(11)}\|l^q\alpha' L'S'\rangle. \end{aligned}\tag{4.29}$$

Reduced matrix elements of the *double tensor operator* $\mathbf{V}^{(11)}$, given by Wybourne (1965),

$$\begin{aligned}&\langle l^q\gamma LS\|\mathbf{V}^{(11)}\|l^q\gamma' L'S'\rangle\\ &= q\sum_{\gamma_1L_1S_1}\langle l\gamma LS\{|l^{q-1}(\gamma_1L_1S_1)lLS\rangle\langle l^{q-1}(\gamma_1L_1S_1)lL'S'|\}l\gamma' L'S'\rangle(-1)^{L_1+S_1+l+3/2+L'+S'}\\ &\times\sqrt{\tfrac{3}{2}(2L+1)(2S+1)(2L'+1)(2S'+1)}\,W(LL'll;1L_1)W(SS'\tfrac{1}{2}\tfrac{1}{2};1S_1),\end{aligned}\tag{4.30}$$

are tabulated for p^q, d^q and f^q configurations [Nielson and Koster (1963)].

The spin–orbit interaction for a hole is the negative of that for an electron; consequently, the spin–orbit constants for corresponding terms of conjugate configurations have opposite signs,

$$\lambda(l^q\gamma LS) = -\lambda(l^{2(2l+1)-q}\gamma LS). \tag{4.31}$$

Spin–orbit constants are positive for less-than-half-filled shells and negative for more-than-half-filled shells (*inverted multiplets*). They vanish for half-filled shells, $q = 2l+1$. The fine-structure splitting of an $(np)^2$ configuration is illustrated in Fig. 4.1; spin–orbit interaction occurs only in the Hund's rule ground term, ^3P, in this configuration. The order of the terms would be the same in an $(np)^4$ configuration, but the order of the fine-structure levels would be inverted.

4.3 Crystal potential

4.3.1 Point-ion crystal–potential expansion

The effect of the crystalline environment is represented by an electrostatic potential which satisfies the Laplace equation except at the positions of surrounding ions. Within the first shell of ions, the crystal potential term in Eq. (4.1) can be expanded in spherical harmonics in the form

$$V_c(\mathbf{r}) = \sum_{k=0}^{\infty}\sum_{q=-k}^{k} B_k^q r^k Y_k^q(\theta,\phi). \tag{4.32}$$

Although this potential expansion involves an infinite sum, the only terms with non-vanishing matrix elements in an l^q configuration are those with even values of $k \le 2l$.

In the *point-ion approximation* in which the charge distribution of surrounding ions is represented by an array of point charges, the coefficients in Eq. (4.32) are given by

$$B_k^q = -\sum_\alpha \frac{4\pi e q_\alpha}{(2k+1)4\pi\varepsilon_0 r_\alpha^{k+1}} Y_k^q(\theta_\alpha, \phi_\alpha)^*, \tag{4.33}$$

where the sum is over point ions α with charge q_α. In practice, the point-ion approximation is unreliable, and the coefficients B_k^q are commonly treated as adjustable parameters, appropriately restricted by symmetry. The point-ion approximation is more reliable for electron-excess colour centres, because of their more diffuse wavefunctions. For that case, the potential expansion must be extended beyond the first shell of ions,

$$V_c(\mathbf{r}) = \sum_{k=0}^{\infty} \sum_{q=-k}^{k} \left[B_k^q r^k + \sum_{\alpha<} C_{k\alpha}^q \left(r_\alpha^k r^{-k-1} - r^k r_\alpha^{-k-1} \right) \right] Y_k^q(\theta, \phi), \tag{4.34}$$

$$C_{k\alpha}^q = -\frac{4\pi e q_\alpha}{(2k+1)4\pi\varepsilon_0} Y_k^q(\theta_\alpha, \phi_\alpha)^*, \tag{4.35}$$

where $\sum_{\alpha<}$ denotes that the sum is restricted to ions for which $r_\alpha < r$.

4.3.2 *Operator equivalents*

As noted previously, the crystal-field term in the Hamiltonian, Eq. (4.1), is roughly comparable to the mutual electrostatic interaction of the electrons for transition-metal ions. It is nonetheless expedient and instructive to distinguish cases where one or the other can be considered as the dominant perturbation to a first approximation. The *strong-field approximation*, appropriate when the crystal-field is dominant, is applicable to transition-metal complexes of the palladium and platinum groups. For many compounds of iron-group transition-metals, the electrostatic interaction is dominant and the crystal-field can be treated as a perturbation within each term (*intermediate-field approximation*). The designation *weak-field approximation* is reserved for rare-earth compounds in which the crystal-field is weaker than the spin–orbit interaction, and can be treated as a perturbation within each fine-structure level.

In order to facilitate the evaluation of matrix elements of the crystal potential in each of these approximations, it is convenient to exploit the replacement theorem corollary of the Wigner–Eckart theorem, §2.4.5. The crystal potential is expressed in Eq. (4.32) as a sum of spherical tensor operators $V_{kq}(r) = B_k^q r^k Y_k^q(\theta, \phi)$, which are homogeneous polynomials in the components of $\mathbf{r}$. One can generate a corresponding tensor operator $T_{kq}(\mathbf{K})$, where $\mathbf{K}$ is an angular momentum operator, by repeated application of the scalar operator $\mathbf{K} \cdot \nabla$ [Rose (1957)]:

$$T_{kq}(\mathbf{K}) = (\mathbf{K} \cdot \nabla)^k V_{kq}(\mathbf{r}). \tag{4.36}$$

This new tensor operator is not quite the same polynomial, however, since the components of $\mathbf{r}$ commute with one another but the components of $\mathbf{K}$ do not. The utility of this transformation lies in the fact that matrix elements of $T_{kq}(\mathbf{K})$ and $V_{kq}(\mathbf{r})$ are

Table 4.1. *Selected polynomials, operator equivalents and conversion factors, after Stevens (1952) and Abragam and Bleaney (1970, 1986)*

Polynomials
$P_2^0(\mathbf{r}) = 3z^2 - r^2$
$P_4^0(\mathbf{r}) = 35z^4 - 30r^2z^2 + 3r^4$
$P_4^3(\mathbf{r}) = xz(x^2 - 3y^2)$
$P_4^4(\mathbf{r}) = x^4 - 6x^2y^2 + y^4$
Operator equivalents
$O_2^0(\mathbf{K}) = 3K_z^2 - K(K+1)$
$O_4^0(\mathbf{K}) = 35K_z^4 - 30K(K+1)K_z^2 + 25K_z^2 - 6K(K+1) + 3K^2(K+1)^2$
$O_4^3(\mathbf{K}) = \frac{1}{4}\left[K_z(K_+^3 + K_-^3) + (K_+^3 + K_-^3)K_z\right]$
$O_4^4(\mathbf{K}) = \frac{1}{2}(K_+^4 + K_-^4)$
Conversion factors
$A_2^0 = \frac{1}{\sqrt{2\pi}}\sqrt{\frac{5}{8}}B_2^0$
$A_4^0 = \frac{1}{\sqrt{2\pi}}\frac{3\sqrt{2}}{16}B_4^0$
$A_4^3 = \frac{1}{\sqrt{2\pi}}\frac{3\sqrt{70}}{4}B_4^3$
$A_4^4 = \frac{1}{\sqrt{2\pi}}\frac{3\sqrt{35}}{8}B_4^4$

proportional in a $|KM_K\rangle$ representation, and that the former are much easier to evaluate. The angular momentum operator $\mathbf{K}$ can be identified with $\mathbf{l}$ in the strong-field approximation, with $\mathbf{L}$ in the intermediate-field approximation and with $\mathbf{J}$ in the weak-field approximation. It remains only to determine the constants of proportionality in each case.

Conventional operator equivalents $O_k^q(\mathbf{K})$ are derived instead from real polynomials $P_k^q(\mathbf{r})$, which are proportional to $r^k[Y_k^q(\theta,\phi) + Y_k^q(\theta,\phi)^*]$ but have the smallest integral coefficients [Stevens (1952)]. The crystal potential can be written in terms of these polynomials as

$$V_c(\mathbf{r}) = \sum_{k=0}^{\infty}\sum_{q=0}^{k} A_k^q P_k^q(\mathbf{r}). \tag{4.37}$$

The polynomials $P_k^q(\mathbf{r})$ and $O_k^q(\mathbf{K})$ are tabulated by Abragam and Bleaney (1970, 1986), together with the constants of proportionality between them, as well as conversion factors relating coefficients A_k^q and B_k^q. Selected examples are listed in Table 4.1.

The constants of proportionality between the polynomials and their operator equivalents are defined by

$$P_k^q(\mathbf{r}) = \langle r^k\rangle_{nl}\langle K\|a_k\|K\rangle O_k^q(\mathbf{K}), \tag{4.38a}$$

$$\langle r^k\rangle_{nl} = \int_0^{\infty} P_{nl}(r)^2 r^k dr. \tag{4.38b}$$

With the conventional notation $a_2 \equiv \alpha$, $a_4 \equiv \beta$ and $a_6 \equiv \gamma$, some explicit formulas for reduced matrix elements are

$$\langle l\|\alpha\|l\rangle = \frac{-2}{(2l-1)(2l+3)}, \tag{4.39a}$$

$$\langle l\|\beta\|l\rangle = \frac{6}{(2l-1)(2l-3)(2l+3)(2l+5)}, \tag{4.39b}$$

$$\langle l\|\gamma\|l\rangle = \frac{-20}{(2l-1)(2l-3)(2l-5)(2l+3)(2l+5)(2l+7)}, \tag{4.39c}$$

and, for a Hund's rule ground term,

$$\langle L\|\alpha\|L\rangle = \mp\frac{2(2l+1-4S)}{(2l-1)(2l+3)(2L-1)}, \tag{4.39d}$$

$$\langle L\|\beta\|L\rangle = \langle L\|\alpha\|L\rangle\frac{3[3(l-1)(l+2)-7(l-2S)(l+1-2S)]}{2(2l-3)(2l+5)(L-1)(2L-3)}, \tag{4.39e}$$

where the upper sign in Eq. (4.39d) applies for a less-than-half-filled shell, and the lower sign for a more-than-half-filled shell.

4.3.3 *Explicit formulas for crystal–potential matrix elements*

An alternative approach to the evaluation of crystal–potential matrix elements relies directly on the Wigner–Eckart theorem, Eq. (2.63), rather than on its replacement-theorem corollary [Judd (1963), Wybourne (1965)]. Within an l^q configuration, components of the crystal potential can be expressed in terms of unit tensor operators defined by Eqs. (4.22),

$$V_{kq}(\mathbf{r}) = B_k^q\langle r^k\rangle_{nl}\langle l\|Y_k\|l\rangle\mathbf{u}_q^{(k)}, \tag{4.40a}$$

$$\langle l\|Y_k\|l\rangle = (-1)^k\sqrt{\frac{(2k+1)(2l+1)}{4\pi}}\langle lk00|l0\rangle. \tag{4.40b}$$

In the strong-field approximation, the required matrix elements are then given by

$$\langle l^q nlm_lm_s|V_{kq}|l^q nlm_l'm_s'\rangle = B_k^q\langle r^k\rangle_{nl}\langle l\|Y_k\|l\rangle\langle lkm_l'q|lm_l\rangle\delta_{m_sm_s'}. \tag{4.41}$$

In the intermediate-field approximation, they are given by

$$\left\langle l^q\alpha LSM_LM_S\left|\sum_i V_{kq}(i)\right|l^q\alpha' L'S'M_L'M_S'\right\rangle = B_k^q\langle r^k\rangle_{nl}\langle l\|Y_k\|l\rangle\langle l^q\alpha LS\|\mathbf{U}^{(k)}\|l^q\alpha' L'S'\rangle\langle L'kM_L'q|LM_L\rangle\delta_{M_SM_S'}, \tag{4.42}$$

where the reduced matrix element of $\mathbf{U}^{(k)}$ is given by Eq. (4.23). Finally, in the weak-field approximation, they are given by

$$\left\langle l^q\alpha LSJM_J\left|\sum_i V_{kq}(i)\right|l^q\alpha' L'S'J'M_J'\right\rangle = B_k^q\langle r^k\rangle_{nl}\langle l\|Y_k\|l\rangle\langle l^q\alpha LSJ\|\mathbf{U}^{(k)}\|l^q\alpha' L'S'J'\rangle\langle J'kM_J'q|JM_J\rangle, \tag{4.43}$$

where the doubly reduced matrix element of $\mathbf{U}^{(k)}$ can be evaluated with the help of the generalized projection theorem, Eq. (2.69),

$$\langle l^q \alpha LSJ \| \mathbf{U}^{(k)} \| l^q \alpha' L'S'J' \rangle$$
$$= \delta_{SS'}(-1)^{S+k-L-J}\sqrt{(2L+1)(2J'+1)}W(L'J'LJ;Sk)\langle l^q \alpha LS \| \mathbf{U}^k \| l^q \alpha' L'S \rangle. \quad (4.44)$$

4.3.4 Dominant symmetry

The crystal–potential term in the Hamiltonian entails a reduction in symmetry from the full rotation group to one of the crystallographic point groups. For most transition-metal and rare-earth complexes of interest, it is useful to recognize a dominant symmetry determined by the coordination. The dominant symmetry for both octahedral (six-fold) and cubic (eight-fold) coordination is described by point group O_h, the group of proper rotations which leave an octahedron or a cube invariant, and that for tetrahedral (four-fold) coordination, by point group T_d, the group of proper and improper rotations which leave a tetrahedron invariant. A procedure is then adopted which greatly diminishes the complexity of the problem. The crystal-field is divided into two parts, a dominant-symmetry part and a part of lower symmetry. The dominant-symmetry part is first treated by degenerate perturbation theory within the appropriate free-ion level, and the lower-symmetry part, which reflects the extended crystal structure, is subsequently treated as a perturbation within each dominant-symmetry level.

The dominant crystal field splits each free-ion level into a set of crystal-field levels whose number and degeneracies are determined both by the point-group irreducible representations contained in the subduced representation of the full rotation group (§2.5.3) and by the effects of time-reversal invariance (§2.5.4). The numbers of crystal-field levels for each value of J are listed for all of the crystallographic point groups in Tables 2.19 and 2.20.

Subduced representations also constrain the crystal potential. As an example, consider the form of the crystal potential for cubic or octahedral coordination. Since the crystal potential must transform as the identity representation, A_{1g}, of O_h, its expansion in spherical harmonics, Eq. (4.32), can include only linear combinations of spherical harmonics which transform as bases for that representation. It is evident from the rotation group compatibility table for the octahedral group, Table 2.12, that there is one such combination for $J=0$, one for $J=4$ and one for $J=6$. The $J=0$ term in the potential expansion, which is spherically symmetrical and shifts all energy levels equally, is of little interest for transition-metal and rare-earth complexes, but is the dominant term for colour centres. Only the $J=4$ term contributes to the cubic or octahedral crystal-field splitting in transition-metal complexes and only the $J=4$ and $J=6$ terms contribute in rare-earth complexes. The corresponding symmetry-adapted combinations of spherical harmonics are:

$$Y_4^0(\theta,\phi)+\frac{\sqrt{5}}{\sqrt{14}}\left[Y_4^4(\theta,\phi)+Y_4^{-4}(\theta,\phi)\right], \quad (4.45a)$$

$$Y_6^0(\theta,\phi)-\frac{\sqrt{14}}{2}\left[Y_6^4(\theta,\phi)+Y_6^{-4}(\theta,\phi)\right]. \quad (4.45b)$$

The full rotation group compatibility table for the tetrahedral group T_d is listed in Table 2.13. Since the spherical harmonics which appear in the potential expansion, $Y_k^q(\theta,\phi)$, have parity $(-1)^k$, only even-parity subduced representations with even values of $J \leq 6$ have non-vanishing matrix elements in an l^q configuration. Consequently, the same symmetry-adapted combinations of spherical harmonics, Eqs. (4.45), appear in the potential expansion for tetrahedral coordination. The odd parity combination for $J=3$ is not without effect, however. Although it makes no contribution to the crystal-field splitting within the ground configuration, it can mix wavefunctions of opposite parity (configuration mixing), and so relax the selection rule which forbids optical transitions within a single configuration. This point is discussed further in Chapter 5.

The fine-structure levels of rare-earth ions with an odd number of 4*f* electrons present a special case. They are labeled by half-integral values of *J*, corresponding to double-valued irreducible representations of the rotation group which are single-valued irreducible representations of the unitary, unimodular group $SU(2)$. The crystal-field levels derived from each fine-structure level are then labeled by the irreducible representations of the double point group contained in the subduced representation of $SU(2)$. The characters of the additional representations of the double point groups O^* and T_d^* are listed in Table 2.14, and the rotation-group compatibility tables are displayed in Tables 2.15 and 2.16. Time-reversal invariance ensures at least double degeneracy (Kramers degeneracy) of crystal-field levels derived from fine-structure levels with half-integral *J*, as discussed in §2.5.4 and indicated in Tables 2.19 and 2.20.

4.4 Transition metals

4.4.1 Free-ion energy levels

Transition-metal ions are characterized by ground configurations with open *nd* shells, where the principal quantum number *n* equals 3 for the iron group, 4 for the palladium group and 5 for the platinum group. The three groups are essentially similar except that the relative strength of the spin–orbit interaction increases rapidly with *n*. In this section it is assumed that the fine-structure splitting is small compared with the term separation, an assumption which is most appropriate for the iron group. Term energies for d^q configurations were derived by Racah (1942), who found it expedient to introduce the following linear combinations of Slater integrals (*Racah parameters*):

$$A \equiv F^0 - F^4/9, \tag{4.46a}$$

$$B \equiv (9F^2 - 5F^4)/441, \tag{4.46b}$$

$$C \equiv 5F^4/63. \tag{4.46c}$$

Term energies for d^2 and d^3 configurations, expressed in terms of Racah parameters, are listed in Table 4.2.

Since all of the terms within each configuration have the same dependence on *A*, that parameter contributes nothing to the term separations and it is feasible to plot the relative term energies in units of *C* as functions of B/C [Griffith (1961)]. One can then fit the free-ion parameters *B* and *C* by matching the term energies inferred from optical

Table 4.2. *Term energies of d^q configurations. (After Racah* (1942))

d^2
$E(^1S)=A+14B+7C$
$E(^3P)=A+7B$
$E(^1D)=A-3B+2C$
$E(^3F)=A-8B$
$E(^1G)=A+4B+2C$
d^3
$E(^2P)=3A-6B+3C$
$E(^4P)=3A$
$E(^2D)=3A+5B+5C\pm\sqrt{193B^2+8BC+4C^2}$
$E(^2F)=3A+9B+3C$
$E(^4F)=3A-15B$
$E(^2G)=3A-11B+3C$
$E(^2H)=3A-6B+3C$
d^4
$E(^1S)=6A+10B+10C\pm 2\sqrt{193B^2+8BC+4C^2}$
$E(^3P)=6A-5B+\frac{11}{2}C\pm\frac{1}{2}\sqrt{912B^2-24BC+9C^2}$
$E(^1D)=6A+9B+\frac{15}{2}C\pm\frac{3}{2}\sqrt{144B^2+8BC+C^2}$
$E(^3D)=6A-5B+4C$
$E(^5D)=6A-21B$
$E(^1F)=6A+6B$
$E(^3F)=6A-5B+\frac{11}{2}C\pm\frac{3}{2}\sqrt{68B^2+4BC+C^2}$
$E(^1G)=6A-5B+\frac{15}{2}C\pm\frac{1}{2}\sqrt{708B^2-12BC+9C^2}$
$E(^3G)=6A-12B+4C$
$E(^3H)=6A-17B+4C$
$E(^1I)=6A-15B+6C$
d^5
$E(^2S)=10A-3B+8C$
$E(^6S)=10A-35B$
$E(^2P)=10A+20B+10C$
$E(^4P)=10A-28B+7C$
$E(^2D)=10A-3B+11C\pm 3\sqrt{57B^2+2BC+C^2}$
$E(^2D')=10A-4B+10C$
$E(^4D)=10A-18B+5C$
$E(^2F)=10A-9B+8C$
$E(^2F')=10A-25B+10C$
$E(^4F)=10A-13B+7C$
$E(^2G)=10A-13B+8C$
$E(^2G')=10A+3B+10C$
$E(^4G)=10A-25B+5C$
$E(^2H)=10A-22B+10C$
$E(^2I)=10A-24B+8C$

spectra. The fit is never perfect because of the neglect of configuration mixing in the theoretical curves. The empirical parameter values depend on the ion as well as its configuration, and are modified by its surroundings when the ion is embedded in a solid; the latter effect is discussed more fully in Chapters 8 and 9. Free-ion empirical values of B and C for divalent, trivalent and tetravalent iron-group transition-metal

Table 4.3. *Free-ion empirical values, in* cm^{-1}, *of Racah parameters and spin–orbit constants (after Griffith* (1961), *Tanabe and Sugano* (1954)* *and Burns* (1993)**) *for divalent, trivalent and tetravalent iron-group transition-metal ions*

Config.	M^{2+}	M^{3+}	M^{4+}	B	C	C/B	ζ
$3d^1$		Ti^{3+}					154
			V^{4+}				248
$3d^2$	Ti^{2+}			718	2629	3.66	121
		V^{3+}		861	4165	4.84	209
			Cr^{4+}	1039	4238	4.08	327
$3d^3$	V^{2+}			766	2855	3.73	167
		Cr^{3+}		918*	3850	4.19	273
			Mn^{4+}	1064**			402
$3d^4$	Cr^{2+}			830	3430	4.13	230
		Mn^{3+}		965*	3675	3.81	352
			Fe^{4+}	1144	4459	3.90	514
$3d^5$	Mn^{2+}			960	3325	3.46	347
		Fe^{3+}		1015*	4800*	4.73	
$3d^6$	Fe^{2+}			1058	3901	3.69	410
		Co^{3+}		1065*	5120*	4.81	
$3d^7$	Co^{2+}			971*	4366	4.50	533
		Ni^{3+}		1115*	5450*	4.89	
$3d^8$	Ni^{2+}			1041**	4831	4.64	649
$3d^9$	Cu^{2+}			1238	4659	3.76	829

ions are listed in Table 4.3. It is evident from this table that the ratio C/B has approximately the same value, 4.5, for all of these ions. The one-electron spin–orbit constant ζ is also listed in Table 4.3.

A more complete tabulation, including other ionization states and all three transition series, is provided by Griffith (1961), together with primary references. Nearly all of the parameters are derived from experimental data compiled by Moore (1952), but different fitting criteria are employed with somewhat variable results.

4.4.2 *One-electron configuration*

For an ion with a d^1 configuration, e.g. Ti^{3+}, the distinction between the strong-field and the intermediate-field approximations disappears. The degeneracy of the free-ion configuration is $(2s+1)(2l+1)=10$, corresponding to the two allowed values of m_s and the five allowed values of m_l. The five-fold orbital degeneracy is partially removed by the crystal-field, to an extent which depends on its symmetry and on time-reversal invariance, as discussed in §2.5.4. It is evident from Tables 2.12, 2.13, 2.19 and 2.20 that the ^2D ground state splits into two crystal-field states, ^2E and $^2T_2$, for the case of dominant O_h or T_d symmetry. Real linear combinations of d-orbitals which transform as bases for the E and T_2 representations of O or T_d are listed in Table 4.4. Combinations which transform as bases for the E representation are called e orbitals, and those which transform as bases for the T_2 representation, t_2 orbitals. All of the orbitals are even under inversion. These orbitals are illustrated in Fig. 4.2.

Table 4.4. *Real linear combinations of d-orbitals which transform as bases for the E and T_2 representations of O or T_d*

Rep.	Combinations of *d*-orbitals	Polynomials
$E\theta$	$r^{-1}P_{3d}(r)Y_2^0(\theta,\phi)$	$r^{-3}P_{3d}(r)\sqrt{5/4\pi}\frac{1}{2}(3z^2-r^2)$
$E\varepsilon$	$r^{-1}P_{3d}(r)\frac{1}{\sqrt{2}}[Y_2^2(\theta,\phi)+Y_2^{-2}(\theta,\phi)]$	$r^{-3}P_{3d}(r)\sqrt{5/4\pi}\frac{\sqrt{3}}{2}(x^2-y^2)$
$T_2\xi$	$r^{-1}P_{3d}(r)\frac{(-1)}{i\sqrt{2}}[Y_2^1(\theta,\phi)+Y_2^{-1}(\theta,\phi)]$	$r^{-3}P_{3d}(r)\sqrt{5/4\pi}\sqrt{3}yz$
$T_2\eta$	$r^{-1}P_{3d}(r)\frac{(-1)}{\sqrt{2}}[Y_2^1(\theta,\phi)-Y_2^{-1}(\theta,\phi)]$	$r^{-3}P_{3d}(r)\sqrt{5/4\pi}\sqrt{3}xz$
$T_2\zeta$	$r^{-1}P_{3d}(r)\frac{1}{i\sqrt{2}}[Y_2^2(\theta,\phi)-Y_2^{-2}(\theta,\phi)]$	$r^{-3}P_{3d}(r)\sqrt{5/4\pi}\sqrt{3}xy$

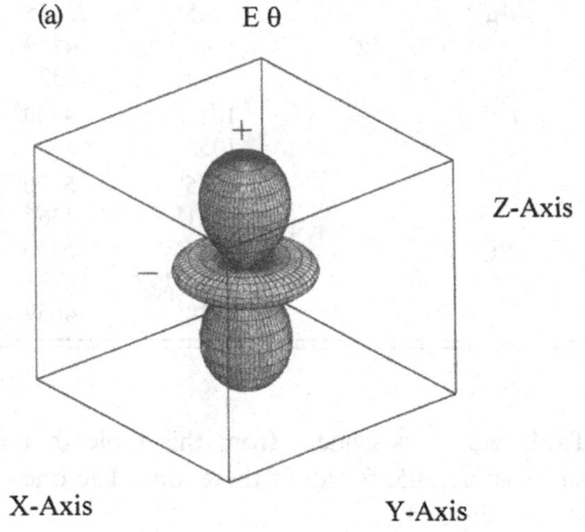

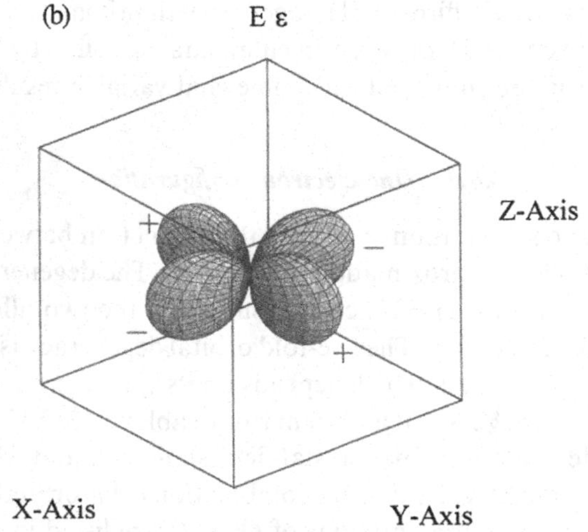

Figure 4.2. Angular dependences of the real linear combinations of *d*-orbitals defined in Table 4.4.(a) $E\theta[d(z^2)]$, (b) $E\varepsilon[d(x^2-y^2)]$, (c) $T_2\xi[d(yz)]$, (d) $T_2\eta[d(xz)]$ and (e) $T_2\zeta[d(xy)]$.

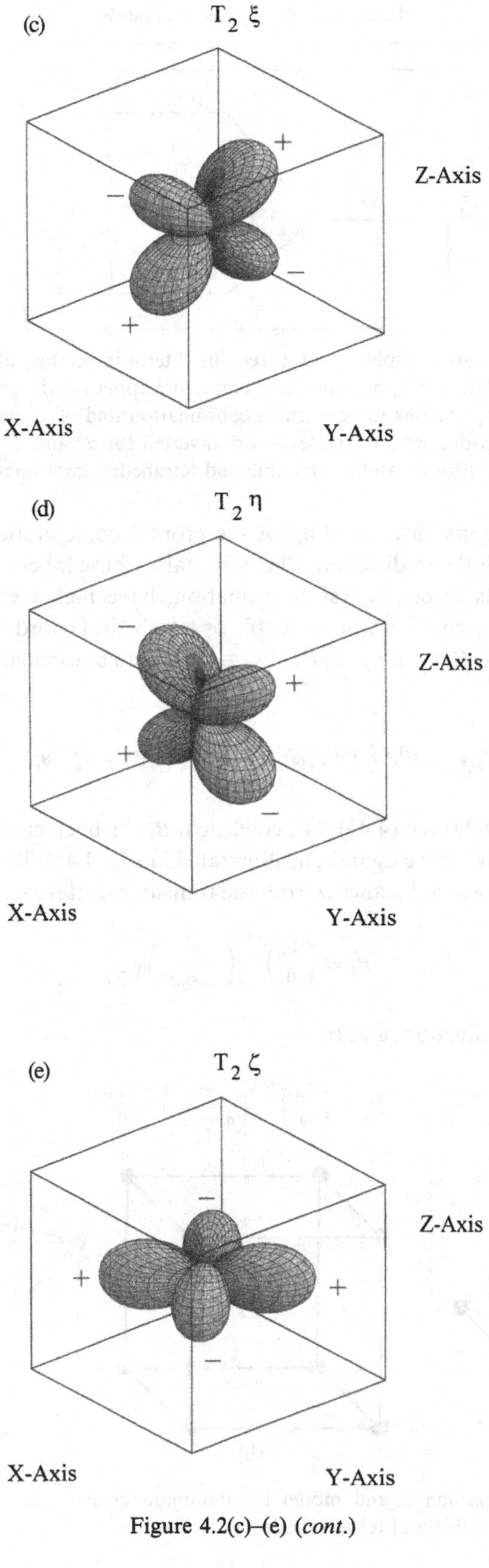

Figure 4.2(c)–(e) (*cont.*)

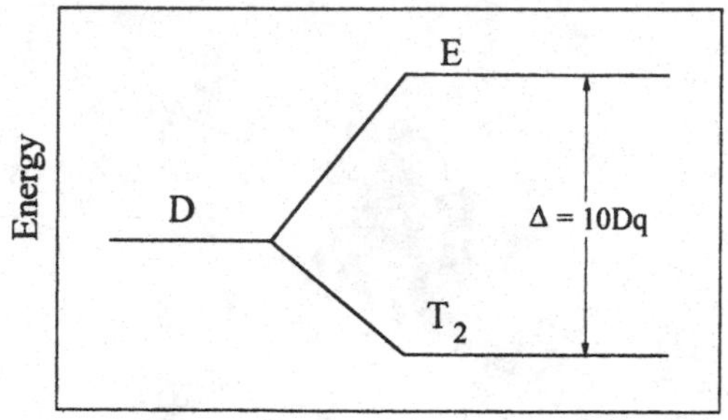

Figure 4.3. Crystal-field splitting of a free ion D term in octahedral coordination in the intermediate field approximation. This figure applies to the ground terms of d^1 and d^6 configurations in octahedral coordination and of d^4 and d^9 in cubic or tetrahedral coordination. The levels are inverted for d^4 and d^9 in octahedral coordination and for d^1 and d^6 in cubic and tetrahedral coordination.

The dominant crystal-field splitting of the ground configuration is illustrated in Fig. 4.3 for octahedral coordination. The e-orbitals, whose lobes are directed toward the negative ligands in octahedral coordination, have higher energy than the t_2-orbitals, while the opposite is true for cubic or tetrahedral coordination. The energy level separation, $\Delta \equiv E(d^1, {}^2E) - E(d^1, {}^2T_2) \equiv 10Dq$, can be calculated from the crystal potential

$$V_c(\mathbf{r}) = B_4^0 r^4 \left\{ Y_4^0(\theta, \phi) + \frac{\sqrt{5}}{\sqrt{14}} \left[Y_4^4(\theta, \phi) + Y_4^{-4}(\theta, \phi) \right] \right\}, \tag{4.47}$$

derived from Eqs. (4.32) and (4.45). The coefficient B_4^0 can be calculated from Eq. (4.33) for a point-ion model of the ligands, as illustrated in Fig. 4.4. With the assumption of ligands of charge $-e$ at a distance a from the transition-metal ion, it is given by

$$B_4^0 = \left(\frac{4\pi}{9}\right)^{1/2} \left(\frac{e^2}{4\pi\varepsilon_0 a^5}\right) \left(\frac{7}{2}\right) \tag{4.48a}$$

for octahedral coordination, and by

$$B_4^0 = \left(\frac{4\pi}{9}\right)^{1/2} \left(\frac{e^2}{4\pi\varepsilon_0 a^5}\right) \left(\frac{-28}{9}\right) \tag{4.48b}$$

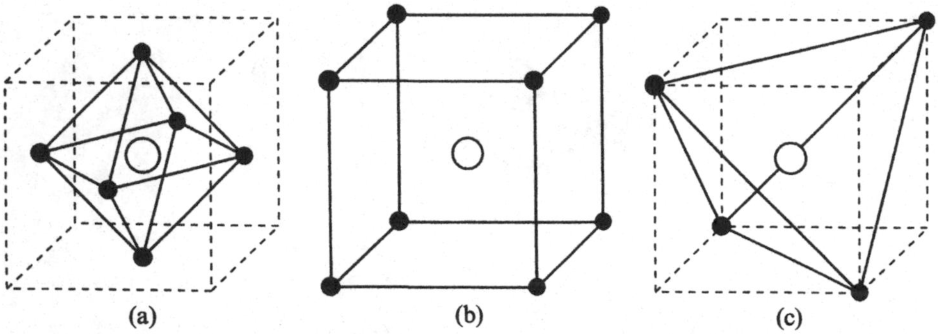

Figure 4.4. Point-ion ligand model for dominant crystal-field splitting in (a) octahedral, (b) cubic and (c) tetrahedral coordination.

for cubic coordination. The value for tetrahedral coordination is just half of that for cubic coordination. The expression for the crystal potential in terms of operator equivalents which corresponds to Eq. (4.47) is

$$V_c(\mathbf{l}) = A_4^0\langle r^4\rangle_{3d}\langle l\|\beta\|l\rangle\left[O_4^0(\mathbf{l}) + 5O_4^4(\mathbf{l})\right]. \tag{4.49}$$

The expectation value of $V_c(\mathbf{l})$ can be evaluated in one crystal-field state, say $E\theta$, with the help of Table 4.1, to find the crystal-field splitting. The resulting expressions are

$$\Delta = \left(\frac{5}{3}\right)\left(\frac{e^2}{4\pi\varepsilon_0 a^5}\right)\langle r^4\rangle_{3d} \tag{4.50a}$$

for octahedral coordination, and

$$\Delta = -\left(\frac{40}{27}\right)\left(\frac{e^2}{4\pi\varepsilon_0 a^5}\right)\langle r^4\rangle_{3d} \tag{4.50b}$$

for cubic coordination. As noted previously, the point-ion approximation is unreliable, and the strength of the crystal-field is ordinarily treated as an adjustable parameter. In that case, it is really the combination $A_4^0\langle r^4\rangle_{3d}$ in Eq. (4.49) which is adjustable, since the 3*d*-atomic orbitals are modified in the solid. Further discussion of this point is deferred to Chapter 8.

4.4.3 *Intermediate-field approximation*

In the intermediate-field approximation, the crystal potential is treated as a small perturbation within each term of a d^q configuration. In the present section, we will consider only octahedral or cubic symmetry and restrict consideration to the Hund's rule ground terms, which are listed in Table 4.5 together with the irreducible representations of point group O_h contained in their respective subduced representations.

Since the ^6S term is not orbitally degenerate, there are really only two distinct cases of interest: a D term, already considered in the d^1 configuration, and an F term. For configurations with more than one electron, the crystal potential can be written in the form

$$V_c(\mathbf{L}) = A_4^0\langle r^4\rangle_{3d}\langle L\|\beta\|L\rangle\left[O_4^0(\mathbf{L}) + 5O_4^4(\mathbf{L})\right], \tag{4.51}$$

where the coefficient $\langle L\|\beta\|L\rangle$ is given by Eqs. (4.39d) and (4.39e). It is evident from these equations that the crystal-field levels are inverted for Hund's rule ground terms of conjugate configurations, d^q and $d^{2(2l+1)-q}$. They are also inverted for Hund's rule

Table 4.5. *Hund's rule ground terms of d^q configurations and representations of the octahedral group contained in their subduced representations*

Configuration	d^1, d^9	d^2, d^8	d^3, d^7	d^4, d^6	d^5
Ground term	^2D	^3F	^4F	^5D	^6S
reps. of O_h	$E_g + T_{2g}$	$A_{1g} + T_{1g} + T_{2g}$	$A_{1g} + T_{1g} + T_{2g}$	$E_g + T_{2g}$	A_{1g}

Table 4.6. *Symmetry-adapted combinations of eigenkets $|LM_L\rangle$, with $L=3$, which transform as bases for irreducible representations of O_h*

$\lvert A_2\rangle = \frac{1}{\sqrt{2}}\lvert 32\rangle - \frac{1}{\sqrt{2}}\lvert 3-2\rangle$
$\lvert T_1 1\rangle = -\frac{\sqrt{5}}{2\sqrt{2}}\lvert 3-3\rangle - \frac{\sqrt{3}}{2\sqrt{2}}\lvert 31\rangle$
$\lvert T_1 0\rangle = \lvert 30\rangle$
$\lvert T_1 -1\rangle = -\frac{\sqrt{5}}{2\sqrt{2}}\lvert 33\rangle - \frac{\sqrt{3}}{2\sqrt{2}}\lvert 3-1\rangle$
$\lvert T_2 1\rangle = -\frac{\sqrt{3}}{2\sqrt{2}}\lvert 33\rangle + \frac{\sqrt{5}}{2\sqrt{2}}\lvert 3-1\rangle$
$\lvert T_2 0\rangle = \frac{1}{\sqrt{2}}\lvert 32\rangle + \frac{1}{\sqrt{2}}\lvert 3-2\rangle$
$\lvert T_2 -1\rangle = -\frac{\sqrt{3}}{2\sqrt{2}}\lvert 3-3\rangle + \frac{\sqrt{5}}{2\sqrt{2}}\lvert 31\rangle$

ground terms of configurations d^q and d^{2l+1-q} for $q<5$, for which all of the spins are parallel. These results are to be expected, since the interactions of electrons and holes with the crystal-field have opposite signs. Thus the only new case is presented by the d^2 configuration with a ^3F ground term.

Symmetry-adapted combinations of eigenkets $|LM_L\rangle$ which transform as bases for irreducible representations of O_h are listed in Table 4.6 for $L=3$. The complex representations $T_1(1,0,-1)$ and $T_2(1,0,-1)$ in this table are related, respectively, to equivalent real representations $T_1(\alpha,\beta,\gamma)$ and $T_2(\xi,\eta,\zeta)$ by the unitary transformation

$$\mathbf{U} = \begin{bmatrix} -\frac{i}{\sqrt{2}} & \frac{1}{\sqrt{2}} & 0 \\ 0 & 0 & i \\ \frac{i}{\sqrt{2}} & \frac{1}{\sqrt{2}} & 0 \end{bmatrix}. \tag{4.52}$$

The matrix of the crystal-potential operator, $V_c(\mathbf{L})$, is a constant matrix within each representation, and the energy differences of the crystal-field levels are

$$E(d^2,{}^3A_2) - E(d^2,{}^3T_2) = \Delta = 10Dq, \tag{4.53a}$$

$$E(d^2,{}^3T_2) - E(d^2,{}^3T_1) = \frac{4}{5}\Delta = 8Dq. \tag{4.53b}$$

Crystal-field splitting of the ^3F Hund's rule ground term for the d^2 configuration in octahedral coordination is illustrated in Fig. 4.5. This figure applies as well to the ^4F ground term of configuration d^7 in octahedral coordination and to the ^4F and ^3F ground terms of the d^3 and d^8 configurations, respectively, in cubic or tetrahedral coordination. The levels are inverted for the d^3 and d^8 configurations in octahedral coordination and the d^2 and d^7 configurations in cubic or tetrahedral coordination. Similarly, Fig. 4.3 applies not only to the ^2D ground term of the d^1 configuration in octahedral coordination, but also to the ^5D ground term of the d^6 configuration in octahedral coordination and to the ^5D and ^2D ground terms of the d^4 and d^9 configurations, respectively, in cubic or tetrahedral coordination. The levels are

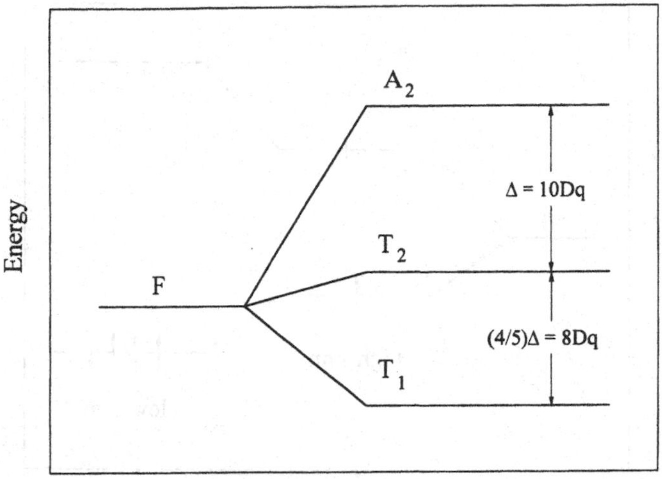

Figure 4.5. Crystal-field splitting of a free ion F term in octahedral coordination in the intermediate field approximation. This figure applies to the ground terms of d^2 and d^7 configurations in octahedral coordination and of d^3 and d^8 in cubic and tetrahedral coordination. The levels are inverted for d^3 and d^8 in octahedral coordination and for d^2 and d^7 in cubic and tetrahedral coordination.

inverted for the d^4 and d^9 configurations in octahedral coordination and the d^1 and d^6 configurations in cubic or tetrahedral coordination. Although crystal-field levels derived from the Hund's rule ground term are of special interest in paramagnetic resonance [Abragam and Bleaney (1970, 1986)], they provide at best an incomplete description of optical properties of iron-group transition-metal complexes.

4.4.4 Strong-field approximation

The strong-field approximation proceeds from the assumption that the crystal potential exceeds the mutual electrostatic interaction of the electrons. In the strong-field limit, a free-ion configuration d^q is split by the dominant crystal-field into strong-field configurations $t_2^n e^m$, with energies given by

$$E(t_2^n e^m) = \left(-\frac{2}{5}n + \frac{3}{5}m\right)\Delta = (-4n + 6m)Dq, \tag{4.54}$$

where $n + m = q$. The configuration of lowest energy is predicted to be that with the most electrons in the lower energy level, which is the t_2 level for octahedral coordination and the e level for cubic or tetrahedral coordination. As a consequence, the total spin S in the ground state is predicted to be less than maximum for values of q between 4 and 7 in octahedral coordination and between 3 and 6 in cubic and tetrahedral coordination, contrary to Hund's rule for the free ion. The distinction between high and low spin states is illustrated in Fig. 4.6.

Each strong-field configuration is split into crystal-field terms by the electrostatic interaction, which may be treated by degenerate, first-order perturbation theory. The number of possible terms and their symmetry designations are determined by successive reductions of the Kronecker product representation. The irreducible

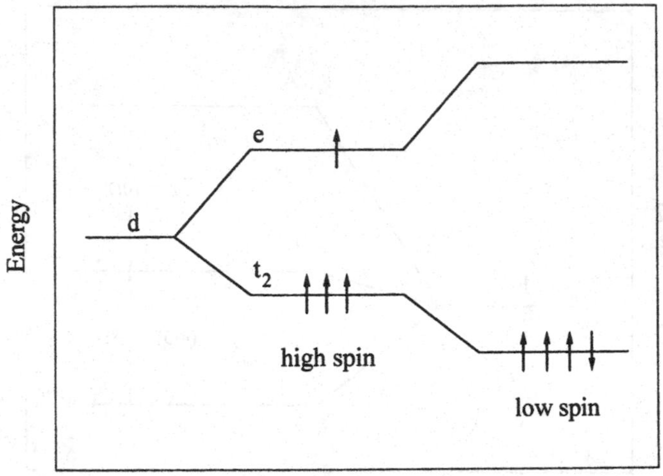

Figure 4.6. Illustration of the distinction between high and low spin states of a d^4 configuration, reflecting competition between exchange interaction and crystal-field splitting.

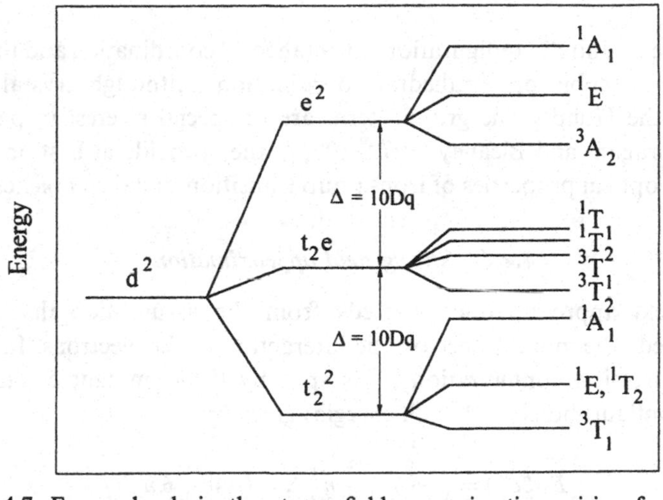

Figure 4.7. Energy levels in the strong-field approximation arising from a d^2 configuration in octahedral coordination.

representations of point-groups O and T_d contained in each Kronecker product representation are displayed in Table 2.21

For example, the terms in a t_2e configuration are found from the reduction of the Kronecker product representation $T_2 \times E = T_1 + T_2$, each representation appearing as both a singlet and a triplet; thus the terms are ${}^1T_1, {}^1T_2, {}^3T_1, {}^3T_2$. However, anti-symmetry imposes restrictions in the case of equivalent electrons, e.g. a t_2^2 configuration, so the actual number of terms is less than reduction of the Kronecker product would suggest. The energy levels in the strong-field approximation for a d^2 configuration in octahedral coordination are illustrated in Fig. 4.7.

4.4.5 Tanabe–Sugano theory

Crystal-field theory was developed in the strong-field representation by Tanabe and Sugano (1954) and was presented in detail by Sugano, Tanabe and Kamimura (1970). It is actually an intermediate coupling theory, since it includes matrix elements of the electrostatic interaction between strong-field configurations, but it contains the strong-field approximation as a limiting case.

Although it is feasible to define coefficients of fractional parentage for many-electron wavefunctions which transform as bases for point group representations [Griffith (1961)], Tanabe and Sugano construct symmetry-adapted linear combinations of Slater determinants instead. For example, an antisymmetric, symmetry-adapted basis function for terms of the t_2e configuration can be constructed as follows:

$$|t_2e\Gamma SM_\Gamma M_s\rangle = \sum_{\gamma_1}\sum_{\gamma_2}\sum_{m_{s1}}\sum_{m_{s2}} \frac{1}{\sqrt{2}} \det|\phi(t_2\gamma_1 m_{s1})\phi(e\gamma_2 m_{s2})| \times \langle t_2e\gamma_1\gamma_2|\Gamma M_\Gamma\rangle \left\langle \frac{1}{2}\frac{1}{2}m_{s1}m_{s2}\middle|SM_S\right\rangle, \tag{4.55}$$

where Γ denotes either T_1 or T_2, and the rows of the coupled representation are labeled by M_Γ. A three-electron basis function can be constructed by coupling an additional spin-orbital, and subsequently applying the antisymmetrizer, $\mathscr{A}$, defined by Eq. (4.9b). Duplicated terms in this method are distinguished by their parent terms. A complication arises when two or more orbitals transform as bases for the same irreducible representation. Then some Slater determinants vanish and others are equal within a sign. Some potential terms are thereby eliminated, and basis functions for the surviving terms must be renormalized.

Matrix elements of the electrostatic interaction were calculated with these many-electron wavefunctions, including off-diagonal elements not only between duplicated terms of the same strong-field configuration, but also between strong-field

Table 4.7. *Matrix elements of the coulomb interaction for a d^2 configuration in the strong-field representation [Sugano et al. (1970)]*

Strong-field term (free-ion terms)	Matrices of the Coulomb interaction
$^1A_1(^1G, ^1S)$	$\begin{matrix} t_2^2 \\ e^2 \end{matrix}\begin{pmatrix} 10B+5C & \sqrt{6}(2B+C) \\ \sqrt{6}(2B+C) & 8B+4C \end{pmatrix}$
$^1E(^1D, ^1G)$	$\begin{matrix} t_2^2 \\ e^2 \end{matrix}\begin{pmatrix} B+2C & -2\sqrt{3}B \\ -2\sqrt{3}B & 2C \end{pmatrix}$
$^1T_2(^1D, ^1G)$	$\begin{matrix} t_2^2 \\ t_2e \end{matrix}\begin{pmatrix} B+2C & 2\sqrt{3}B \\ 2\sqrt{3}B & 2C \end{pmatrix}$
$^3T_1(^3F, ^3P)$	$\begin{matrix} t_2^2 \\ t_2e \end{matrix}\begin{pmatrix} -5B & 6B \\ 6B & 4B \end{pmatrix}$
$^1T_1(^1G)$	$t_2e\ (4B+2C)$
$^3T_2(^3F)$	$t_2e\ (-8B)$
$^3A_2(^3F)$	$e^2\ (-8B)$

configurations, as shown in Table 4.7 for the d^2 configuration. Energy levels in intermediate coupling were then obtained by diagonalizing the combined matrices of the electrostatic interaction and of the crystal-field, Eq. (4.54), which is diagonal in this representation. Eigenvectors provide the expansion coefficients for expressing eigenfunctions in terms of the constructed many-electron functions. Matrix elements of the electrostatic interaction are the same for conjugate configurations (*corresponding states*) and $t_2^n e^m$ and $t_2^{6-n} e^{4-m}$, since the coulomb interaction of two holes is the same as that of two electrons, but the signs of the crystal-field matrix elements are reversed. Energy levels for the d^2 configuration are listed in Table 4.8 as functions of the octahedral crystal-field parameter $\Delta(= 10Dq)$ and the Racah parameters B and C. This table also applies to the d^8 configuration with the sign of Δ reversed. Since larger matrices are involved in the remaining many-electron configurations, they must be diagonalized numerically; the resulting energy levels are plotted in units of Racah parameter B as functions of Dq/B for fixed values of C/B (*Tanabe–Sugano diagrams*) for several d^q configurations in Fig. 4.8. The ground-state energy is taken as the zero of energy in these energy-level diagrams; thus the diagram for the d^7 configuration exhibits discontinuous slopes where the lowest energy levels cross, marking the transition between high and low spin states.

4.4.6 Spin–orbit interaction

The crystal-field levels are further split into fine-structure levels, which can be determined to a good approximation by treating the spin–orbit interaction as a small perturbation within each crystal-field level. In the intermediate-field approximation, L and S are good quantum numbers and it follows readily from Eqs. (4.27) that the spin–orbit constant in the Hund's rule ground term of an l^q configuration is

$$\lambda(L, S) = \pm\frac{\zeta_{nl}}{2S}, \tag{4.56}$$

where the + sign applies for a less-than-half-filled shell, and the − sign for a more-than-half-filled shell. Matrix elements of the orbital angular momentum operator, **L**,

Table 4.8. *Energy levels of a d^2 configuration as a function of the octahedral crystal-field parameter Δ ($= 10Dq$) and the Racah parameters B and C. The energy levels of the d^8 configuration are the same except that Δ is replaced by $-\Delta$*

Strong-field term (free-ion terms)	Term energies
$^1A_1(^1G, {}^1S)$	$\Delta + 9B + \frac{9}{2}C \pm \sqrt{(\Delta - B - \frac{1}{2}C)^2 + 6(2B + C)^2}$
$^1E(^1D, {}^1G)$	$\Delta + \frac{1}{2}B + 2C \pm \sqrt{(\Delta - \frac{1}{2}B)^2 + 12B^2}$
$^1T_2(^1D, {}^1G)$	$\frac{1}{2}\Delta + \frac{1}{2}B + 2C \pm \sqrt{\frac{1}{4}(\Delta - B)^2 + 12B^2}$
$^3T_1(^3F, {}^3P)$	$\frac{1}{2}\Delta - \frac{1}{2}B \pm \sqrt{\frac{1}{4}(\Delta + 9B)^2 + 36B^2}$
$^1T_1(^1G)$	$\Delta + 4B + 2C$
$^3T_2(^3F)$	$\Delta - 8B$
$^3A_2(^3F)$	$2\Delta - 8B$

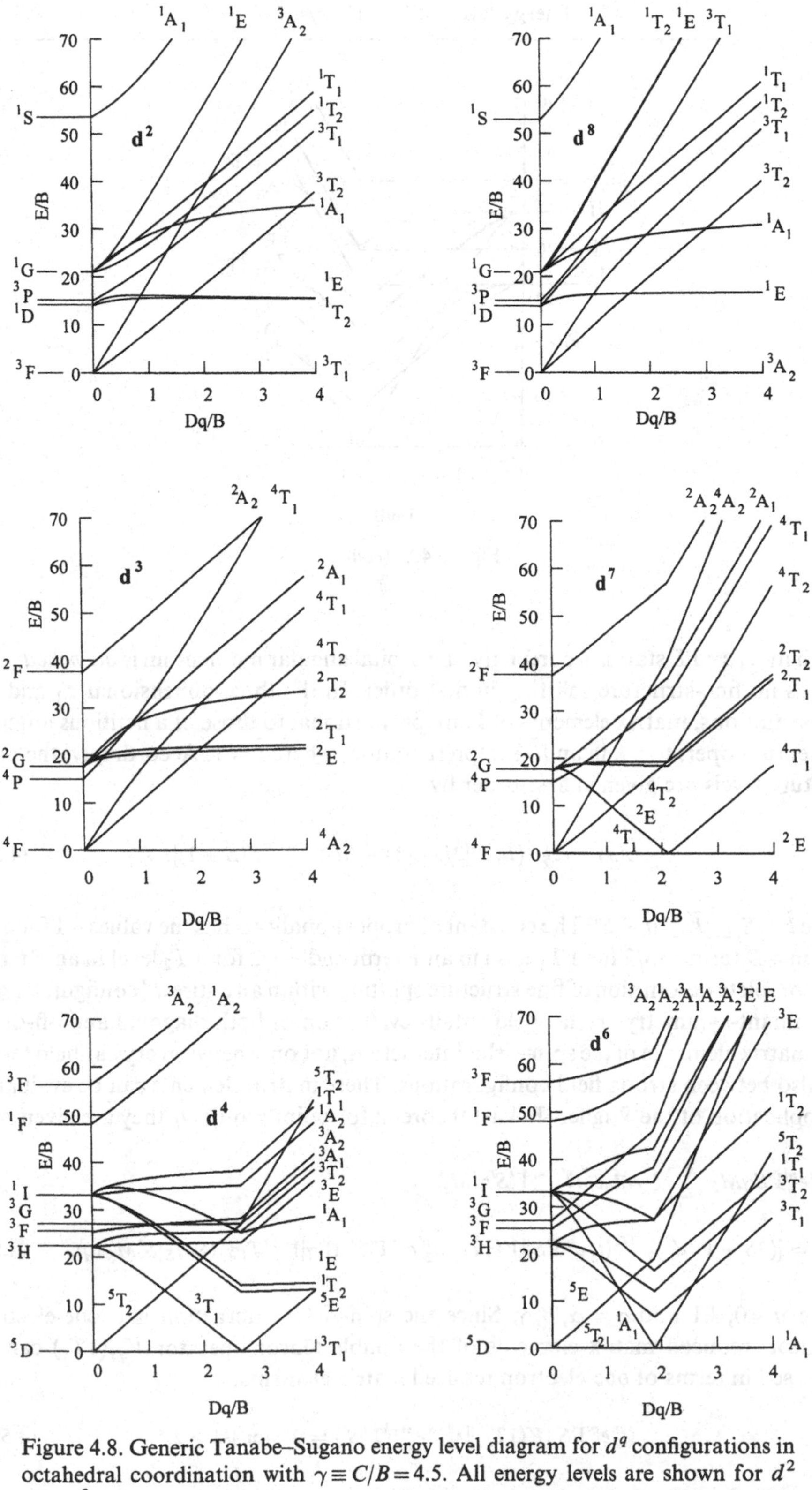

Figure 4.8. Generic Tanabe–Sugano energy level diagram for d^q configurations in octahedral coordination with $\gamma \equiv C/B = 4.5$. All energy levels are shown for d^2 and d^8, but only selected levels, including the lowest level for each irreducible representation, are shown for the remaining configurations. (*Continued overleaf.*)

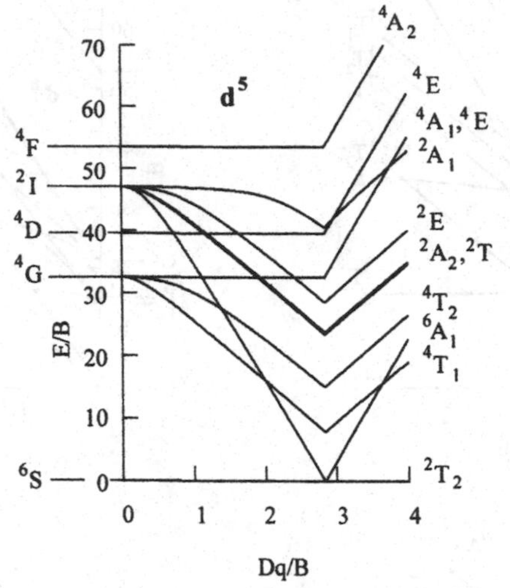

Figure 4.8. (*cont.*)

vanish in A_1 and E states; accordingly, the orbital angular momentum is *quenched*, and there is no fine-structure splitting in first-order. In the three-dimensional T_1 and T_2 representations, matrix elements of $\mathbf{L}$ are proportional to those of a fictitious angular momentum operator, $\tilde{\mathbf{l}}$, in an $|\tilde{l}\tilde{m}\rangle$ representation, where $\tilde{l} = 1$. Accordingly, the fine-structure levels are given in first-order by

$$E(\tilde{J}) = \alpha\tfrac{1}{2}\lambda(L,S)[\tilde{J}(\tilde{J}+1) - \tilde{l}(\tilde{l}+1) - S(S+1)], \tag{4.57}$$

where $\tilde{l} + S \geq \tilde{J} \geq |\tilde{l} - S|$. The constant of proportionality α has the values -1 for a T_2 level in a D term, $-3/2$ for a T_1 level in an F term and $+1/2$ for a T_2 level in an F term.

A complete description of fine structure splitting within an entire d^q configuration in a dominant-symmetry crystal-field entails evaluation of both diagonal and off-diagonal matrix elements of the spin–orbit interaction, not only between crystal-field terms but also between strong-field configurations. These matrix elements can be evaluated by application of the Wigner–Eckart theorem; for point group O_h they are given by

$$\langle t_2^n e^m \Gamma S \gamma M_S | \sum_i \xi(r_i)\mathbf{l}_i\cdot\mathbf{s}_i | t_2^{n'} e^{m'} \Gamma' S' \gamma' M'_S \rangle$$

$$= [(2S+1)(d_\Gamma)]^{-1/2} \langle t_2^n e^m \Gamma S \| V(1T_1) \| t_2^{n'} e^{m'} \Gamma' S' \rangle \langle \Gamma\gamma | \Gamma'\gamma' T_1 \bar{\gamma} \rangle \langle S M_S | S' M'_S 1 q \rangle, \tag{4.58}$$

where $q = 0, \pm 1$ and $\bar{\gamma} = \alpha, \beta, \gamma$. Since the spin–orbit interaction is a one-electron operator, reduced matrix elements of the double tensor operator $V_{q\bar{\gamma}}(1T_1)$ can be expressed in terms of one-electron reduced matrix elements,

$$\langle t_2^n e^m \Gamma S \| V(1T_1) \| t_2^{n-k} e^{m+k} \Gamma' S' \rangle = 0 \quad \text{for } |k| \geq 2, \tag{4.59a}$$

$$\langle t_2^n e^m \Gamma S \| V(1T_1) \| t_2^{n-1} e^{m+1} \Gamma' S' \rangle = C_0 \langle t_2 \| v(1T_1) \| e \rangle, \tag{4.59b}$$

$$\langle t_2^n e^m \Gamma S \| V(1T_1) \| t_2^n e^m \Gamma' S' \rangle = C_1 \langle t_2 \| v(1T_1) \| t_2 \rangle, \tag{4.59c}$$

$$\langle t_2 \| v(1T_1) \| e \rangle = -3\sqrt{2} i \zeta_{nd}, \tag{4.59d}$$

$$\langle t_2 \| v(1T_1) \| t_2 \rangle = 3 i \zeta_{nd}. \tag{4.59e}$$

Values of the numerical constants C_0 and C_1, calculated from symmetry-adapted many-electron basis functions, are tabulated by Sugano, Tanabe and Kamimura (1970). The values for a d^2 configuration are listed in Table 4.9. Off-diagonal matrix elements of the spin–orbit interaction for which $S' \neq S$ mediate states of mixed multiplicity, which are essential for explaining the occurrence of spin-forbidden transitions such as the ruby R-lines.

Table 4.9. *Spin–orbit parameters C_0 and C_1 for a d^2 configuration* [*Sugano et al.* (1970)]

$$C_1 = \frac{\langle t_2^2 S\Gamma \| V(1T_1) \| t_2^2 S'\Gamma' \rangle}{\langle t_2 \| v(1T_1) \| t_2 \rangle}$$

$S\Gamma \backslash S'\Gamma'$	$^1A_1$	$^3T_1$	^1E	$^1T_2$
$^1A_1$		$-\sqrt{2}/\sqrt{3}$		
$^3T_1$	$\sqrt{2}/\sqrt{3}$	-1	$-1/\sqrt{3}$	$-1/\sqrt{2}$
^1E		$1/\sqrt{3}$		
$^1T_2$		$1/\sqrt{2}$		

$$C_1 = \frac{\langle t_2 e S\Gamma \| V(1T_1) \| t_2 e S'\Gamma' \rangle}{\langle t_2 \| v(1T_1) \| t_2 \rangle}$$

$S\Gamma \backslash S'\Gamma'$	$^3T_1$	$^3T_2$	$^1T_1$	$^1T_2$
$^3T_1$	$1/2$	$\sqrt{3}/2$	$1/2\sqrt{2}$	$\sqrt{3}/2\sqrt{2}$
$^3T_2$	$\sqrt{3}/2$	$-1/2$	$\sqrt{3}/2\sqrt{2}$	$-1/2\sqrt{2}$
$^1T_1$	$-1/2\sqrt{2}$	$-\sqrt{3}/2\sqrt{2}$		
$^1T_2$	$-\sqrt{3}/2\sqrt{2}$	$1/2\sqrt{2}$		

$$C_0 = \frac{\langle t_2^2 S\Gamma \| V(1T_1) \| t_2 e S'\Gamma' \rangle}{\langle t_2 \| v(1T_1) \| e \rangle}$$

$S\Gamma \backslash S'\Gamma'$	$^3T_1$	$^3T_2$	$^1T_1$	$^1T_2$
$^1A_1$	$1/\sqrt{3}$			
$^3T_1$	$-1/\sqrt{2}$	$\sqrt{3}/\sqrt{2}$	$1/2$	$-\sqrt{3}/2$
^1E	$\sqrt{2}/\sqrt{3}$			
$^1T_2$	$-1/2$	$\sqrt{3}/2$		

$$C_0 = \frac{\langle t_2 e S\Gamma \| V(1T_1) \| e^2 S'\Gamma' \rangle}{\langle t_2 \| v(1T_1) \| e \rangle}$$

$S\Gamma \backslash S'\Gamma'$	$^1A_1$	$^3A_2$	^1E
$^3T_1$	$-1/\sqrt{2}$		$-1/\sqrt{2}$
$^3T_2$		1	$1/\sqrt{2}$
$^1T_1$			
$^1T_2$		$-1/\sqrt{2}$	

4.4.7 *Lower symmetry fields*

Crystal-field components of lower than cubic or tetrahedral symmetry are treated as small perturbations within each level in the dominant-symmetry approximation. These fields arise from the crystalline environment beyond the ligands, either directly from the charge distribution of more remote ions or indirectly through distortion of ligand positions. They effect a reduction in symmetry from the dominant-symmetry point group, O_h or T_d, to one of their respective subgroups. We will restrict consideration to fields of axial symmetry. The operator-equivalent crystal potential for a tetragonal field, e.g., point group C_{4v}, Table 2.22, becomes

$$\begin{aligned} V_c(\mathbf{K}) &= A_4^0\langle r^4\rangle_{3d}\langle K|\beta|K\rangle\left[O_4^0(\mathbf{K}) + 5O_4^4(\mathbf{K})\right] \\ &\quad + A_2^0\langle r^2\rangle_{3d}\langle K|\alpha|K\rangle O_2^0(\mathbf{K}) + \delta A_4^0\langle r^4\rangle_{3d}\langle K|\beta|K\rangle O_4^0(\mathbf{K}), \end{aligned} \tag{4.60}$$

where $\mathbf{K}$ is replaced by $\mathbf{L}$ in the intermediate-field approximation and by $\mathbf{l}$ in the strong-field approximation. The first term on the right-hand side of Eq. (4.60) is the dominant symmetry part of the crystal potential, characterized by a single parameter. Two additional parameters are required to characterize the tetragonal part of the crystal potential, whose diagonal matrix elements in a symmetry-adapted basis determine the additional energy-level splitting.

In the case of a trigonal distortion of the crystal-field, e.g., point group C_3, Table 2.18, it is expedient to reorient the coordinate axes so that the z-axis is parallel to a body diagonal of the cube or perpendicular to a triangular face of the octahedron. The operator-equivalent crystal potential then takes the form

$$\begin{aligned} V_c(\mathbf{K}) &= A_4^{0'}\langle r^4\rangle_{3d}\langle K|\beta|K\rangle\left[O_4^0(\mathbf{K}) + 20\sqrt{2}O_4^3(\mathbf{K})\right] \\ &\quad + A_2^{0'}\langle r^2\rangle_{3d}\langle K|\alpha|K\rangle O_2^0(\mathbf{K}) + \delta A_4^{0'}\langle r^4\rangle_{3d}\langle K|\beta|K\rangle O_4^0(\mathbf{K}), \end{aligned} \tag{4.61a}$$

$$A_4^{0'} = -\frac{2}{3}A_4^0. \tag{4.61b}$$

An example of trigonal distortion in octahedral coordination is the aluminum site in corundum, αAl_2O_3, described in §3.3.1 and illustrated in Figs. 2.5b and 3.1b. Two important laser materials based on this host mineral are Ti-sapphire ($Ti^{3+}:Al_2O_3$) and ruby ($Cr^{3+}:Al_2O_3$), with d^1 and d^3 configurations, respectively.

In general, low symmetry components of the crystal-field are comparable in strength with the spin–orbit interaction, so these two perturbations must be considered simultaneously in a complete description of the fine structure. Examples are presented in Chapter 5.

4.4.8 *Empirical parameters*

Typical values of the crystal-field splitting parameter $\Delta = 10\,Dq$ range from 10 000 to 20 000 cm^{-1} for octahedral complexes of iron-group transition-metals, and increase by one-third to one-half for each subsequent transition series. Values of Δ depend on a number of variables, including the number of d electrons, the ionization state of the transition-metal ion, the ligand charge and distance, and the nature of the ligand. As noted previously, the point-ion model is not quantitatively reliable; accordingly, empirical rules have been developed for both cations and anions [Burns (1993)] called *spectrochemical series*.

The Racah parameter B is invariably diminished from its free-ion value when a transition-metal ion is incorporated in a crystal, while the ratio C/B is approximately preserved. The *nephelauxetic* (cloud-expanding) ratio, $\beta = B/B_0$, is a measure of the strength of covalent bonding. *Nephelauxetic series* have been developed for both cations and anions, analogous to but different from the spectrochemical series.

The spectrochemical and nephelauxetic series, which are presented in §9.6.2 and §9.6.3, respectively, provide qualitative guidance, but have limited predictive capability. Ultimately, one must fit Racah parameters B and C and the crystal-field splitting parameter Δ to optical spectra for each transition-metal complex in each host crystal. Empirical values of Racah parameters, spin–orbit constants and crystal-field parameters have been compiled by Morrison (1992) for transition-metal complexes in a number of ionic crystals of potential interest as laser materials. Crystal-field energy levels of transition-metal ions in laser materials have been compiled by Kaminskii (1996).

4.5 Rare earths

4.5.1 Free-ion energy levels

Rare-earth ions are characterized by ground configurations with an open $4f$ shell. Electrostatic matrix elements for f^q configurations are tabulated by Nielson and Koster (1963) in terms of linear combinations, E^k, of Slater integrals, F^k, introduced by Racah (1949),

$$E^0 \equiv F^0 - 2F^2/45 - F^4/33 - 50F^6/1287, \tag{4.62a}$$

$$E^1 \equiv 14F^2/405 + 7F^4/297 + 350F^6/11\,583, \tag{4.62b}$$

$$E^2 \equiv F^2/2025 - F^4/3267 + 175F^6/1\,656\,369, \tag{4.62c}$$

$$E^3 \equiv F^2/135 + 2F^4/1089 - 175F^6/42\,471. \tag{4.62d}$$

There are no off-diagonal electrostatic matrix elements between terms which differ in L or S. However, since there are as many as ten duplicated terms in some configurations, it is no longer feasible to derive explicit expressions for term energies as was done for transition metals. In addition, it is essential to include off-diagonal matrix elements of the spin–orbit interaction between terms. These matrix elements can be calculated by means of Eq. (4.29) in an L, S, J, M_J basis in terms of tabulated reduced matrix elements of $\mathbf{V}^{(11)}$ [Nielson and Koster (1963)]. The procedure is then to diagonalize numerically the combined electrostatic and spin–orbit matrices within the entire ground configuration, and to adjust the Racah parameters, E^k, and the spin–orbit constant, ζ, for a least-squares fit to free-ion optical data.

Alternative reduced Slater integrals, F_k, introduced by Condon and Shortley (1935) to eliminate large denominators, are related to F^k and E^k by

$$F_0 \equiv F^0 = (7E^0 + 9E^1)/7, \tag{4.63a}$$

$$F_2 \equiv F^2/225 = (E^1 + 143E^2 + 11E^3)/42, \tag{4.63b}$$

$$F_4 \equiv F^4/1089 = (E^1 - 130E^2 + 4E^3)/77, \tag{4.63c}$$

$$F_6 \equiv F^6(25/184\,041) = (E^1 + 35E^2 - 7E^3)/462, \tag{4.63d}$$

Table 4.10. *Empirical values, in* cm^{-1}, *of Racah parameters and spin–orbit constants for trivalent rare-earth ions in* $LaCl_3$ *(after Hüfner (1978))*

Config.	Ion	F_2	F_4	F_6	ζ
$4f^1$	Ce^{3+}	—	—	—	625
$4f^2$	Pr^{3+}	304	45.9	4.45	744
$4f^3$	Nd^{3+}	319	47.9	4.82	880
$4f^4$	Pm^{3+}	337	49.9	5.27	1022
$4f^5$	Sm^{3+}	347	52.2	5.45	1168
$4f^6$	Eu^{3+}	375	55.4	5.65	1331
$4f^7$	Gd^{3+}	379	55.5	6.09	1513
$4f^8$	Tb^{3+}	405	59.1	5.83	1707
$4f^9$	Dy^{3+}	412	60.3	6.19	1920
$4f^{10}$	Ho^{3+}	424	61.7	6.35	2137
$4f^{11}$	Er^{3+}	436	64.0	6.67	2370
$4f^{12}$	Tm^{3+}	—	—	—	—
$4f^{13}$	Yb^{3+}	—	—	—	—

The parameter F_0 makes no contribution to splitting of the ground configuration. Values of F_2, F_4, F_6 and ζ, optimized to fit optical spectra, are listed in Table 4.10 for trivalent rare-earth ions in $LaCl_3$. A number of unlisted additional parameters, representing higher order free-ion interactions as well as the crystal-field, were optimized simultaneously to ensure a precise fit. A total of nineteen free-ion parameters was employed.

The free-ion energy level structures of rare-earth ions are, in general, much more complex than those of transition-metal ions; they are relatively simple only for the $4f^1$, $4f^2$, $4f^{12}$ and $4f^{13}$ configurations. Configuration $4f^1$, exemplified by Ce^{3+}, has only a single term, split into two fine-structure levels, $^2F_{5/2,7/2}$. Configuration $4f^2$, exemplified by Pr^{3+}, has the following thirteen fine-structure levels: $^1S_0$, $^3P_{0,1,2}$, $^1D_2$, $^3F_{2,3,4}$, $^1G_4$, $^3H_{4,5,6}$, $^1I_6$. The fine-structure levels are labeled not only by their J values, but also, conventionally, by the values of L and S which would be appropriate in the limit of vanishing spin–orbit interaction. The complementary configurations $4f^{13}$ and $4f^{12}$ have the same multiplet structures as $4f^1$ and $4f^2$, respectively, but inverted fine structure.

4.5.2 *Crystal-field splitting of fine-structure levels*

The free-ion parameters for rare-earth ions have a relatively invariant significance, in contrast with those for transition-metal ions, since they are nearly the same as empirical parameters derived for rare-earth ions in aqueous solution and in various ionic host crystals. Thus the parameter values listed in Table 4.10 provide a nearly complete set, for illustrative purposes, of those which have approximate universal applicability. Crystal-field splitting of fine-structure levels is only of the order of 100–500 cm^{-1} in rare-earth complexes. Accordingly, it is useful to present an energy-level diagram, due to Dieke (1968), for all fourteen trivalent rare-earth ions in $LaCl_3$, reproduced in Fig. 4.9. Crystal-field splitting is unresolved, but its extent is indicated by the width of the line representing each fine-structure level. Crystal-field split rare-earth energy levels

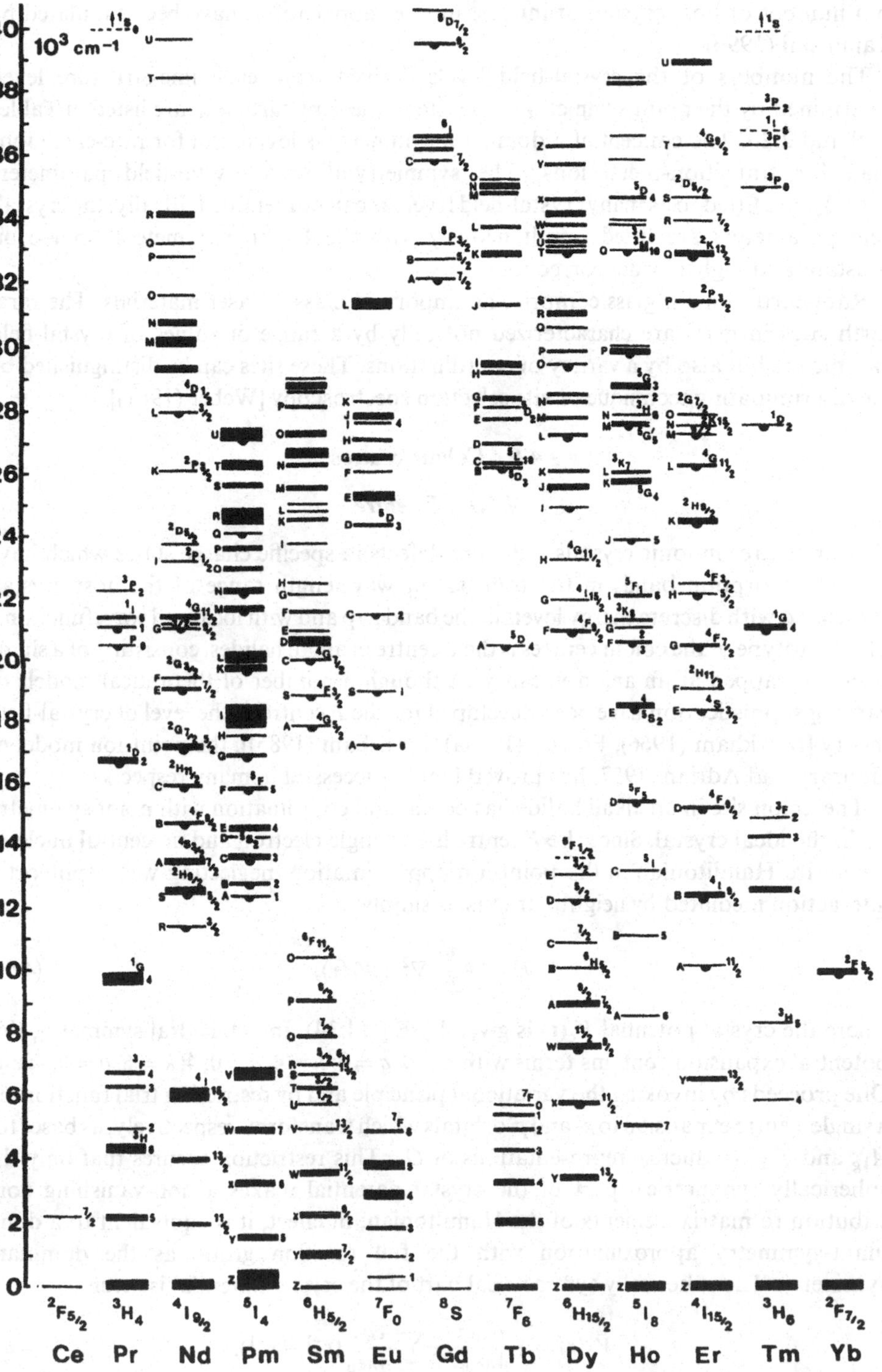

Figure 4.9. Dieke diagram for trivalent rare-earth ions in $LaCl_3$. The width of each level is a measure of its crystal-field splitting [after Henderson and Imbusch (1989)].

in a number of host crystals of interest in laser applications have been tabulated by Kaminskii (1996).

The numbers of the crystal-field levels derived from each fine-structure level, determined by the point symmetry at the site of the rare-earth ion, are listed in Tables 2.19 and 2.20. The concept of a dominant symmetry is less useful for rare-earth ions than for transition-metal ions. The symmetry-allowed crystal-field parameters $A_k^q\langle r^k\rangle_{4f}$ are fitted to as many crystal-field levels as can be identified. Ideally, the crystal-field parameters are fitted simultaneously with the Racah parameters, spin–orbit constant and higher-order corrections.

Rare-earth ions in glass comprise an important class of laser materials. The rare-earth sites in glass are characterized not only by a range of values of crystal-field parameters but also by a variety of coordinations. These sites can be distinguished by the discriminating technique of site selection spectroscopy [Weber (1981)].

4.6 Colour centres

4.6.1 F centre

Colour centres in ionic crystals are point defects in specific charge states which have optical absorption bands in the transparent wavelength range of the host crystal, associated with discrete energy levels in the band gap and with localized wavefunctions. The prototype of the colour centres is the *F* centre in alkali halides, consisting of a single electron trapped at an anion vacancy. Although a number of theoretical models of varying sophistication have been developed for the *F* centre at the level of crystal-field theory [Markham (1966), Fowler (1968a), Stoneham (1985)], the point-ion model of Gourary and Adrian (1957) has proved highly successful in many respects.

The anion site in an alkali halide has octahedral coordination with point symmetry O_h in the ideal crystal. Since the *F* centre has a single electron and no central nucleus, the entire Hamiltonian in the point-ion approximation, neglecting weak spin–orbit interaction mediated by neighbour ions, is simply

$$H = -\frac{\hbar^2}{2m}\nabla^2 + V_c(\mathbf{r}), \tag{4.64}$$

where the crystal potential $V_c(\mathbf{r})$ is given by Eq. (4.34). In octahedral symmetry, this potential expansion contains terms with $k=0, q=0$; $k=4, q=0,4$; $k=6, q=0,4$; etc. One proceeds by invoking the variational principle and by restricting trial functions in a single-centre expansion to *s*- and *p*-orbitals which transform, respectively, as bases for A_{1g} and T_{1u} irreducible representations of O_h. This restriction ensures that only the spherically symmetrical part of the crystal potential makes a non-vanishing contribution to matrix elements of the Hamiltonian; in effect, it is equivalent to a dominant-symmetry approximation with the full rotation group as the dominant symmetry. The spherically symmetrical part of the crystal potential is given by

$$\bar{V}_c(r) = -\frac{\alpha_M e^2}{4\pi\varepsilon_0 a} - \sum_{\alpha<}\frac{eq_\alpha}{4\pi\varepsilon_0}(r^{-1} - r_\alpha^{-1}), \tag{4.65}$$

where α_M is the Madelung constant and a is the nearest-neighbour distance. This potential is plotted in Fig. 4.10 for alkali halides with rock-salt structure, for which $\alpha_M = 1.747\,565\ldots$.

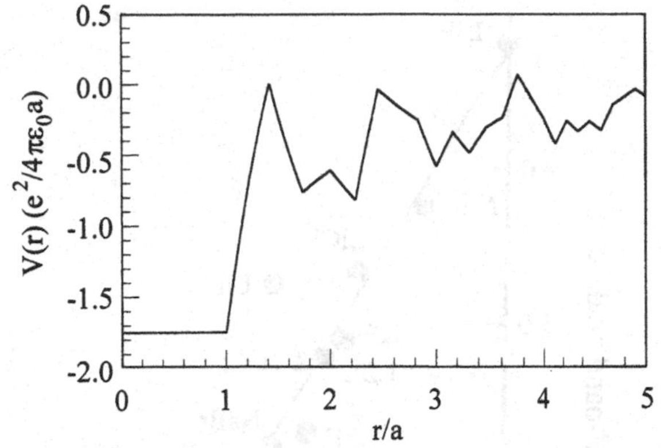

Figure 4.10. Spherically symmetric component of the point-ion potential for an F centre in an alkali halide.

The single optical absorption band associated with the F centre (*F band*) is identified with an electric-dipole allowed transition between the A_{1g} $(1s)$ ground state and the T_{1u} $(2p)$ excited state. It can be seen that the spherically symmetrical part of the point-ion potential shown in Fig. 4.10 bears some resemblance to a spherical square-well potential which confines the electron to the anion vacancy; a square-well potential with the depth and range appropriate to the F centre would support just the two bound states cited. However, the extended portion of the point-ion potential for $r > a$ binds additional states with diffuse wavefunctions, giving rise to additional weak absorption bands at higher energies, the K and L bands.

The point-ion approximation is quite successful in predicting the dependence of the transition energy on lattice parameter, as summarized by the empirical Mollwo–Ivey relation [Ivey (1947), Dawson and Pooley (1969)],

$$\Delta E = 16.8[a(A)]^{-1.772} = 0.97[2a(nm)]^{-1.772}, \tag{4.66}$$

where ΔE is the F-band energy in electron volts, $a(A)$ is the nearest-neighbour distance in Ångstrom units and $2a(nm)$ is the lattice parameter in nanometres. This relation is plotted in Fig. 4.11, together with measured F-band energies. However, experiments by Buchenauer and Fitchen (1968) revealed a systematic dependence of ΔE on the ratio of ionic radii as well, attributed to ion-size effects not reflected in the point-ion potential. These ion-size effects were subsequently explained by Bartram, Stoneham and Gash (BSG) (1968) in terms of a local model pseudopotential,

$$V_{\mathrm{p}}^{\mathrm{BSG}}(\mathbf{r}) = V_{\mathrm{PI}}(\mathbf{r}) + \sum\nolimits_{\gamma} C_{\gamma}\delta(\mathbf{r} - \mathbf{r}_{\gamma}), \tag{4.67a}$$

$$C_{\gamma} = A_{\gamma} + (E - U_{\gamma})B_{\gamma}, \tag{4.67b}$$

where A_{γ} and B_{γ} are properties of the ions, calculated and tabulated by BSG, and U_{γ} is the point-ion potential at the site of ion γ due to the remaining ions. A pseudopotential replaces the kinetic energy associated with orthogonalization of the valence wavefunction to occupied ion-core orbitals by an effective potential for use with a smooth pseudo-wavefunction [Phillips and Kleinman (1959)]. In the original BSG

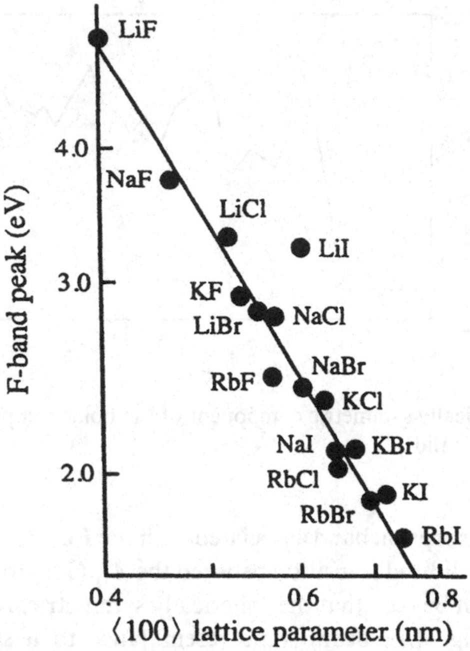

Figure 4.11. Mollwo–Ivey relation, Eq. (4.66), and experimental *F*-band energies for alkali halides [after Dawson and Pooley (1969)].

formulation, the average pseudopotential $\bar{V}$ appeared in Eq. (4.67b) instead of the total energy E, which has been used in subsequent work; they are equivalent within the approximation of a smooth pseudo-wavefunction, but the latter is more convenient. The point-ion potential itself has been interpreted as a crude pseudopotential [Gourary and Fein (1962)]. Consequently, the pseudo-wavefunctions for both the point-ion and BSG models must be used with caution in calculating properties other than transition energies; e.g., orthogonalization to occupied core orbitals is an essential prerequisite for calculation of transferred hyperfine interactions. A more thorough discussion of the pseudopotential theorem is presented in §8.4.1 in the context of effective core potentials for molecular-orbital calculations.

Lattice relaxation and polarization have a major influence on the optical properties of *F* centres. In the 1*s* ground state, the removal of repulsive forces associated with the missing anion is largely offset by the reduction of electron charge in the vacancy, but a much larger relaxation occurs in the 2*p* excited state, resulting in very diffuse wavefunctions and a crossing of 2*s* and 2*p* energy levels [Bogan and Fitchen (1970), Ham and Grevsmühl (1973)]. Competing models for the *F* centre which emphasize polarization effects include continuum and semi-continuum models. In the simplest form of the continuum model, the anion vacancy is represented by a point positive charge embedded in a dielectric continuum with dielectric constant K, and the electron is assigned an effective mass m^*. The Hamiltonian is that for a hydrogen atom with scaled wavefunctions and energy levels,

$$H = -\frac{\hbar^2}{2m^*}\nabla^2 - \frac{e^2}{(4\pi\varepsilon_0)Kr}, \qquad (4.68)$$

and either m^* or K is adjusted to make the energy difference between the $1s$ and $2p$ states equal to the F-band energy,

$$\Delta E = \frac{3(m^*/m)}{4K^2}\,\mathrm{Ry}. \tag{4.69}$$

Refinements of the continuum model include polarization by the electron charge-density distribution [Simpson (1949)] and explicit coupling to vibrational modes [Pekar (1954)]. Although the continuum model is appropriate for shallow donor states in semiconductors [Kohn (1957)], it is manifestly inconsistent for F centres, since it predicts that 90% of the ground-state electronic charge is contained in a sphere whose radius is the nearest-neighbour distance; nevertheless, it remains popular by virtue of its simplicity. The semi-continuum model, which combines a spherical cavity in the dielectric continuum with a Coulomb-potential tail, provides a much more realistic representation of the F centre [Simpson (1949), Krumhansl and Schwartz (1953)].

4.6.2 Laser-active colour centres

Colour centres in alkali halides have been exploited successfully in tunable, optically pumped infrared lasers in the wavelength range 0.8–4 μm. Although the F centre is the prototype of the colour centres, it is not laser-active. Stability against ionization or reorientation is a critical requirement of colour centres for laser applications. Colour centres which have been employed successfully in lasers include perturbed F centres, F-aggregate centres and $Tl^0(1)$ centres [Mollenauer (1987, 1992)]. These centres are illustrated schematically in Fig. 4.12, and their electronic structures are considered in subsequent sections. In contrast with the F centre, these complex colour centres do not necessarily preserve the point symmetry of their site in the host crystal.

4.6.3 Perturbed F centres

F centres in alkali halides can be stabilized by intimate association with one or two monovalent cation impurities or with two distinct radiation-induced defects (§9.2.1), giving rise to F_A, F_B, F^* and F^{**} centres, respectively [Mollenauer (1987, 1992)]. The F_A centre will be considered here as an example. The character table for point group C_{4v}, which describes the symmetry of the F_A centre, is displayed in Table 2.22, and the compatibility table of O_h with C_{4v} in Table 2.23. It is evident from these tables, together with Table 2.20, that the perturbing cation impurity in the F_A centre splits the triply degenerate T_{1u} $(2p)$ excited state of the F centre into a non-degenerate A_2 state and a doubly degenerate E state. The wavefunction for the A_2 state is the $2p$ orbital oriented in the direction of the perturbing cation, and the wavefunctions for the E state are the two remaining $2p$ orbitals, with perpendicular orientations. The splitting of the excited-state energy level is very evident in optical absorption spectra, and is typically about 0.2 eV, or about 10% of the F-band energy [Lüty (1968)].

At the level of the point-ion model in an undistorted lattice, there is no distinction between the F centre and the F_A centre; accordingly, one must look deeper for an explanation of the energy-level splitting. Part of the explanation may lie in differential distortion accompanying lattice relaxation in the ground state. Since the perturbing cation is typically smaller than the cations of the host crystal; e.g., KCl : Na^+ and

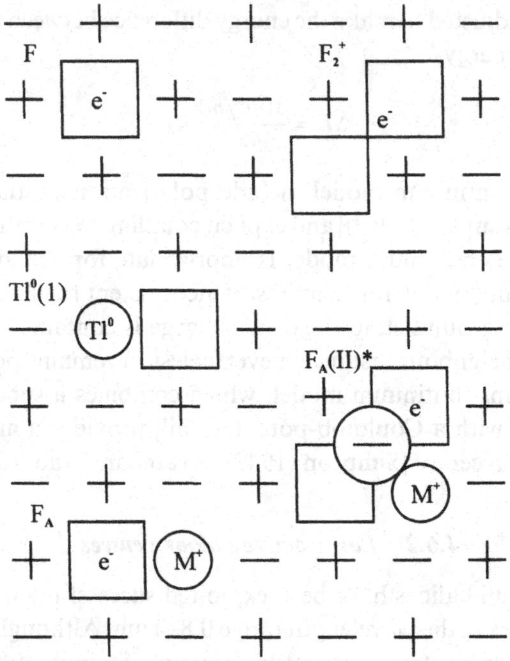

Figure 4.12. Schematic representation of the F centre and selected laser-active colour centres in alkali halides.

KCl : Li^+, its displacement from the vacancy may be different and the resulting dipole moment may give rise to an axial component of the crystal-field. However, it appears that the dominant mechanism is an ion-size effect, which has been explained satisfactorily [Weber and Dick (1969), Alig (1970)] in terms of the BSG local model pseudopotential, Eqs. (4.67).

F_A centres are classified as type I or II, depending on the nature of their relaxed excited states [Lüty (1968)]. The relaxed excited state of the F_A(I) centre, exemplified by KCl : Na, is similar to that of the F centre, but that of the F_A(II) centre, exemplified by KCl : Li, exhibits a much more extreme behaviour. In the latter case, a neighbouring anion occupies a saddle-point position leaving a split anion vacancy, as illustrated in Fig. 4.12, with a reduction in symmetry to C_{2v} and a pronounced reduction in optical transition energy. The character table of point group C_{2v} is displayed in Table 2.24, and its compatibility table with point group O_h in Table 2.23; these tables suggest, and Table 2.20 confirms, that no orbital degeneracy remains. Only F_A(II) centres are laser active.

4.6.4 F_2^+ centre

The F_2^+ centre in alkali halides consists of a single electron trapped at a pair of adjacent anion vacancies oriented along a ⟨110⟩ direction, with point symmetry D_{2h}. Although other F-aggregate centres have been modeled successfully by a single-centre expansion with a point-ion potential [Kern *et al.* (1981), DeLeo *et al.* (1981)], the preferred model for the F_2^+ centre appears to be a continuum model [Herman *et al.* (1956), Aegerter and Lüty (1971)]. It exploits the exact solution for the H_2^+ molecular ion in spheroidal coordinates by Bates, Ledsham and Stewart (BLS) (1953) by embedding the molecule ion in a dielectric continuum and by adjusting the dielectric constant K and internuclear separation r_{12} to fit the two absorption bands accessible from the ground state, $1s\sigma_g \to 2p\sigma_u, 2p\pi_u$. The energy levels are calculated from the recipe

$$E_{F_2^+} = K^{-2} E_{H_2^+}(r_{12}), \tag{4.70a}$$

$$r_{12} = K^{-1} R_{12}, \tag{4.70b}$$

where R_{12} is the actual separation of the two vacancies. The effective mass ratio, m^*/m, is assumed to be unity in this model. The parameter r_{12} is readjusted in the relaxed excited state to fit the energy of the emission band, $2p\sigma_u \to 1s\sigma_g$. With this readjusted value, three additional bands observed in excited-state absorption in NaF and NaCl, $2p\sigma_u \to 3d\sigma_g, 2s\sigma_g, 3d\pi_g$, were found in nearly the predicted energy range, but not in the predicted order [Mollenauer (1979)].

The energy levels in the continuum model are labeled by the united-atom designation; i.e., atomic quantum numbers of the He^+ ion together with irreducible representations of $D_{\infty h}$, the group of the molecule ion, for which the character table is displayed in Table 2.25. The latter has both one- and two-dimensional representations, whereas the exact point group, D_{2h}, has only one-dimensional representations; thus the predicted double degeneracy of the π states must be removed by components of the crystal field, but, presumably, this crystal-field splitting is not resolved in the optical spectra. Point group D_{2h} is a direct product of its subgroups, $D_{2h} = D_2 \otimes C_i$; the character table of point group D_2 is displayed in Table 2.26 and states of the F_2^+ centre, labeled by the irreducible representations of both $D_{\infty h}$ and D_{2h}, are listed in Table 4.11.

Both the pumping and lasing transitions of the F_2^+-centre laser utilize the two lowest states, $2p\sigma_u \leftrightarrow 1s\sigma_g$, as shown in Fig. 4.13, and the set of alkali halides provides a wide range of transition energies in the infrared.

4.6.5 *$Tl^0(1)$ centre*

The $Tl^0(1)$ centre in alkali halides, consisting of a neutral thallium atom at a cation site adjacent to an anion vacancy, was first identified in electron spin resonance [Goovaerts *et al.* (1981)]. A stable laser-active centre in KCl : *Tl*, at first identified as a *Tl*-F_A centre [Gellermann *et al.* (1981a,b)], was subsequently shown by magneto-optical techniques

Table 4.11. *Symmetry designations of states of the F_2^+ centre in ascending order*

$D_{\infty h}$	D_{2h}
$1s\sigma_g$	A_{1g}
$2p\sigma_u$	B_{1u}
$3d\sigma_g$	A_{1g}
$2p\pi_u$	B_{2u}, B_{3u}
$2s\sigma_g$	A_{1g}
$3d\pi_g$	B_{2g}, B_{3g}

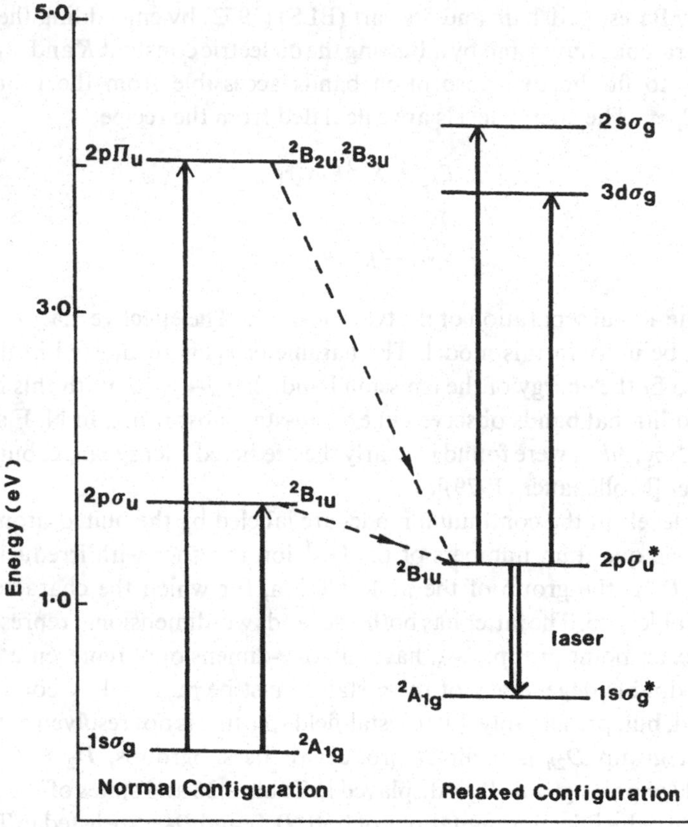

Figure 4.13. Energy levels and transitions of the laser-active F_2^+ centre in NaF [adapted from Mollenauer (1979)].

to be identical with the $Tl^0(1)$ centre [Ahlers *et al.* (1983)]. The distinction is somewhat semantic, since the wavefunction of the valence electron is not precisely localized [Fockele *et al.* (1985)], but the $Tl^0(1)$-centre designation provides the more useful description.

The strong- and intermediate-field approximations have been considered previously for transition-metal ions, and the weak-field approximation for rare-earth ions. Heavy-metal-ion impurities require still another approximation regime in which the spin–orbit, crystal-field and electrostatic interactions are all comparable. Although the latter interaction is not involved in the one-electron configuration of the $Tl^0(1)$ centre, it is involved, for example, in the two-electron configuration of the $Pb^0(2)$ centre in SrF_2 [Bartram *et al.* (1989)] The electronic structure of the $Tl^0(1)$ centre has been elucidated successfully at the level of crystal-field theory [Mollenauer *et al.* (1983b)] by simultaneous diagonalization of the crystal-potential and spin–orbit matrices within the $6p^1$ ground configuration of Tl^0. The point symmetry of the centre is C_{4v}, the same as that of the F_A centre, but the prominence of spin–orbit interaction necessitates the use of double-group representations, displayed in Table 2.27. The full rotation group compatibility table for double point group C_{4v}^* is displayed in Table 2.28.

The only non-constant term in the potential expansion, Eq. (4.37), which transforms as the identity representation of point group C_{4v} and has non-vanishing matrix elements within the $6p^1$ ground configuration of Tl^0, is the term with $k=2$ and $q=0$. This term may be replaced by its operator equivalent from Table 4.1,

$$V_{20}(\mathbf{r}) \to \gamma O_2^0(\mathbf{l}) = \gamma[3l_z^2 - l(l+1)], \tag{4.71}$$

where $l=1$ and the constant γ is retained as an adjustable parameter.

It is evident from Eq. (4.71) that the crystal potential is diagonal in an $|lsm_lm_s\rangle$ representation, with diagonal matrix elements $\gamma[3m_l^2 - l(l+1)]$, whereas, from Eq. (4.25), the spin–orbit interaction is diagonal in an $|lsjm_j\rangle$ representation with diagonal matrix elements $\zeta_{6p}[j(j+1) - l(l+1) - s(s+1)]$. Matrix elements of the crystal potential, transformed to an $|lsjm_j\rangle$ representation by application of Eq. (4.24), are then given by

$$\langle lsjm_j|V_c|lsj'm_j'\rangle = \delta_{m_jm_j'}\gamma \sum_{m_l=-l}^{l}\sum_{m_s=-s}^{s}[3m_l^2 - l(l+1)]\langle jm_j|lsm_lm_s\rangle\langle lsm_lm_s|j'm_j\rangle. \tag{4.72}$$

Clebsch–Gordon coefficients for $l=1$ and $s=1/2$ are listed in Table 2.9.

The matrix of the combined spin–orbit and crystal-field interactions within the ground configuration is partitioned into two identical blocks of the form

$$H = \begin{pmatrix} \Delta+\gamma & 0 & 0 \\ 0 & \Delta-\gamma & \sqrt{2}\gamma \\ 0 & \sqrt{2}\gamma & 0 \end{pmatrix}, \tag{4.73}$$

for positive and negative values of m_j, respectively, where $\Delta \equiv \frac{3}{2}\zeta_{6p}$ is the fine-structure splitting in the absence of the crystal-field. The resulting energy levels, labeled by irreducible representations of double point group C_{4v}, are

$$E(E_{3/2}) = \Delta + \gamma, \tag{4.74a}$$

$$E(E_{1/2}) = \tfrac{1}{2}(\Delta-\gamma) \pm \tfrac{1}{2}\sqrt{(\Delta-\gamma)^2 + 8\gamma^2}, \tag{4.74b}$$

and each level retains a two-fold Kramers degeneracy. The parameters Δ and γ are fitted to optical transitions enabled by odd-parity components of the crystal potential. In KCl : Tl, the value of Δ is found to be 5/6 of its free-atom value, 0.97 eV, for reasons analogous to the nephelauxetic effect, and the ratio γ/Δ has the value 0.5 in absorption and half that in emission [Mollenauer *et al.* (1983b)]. The optical transition between the two lowest states, with energy

$$h\nu = \sqrt{(\Delta-\gamma)^2 + 8\gamma^2}, \tag{4.75}$$

is utilized in the $Tl^0(1)$-centre laser.

5

Spectra of ions in crystals

5.1 Theory of optical transitions

Optical spectra of ions in crystals share features with the free-ion spectra from which they are derived, but are substantially modified by the crystalline environment in a variety of ways. These include the effect of the dielectric medium on the electromagnetic field, effects of the crystal-field on selection rules and intensities, and effects of electron–lattice coupling on spectral line shapes and intensities.

We begin with the semi-classical approximation in which the electronic system is described quantum-mechanically but the time-dependent electromagnetic field which induces transitions between its energy levels is described classically. Quantization of the electromagnetic field is considered subsequently in connection with spontaneous emission. Electron–lattice coupling and spectral intensities are the topics of §5.2 and §5.3, respectively. Examples are presented in §5.4. Line-shape functions are discussed in §5.5 and nonlinear susceptibilities in §5.6.

5.1.1 Free-ion transition probabilities

The Hamiltonian of the ion in an electromagnetic field is

$$H = H_0 + H'(t), \tag{5.1}$$

where H_0 is the Hamiltonian of the ion in the absence of electromagnetic radiation, given by Eq. (4.1), and $H'(t)$ is given approximately, in a Coulomb gauge with scalar potential ϕ chosen to vanish, by

$$H'(t) \cong \frac{e}{m}\sum_{i=1}^{N}[\mathbf{A}(\mathbf{r}_i, t)\cdot\mathbf{p}_i + \mathbf{B}(\mathbf{r}_i, t)\cdot\hbar\mathbf{s}_i], \tag{5.2a}$$

$$\mathbf{p}_i = -i\hbar\nabla_i, \tag{5.2b}$$

$$\mathbf{B} = \nabla\times\mathbf{A}, \tag{5.2c}$$

$$\mathbf{E} = -\frac{\partial\mathbf{A}}{\partial t}, \tag{5.2d}$$

$$\nabla\cdot\mathbf{A} = 0, \tag{5.2e}$$

$$\nabla^2\mathbf{A} - \frac{n^2}{c^2}\frac{\partial^2\mathbf{A}}{\partial t^2} = 0, \tag{5.2f}$$

where **B** is the magnetic flux density, **E** is the electric field intensity, **A** is the magnetic vector potential and $n \equiv \sqrt{\kappa}$ is the index of refraction.

In the special case of a harmonic perturbation, defined by

$$H'(t) = H^{(0)} \exp(-i\omega t) + c.c., \tag{5.3}$$

first-order time-dependent perturbation theory yields the following expression for the transition probability per unit time from state m to state k:

$$w_{m\to k} = \frac{2\pi}{\hbar} |\langle k|H^{(0)}|m\rangle|^2 \delta(E_k - E_m \mp \hbar\omega). \tag{5.4}$$

Equation (5.4), which contains a δ-function ensuring energy conservation, may be employed either for monochromatic light and a continuum of final energy states, or for discrete energy levels and a continuous frequency distribution. Only first-order time-dependent perturbation theory is employed in the present chapter, except for §5.6.

We assume a plane electromagnetic wave,

$$\mathbf{A}(\mathbf{r}, t) = \mathbf{A}^{(0)} \exp[i(\mathbf{k}\cdot\mathbf{r} - \omega t)] + c.c., \tag{5.5a}$$

$$\mathbf{B}(\mathbf{r}, t) = \mathbf{B}^{(0)} \exp[i(\mathbf{k}\cdot\mathbf{r} - \omega t)] + c.c., \tag{5.5b}$$

$$\mathbf{E}(\mathbf{r}, t) = \mathbf{E}^{(0)} \exp[i(\mathbf{k}\cdot\mathbf{r} - \omega t)] + c.c., \tag{5.5c}$$

and proceed by expanding the exponential function in powers of $\mathbf{k}\cdot\mathbf{r}$, which is of order 10^{-3} since the dimensions of the ion are very small compared with the wavelength of visible light. The contribution of the leading term in the expansion to the perturbation $H'(t)$ is then

$$H_1'(t) = \frac{e}{m} \sum_{i=1}^{N} \mathbf{A}^{(0)} \cdot \mathbf{p}_i \exp(-i\omega t) + c.c. \tag{5.6}$$

It follows from Eq. (4.1) and from commutation relations of the form

$$[x, p_x] = i\hbar \tag{5.7}$$

that $\mathbf{r}_i$ and $\mathbf{p}_i$ are related by

$$\mathbf{p}_i = -\frac{im}{\hbar}[\mathbf{r}_i, H], \tag{5.8}$$

and their matrix elements by

$$\langle k|\mathbf{p}_i|m\rangle = im\omega_{km}\langle k|\mathbf{r}_i|m\rangle = \pm im\omega\langle k|\mathbf{r}_i|m\rangle, \tag{5.9a}$$

$$\omega_{km} \equiv \frac{E_k - E_m}{\hbar}, \tag{5.9b}$$

where the latter equality in Eq. (5.9a) is a consequence of energy conservation; the upper sign refers to absorption and the lower sign to emission. In addition, from Eq. (5.2) we obtain

$$\mathbf{E}^{(0)} = i\omega\mathbf{A}^{(0)}. \tag{5.10}$$

With the help of Eqs. (5.9) and (5.10), Eq. (5.6) may be replaced by

$$H_1'(t) = \sum_{i=1}^{N} e\mathbf{E}^{(0)} \cdot \mathbf{r}_i \exp(-i\omega t) + c.c. \tag{5.11}$$

This expression defines the *electric-dipole approximation.*

We now consider the second term in the series expansion of $\exp(i\mathbf{k}\cdot\mathbf{r})$ in Eqs. (5.5). Its contribution to the perturbation $H'(t)$ is

$$H_2'(t) = \frac{ie}{m} \sum_{i=1}^{N} (\mathbf{k}\cdot\mathbf{r}_i)(\mathbf{A}^{(0)}\cdot\mathbf{p}_i) \exp(-i\omega t) + c.c. \tag{5.12}$$

This expression can be transformed as follows:

$$\begin{aligned}(\mathbf{k}\cdot\mathbf{r}_i)&(\mathbf{A}^{(0)}\cdot\mathbf{p}_i)\\ &= \frac{1}{2}\left[(\mathbf{k}\cdot\mathbf{r}_i)(\mathbf{A}^{(0)}\cdot\mathbf{p}_i) - (\mathbf{k}\cdot\mathbf{p}_i)(\mathbf{A}^{(0)}\cdot\mathbf{r}_i)\right] + \frac{1}{2}\left[(\mathbf{k}\cdot\mathbf{r}_i)(\mathbf{A}^{(0)}\cdot\mathbf{p}_i) + (\mathbf{k}\cdot\mathbf{p}_i)(\mathbf{A}^{(0)}\cdot\mathbf{r}_i)\right]\\ &= \frac{1}{2}(\mathbf{k}\times\mathbf{A}^{(0)})\cdot(\mathbf{r}_i\times\mathbf{p}_i) - \frac{im}{2\hbar}\mathbf{k}\cdot[\mathbf{r}_i\mathbf{r}_i, H]\cdot\mathbf{A}^{(0)},\end{aligned} \tag{5.13}$$

where we have applied a vector identity and Eq. (5.8). The matrix element of the commutator in the last line satisfies

$$\langle k|[\mathbf{r}_i\mathbf{r}_i, H]|m\rangle = -\hbar\omega_{km}\langle k|\mathbf{r}_i\mathbf{r}_i|m\rangle = \mp\hbar\omega\langle k|\mathbf{r}_i\mathbf{r}_i|m\rangle, \tag{5.14}$$

where the latter equality follows from energy conservation. From Eq. (5.2c) we obtain

$$\mathbf{B}^{(0)} = i\mathbf{k}\times\mathbf{A}^{(0)}. \tag{5.15}$$

The second term in brackets in Eq. (5.2a) must also be incorporated in the perturbation at this level of approximation, with only the leading term retained in the expansion of $\exp(i\mathbf{k}\cdot\mathbf{r})$. With the help of the foregoing relations, the additional contributions to the time-dependent perturbation can be expressed in the form

$$H_2'(t) = \sum_{i=1}^{N}\left[\frac{e\hbar}{2m}\mathbf{B}^{(0)}\cdot(\mathbf{l}_i + 2\mathbf{s}_i) \pm \frac{ie}{2}\mathbf{k}\cdot\mathbf{r}_i\mathbf{r}_i\cdot\mathbf{E}^{(0)}\right]\exp(-i\omega t) + c.c., \tag{5.16}$$

where the two terms in brackets define, respectively, the *magnetic-dipole* and *electric-quadrupole* contributions. Terms of still higher degree in the expansion of $\exp(i\mathbf{k}\cdot\mathbf{r})$ are too small to be of practical importance.

5.1.2 *Free-ion selection rules*

Symmetry imposes severe constraints, called *selection rules*, on the matrix elements of the time-dependent perturbation $H'(t)$. The Wigner–Eckart theorem, discussed in §2.4.5, provides a convenient tool for investigating these constraints. In the present section, selection rules in the absence of the crystal field are considered for each contribution to $H'(t)$ at successive levels of approximation of the atomic structure.

The relevant matrix element for the electric-dipole part of the perturbation, Eq. (5.11), is $\langle k|\mathbf{r}_i|m\rangle$. We can deduce from Eqs. (2.60) that the operator $\mathbf{r}_i$ is a spherical tensor operator of the first rank with respect to the orbital angular momentum operator $\mathbf{l}_i$, with spherical components r_{iM}, $M=0, \pm 1$, defined by

$$r_{i,\pm 1} \equiv \mp \frac{1}{\sqrt{2}}(x_i \pm iy_i), \tag{5.17a}$$

$$r_{i0} \equiv z_i, \tag{5.17b}$$

and that $\mathbf{r}_i$ is a scalar with respect to $\mathbf{s}_i$, with which it commutes. Since only one-electron operators are involved, antisymmetrization is irrelevant at the level of the central-field approximation and the matrix element is non-vanishing only between wavefunctions which differ in a single spin-orbital. From the Wigner–Eckart theorem, Eq. (2.63), the matrix element is given by

$$\langle n'l'm_l'm_s'|r_{iM}|nlm_lm_s\rangle = \frac{1}{\sqrt{2l'+1}}\langle n'l'\|r_i\|nl\rangle \delta_{m_s'm_s}\langle l1m_lM|l'm_l'\rangle. \tag{5.18}$$

The Clebsch–Gordon coefficient $\langle l1m_lM|l'm_l'\rangle$ vanishes unless the triangle condition, $\Delta(l1l')$, is satisfied. In addition, the orbitals must have opposite parity, since $\mathbf{r}_i$ is a polar vector,

$$(-1)^{l'} = -(-1)^l. \tag{5.19}$$

These constraints are summarized by the following selection rule for electric-dipole-allowed transitions in the central-field approximation between configurations of opposite parity:

$$\Delta l = \pm 1, \tag{5.20}$$

The operator $\mathbf{r}_i$ is also a spherical tensor operator of the first rank with respect to the total orbital angular momentum operator $\mathbf{L}$, and a scalar with respect to the total spin angular momentum operator $\mathbf{S}$. Application of the Wigner–Eckart theorem then provides the following selection rules for electric-dipole-allowed transitions between terms in Russell–Saunders coupling, which must belong to configurations of opposite parity:

$$\Delta L = 0, \pm 1, \text{ excluding } 0 \rightarrow 0, \tag{5.21a}$$

$$\Delta S = 0, \tag{5.21b}$$

The selection rule in Eq. (5.20) is still nearly satisfied at this level of approximation. However, the parity selection rule for electric-dipole-allowed transitions (*Laporte's rule*) is rigorous for the free ion and applies at all levels of approximation, since all of the states of a configuration have the same parity and configuration interaction is restricted to configurations of the same parity.

Finally, $\mathbf{r}_i$ is a spherical tensor operator of the first rank with respect to the total angular momentum operator $\mathbf{J}$, and the selection rule for electric-dipole-allowed transitions between fine-structure levels belonging to configurations of opposite parity is:

$$\Delta J = 0, \pm 1, \text{ excluding } 0 \rightarrow 0. \tag{5.22}$$

Again, the preceding selection rules, Eqs. (5.19) and (5.20), are nearly satisfied; however, the *multiplicity selection rule*, Eq. (5.21b), is a poor approximation for rare-earth and heavy-metal ions, by virtue of their strong spin–orbit interactions.

The relevant matrix elements for the magnetic-dipole part of the perturbation in Eq. (5.16) are $\langle k|\mathbf{l}_i|m\rangle$ and $\langle k|\mathbf{s}_i|m\rangle$, and the corresponding angular momentum operators, $\mathbf{l}_i$ and $\mathbf{s}_i$, are spherical tensor operators of the first rank. However, they differ from $\mathbf{r}_i$ in one essential feature; they are both axial vectors or pseudo-vectors, which are even under inversion. Accordingly, they mediate transitions only between configurations of the same parity and the selection rule for magnetic-dipole-allowed transitions in the central-field approximation is

$$\Delta l = 0, \text{ excluding } 0 \rightarrow 0. \tag{5.23}$$

Since $\mathbf{l}_i$ is a spherical tensor of the first rank with respect to $\mathbf{L}$, and $\mathbf{s}_i$ with respect to $\mathbf{S}$, the selection rules for magnetic-dipole-allowed transitions between terms in Russell–Saunders coupling belonging to configurations of the same parity are

$$\Delta L = 0, \pm 1, \text{ excluding } 0 \rightarrow 0, \tag{5.24a}$$

$$\Delta S = 0, \pm 1, \text{ excluding } 0 \rightarrow 0. \tag{5.24b}$$

As before, Eq. (5.23) is still nearly satisfied.

Finally, since both $\mathbf{l}_i$ and $\mathbf{s}_i$ are spherical tensors of the first rank with respect to $\mathbf{J}$, the selection rules for magnetic-dipole-allowed transitions between fine-structure levels belonging to configurations of the same parity are given by Eq. (5.22).

The relevant matrix element for the electric-quadrupole part of the perturbation in Eq. (5.16) is $\langle k|\mathbf{r}_i\mathbf{r}_i|m\rangle$. One linear combination of the six distinct components of the operator $\mathbf{r}_i\mathbf{r}_i$ can be eliminated from consideration, since

$$r^2\mathbf{k} \cdot \mathbf{E}^{(0)} = 0 \tag{5.25}$$

by virtue of transverse polarization. The remaining five linearly independent combinations then transform as a spherical tensor operator of the second rank with respect to $\mathbf{l}_i$ (see Table 4.4) and as a scalar with respect to $\mathbf{s}_i$. Since the operator is even under inversion, it mediates transitions only between configurations of the same parity. Application of the Wigner–Eckart theorem provides the following selection rule for electric-quadrupole-allowed transitions between configurations of the same parity in the central-field approximation:

$$\Delta l = 0, \pm 2, \text{ excluding } 0 \rightarrow 0, \tag{5.26}$$

Selection rules for electric-quadrupole-allowed transitions between terms in Russell–Saunders coupling belonging to configurations of the same parity are

$$\Delta L = 0, \pm 1, \pm 2, \text{ excluding } 0 \rightarrow 0, 0 \rightarrow 1, 1 \rightarrow 0, \tag{5.27a}$$

$$\Delta S = 0, \tag{5.27b}$$

and Eq. (5.26) is nearly satisfied.

Finally, the selection rule for electric-quadrupole-allowed transitions between fine-structure levels belonging to configurations of the same parity is

$$\Delta J = 0, \pm 1, \pm 2, \text{ excluding } 0 \rightarrow 0, 0 \rightarrow 1, 1 \rightarrow 0, \tag{5.28}$$

and Eqs. (5.26) and (5.27) are nearly satisfied.

Since the transition probability for electric-dipole transitions is five or six orders of magnitude greater than for magnetic-dipole and electric-quadrupole transitions, the latter are ordinarily considered only when the former are rigorously forbidden.

5.1.3 Crystal-field selection rules

The point-group analogue of the Wigner–Eckart theorem [Koster (1958)] is

$$\langle \lambda k | T_i^{(\mu)} | \nu j \rangle = d_\lambda^{-1/2} \sum_{\tau_\lambda} \langle \lambda \tau_\lambda \| T^{(\mu)} \| \nu \rangle \langle \mu \nu i j | \lambda \tau_\lambda k \rangle^*, \tag{2.71}$$

where the tensor operator $T_i^{(\mu)}$ is defined by

$$O_R T_i^{(\mu)} O_R^{-1} = \sum_{j=1}^{d_\mu} T_j^{(\mu)} D_{ji}^{(\mu)}(R), \tag{2.72}$$

the coupling coefficient $\langle \mu \nu i j | \lambda \tau_\lambda k \rangle$ is an element of the unitary matrix $\mathbf{U}^+$, defined in §2.2.5, which effects the reduction of the Kronecker product representation $\mathbf{D}^{(\mu \times \nu)}(R)$, and the index τ_λ distinguishes multiple occurrences of representation λ in that reduction. Selection rules are then derived from the condition that the coupling coefficient vanishes unless representation λ is contained at least once in the reduction of the Kronecker product representation.

We proceed to derive selection rules for dominant-symmetry crystal fields. There is a fundamental difference between octahedral and cubic coordinations (point group O_h), which preserve inversion symmetry, and tetrahedral coordination (point group T_d), which removes it. Laporte's parity selection rule continues to apply in perfect octahedral or cubic coordination, so electric-dipole transitions within the ground configuration are possible only if additional odd components of the crystal field are provided by local distortions, both static and dynamic, or by more remote ions. Since odd crystal-field components are inherent in tetrahedral coordination, no additional perturbations are required, and electric-dipole transitions within the ground configuration are more strongly allowed than for octahedral or cubic coordination. However, in both cases electric-dipole transitions within the ground configuration are appreciably weaker than those between configurations of opposite parity, since the crystal potential is a small perturbation on the central-field model in all cases and Eq. (5.20) is still nearly satisfied.

It is apparent from Tables 2.12 and 2.13 that the components of the electric-dipole operator $\mathbf{r}_i$ transform as bases for the T_{1u} representation of O_h and the T_2 representation of T_d. Table 2.21 then provides the necessary information concerning the irreducible representations contained in the reduction of the Kronecker product representations. Since both groups under consideration are simply reducible, the index τ_λ in Eq. (2.71) is irrelevant. In the present section, consideration of electric-dipole selection rules is restricted to point group T_d.

In the strong-field approximation, we are concerned first with transitions between orbitals. The transition $t_2 \leftrightarrow e$ is allowed, since

$$T_2 \times E = T_1 + T_2. \tag{5.29}$$

The transition $t_2 \leftrightarrow t_2$ is also allowed, but not $e \leftrightarrow e$. Transitions are possible only between strong-field configurations which differ in at most one spin-orbital; e.g., for a d^3 configuration, $t_2^3 \leftrightarrow t_2^2 e \leftrightarrow t_2 e^2 \leftrightarrow e^3$ transitions are allowed, but not $t_2^3 \leftrightarrow t_2 e^2$, $t_2^2 e \leftrightarrow e^3$ or $t_2^3 \leftrightarrow e^3$. Transitions between terms in the strong-field approximation are also constrained by the Kronecker-product rule. In addition, the multiplicity selection rule is operative,

$$\Delta S = 0. \tag{5.30}$$

In the intermediate-field approximation, Eqs. (5.20) and (5.21) are nearly satisfied in addition to the Kronecker-product rule for transitions between crystal-field levels. In the weak-field approximation, applicable to rare-earth ions, Eqs. (5.20), (5.21) and (5.22) are all nearly satisfied in addition to the Kronecker-product rule. For configurations with an odd number of electrons, crystal-field levels split from fine-structure levels are labeled by double-group representations; the reductions of their Kronecker product representations are tabulated by Koster *et al.* (1963).

It can be seen from Tables 2.12 and 2.13 that components of the magnetic-dipole operator $\mathbf{l}_i$ transform as bases for the T_{1g} representation of O_h and the T_1 representation of T_d, while components of the electric quadrupole operator $\mathbf{r}_i\mathbf{r}_i$ transform as bases for the E_g and T_{2g} representations of O_h and the E and T_2 representations of T_d. Since both operators are even under inversion, they mediate transitions between states of the same parity, including transitions within the ground configuration. Selection rules for these operators can be derived from the Kronecker-product rule in a fashion analogous to that for electric-dipole operators. Eq. (5.24b) remains the selection rule for transitions mediated by the magnetic-dipole operator $\mathbf{s}_i$.

5.1.4 *Electric-dipole transitions*

Symmetry considerations provide selection rules, but not the absolute rates of allowed transitions. Electric-dipole-allowed transitions are assumed in the present section. The reduced matrix elements in Eqs. (5.18) and (2.71) are determined by evaluating the left-hand side of each equation for chosen values of the quantum numbers m_l, M, m_l' and i, j, k, respectively, utilizing tabulated Clebsch–Gordon coefficients [Rotenberg *et al.* (1959), Weissbluth (1978)] and tabulated point-group coupling coefficients [Koster *et al.* (1963)]. The generalized projection theorem, Eq. (2.69), is useful for evaluating reduced matrix elements for transitions between fine-structure levels,

$$\begin{aligned}&\langle L'S'J'\|r_i\|LSJ\rangle \\ &\quad = \delta_{SS'}(-1)^{S+1-L-J'}[(2L'+1)(2J+1)]^{1/2}W(LJL'J';S1)\langle L'\|r_i\|L\rangle,\end{aligned} \tag{5.31}$$

and for transitions between terms of a two-electron system,

$$\begin{aligned}&\langle l_1'l_2'L'\|r_1\|l_1l_2L\rangle \\ &\quad = \delta_{l_2'l_2}(-1)^{l_2+1-l_1-L'}[(2l'+1)(2L+1)]^{1/2}W(l_1Ll_1'L';l_21)\langle l_1'\|r_i\|l_1\rangle.\end{aligned} \tag{5.32}$$

The full matrix elements,

$$\mathbf{r}_{ikm} \equiv \langle k|\mathbf{r}_i|m\rangle, \tag{5.33}$$

are then utilized in Eqs. (5.4) and (5.11) to obtain the transition probability in the electric-dipole approximation,

$$w_{m\to k} = \frac{2\pi e^2}{\hbar}\left|\mathbf{E}^{(0)}\cdot\sum_{i=1}^{N}\mathbf{r}_{ikm}\right|^2 \delta(E_k - E_m \pm \hbar\omega), \tag{5.34}$$

The intensity of the radiation, I, can be related to $\mathbf{E}^{(0)}$ by calculating the time averaged magnitude of the Poynting vector, $\mathbf{P} = \mathbf{E}\times\mathbf{B}/\mu_0$, with the result

$$I = 2\varepsilon_0 nc|\mathbf{E}^{(0)}|^2. \tag{5.35}$$

However, the local field is altered by surface charges due to polarization of the medium on the spherical cavity occupied by the ion, necessitating the *Lorentz correction*,

$$\frac{E^{(0)}_{\text{local}}}{E^{(0)}} = \frac{n^2+2}{3}. \tag{5.36}$$

In the case of isotropic unpolarized light, the transition probability can be expressed in the form

$$w_{m\to k} = \frac{\pi e^2}{3\hbar\varepsilon_0 c}\left(\frac{(n^2+2)^2}{9n}\right)\left|\sum_{i=1}^{N}\mathbf{r}_{ikm}\right|^2 I\delta(E_k - E_m \pm \hbar\omega). \tag{5.37}$$

A property of the ion is the oscillator strength of a transition, defined in analogy with a corresponding classical quantity for an oscillating charge,

$$f_{mk} \equiv \frac{2m\omega_{km}}{3\hbar}\left|\sum_{i=1}^{N}\mathbf{r}_{ikm}\right|. \tag{5.38}$$

The oscillator strengths satisfy the Kuhn–Thomas sum rule,

$$\sum_k f_{mk} = N, \tag{5.39}$$

where N is the number of electrons. The transition probability for isotropic, unpolarized light can then be expressed in the form

$$w_{m\to k} = \frac{\pi e^2}{2\varepsilon_0 mc\omega_{km}}\left(\frac{(n^2+2)^2}{9n}\right) f_{mk} I\delta(E_k - E_m \pm \hbar\omega). \tag{5.40}$$

A quantity related to the oscillator strength is the absorption cross section, $\sigma_{m\to k}$, defined as the number of photons absorbed per unit time divided by the number of incident photons per unit area per unit time,

$$\sigma_{m\to k} \equiv \frac{w_{m\to k}}{I/\hbar\omega_{km}} = \frac{\pi\hbar e^2}{2\varepsilon_0 mc}\left(\frac{(n^2+2)^2}{9n}\right) f_{mk}\delta(E_k - E_m - \hbar\omega). \tag{5.41}$$

The absorption coefficient, $\alpha_{m\to k}$, is defined by

$$\alpha_{m\to k} \equiv (N_i/V)\sigma_{m\to k}, \tag{5.42}$$

where N_i/V is the density of ions. The intensity of radiation traversing a distance x through ions of this density, which are initially in state m, is diminished according to the relation

$$I(x) = I_0 \exp(-\alpha_{m\to k}\, x). \tag{5.43}$$

5.1.5 Spontaneous emission

The semi-classical approximation accounts only for stimulated absorption and emission. In order to explain spontaneous emission, it is necessary to quantize the electromagnetic field [Loudon (1983)]. We proceed by generalizing Eq. (5.5a) to a superposition of modes in a cubic cavity of edge length L with periodic boundary conditions,

$$\mathbf{A}(\mathbf{r},t) = \sum_{\mathbf{k},\varepsilon_\mathbf{k}} \mathbf{A}^{(0)}_{\mathbf{k},\boldsymbol{\varepsilon}_\mathbf{k}} \exp[i(\mathbf{k}\cdot\mathbf{r} - \omega t)] + c.c., \tag{5.44a}$$

$$k_i = 2\pi\nu_i/L; \quad i = x, y, z, \tag{5.44b}$$

$$\nu_i = 0, \pm 1, \pm 2, \ldots, \tag{5.44c}$$

where the sum is over allowed wave vectors, $\mathbf{k}$, and over the two independent transverse polarization directions associated with each wave vector, specified by the unit vector $\varepsilon_\mathbf{k}$. One can deduce from the periodic boundary conditions, and from the relation

$$\omega = (c/n)k, \tag{5.45}$$

that the density of modes per unit volume per unit angular frequency is given by

$$\rho_\omega = \omega^2 n^3/\pi^2 c^3. \tag{5.46}$$

The cycle-averaged energy associated with mode $\mathbf{k}$, $\boldsymbol{\varepsilon}_\mathbf{k}$ in a cavity of volume V is

$$\bar{U}_{\mathbf{k},\boldsymbol{\varepsilon}_\mathbf{k}} = \frac{1}{2}\int_{\text{cavity}} (\varepsilon_0 \kappa \bar{E}^2_{\mathbf{k},\boldsymbol{\varepsilon}_\mathbf{k}} + \mu_0^{-1}\bar{B}^2_{\mathbf{k},\boldsymbol{\varepsilon}_\mathbf{k}})\, d\tau = \varepsilon_0 \kappa V \omega_\mathbf{k}^2 \mathbf{A}^{(0)}_{\mathbf{k},\boldsymbol{\varepsilon}_\mathbf{k}} \cdot \mathbf{A}^{(0)^*}_{\mathbf{k},\boldsymbol{\varepsilon}_\mathbf{k}}, \tag{5.47}$$

where the second equality follows from Eqs. (5.10) and (5.15). This energy can be expressed in the form

$$\bar{U}_{\mathbf{k},\boldsymbol{\varepsilon}_\mathbf{k}} \equiv \tfrac{1}{2}(P^2_{\mathbf{k},\boldsymbol{\varepsilon}_\mathbf{k}} + \omega_\mathbf{k}^2 Q^2_{\mathbf{k},\boldsymbol{\varepsilon}_\mathbf{k}}), \tag{5.48}$$

in terms of real, mass-weighted harmonic oscillator coordinates and momenta defined by

$$\mathbf{A}^{(0)}_{\mathbf{k},\boldsymbol{\varepsilon}_\mathbf{k}} \equiv (4\varepsilon_0 \kappa V \omega_\mathbf{k}^2)^{-1/2}(\omega_\mathbf{k} Q_{\mathbf{k},\boldsymbol{\varepsilon}_\mathbf{k}} + iP_{\mathbf{k},\boldsymbol{\varepsilon}_\mathbf{k}})\boldsymbol{\varepsilon}_\mathbf{k}. \tag{5.49}$$

Quantization of the electromagnetic field is accomplished by replacing the harmonic oscillator coordinates and momenta with non-commuting operators; Eq. (5.48) is then replaced by the Hamiltonian operator

$$\hat{H}_{\mathbf{k},\varepsilon_{\mathbf{k}}} = \tfrac{1}{2}(\hat{p}^2_{\mathbf{k},\varepsilon_{\mathbf{k}}} + \omega^2_{\mathbf{k}}\hat{q}^2_{\mathbf{k},\varepsilon_{\mathbf{k}}}), \tag{5.50a}$$

$$\left[\hat{q}_{\mathbf{k},\varepsilon_{\mathbf{k}}}, \hat{p}_{\mathbf{k},\varepsilon_{\mathbf{k}}}\right] = i\hbar. \tag{5.50b}$$

This Hamiltonian operator is diagonal in a *number state*,

$$\hat{H}_{\mathbf{k},\varepsilon_{\mathbf{k}}}\left|n_{\mathbf{k},\varepsilon_{\mathbf{k}}}\right\rangle = (n_{\mathbf{k},\varepsilon_{\mathbf{k}}} + \tfrac{1}{2})\hbar\omega_{\mathbf{k}}\left|n_{\mathbf{k},\varepsilon_{\mathbf{k}}}\right\rangle, \tag{5.51}$$

and the total cycle-averaged energy in the cavity in this state is given by

$$\bar{U} = \sum_{\mathbf{k},\varepsilon_{\mathbf{k}}}(n_{\mathbf{k},\varepsilon_{\mathbf{k}}} + \tfrac{1}{2})\hbar\omega_{\mathbf{k}}, \tag{5.52}$$

where the quantum number $n_{\mathbf{k},\varepsilon_{\mathbf{k}}}$ is an integer interpreted as the number of photons associated with mode $\mathbf{k}$, $\varepsilon_{\mathbf{k}}$.

Although the photon quantum numbers in Eq. (5.52) can have any values, depending on the nature of the electromagnetic radiation, it is of interest for the present purpose to determine their average values for a cavity in thermal equilibrium at a fixed temperature T. The density operator for a canonical ensemble of cavities is

$$\hat{\rho} = \exp(-\beta\hat{H}), \tag{5.53a}$$

$$\hat{H} \equiv \sum_{\mathbf{k},\varepsilon_{\mathbf{k}}} \hat{H}_{\mathbf{k},\varepsilon_{\mathbf{k}}}, \tag{5.53b}$$

$$\beta \equiv 1/k_{\mathrm{B}}T, \tag{5.53c}$$

where k_{B} is the Boltzmann constant, and the ensemble average of the internal energy is

$$U = \frac{\mathrm{Tr}(\hat{\rho}\hat{H})}{Q} = -\frac{\partial \ln Q}{\partial \beta}, \tag{5.54a}$$

$$Q \equiv \mathrm{Tr}(\hat{\rho}). \tag{5.54b}$$

Since the partition function Q for the ensemble of cavities is

$$Q = \prod_{\mathbf{k},\varepsilon_{\mathbf{k}}}\sum_{n} \exp\left[-\left(n + \frac{1}{2}\right)\beta\hbar\omega_{\mathbf{k}}\right] = \prod_{\mathbf{k},\varepsilon_{\mathbf{k}}} \frac{\exp(-\frac{1}{2}\beta\hbar\omega_{\mathbf{k}})}{1 - \exp(-\beta\hbar\omega_{\mathbf{k}})}, \tag{5.55}$$

the internal energy is given by

$$U = \sum_{\mathbf{k},\varepsilon_{\mathbf{k}}}\left[\frac{1}{\exp(\beta\hbar\omega_{\mathbf{k}}) - 1} + \frac{1}{2}\right]\hbar\omega_{\mathbf{k}}, \tag{5.56}$$

and it follows from comparison with Eq. (5.52) that the average photon quantum numbers are given by

$$\bar{n}_{\mathbf{k},\varepsilon_{\mathbf{k}}} = \frac{1}{\exp(\beta\hbar\omega_{\mathbf{k}}) - 1}. \tag{5.57}$$

This result can be combined with the density of modes per unit volume per unit angular frequency, Eq. (5.46), to yield the Planck radiation law for the energy density per unit

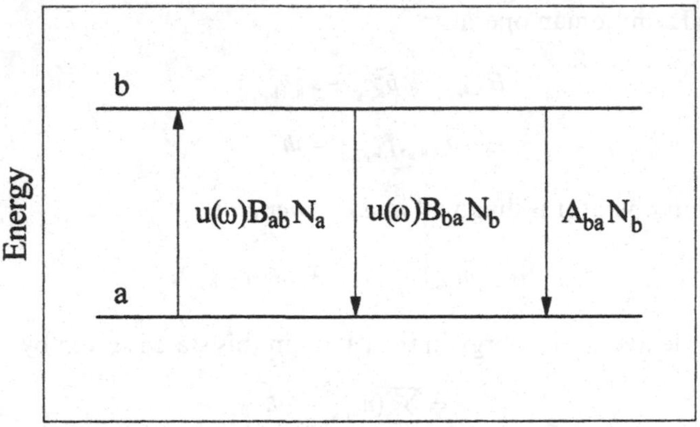

Figure 5.1. Stimulated absorption, stimulated emission and spontaneous emission of a concentration of two-level systems interacting with electromagnetic radiation in thermal equilibrium.

frequency range in thermal equilibrium,

$$u(\omega) = \rho_\omega \bar{n}\hbar\omega = \left(\frac{\hbar\omega^3}{\pi^2 c^3}\right)\frac{1}{\exp(\beta\hbar\omega) - 1}, \tag{5.58}$$

and conversion of the sum in Eq. (5.56) to an integral yields

$$U = \int_{\text{cavity}} d\tau \int_0^\infty u(\omega)\, d\omega = \frac{V\pi^2(k_B T)^4}{15c^3\hbar^3}. \tag{5.59}$$

One could proceed to derive the transition probability for spontaneous emission by application of time-dependent perturbation theory to the quantized electromagnetic field. However, it is more profitable for present purposes to develop a general relation between spontaneous and stimulated transitions, following Einstein (1917). Consider radiation in thermal equilibrium interacting with a concentration N of two-level atoms with ground-state and excited state energies, degeneracies and population densities E_a, g_a, N_a and E_b, g_b, N_b, respectively, as illustrated in Fig. 5.1. At the level of first-order time-dependent perturbation theory, the rate equation for this system is

$$\frac{dN_a}{dt} = -\frac{dN_b}{dt} = A_{ba}N_b + u(\omega)(B_{ba}N_b - B_{ab}N_a), \tag{5.60}$$

where A_{ba}, B_{ba} and B_{ab} are, respectively, rate constants for spontaneous emission, stimulated emission and stimulated absorption, and conservation of energy constrains the atoms to interact only with radiation of angular frequency near ω_{ba}, defined by

$$\hbar\omega_{ba} = E_b - E_a. \tag{5.61}$$

Since the population ratio in thermal equilibrium is

$$\frac{N_b}{N_a} = \frac{g_b}{g_a}\exp(-\beta\hbar\omega_{ba}), \tag{5.62}$$

it follows from Eqs. (5.58) and (5.60), with $dN_a/dt = 0$, that

$$A_{ba} = (\hbar\omega_{ba}^3/\pi^2c^3)B_{ba}, \tag{5.63a}$$

$$B_{ba} = (g_a/g_b)B_{ab}. \tag{5.63b}$$

The spontaneous emission rate A_{ba}, determined readily from Eqs. (5.63a), (5.37) and the relation

$$I(\omega) = \left(\frac{c}{n}\right)u(\omega), \tag{5.64}$$

is then given by

$$A_{ba} = \frac{\omega_{ba}^3 e^2}{3\pi\varepsilon_0\hbar c^3}\left[\frac{(n^2+2)^2}{9n^2}\right]\frac{1}{g_b}\sum_{j,k}\left|\sum_{i=1}^{N}\langle aj|\mathbf{r}_i|bk\rangle\right|^2, \tag{5.65}$$

where j and k distinguish degenerate states.

In the absence of ambient radiation, the rate equation (5.60) reduces to

$$\frac{dN_b}{dt} = -A_{ba}N_b, \tag{5.66}$$

with solution

$$N_b(t) = N_b(0)\exp(-t/\tau_R), \tag{5.67}$$

where $\tau_R \equiv A_{ba}^{-1}$ is the radiative lifetime. It follows that the electric-field intensity of the emitted radiation has time dependence, in complex notation,

$$\mathbf{E}(t) = \mathbf{E}^{(0)}\exp(-i\omega_{ba}t)\exp(-t/2\tau_R), \tag{5.68}$$

and the square of the absolute value of the Fourier transform of $\mathbf{E}(t)$ yields the Lorentzian homogeneous line-shape function

$$I(\omega) = \frac{I^{(0)}}{(\omega_{ba}-\omega)^2 + 1/(2\tau_R)^2}, \tag{5.69}$$

with full width at half maximum given by

$$\Delta\omega = 1/\tau_R = A_{ba}. \tag{5.70}$$

A typical radiative lifetime for a fully electric-dipole-allowed transition is $\tau_R \cong 10^{-8}$ s, with corresponding line-width $\Delta\omega/2\pi c \cong 5\times10^{-4}\,\text{cm}^{-1}$. Since radiative lifetimes of crystal-field transitions are typically of the order of microseconds, lifetime broadening is much less for them.

Thus far, only transitions of an isolated ion embedded in a rigid lattice have been considered, with only electronic polarization included. Transition line shapes of ions in actual crystals are very much broader as a consequence of electron-lattice coupling, which is the topic of the following section.

5.2 Electron–lattice coupling

5.2.1 *Born–Oppenheimer approximation*

Only the electronic coordinates were treated quantum-mechanically in the Hamiltonian of Eq. (4.1), while the nuclear coordinates were assumed to be fixed. A nuclear kinetic energy operator must also be included in a more complete and rigorous formulation; in a concise notation, the revised Hamiltonian is then

$$H = T_{\mathrm{E}} + T_{\mathrm{N}} + V(\mathbf{r}, \mathbf{Q}), \tag{5.71}$$

where T_{E} is the electronic kinetic energy operator, T_{N} is the nuclear kinetic energy operator, $V(\mathbf{r}, \mathbf{Q})$ includes all of the interactions among electrons and nuclei, and $\mathbf{r}$ and $\mathbf{Q}$ respectively denote all electronic coordinates and all nuclear coordinates. Approximate eigenfunctions of H are *Born–Oppenheimer* states [Born and Oppenheimer (1927)]

$$\psi_{nv}^{\mathrm{BO}}(\mathbf{r}, \mathbf{Q}) = \phi_n(\mathbf{r}, \mathbf{Q})\theta_{nv}(\mathbf{Q}), \tag{5.72}$$

where $\phi_n(\mathbf{r}, \mathbf{Q})$ is an eigenfunction of the electronic Hamiltonian $H_{\mathrm{e}}(\mathbf{Q})$ for fixed nuclear coordinates,

$$H_{\mathrm{e}}(\mathbf{Q}) \equiv T_{\mathrm{E}} + V(\mathbf{r}, \mathbf{Q}), \tag{5.73}$$

$$H_{\mathrm{e}}(\mathbf{Q})\phi_n(\mathbf{r}, \mathbf{Q}) = U_n(\mathbf{Q})\phi_n(\mathbf{r}, \mathbf{Q}); \tag{5.74}$$

the electronic energy eigenvalue $U_n(\mathbf{Q})$ then serves as the potential energy of interaction of the nuclei in the nth electronic state,

$$[T_{\mathrm{N}} + U_n(\mathbf{Q})]\theta_{nv}(\mathbf{Q}) = E_{nv}\theta_{nv}(\mathbf{Q}), \tag{5.75}$$

where E_{nv} is the approximate total energy eigenvalue of the Born–Oppenheimer state. Neglected terms in the Schrödinger equation mediate nonradiative transitions between Born–Oppenheimer states, considered in Chapter 6. These terms are small compared with the energy separations of non-degenerate electronic states by virtue of the enormous disparity of electron and nuclear masses. Modification of the Born–Oppenheimer approximation for degenerate electronic states (*Jahn–Teller effect*) is considered below.

5.2.2 *Harmonic approximation*

For small displacements from equilibrium, the electronic energy eigenvalue $U_n(\mathbf{Q})$ can be expanded in symmetry-adapted, mass-weighted normal coordinates Q_k,

$$U_n(\mathbf{Q}) \cong U_n(\mathbf{Q}_0^{(n)}) + \frac{1}{2}\sum_k \omega_k^{(n)2}(Q_k - Q_{0k}^{(n)})^2, \tag{5.76}$$

and the nuclear kinetic energy operator can be expressed in terms of their conjugate momenta. Equation (5.75) is then replaced by

$$\frac{1}{2}\sum_k (P_k^2 + \omega_k Q_k^2)\theta_{nv}(\mathbf{Q}) = \left[E_{nv} - U_n(\mathbf{Q}_0^{(n)})\right]\theta_{nv}(\mathbf{Q}), \tag{5.77}$$

$$[Q_k, P_{k'}] = i\hbar\delta_{kk'}. \tag{5.78}$$

As in the case of quantized electromagnetic radiation, we again encounter a collection of quantum-mechanical harmonic oscillators with energy eigenvalue

$$E_{n\nu} \equiv \hbar\Omega_{n\nu} = U_n(\mathbf{Q}_0^{(n)}) + \sum_k \left(\nu_k + \frac{1}{2}\right)\hbar\omega_k^{(n)}, \tag{5.79}$$

where the quantum numbers ν_k for quantized lattice vibrations are interpreted as the numbers of *phonons* associated with each normal mode, and ν denotes the set of quantum numbers $\{\nu_k\}$. The harmonic oscillator wavefunctions are also of interest, for reasons which will become apparent,

$$\theta_{n\nu}(\mathbf{Q}) = \prod_k \chi_{\nu_k}^{(n)}(Q_k - Q_{0k}^{(n)}), \tag{5.80a}$$

$$\chi_m(Q) = \left(\frac{\alpha}{\sqrt{\pi}2^m m!}\right)^{1/2} H_m(\alpha Q)\exp\left(-\frac{1}{2}\alpha^2 Q^2\right), \tag{5.80b}$$

$$\alpha = \sqrt{\frac{\omega}{\hbar}}, \tag{5.80c}$$

and $H_m(\xi)$ is a *Hermite polynomial*, defined by

$$H_m(\xi) \equiv (-1)^m \exp(\xi^2)\frac{\partial^m}{\partial\xi^m}\exp(-\xi^2). \tag{5.80d}$$

It is also useful to introduce phonon annihilation and creation operators, defined respectively by

$$b_k \equiv \frac{1}{\sqrt{2\hbar\omega_k}}(\omega_k Q_k + iP_k), \tag{5.81a}$$

$$b_k^+ \equiv \frac{1}{\sqrt{2\hbar\omega_k}}(\omega_k Q_k - iP_k), \tag{5.81b}$$

with the following properties:

$$[b_k, b_{k'}^+] = \delta_{kk'}, \tag{5.82a}$$

$$b\chi_m = \sqrt{m}\chi_{m-1}, \tag{5.82b}$$

$$b^+\chi_m = \sqrt{m+1}\chi_{m+1}, \tag{5.82c}$$

$$b^+b\chi_m = m\chi_m. \tag{5.82d}$$

5.2.3 *Electric-dipole transitions between Born–Oppenheimer states*

For electric-dipole transitions between Born–Oppenheimer states, Eq. (5.72), the matrix element of Eq. (5.33) is replaced by

$$\begin{aligned} \mathbf{r}_{ib\beta,a\alpha} &= \int\int \psi_{b\beta}^{\mathrm{BO}}(\mathbf{r},\mathbf{Q})^* \, \mathbf{r}_i \psi_{a\alpha}^{\mathrm{BO}}(\mathbf{r},\mathbf{Q})\, d\mathbf{r}\, d\mathbf{Q} \\ &= \int \theta_{b\beta}(\mathbf{Q})^* \left[\int \phi_b(\mathbf{r},\mathbf{Q})^* \, \mathbf{r}_i \phi_a(\mathbf{r},\mathbf{Q})\, d\mathbf{r}\right] \theta_{a\alpha}(\mathbf{Q})\, d\mathbf{Q}. \end{aligned} \tag{5.83}$$

An immediate simplification of this expression is provided by the *Condon approximation*,

$$\int \phi_b(\mathbf{r},\mathbf{Q})^* \mathbf{r}_i \phi_a(\mathbf{r},\mathbf{Q})\, d\mathbf{r} \cong \mathbf{r}_{iba}, \tag{5.84}$$

where the right-hand side of Eq. (5.84) is assumed to be independent of nuclear coordinates $\mathbf{Q}$. It is also convenient to adopt the *mean-value* approximation,

$$E_{b\beta} - E_{a\alpha} \cong E_{ba} \equiv \hbar\omega_{ba}, \tag{5.85}$$

although departures from this approximation become significant for broad lines such as those of relevance to tunable lasers [Wojtowicz *et al.* (1989)].

If one is interested only in a transition between two electronic states without regard for vibrational substates, one must average statistically over initial vibrational states and sum over final vibrational states. The absorption coefficient for unpolarized light, defined by Eq. (5.42), is then given approximately by

$$\alpha_{a\to b}(\omega) \cong \left(\frac{N_i}{V}\right)\left(\frac{\pi e^2}{3\hbar\varepsilon_0 c}\right)\left(\frac{(n^2+2)^2}{9n}\right)\left|\sum_{i=1}^{N} \mathbf{r}_{iba}\right|^2 \omega_{ba} G(\Omega), \tag{5.86a}$$

where $G(\Omega)$ is a normalized line-shape function defined by

$$G(\Omega) \equiv \sum_\alpha \sum_\beta P_\alpha \left| \int \theta_{b\beta}(\mathbf{Q})^* \theta_{a\alpha}(\mathbf{Q})\, d\mathbf{Q}\right|^2 \delta(\Omega_{b\beta} - \Omega_{a\alpha} - \Omega_0 - \Omega), \tag{5.86b}$$

$$P_\alpha = \frac{\exp(-\hbar\Omega_{a\alpha}/k_B T)}{\sum_{\alpha'} \exp(-\hbar\Omega_{a\alpha'}/k_B T)}, \tag{5.86c}$$

$$\hbar\Omega_0 = U_b(\mathbf{Q}_0^{(b)}) - U_a(\mathbf{Q}_0^{(a)}), \tag{5.86d}$$

with photon energy

$$h\nu = |\hbar(\Omega_0 + \Omega)|. \tag{5.87}$$

Finally, if one assumes the harmonic approximation, Eq. (5.80a), the line-shape function, $G(\Omega)$, becomes

$$G(\Omega) = \sum_\alpha \sum_\beta P_\alpha \prod_k \left| \int \chi_{\beta_k}^{(b)}(Q_k - Q_{0k}^{(b)})^* \chi_{\alpha_k}^{(a)}(Q_k - Q_{0k}^{(a)})\, dQ_k\right|^2 \delta(\Omega_{b\beta} - \Omega_{a\alpha} - \Omega_0 - \Omega), \tag{5.88}$$

where $\Omega_{a\alpha}$ and $\Omega_{b\beta}$ are given by Eq. (5.79).

5.2.4 *Configuration-coordinate diagram*

Although harmonic oscillator wavefunctions are orthonormal, the overlap integrals in Eq. (5.88) are non-vanishing, even for different vibrational quantum numbers, α_k and β_k, when the two wavefunctions involved are defined with respect to different origins, $Q_{0k}^{(a)}$ and $Q_{0k}^{(b)}$. The electronic system is said to be *linearly coupled* to modes for which these origins differ in the two electronic states. The line-shape function $G(\Omega)$ may

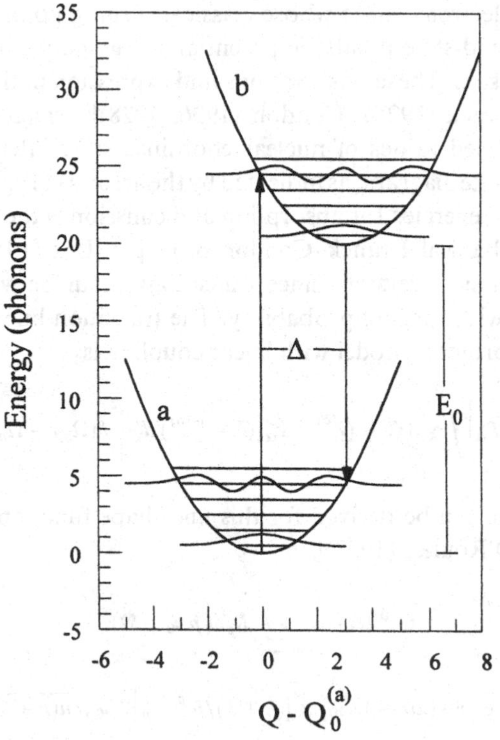

Figure 5.2. Configuration-coordinate diagram for a two-level system with linear coupling to a single mode of vibration, including vibrational energy levels and representative vibrational wavefunctions. Vertical arrows depict optical transitions in accordance with the classical Franck–Condon principle, and their energy difference corresponds to the Stokes shift.

be viewed as a convolution of line-shape functions for the individual coupled modes. Non-degenerate electronic states couple only to fully symmetrical modes.

For a transition-metal or rare-earth dopant in a substitutional site, it is a reasonable approximation to assume that the electronic system is coupled only to a symmetrical (A_{1g}) displacement of the immediate ligands, as illustrated for octahedral coordination in Fig. 3.2. However, such a displacement is not generally a normal coordinate, but rather a *reaction coordinate* which participates in many normal modes with a range of vibration frequencies. It is nevertheless instructive to adopt a *configuration-coordinate* model in which the reaction coordinate is treated as a single normal coordinate Q, associated with a typical vibration frequency. This model would be rigorously correct only if all of the normal modes had identical vibration frequencies in each electronic state [Huang and Rhys (1950)]. *Adiabatic potential-energy curves* $U_n(Q)$ and energy levels E_{nv} are illustrated in a configuration-coordinate diagram in Fig. 5.2. This diagram is drawn for the special case of pure linear coupling, in which the vibration frequency has the same value, ω_0, in both electronic states.

Representative vibrational wavefunctions are also plotted in Fig. 5.2, for cases of maximum overlap. They demonstrate that the most probable absorption transition from the lowest vibrational state of the ground electronic state is to that vibrational state of the excited electronic state whose classical turning point coincides with the minimum of the ground-state adiabatic potential energy curve; an analogous statement applies for emission. These observations find expression in the classical *Franck–Condon principle* [Franck (1925), Condon (1926, 1928)], which holds that optical transitions occur at fixed values of nuclear coordinates; i.e., they are vertical on a configuration coordinate diagram, as indicated by the arrows in Fig. 5.2. The difference between the transition energies for absorption and emission is the *Stokes shift*.

The quantum-mechanical Franck–Condon principle [Lax (1952)] is rather more complex than the classical version, since transitions occur between various Born–Oppenheimer states with varying probability. The transition line-shape function for the configuration-coordinate model with linear coupling is

$$G^{(0)}(\Omega) = \sum_{\alpha}\sum_{\beta} P_\alpha \left| \int \chi_\beta(Q - Q_0^{(b)})^* \chi_\alpha(Q - Q_0^{(a)})\, dQ \right|^2 \delta(\Omega_{b\beta} - \Omega_{a\alpha} - \Omega_0 - \Omega). \quad (5.89)$$

An explicit expression can be derived for this line-shape function [Huang and Rhys (1950), Lax (1952), O'Rourke (1953)],

$$G^{(0)}(\Omega) = \sum_{p=-\infty}^{\infty} R_p \delta(p\omega_0 - \Omega), \quad (5.90a)$$

$$R_p = \exp[-(2\bar{n}+1)S_0] \times [(\bar{n}+1)/\bar{n}]^{p/2} I_p\left[2S_0\sqrt{\bar{n}(\bar{n}+1)}\right], \quad (5.90b)$$

where I_p is a modified Bessel function, $\bar{n}$ is the phonon occupation number given by

$$\bar{n} = [\exp(\hbar\omega_0/k_B T) - 1]^{-1}, \quad (5.91)$$

and S_0 is the zero-temperature *Huang–Rhys factor*, which is a measure of the difference in the equilibrium lattice configuration between the two electronic states,

$$S_0 = \omega_0 \Delta^2/2\hbar, \quad (5.92)$$

$$\Delta = Q_0^{(b)} - Q_0^{(a)}. \quad (5.93)$$

In the low temperature limit, R_p for $p \geq 0$ reduces to a normalized Poisson distribution,

$$R_p = \exp(-S_0) S_0^p / p!. \quad (5.94)$$

Moments of the transition line-shape function $G^{(0)}(\Omega)$ are given by [Lax (1952)]

$$M_1^{(0)} \equiv \langle\Omega\rangle/\omega_0 = S_0, \quad (5.95a)$$

$$M_2^{(0)} \equiv \langle(\Omega - \langle\Omega\rangle)^2\rangle/(\omega_0)^2 = S_0(2\bar{n}+1), \quad (5.95b)$$

$$M_3^{(0)} \equiv \langle(\Omega - \langle\Omega\rangle)^3\rangle/(\omega_0)^3 = S_0. \quad (5.95c)$$

5.2.5 *Linear coupling to many modes*

The transition line-shape for linear coupling to many modes with a range of frequencies has been addressed by several investigators. By exploiting Mehler's formula (1866),

O'Rourke (1953) evaluated its Fourier transform, which may be expressed in the form

$$\Gamma(t) = \int_{-\infty}^{\infty} \exp(i\Omega t) G(\Omega)\, d\Omega = \exp[-S + g(t)], \tag{5.96a}$$

$$S = \int_{0}^{\infty} d\omega\, [2n(\omega) + 1] A(\omega), \tag{5.96b}$$

$$g(t) = \int_{0}^{\infty} d\omega\, \{[n(\omega) + 1] \exp(i\omega t) + n(\omega) \exp(-i\omega t)\} A(\omega), \tag{5.96c}$$

$$A(\omega) = \sum_k S_{0k} \delta(\omega - \omega_k), \tag{5.96d}$$

$$n(\omega) = [\exp(\hbar\omega / k_B T) - 1]^{-1}, \tag{5.96e}$$

where S_{0k} is the zero-temperature, per-mode Huang–Rhys factor and $A(\omega)$ is the zero-temperature, single-phonon-sideband line-shape function. Evaluation of the inverse Fourier transform for the special case of linear coupling to a single mode, or, equivalently, to many modes with the same frequency [Huang and Rhys (1950)], yields Eqs. (5.90). Equations (5.96) also provide the starting point for the further developments of Pryce (1966) and of Weissman and Jortner (1978).

Pryce obtained the transition line-shape function

$$G(\Omega) = \exp(-S) \left[\delta(\Omega) + \sum_{p=1}^{\infty} (S^p / p!) B_p(\Omega) \right], \tag{5.97}$$

where $B_p(\Omega)$ is the p-fold convolution of a normalized single-phonon sideband given by

$$B_1(\Omega) = (S^{-1} / 2\pi) \int_{-\infty}^{\infty} dt\, \exp(-i\Omega t) g(t). \tag{5.98}$$

He then invoked the central-limit theorem to obtain an approximate expression for $B_p(\Omega)$ which exploits the smoothing effect of the p-fold convolution for large values of p. A hypothetical line shape calculated by this algorithm is shown in Fig. 5.3. Note that the zero-phonon line, represented by the δ-function in Eq. (5.97), remains very narrow; at low temperatures, it is subject only to lifetime and inhomogeneous broadening (§5.5.7). Although Pryce's approximation works well in the low-temperature limit, $B_1(\Omega)$ is bimodal at finite temperatures, since it includes both absorption and emission of phonons; consequently, the approximation is adequate only for inconveniently large values of p. This deficiency was remedied by the refinement of Weissman and Jortner (1978), who proceeded from an alternative expansion of the line-shape function which emphasizes the net number of phonons emitted (§5.5.4).

5.2.6 Static Jahn–Teller effect

The Born–Oppenheimer approximation is no longer appropriate when the electronic Hamiltonian $H_e(\mathbf{Q})$ for fixed nuclear coordinates has degenerate eigenvalues in a symmetrical configuration, which we will adopt as the origin of nuclear coordinates, $\mathbf{Q} = \mathbf{0}$. The origin is also assumed to be the equilibrium configuration for purely symmetrical displacements. Symmetry-induced degeneracy is assumed,

$$H_e(\mathbf{0}) \phi_n^{(\Gamma)}(\mathbf{r}, \mathbf{0}) = U^{(\Gamma)}(\mathbf{0}) \phi_n^{(\Gamma)}(\mathbf{r}, \mathbf{0}), \tag{5.99}$$

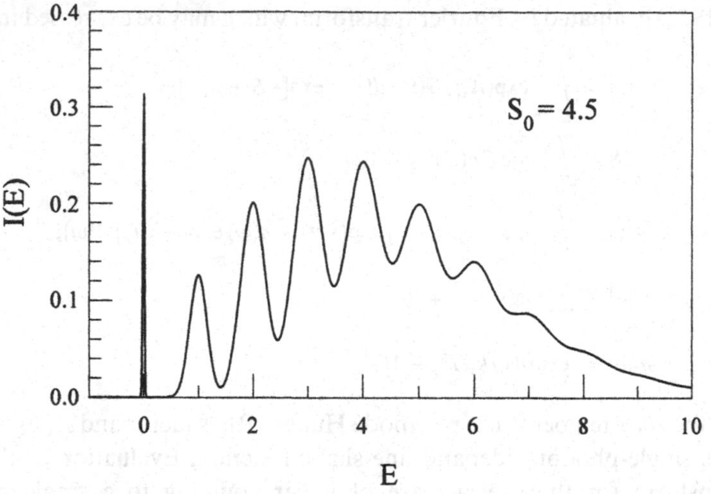

Figure 5.3. Hypothetical optical absorption line-shape for linear coupling to many modes, calculated from Eqs. (5.97) and (5.98). The single-phonon sideband is assumed to be a gaussian function, and the zero-phonon line is represented by a narrow gaussian rather than a δ-function.

where n labels the rows of irreducible representation Γ of the group of the Hamiltonian $H_e(\mathbf{0})$. We proceed by expanding the potential $V(\mathbf{r}, \mathbf{Q})$ in mass-weighted, symmetry-adapted nuclear displacements $Q_j^{(\Gamma')}$, retaining only terms linear and quadratic in these displacements,

$$V(\mathbf{r},\mathbf{Q}) \cong V(\mathbf{r},\mathbf{0}) + \sum_{\Gamma'}\sum_{j} \left[V_j^{(\Gamma')}(\mathbf{r},\mathbf{0})Q_j^{(\Gamma')} + \frac{1}{2}\omega^{(\Gamma')2}Q_j^{(\Gamma')2}\right]. \tag{5.100}$$

The coefficient $V_j^{(\Gamma')}(\mathbf{r},\mathbf{0})$ of $Q_j^{(\Gamma')}$ in the linear term must transform as a tensor operator, in order to ensure the invariance of the potential under simultaneous symmetry operations on electronic and nuclear coordinates,

$$O_R V_j^{(\Gamma')}(\mathbf{r},\mathbf{0})O_R^{-1} = \sum_i V_i^{(\Gamma')}(\mathbf{r},\mathbf{0})D_{ij}^{(\Gamma')}(R). \tag{5.101}$$

Changes in the potential for small displacements from the symmetrical configuration, $\delta V = V(\mathbf{r},\mathbf{Q}) - V(\mathbf{r},\mathbf{0})$, are then treated by degenerate first-order perturbation theory. In the present section, the effect of this perturbation is considered at the level of crystal-field theory by neglecting the nuclear kinetic energy operator, T_N.

Matrix elements $\langle\Gamma m|V_j^{(\Gamma')}|\Gamma n\rangle$ between degenerate unperturbed electronic eigenfunctions survive only if the Kronecker product representation $\Gamma' \times \Gamma$ contains irreducible representation Γ, or equivalently, if the extended Kronecker product representation $\Gamma \times \Gamma' \times \Gamma$ contains the identity representation Γ_1. Jahn and Teller (1937) have shown that each of the thirty-two crystallographic point groups contains at least one Γ' for every Γ for which this condition is satisfied. Their theorem applies to crystallographic double groups as well, with the exception of Kramers degeneracy, but not to linear molecules. The perturbed energy levels corresponding to a displacement $Q_j^{(\Gamma')}$ are then obtained by diagonalizing the matrix $\langle\Gamma m|V_j^{(\Gamma')}|\Gamma n\rangle$. They also demonstrated that the trace of this matrix vanishes for all Γ' except Γ_1,

$$\sum_n \langle \Gamma n | V_j^{(\Gamma')} | \Gamma n \rangle = 0. \tag{5.102}$$

It follows that, if there is a displacement $Q_j^{(\Gamma')}$ which splits the degeneracy of the ground electronic state, at least one component is reduced in energy and the complex distorts spontaneously; this distortion is limited by the quadratic terms in Eq. (5.100). In fact, there are $d_{\Gamma'}$ equivalent distortions.

Although the Jahn–Teller theorem predicts instability of degenerate electronic states against low-symmetry distortions, it makes no prediction concerning the magnitude of the effect. For that, one must rely on explicit evaluation of matrix elements, which is facilitated by the point-group analogue of the Wigner–Eckart theorem, Eq. (2.71),

$$\langle \Gamma m | V_j^{(\Gamma')} | \Gamma n \rangle = d_\Gamma^{-1/2} \sum_{\tau_\Gamma} \langle \Gamma \tau_\Gamma \| V^{(\Gamma')} \| \Gamma \rangle \langle \Gamma' \Gamma j n | \Gamma \tau_\Gamma m \rangle^*. \tag{5.103}$$

The d_Γ-dimensional matrix of the electronic Hamiltonian within the degenerate manifold of unperturbed eigenfunctions can then be expressed in the form

$$\begin{aligned}\mathbf{H}_e = {} & \left[U^{(\Gamma)}(\mathbf{0}) + \sum_{\Gamma'} \frac{1}{2} \omega^{(\Gamma')2} \sum_j Q_j^{(\Gamma')2} \right] \mathbf{I} \\ & + \sum_{\Gamma'} d_\Gamma^{-1/2} \sum_{\tau_\Gamma} \langle \Gamma \tau_\Gamma \| V^{(\Gamma')} \| \Gamma \rangle \sum_j Q_j^{(\Gamma')} \mathbf{U}_j(\Gamma', \Gamma, \tau_\Gamma),\end{aligned} \tag{5.104}$$

where $\mathbf{I}$ is the identity matrix and the elements of the matrices $\mathbf{U}_j(\Gamma', \Gamma, \tau_\Gamma)$ are the coupling coefficients $\langle \Gamma' \Gamma j n | \Gamma \tau_\Gamma m \rangle^*$. These matrices have been tabulated for a number of cases of interest by F.S. Ham (1972). Adiabatic potential-energy surfaces as functions of the coordinates $Q_j^{(\Gamma')}$ are then obtained by diagonalization of $\mathbf{H}_e$; such surfaces have been investigated exhaustively [Liehr (1963)]. Evaluation of reduced matrix elements for specific cases can be accomplished either empirically or by detailed calculation.

As an example of the static Jahn–Teller effect, consider a doubly degenerate (E_g or E_u) electronic state of a substitutional impurity or point defect in octahedral coordination. Jahn–Teller-active displacements, which must be even under inversion, include e_g and t_{2g} distortions of the octahedron, illustrated in Fig. 3.2. However, only the e_g distortion couples to an E_g or E_u electronic state, since repeated application of Table 2.21 yields the relations

$$E \times E \times E = A_1 + A_2 + 3E, \tag{5.105a}$$

$$E \times T_2 \times E = 2T_1 + 2T_2. \tag{5.105b}$$

Equation (5.104) then reduces to

$$\mathbf{H}_e = [U(\mathbf{0}) + \tfrac{1}{2}\omega^2 (Q_\theta^2 + Q_\varepsilon^2)]\mathbf{I} + V(Q_\theta \mathbf{U}_\theta + Q_\varepsilon \mathbf{U}_\varepsilon), \tag{5.106a}$$

$$V \equiv \langle E_{g,u} \| V^{(e_g)} \| E_{g,u} \rangle / \sqrt{2}, \tag{5.106b}$$

$$\mathbf{U}_\theta = \begin{pmatrix} -1 & 0 \\ 0 & 1 \end{pmatrix}, \tag{5.106c}$$

$$\mathbf{U}_\varepsilon = \begin{pmatrix} 0 & 1 \\ 1 & 0 \end{pmatrix}, \tag{5.106d}$$

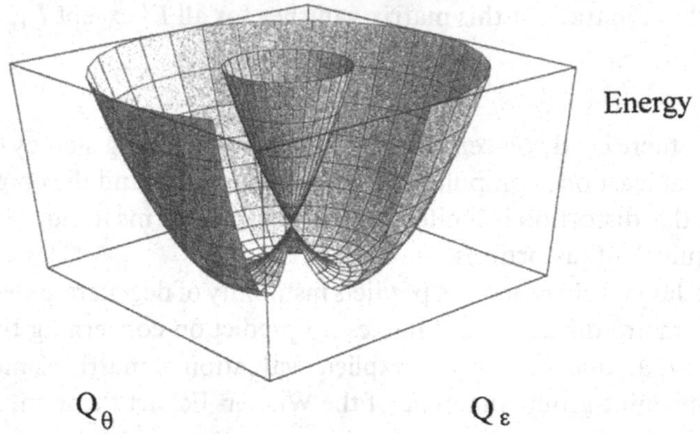

Figure 5.4. Adiabatic potential energy surfaces for an $E \times e$ Jahn–Teller system, the 'Mexican-hat potential'. Configuration coordinates Q_θ and Q_ε may be identified with the $E_g\theta$ and $E_g\varepsilon$ distortions of an octahedron in Fig. 3.2.

with eigenvalues

$$E_\pm = U(\mathbf{0}) \pm V\rho + \tfrac{1}{2}\omega^2\rho^2, \tag{5.107a}$$

$$Q_\theta = \rho\cos\phi, \tag{5.107b}$$

$$Q_\varepsilon = \rho\sin\phi. \tag{5.107c}$$

The corresponding adiabatic-potential-energy surfaces, metaphorically designated 'the Mexican-hat potential', are plotted in Fig. 5.4. There is a stable value of the radial coordinate ρ,

$$\rho_0 = \frac{V}{\omega^2}, \tag{5.108}$$

corresponding to the Jahn–Teller stabilization energy,

$$E_{\mathrm{JT}} = \frac{V^2}{2\omega^2}, \tag{5.109}$$

but no stable value of the azimuthal coordinate ϕ. The latter result is an artifact of the approximation, removed by inclusion of bilinear and cubic terms which warp the adiabatic-potential-energy surfaces and introduce three stable minima, separated by 120° in ϕ [Englman (1972)].

5.2.7 *Dynamic Jahn–Teller effect*

Inclusion of the kinetic energy operator T_{N} in the Hamiltonian, important in the weak-coupling regime, leads to the *dynamic Jahn–Teller effect* [Moffitt and Liehr (1957), Moffitt and Thorson (1957), Longuet-Higgins, Öpik, Pryce and Sack (L-HOPS) (1958)]. The matrix of the Hamiltonian within the degenerate electronic manifold then becomes

$$\mathbf{H} = \left[U^{(\Gamma)}(\mathbf{0}) + \sum_{\Gamma'}\left(-\frac{1}{2}\hbar^2\sum_j \frac{\partial^2}{\partial Q_j^{(\Gamma')2}} + \frac{1}{2}\omega^{(\Gamma')2}\sum_j Q_j^{(\Gamma')2}\right)\right]\mathbf{I}$$
$$+ \sum_{\Gamma'} d_\Gamma^{-1/2}\sum_{\tau_\Gamma}\left\langle \Gamma\tau_\Gamma \middle\| V^{(\Gamma')} \middle\| \Gamma\right\rangle \sum_j Q_j^{(\Gamma')}\mathbf{U}_j(\Gamma',\Gamma,\tau_\Gamma). \qquad (5.110)$$

It no longer suffices to diagonalize **H** for a fixed configuration of the nuclei, since the matrices $\mathbf{U}_j(\Gamma',\Gamma,\tau_\Gamma)$ do not generally commute and cannot be diagonalized simultaneously. Consequently, the electronic and vibrational motions are inextricably linked in a *vibronic* wavefunction of the form

$$\psi_{nv}^{(\Gamma')}(\mathbf{r},\mathbf{Q}) = \sum_m \phi_m^{(\Gamma)}(\mathbf{r},\mathbf{0})\theta_{mnv}^{(\Gamma,\Gamma')}(\mathbf{Q}). \qquad (5.111)$$

A ubiquitous feature of vibronic wavefunctions is that they reflect the symmetry of the symmetrical configuration rather than the reduction in symmetry inherent in the static Jahn–Teller effect, since probability is shared among equivalent stable distortions; consequently, manifestations of the dynamic Jahn–Teller effect are more subtle than those of the static effect [Ham (1972)].

We again consider the $E \times e$ system (Mexican-hat potential) as an example [Moffitt and Thorson (1957), L-HOPS (1958), Struck and Herzfeld (1966), Ham (1968)],

$$\mathbf{H} = H_0\mathbf{I} + V(Q_\theta\mathbf{U}_\theta + Q_\varepsilon\mathbf{U}_\varepsilon), \qquad (5.112\text{a})$$

$$H_0 = U(\mathbf{0}) - \frac{1}{2}\hbar^2\left(\frac{\partial^2}{\partial Q_\theta^2} + \frac{\partial^2}{\partial Q_\varepsilon^2}\right) + \frac{1}{2}\omega^2(Q_\theta^2 + Q_\varepsilon^2). \qquad (5.112\text{b})$$

The matrix **H** commutes with the operator

$$\mathbf{J} = \hbar^{-1}(Q_\theta P_\varepsilon - Q_\varepsilon P_\theta)\mathbf{I} - \frac{1}{2}\begin{pmatrix} 0 & -i \\ i & 0 \end{pmatrix}, \qquad (5.113)$$

which has eigenvalues $l = m + \frac{1}{2}$, where m is an integer. In a representation in which **J** is diagonal, **H** has the form

$$\mathbf{H} = H_0\mathbf{I} + V\begin{pmatrix} 0 & \rho\exp(-i\phi) \\ \rho\exp(i\phi) & 0 \end{pmatrix}. \qquad (5.114)$$

The vibronic wavefunctions are

$$\psi_{lp}(\mathbf{r},\mathbf{Q}) = \phi_a(\mathbf{r},\mathbf{0})\chi_a(\rho,\phi) + \phi_b(\mathbf{r},\mathbf{0})\chi_b(\rho,\phi), \qquad (5.115\text{a})$$

$$\phi_a = \frac{1}{\sqrt{2}}(\phi_\theta + i\phi_\varepsilon), \qquad (5.115\text{b})$$

$$\phi_b = -\frac{1}{\sqrt{2}}(\phi_\theta - i\phi_\varepsilon), \qquad (5.115\text{c})$$

where χ_a and χ_b are eigenfunctions of **H**. L-HOPS proceeded by expanding χ_a and χ_b as linear combinations of the eigenfunctions $\chi_{nm}(\rho,\phi)$ of H_0, the isotropic two-dimensional harmonic-oscillator Hamiltonian. They tabulated the energy eigenvalues

E_{lp} as functions of the coupling constant $k^2 \equiv 2E_{JT}/\hbar\omega = V^2/\hbar\omega^3$, and they also calculated low-temperature line shapes for optical transitions between degenerate and non-degenerate electronic states as functions of k^2. Line shapes for transitions from degenerate to non-degenerate states are similar to that of Eq. (5.90), but those for transitions from non-degenerate to degenerate states are distinctly bimodal for strong coupling ($k^2 \gg 1$), since the vibronic wavefunctions for the degenerate state are associated primarily with one or the other of the adiabatic potential-energy surfaces in that limit.

Symmetry properties of the vibronic functions of Eqs. (5.115) were investigated by Ham (1968). The topic of Jahn–Teller coupling to many modes of the same symmetry with a range of frequencies, addressed by O'Brien (1972), is omitted here for brevity.

5.3 Spectral intensities

5.3.1 Electric-dipole-allowed transitions

Examples of electric-dipole-allowed transitions of point imperfections in solids include charge-transfer spectra of transition-metal complexes [Burns (1993)]. Intervalence charge transfer (IVCT) occurs between two nearby substitutional impurity ions of the same or different species; for example, the transfer of an electron from Fe^{2+} to Ti^{4+} in corundum (Al_2O_3) produces the broad, long-wavelength optical absorption band responsible for the blue colour of sapphire [Burns and Burns (1984)]. The inverse transition is normally nonradiative as a consequence of very strong electron–lattice coupling. Oxygen–metal charge-transfer spectra (OMCT), involving the transfer of an electron from the ligands to the central ion, occur at shorter wave lengths and are prominent in tetrahedral transition-metal complexes such as the chromate ion $(CrO_4)^{3-}$ and its isoelectronic analogues [Hazenkamp and Güdel (1996)].

Electric-dipole-allowed $4f \rightarrow 5d$ transitions in rare-earth ions occur at lower energies in divalent ions than in trivalent ions. These transitions in Ce^{3+} find application in scintillator materials [Melcher and Schweitzer (1992)].

Optical transitions of interest in colour centres are generally between states of opposite parity and are thus strongly electric-dipole-allowed.

5.3.2 Crystal-field spectra

Nearly all of the transitions in transition-metal, rare-earth and heavy-metal ions which are of interest for laser applications occur within the ground configuration, and so are electric-dipole-forbidden in the free ion by Laporte's parity selection rule. Although magnetic-dipole and electric-quadrupole transitions can sometimes be identified in the spectra of these ions, they are generally too weak to be useful in laser applications. Accordingly, one must rely on environmental effects to relax the parity selection rule for electric-dipole transitions, and the analysis of these effects must go beyond the levels of approximation considered in Chapter 4 to include mixing of configurations of opposite parity by components of the crystal field which are odd under inversion.

Parity mixing can be accomplished both by odd components of the static crystal field and by odd modes of lattice vibration. Since tetrahedral complexes lack inversion symmetry, their crystal fields can mediate parity mixing without further distortion.

Odd distortions of an octahedral complex are illustrated in Fig. 3.2. Rigid translations of the entire complex, which also transform as t_{1u}, participate in low-frequency acoustical modes but are ineffectual in parity mixing, since the electronic wavefunction tends to follow the ligands adiabatically.

Consider an electric-dipole transition between mixed-parity states $|a\rangle$ and $|b\rangle$, derived respectively from states $|a_0\rangle$ and $|b_0\rangle$ of a $3d^N$ configuration by admixture of opposite-parity states $|\phi_{nl}\rangle$ from excited configurations $3d^{N-1}nl$,

$$|a\rangle \cong |a_0\rangle + \sum_{|\phi_{nl}\rangle} \frac{|\phi_{nl}\rangle\langle\phi_{nl}|\sum_i V_c^{(\text{odd})}(\mathbf{r}_i)|a_0\rangle}{E(a_0) - E(\phi_{nl})}, \tag{5.116a}$$

$$|b\rangle \cong |b_0\rangle + \sum_{|\phi_{nl}\rangle} \frac{|\phi_{nl}\rangle\langle\phi_{nl}|\sum_i V_c^{(\text{odd})}(\mathbf{r}_i)|b_0\rangle}{E(b_0) - E(\phi_{nl})}, \tag{5.116b}$$

$$V_c^{(\text{odd})}(\mathbf{r}) = \sum_{k=1(\text{odd})}^{\infty} \sum_{q=-k}^{k} B_k^q r^k Y_k^q(\theta, \phi). \tag{5.117}$$

The required matrix element is then given by

$$\left\langle b\left|\sum_i r_{ip}\right|a\right\rangle \cong \sum_{k=1(\text{odd})}^{\infty} \sum_{q=-k}^{k} B_k^q \sum_{|\phi_{nl}\rangle} \left[\frac{\langle b_0|\sum_i r_i^k Y_k^q(\theta_i,\phi_i)|\phi_{nl}\rangle\langle\phi_{nl}|\sum_i r_{ip}|a_0\rangle}{E(b_0) - E(\phi_{nl})} + \frac{\langle b_0|\sum_i r_{ip}|\phi_{nl}\rangle\langle\phi_{nl}|\sum_i r_i^k Y_k^q(\theta_i,\phi_i)|a_0\rangle}{E(a_0) - E(\phi_{nl})}\right]. \tag{5.118}$$

5.3.3 *Odd modes of vibration*

Odd modes of lattice vibration can also mediate electric-dipole transitions. The odd crystal-field parameters can be expanded in terms of symmetry-adapted, mass-weighted normal coordinates, which can then be expressed in terms of creation and annihilation operators defined by Eqs. (5.81),

$$B_k^q(\mathbf{Q}) \cong B_k^q(\mathbf{0}) + \sum_{\Gamma j} \frac{\partial B_k^q}{\partial Q_j^{(\Gamma)}} \sqrt{\frac{\hbar}{2\omega^{(\Gamma)}}} (b_j^{(\Gamma)} + b_j^{(\Gamma)+}). \tag{5.119}$$

Several simplifying assumptions are introduced for clarity. The constant term $B_k^q(\mathbf{0})$ is assumed to vanish in Eq. (5.119). The electronic system is assumed to be linearly coupled only to fully symmetrical even modes, so that one can disregard any degeneracy of electronic states and consider transitions between Born–Oppenheimer states $|a\alpha\rangle$ and $|b\beta\rangle$. There can be no linear coupling to odd modes, and it is assumed that there is no quadratic coupling. Finally, the crude-adiabatic approximation is adopted,

$$\psi_{nv}^{\text{CA}}(\mathbf{r}, \mathbf{Q}) = \phi_n(\mathbf{r}, \mathbf{0})\theta_{nv}(\mathbf{Q}), \tag{5.120a}$$

$$[T_{\text{N}} + \langle\phi_n(\mathbf{0})|V(\mathbf{Q})|\phi_n(\mathbf{0})\rangle]\theta_{nv}(\mathbf{Q}) = E_{nv}\theta_{nv}(\mathbf{Q}); \tag{5.120b}$$

i.e., the electronic wave function, which satisfies Eq. (5.74) with $\mathbf{Q} = \mathbf{0}$, is assumed to be independent of nuclear coordinates. With this approximation, the matrix element can be factored into electronic and vibrational parts. The electronic part is as in Eq. (5.118) with B_k^q replaced by $\left(\partial B_k^q/\partial Q_j^{(\Gamma)}\right)\sqrt{\hbar/2\omega^{(\Gamma)}}$, provided differences in vibrational energy are neglected in the energy denominators. The vibrational matrix element of the operator in Eq. (5.119) is non-vanishing only if the vibrational wavefunctions differ by one quantum number in just one mode,

$$\left\langle\beta\left|b_j^{(\Gamma)}+b_j^{(\Gamma)+}\right|\alpha\right\rangle=\sqrt{\alpha_j^{(\Gamma)}+\tfrac{1}{2}\pm\tfrac{1}{2}}, \tag{5.121}$$

corresponding to creation or annihilation of one odd-mode phonon. It follows that mode Γ, j contributes Stokes and anti-Stokes lines to the spectrum which are displaced in energy from the magnetic-dipole zero-phonon line by $\pm\hbar\omega^{(\Gamma)}$ in absorption and by $\mp\hbar\omega^{(\Gamma)}$ in emission, and that the sum of their intensities in thermal equilibrium is proportional to

$$2\bar{n}_j^{(\Gamma)}+1=\coth\left[\frac{\hbar\omega^{(\Gamma)}}{2k_{\mathrm{B}}T}\right]. \tag{5.122}$$

This characteristic temperature dependence is a distinguishing feature of phonon-assisted transitions. With the assumption of linear coupling to symmetrical even modes, these odd-mode side bands serve as false origins for progressions of even-mode side bands with intensity distributions given by Eqs. (5.90) or (5.97), and Eq. (5.122) applies to the overall intensity of the spectrum.

5.3.4 *Judd–Ofelt theory*

The complex problem of calculating electric-dipole transition probabilities for crystal-field spectra in rare-earth ions has been simplified by Judd (1962) and Ofelt (1962), in a formal elaboration of an idea due to Van Vleck (1937). Consider an electric-dipole transition between states $|a\rangle$ and $|a'\rangle$ derived from a $4f^N$ configuration by admixture of states $|\phi_{nl}\rangle$ from configurations $4f^{N-1}nl$, for all values of n and l,

$$|a\rangle\cong|f^N JM_J\rangle+\sum_{|\phi_{nl}\rangle}\frac{|\phi_{nl}\rangle\langle\phi_{nl}|\sum_i V_{\mathrm{c}}^{(\mathrm{odd})}(\mathbf{r}_i)|f^N JM_J\rangle}{E(f^N J)-E(\phi_{nl})}, \tag{5.123}$$

where $V_{\mathrm{c}}^{(\mathrm{odd})}(\mathbf{r})$ is defined by Eq. (5.117). The required matrix element is then

$$\left\langle a\left|\sum_i \mathbf{r}_i\right|a'\right\rangle\cong\sum_{|\phi_{nl}\rangle}\left[\frac{\langle f^N JM_J|\sum_i V_{\mathrm{c}}^{(\mathrm{odd})}(\mathbf{r}_i)|\phi_{nl}\rangle\langle\phi_{nl}|\sum_i \mathbf{r}_i|f^N J'M_J'\rangle}{E(f^N J)-E(\phi_{nl})}\right.$$
$$\left.+\frac{\langle f^N JM_J|\sum_i \mathbf{r}_i|\phi_{nl}\rangle\langle\phi_{nl}|\sum_i V_{\mathrm{c}}^{(\mathrm{odd})}(\mathbf{r}_i)|f^N J'M_J'\rangle}{E(f^N J')-E(\phi_{nl})}\right]. \tag{5.124}$$

An enormous simplification is achieved by adopting the drastic approximation that all of the energy denominators are equal,

$$E(f^N J)-E(\phi_{nl})\cong E(f^N J')-E(\phi_{nl})\cong\Delta E. \tag{5.125}$$

Closure can then be invoked with respect to intermediate states,

$$\sum_{|\phi_{nl}\rangle} |\phi_{nl}\rangle\langle\phi_{nl}| = 1, \tag{5.126}$$

leading to the simpler expression

$$\left\langle a\left|\sum_i \mathbf{r}_i\right|a'\right\rangle = \frac{2}{\Delta E}\left\langle f^N JM_J\left|\sum_i V_c^{(\text{odd})}(\mathbf{r}_i)\mathbf{r}_i\right|f^N J'M'_J\right\rangle. \tag{5.127}$$

The matrix element on the right-hand side of Eq. (5.127) is then evaluated by tensor-operator methods, including application of Eqs. (2.62), (2.63) and (2.70). A much less drastic, although less consistent, approximation is obtained by restoring the dependence of radial integrals and energy denominators on the intermediate-state quantum numbers n and l in the final expression [Hüfner (1978)],

$$\begin{aligned}
&\left\langle a\left|\sum_i r_{ip}\right|a'\right\rangle \\
&= \sqrt{\frac{4\pi}{3}} \sum_{k=1(\text{odd})}^{\infty} \sum_{q=-k}^{k} \sum_{\lambda(\text{even})} B_k^q \sqrt{\frac{2\lambda+1}{2J+1}} \sum_{n,l} (-1)^l W(1\lambda lf;kf) \frac{2}{E(f^N J) - E(nl)} \\
&\times \langle 4f|r^k|nl\rangle\langle nl|r|4f\rangle\langle f\|Y_k\|l\rangle\langle l\|Y_1\|f\rangle\left\langle f^N J\|\mathbf{U}^{(\lambda)}\|f^N J'\right\rangle \\
&\times \langle k1qp|\lambda q+p\rangle\langle J'\lambda M'_J q+p|JM_J\rangle.
\end{aligned} \tag{5.128}$$

The reduced matrix element of the sum of unit tensor operators $\mathbf{U}^{(\lambda)}$, defined by Eqs. (4.22), can be evaluated by means of Eq. (4.23) together with the generalized projection theorem, Eq. (2.69). The formula of Eq. (5.128) works best when applied to transitions between fine-structure levels as a whole, with adjustable crystal-field parameters [Krupke (1966)].

5.4 Examples of crystal-field spectra

5.4.1 *Octahedrally-coordinated Cr^{3+}*

Selected examples of crystal-field spectra of transition-metal complexes serve to clarify matters of principle in electron-lattice coupling. Octahedrally-coordinated Cr^{3+}, which has served as the laser-active centre in more than one host lattice, is especially instructive in illustrating how the nature of optical spectra can be inferred from the energy level structure. The Tanabe–Sugano energy-level diagram for a d^3 configuration with $C/B = 4.50$ is shown in Fig. 4.8. The energies of the $^4T_2$ state and lower $^4T_1$ state with respect to that of the $^4A_2$ state are given as functions of the crystal-field parameter Dq and the Racah parameters B and C by

$$E(^4T_2) = 10Dq, \tag{5.129a}$$

$$E(^4T_1) = 15Dq + \tfrac{15}{2}B - \tfrac{1}{2}\sqrt{100(Dq)^2 - 180DqB + 225B^2}. \tag{5.129b}$$

The energies of the lowest ^2E and $^2T_1$ states, which approach the same constant value in the limit of large Dq, were obtained by numerical matrix diagonalization. Approximate second-order perturbation expressions, more convenient in the present context, are

$$E(^2E) \cong 9B + 3C - \frac{72B^2}{10Dq + 14B + 3C} - \frac{18B^2}{10Dq + 5C}, \tag{5.129c}$$

$$E(^2T_1) \cong 9B + 3C - \frac{9B^2}{10Dq + 6B} - \frac{9B^2}{10Dq} - \frac{12B^2}{20Dq + 4B}. \tag{5.129d}$$

The Tanabe–Sugano diagram can be related to a configuration-coordinate diagram by re-plotting the energy levels as functions of the configuration coordinate Q for symmetrical displacements of the octahedron, provided the dependence of Dq on Q is known. With the assumption of a point-charge model for the crystal field, the required dependence is $Dq \propto Q^{-5}$. The elastic energy associated with displacements from the ground-state equilibrium configuration Q_0 must be added in order to obtain the adiabatic-potential-energy curves. If one assumes that this additional elastic energy is the same for all states and depends quadratically on $Q - Q_0$, and if one further neglects that part of the curvature of each excited-state adiabatic-potential-energy curve derived from the Tanabe–Sugano diagram, which is indeed negligible in comparison with the curvature of the elastic energy, then linear coupling is preserved and only the effective phonon energy $\hbar\omega_0$ is required in order to complete the configuration-coordinate diagram.

The Franck–Condon offset Δ, defined by Eq. (5.93), is given by

$$\Delta = -\omega_0^{-2} \left.\frac{dE}{dQ}\right|_{Q=Q_0} = 5\left(\frac{Dq_0}{\omega_0^2 Q_0}\right) \left.\frac{dE}{dDq}\right|_{Q=Q_0} \tag{5.130}$$

for each excited state, and the corresponding zero-temperature Huang–Rhys factor S_0, defined by Eq. (5.91), is given by

$$S_0 = \left(\frac{25}{12}\right) \frac{(Dq_0)^2}{\hbar\omega_0^3 M r_0^2} \left(\left.\frac{dE}{dDq}\right|_{Q=Q_0}\right)^2, \tag{5.131}$$

where $r_0 = Q_0/\sqrt{6M}$ is the equilibrium metal-ligand distance and M is the mass of one ligand ion. The energy gap between each excited state and the ground state, $E_0(= |\hbar\Omega_0|)$, defined by Eq. (5.86d), is reduced by lattice relaxation to

$$E_0 = E - S_0\hbar\omega_0. \tag{5.132}$$

Typical values of the quantities involved in Eqs. (5.129)–(5.132) are listed in Tables 5.1 and 5.2 for $K_2NaGaF_6:Cr^{3+}$, which has the elpasolite structure, in which the chromium ion occupies a substitutional site of rigorous octahedral symmetry [Andrews *et al.* (1986b)]. It is evident from these tables that the nearly linear energy dependence of the quartet excited states of the Dq configuration in Fig. 4.8 is manifest as strong linear coupling to symmetrical displacements, in contrast with the weak coupling of doublet states whose energies are nearly independent of Dq.

The configuration-coordinate diagram corresponding to the Tanabe–Sugano diagram of Fig. 4.8 is shown schematically in Fig. 5.5. This diagram is not drawn to scale; in particular, the energy gaps are greatly reduced for clarity. Since the multiplicity-allowed transitions between the $^4A_2$ ground state and the $^4T_2$ and $^4T_1$ excited states

Table 5.1. *Parameter values for $K_2NaGaF_6:Cr^{3+}$*

Parameter	Value
B	$771\ cm^{-1}$
C/B	4.30
Dq_0	$1600\ cm^{-1}$
$\hbar\omega_0$	$378\ cm^{-1}$
r_0*	2.06 Å
M	3.154×10^{26} kg

*one-quarter of the lattice parameter

Table 5.2. *Franck–Condon offsets, zero-temperature Huang–Rhys factors, vertical energies and energy gaps for excited states of $K_2NaGaF_6:Cr^{3+}$, calculated with the parameter values listed in Table* 5.1

Excited state	$^4T_2$	$^4T_1$	^2E	$^2T_1$
Δ/Q_0	0.0390	0.0449	0.0023	0.0018
S_0	4.13 (3.98*)	5.46	0.0143	0.0084
E (exp.*)	$16\,000\ cm^{-1}$**	$23\,310\ cm^{-1}$**	$15\,460\ cm^{-1}$**	$16\,450\ cm^{-1}$
E_0	$14\,440\ cm^{-1}$	$21\,240\ cm^{-1}$	$15\,455\ cm^{-1}$	$16\,447\ cm^{-1}$

*Experimental value from Andrews *et al.* (1986b)
**Parameters Dq_0, B and C/B in Table 5.1 were adjusted for a precise fit to these experimental energies with full matrix diagonalization

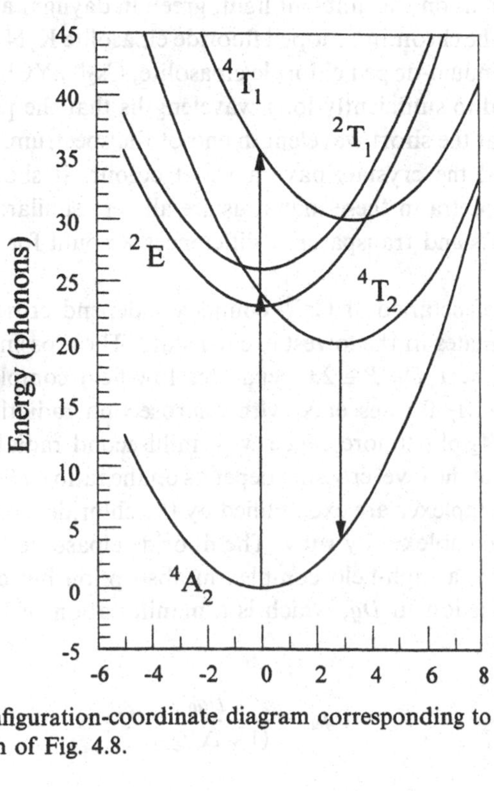

Figure 5.5. Configuration-coordinate diagram corresponding to the d^3 Tanabe–Sugano diagram of Fig. 4.8.

involve strong coupling to symmetrical modes of vibration, the absorption spectrum is dominated by two partially overlapping broad bands which span the visible spectrum. Multiplicity-forbidden transitions, enabled by spin–orbit interaction [Yamaga *et al.* (1989a)], are also evident in the spectrum as weak, narrow features. The general appearance of the absorption spectrum is a common characteristic of Cr^{3+} in octahedral coordination, but the peak energies of the broad bands depend nearly linearly on the crystal-field parameter Dq_0, which exhibits substantial variation, while the weak, narrow features remain at nearly fixed energies. For example, consider substitutional Cr^{3+} in the following sequence of host crystals, listed in order of increasing values of Dq_0: halide elpasolites Cs_2NaYCl_6 and K_2NaGaF_6, beryl ($Be_3Al_2Si_6O_{18}$), chrysoberyl ($BeAl_2O_4$) and corundum (Al_2O_3). The elpasolites are of interest for their rigorous octahedral site symmetry, but are of poor mechanical and optical quality, unsuitable for laser applications. Both magnetic-dipole and phonon-assisted transitions contribute to their optical absorption spectra [Knochenmüss *et al.* (1986)]. In the remaining crystals, which have all been utilized as laser materials as well as precious gems, electric-dipole transitions are mediated by odd-symmetry crystal-field components. In chromium-doped corundum (ruby), which was utilized in the first laser, the absorption bands are shifted to sufficiently short wavelengths that the principal window of transparency occurs at the long wavelength end of the visible spectrum, before the first absorption band, accounting for the dominant red colour, with a secondary window in the blue between the two bands. In chromium-doped beryl (emerald), the principal window of transparency occurs in the green between the two bands. Chromium-doped chrysoberyl (alexandrite) is an intermediate case with two comparable transparency windows, accounting for the alexandrite effect in which the apparent colour depends on the ambient light, green in daylight and red in incandescent or candle light. The chromium-doped fluoride elpasolite $K_2NaGaF_6:Cr^{3+}$ is also green, but in the chromium-doped chloride elpasolite, $Cs_2NaYCl_6:Cr^{3+}$, the absorption bands are shifted to sufficiently long wavelengths that the principal window of transparency occurs at the short wavelength end of the spectrum, beyond the second absorption band, and the crystals have a violet colour. It should be emphasized that the absorption spectra in these materials are all very similar, and that relatively small wavelength shifts and transparency differences account for pronounced colour changes.

Emission spectra of octahedral Cr^{3+} complexes depend critically on Dq_0, since optical emission originates in the lowest excited state. The crossing of the ${}^4T_2$ and ${}^2E$ excited energy levels near $Dq/B \cong 2.0$ separates low-field complexes, characterized by broadband ${}^4T_2 \rightarrow {}^4A_2$ fluorescence with microsecond radiative life times, from narrow-band ${}^2E \rightarrow {}^4A_2$ phosphorescence with millisecond radiative life times. The precise value of Dq/B at the level crossing depends on the ratio C/B, which is somewhat variable. Low-field complexes are exemplified by the chloride elpasolite $Cs_2NaYCl_6:Cr^{3+}$, and high-field complexes by ruby. The fluoride elpasolite $K_2NaGaF_6:Cr^{3+}$ is a border-line case; it is a high-field complex in absorption but a low-field complex in emission. The reduction in Dq, which is a manifestation of lattice relaxation, is given by

$$Dq_{\text{em}} = \frac{Dq_0}{(1 + \Delta/Q_0)^5}. \tag{5.133}$$

However, the nature of the emission spectrum is determined by neither Dq_0 nor Dq_{em}, but rather by the energy gaps, E_0. The criterion for broadband fluorescence at $T=0$ is

$$E_0(^2E) > E_0(^4T_2). \tag{5.134}$$

Low-field complexes are of special interest for wavelength-tunable lasers both because of their broad emission bands and because they behave like four-level systems by virtue of their large Stokes shift, which detunes absorption from emission. Although emerald and alexandrite are examples of high-field complexes, their values of Dq_0 are sufficiently close to the critical value that their $^4T_2$ states are thermally occupied at room temperature, enabling them to function as tunable laser materials. The criterion for broadband fluorescence at finite temperature is

$$\frac{\tau_R(^2E)}{\tau_R(^4T_2)}\exp\left\{-\left[E_0(^4T_2) - E_0(^2E)\right]\Big/k_BT\right\} > 1.0, \tag{5.135}$$

where τ_R is the radiative lifetime. Eq. (5.135) includes Eq. (5.134) as a limiting case.

Although the chromium-doped halide elpasolites are examples of low-field complexes in emission, it has been demonstrated that they can be converted to high-field complexes by application of hydrostatic pressure, which increases Dq_0 by diminishing r_0 [Dolan *et al.* (1986, 1992), Rinzler *et al.* (1993)]. This pressure-induced crossing of excited energy levels is accompanied by a transition from low-field behaviour (broadband fluorescence) to high-field behaviour (narrow-band phosphorescence). The emission spectrum exhibits a pronounced temperature dependence near the level crossing.

Similar behaviour can be observed in a series of chromium-doped compounds with a range of crystal fields which bracket the level crossing. The fact that the transition from low- to high-field behaviour is not as abrupt at low temperature as Eq. (5.134) would imply is explained by a phenomenological model which postulates mixing of the $^4T_2$ and ^2E states, attributed to combined spin–orbit interaction and zero-point vibrations of low-symmetry modes [Henderson *et al.* (1988), Yamaga *et al.* (1990a,b,c)]. Mixed vibronic states are described by the eigenvalues and eigenvectors of the matrix

$$\mathbf{H} = \begin{pmatrix} E(^2E) & \delta \\ \delta & E(^4T_2) \end{pmatrix}, \tag{5.136}$$

where δ is the adjustable mixing parameter. This mixing is manifest as an avoided crossing of excited energy levels, illustrated in Fig. 5.6. The model predicts the dependence of the intensity ratio of broadband to structured emission, I_T/I_E, and the radiative transition rate, $1/\tau$, on both temperature T and energy difference $\Delta E \equiv E(^4T_2) - E(^2E)$. The model was found to work well for the series of host crystals YAG, YGG, GGG, YSGG, GSAG, GSGG and LLGG with adjusted parameters $2\delta = 85\,\text{cm}^{-1}$, $\tau_T = 60\,\mu\text{s}$ and $\tau_E = 4000\,\mu\text{s}$. A more detailed model might include vibronic mixing of fine-structure levels in intermediate coupling.

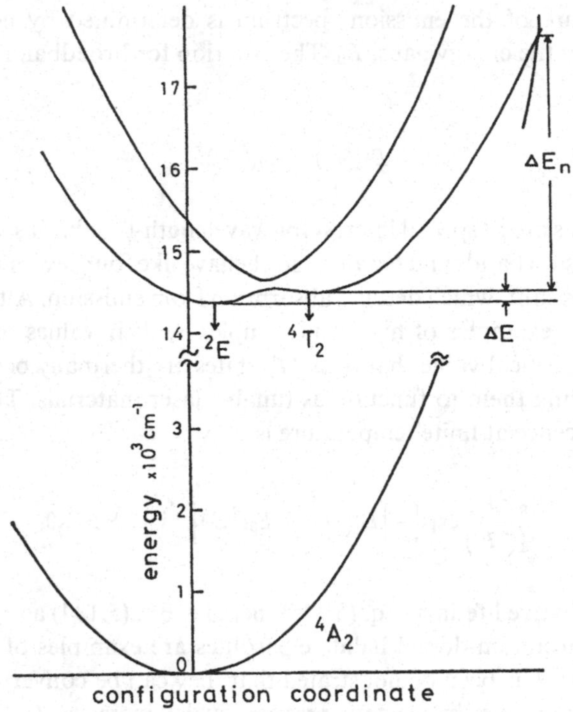

Figure 5.6. Avoided crossing of adiabatic-potential energy curves of octahedrally coordinated Cr^{3+} induced by spin–orbit interaction and zero-point vibrations of low symmetry modes [after Yamaga *et al.* (1990b)].

5.4.2 *Tetrahedrally-coordinated* Cr^{4+}

An example of the effect of low-symmetry crystal-field components is provided by a class of laser materials based on Cr^{4+} in distorted tetrahedral coordination. The crystal-field levels for a d^2 configuration in tetrahedral coordination are the same as those of a d^8 configuration in octahedral coordination, shown in Fig. 4.8. Thus the three lowest triplet states derived from the ^3F ground term of the free ion are ordered as shown schematically at the left-hand side of Fig. 5.7, but their separation is not shown to scale. The electric and magnetic dipole-moment operators transform as the T_2 and T_1 representations of T_d, whose Kronecker products with A_2 are, from Table 2.21,

$$T_2 \times A_2 = T_1, \tag{5.137a}$$

$$T_1 \times A_2 = T_2. \tag{5.137b}$$

Consequently, transitions $^3A_2 \leftrightarrow {}^3T_1$ are only electric-dipole allowed, and transitions $^3A_2 \leftrightarrow {}^3T_2$ are only magnetic-dipole allowed.

Tetragonal distortion of the ligand tetrahedron reduces the symmetry to D_{2d} and splits the crystal-field levels in accordance with Table 2.29, as shown in Fig. 5.7. It is also evident from this table that components of the electric and magnetic dipole-moment operators transform as $B_2 + E$ and as $A_2 + E$, respectively, enabling the

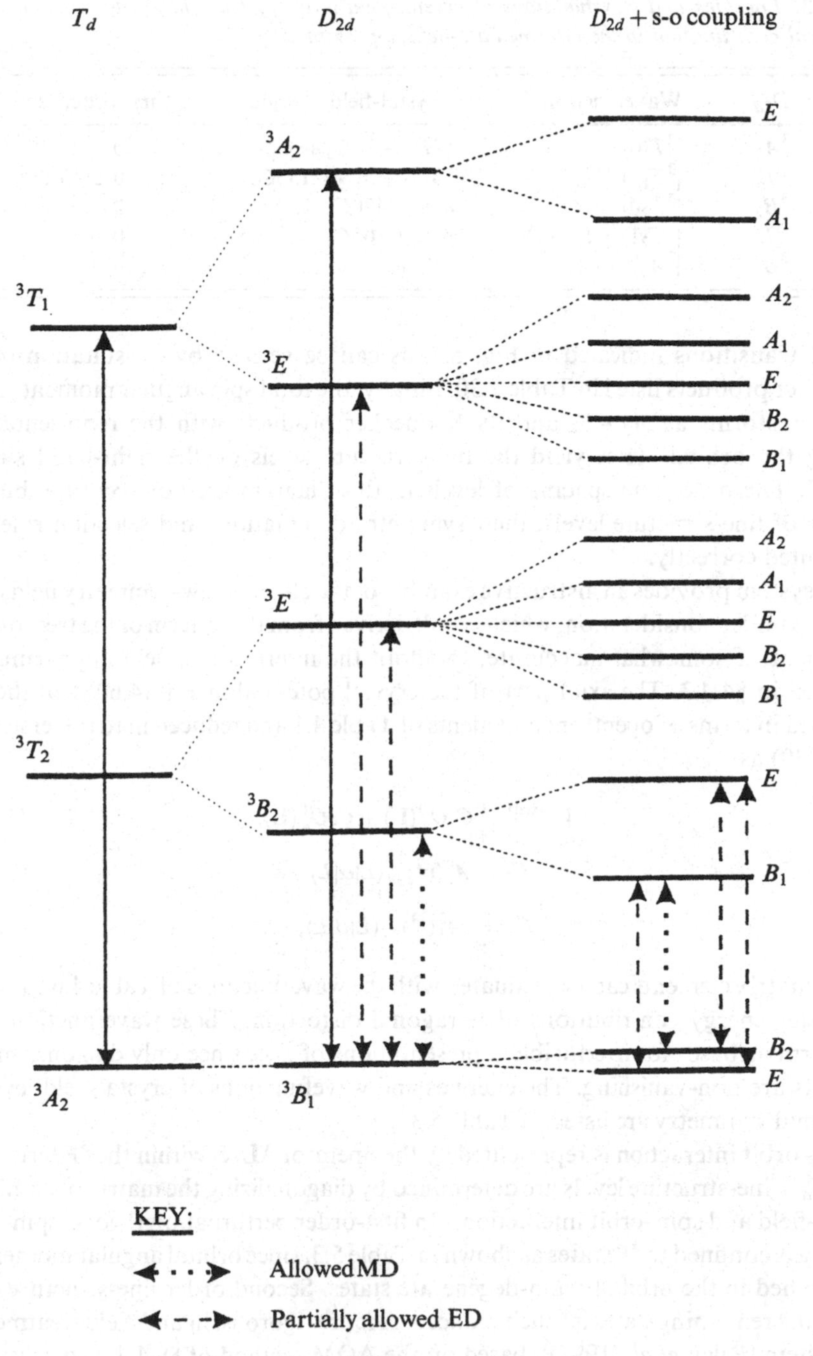

Figure 5.7. Crystal-field and spin–orbit splitting of Cr^{4+} energy levels in tetragonally distorted tetrahedral coordination. Symmetry designations and selection rules are represented correctly, but the order and spacing of energy levels are not.

Table 5.3. *Energies and wavefunctions of crystal-field levels of Cr^{4+} in tetragonally distorted tetrahedral coordination in the intermediate-field approximation*

T_d	D_{2d}	Wavefunction	Crystal-field energies	First-order, spin-orbit
$^3T_1$	$^3A_2$	$\|^3T_1 0\rangle$	$18Dq - 12C_2 + 360C_4$	0
	^3E	$\|^3T_1 1\rangle, \|^3T_1 -1\rangle$	$18Dq + 6C_2 + 135C_4$	$0, \pm 3\lambda/2$
$^3T_2$	$^3B_2$	$\|^3T_2 0\rangle$	$8Dq - 420C_4$	0
	^3E	$\|^3T_2 1\rangle, \|^3T_2 -1\rangle$	$8Dq + 105C_4$	$0, \pm\lambda/2$
$^3A_2$	$^3B_1$	$\|^3A_2\rangle$	$-420C_4$	0

allowed transitions indicated in Fig. 5.7, as can be verified by consultation of the Kronecker products listed in Table 2.30. Finally, the total spin angular momentum for $S=1$ transforms as A_2+E, and its Kronecker products with the representations labeling the orbital states yield the fine-structure levels on the right-hand side of Fig. 5.7. The order and spacing of levels in this diagram are not accurate, but the number of fine-structure levels, their symmetry designations and selection rules are represented correctly.

This system provides an instructive example of the effect of low-symmetry fields. For the states under consideration, which are all derived from the ^3F term of the free ion, it is convenient, if somewhat inaccurate, to adopt the intermediate-field approximation discussed in §4.4.3. The axial part of the crystal potential in Eq. (4.60) can then be expressed in terms of operator equivalents of Table 4.1 and reduced matrix elements of Eqs. (4.39) as

$$V_c^{(\text{axial})} = C_2 O_2^0(\mathbf{L}) + C_4 O_4^0(\mathbf{L}), \tag{5.138a}$$

$$C_2 = A_2^0 \langle r^2 \rangle_{3d} \langle L|\alpha|L\rangle, \tag{5.138b}$$

$$C_4 = \delta A_4^0 \langle r^4 \rangle_{3d} \langle L|\beta|L\rangle, \tag{5.138c}$$

and its matrix elements can be evaluated with the wavefunctions of Table 4.6 to obtain first-order energy contributions of tetragonal distortion. These wavefunctions also transform as bases for irreducible representations of D_{2d}, since only diagonal matrix elements are non-vanishing. The energies and wavefunctions of crystal-field levels in tetragonal symmetry are listed in Table 5.3.

Spin–orbit interaction is represented by the operator $\lambda\mathbf{L}\cdot\mathbf{S}$ within the ^3F term, with $\lambda = \frac{1}{2}\zeta_{3d}$. Fine-structure levels are determined by diagonalizing the matrix of combined crystal-field and spin–orbit interactions. In first-order perturbation theory, spin–orbit splitting is confined to ^3E states as shown in Table 5.3, since orbital angular momentum is quenched in the orbitally non-degenerate states. Second-order fine-structure splitting of the remaining states is much reduced. A more rigorous, ligand-field treatment of this system [Riley *et al.* (1998)], based on the AOM method of §8.4.4, is discussed in §9.3.2 and §10.3.6.

5.4.3 *Octahedrally-coordinated Ti^{3+}*

Octahedrally coordinated Ti^{3+} provides another instructive example. The d^1 ground configuration has only two states in octahedral symmetry, the ground $^2T_2$ state and the excited ^2E state, separated by $10Dq$. Since this energy-level separation has the same

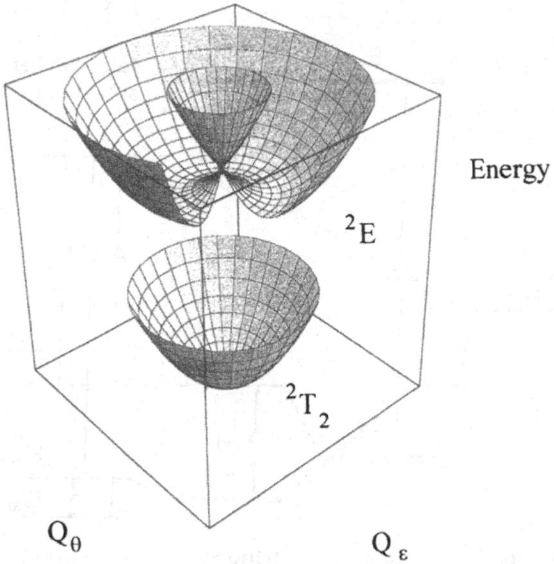

Figure 5.8. Adiabatic potential energy surfaces of the ground and excited states of octahedrally-coordinated Ti^{3+}. Configuration coordinates Q_θ and Q_ε may be identified with the $E_g\theta$ and $E_g\varepsilon$ distortions of an octahedron in Fig. 3.2.

crystal-field dependence as that of the $^4A_2$ and $^4T_2$ states of octahedrally coordinated Cr^{3+}, linear coupling to symmetrical displacements is comparable for the two complexes. In addition, the absence of higher excited states within the ground configuration eliminates excited-state absorption, an attractive feature for tunable-laser applications.

A distinctive feature of this complex is strong Jahn–Teller coupling to modes of e_g symmetry [Macfarlane *et al.* (1968)]. $E \times e$ coupling in the excited state is strongly favored by the shapes of the e_g orbitals, which conform with the symmetry-adapted displacements of the ligands, as is evident from Figs. 4.2 and 3.2 [Sturge (1967)]. $T_2 \times e$ coupling in the ground state is much less favoured, as is coupling in both states to t_{2g} displacements, which involve bending rather than stretching. Accordingly, we will retain only $E \times e$ coupling, and the adiabatic potential energy surfaces in the two-dimensional space of e_g ligand displacements are as shown in Fig. 5.8. Consequently, the absorption spectrum is distinctly bimodal, while the emission spectrum appears as a single broadband [L-HOPS (1958)]. The shape of the emission band is a convolution of the individual line shapes for coupling to a_{1g} and e_g modes, and its effective coupling constant in the strong-coupling limit is given approximately by

$$S_{0\text{eff}} \cong S_0 + \tfrac{1}{2}k^2. \tag{5.139}$$

The additional broadening of the emission spectrum by coupling to e_g modes serves to extend the tuning range in tunable laser applications, exemplified by Ti-sapphire ($Al_2O_3 : Ti^{3+}$) [Moulton (1986a)].

The effect of low-symmetry distortions on this system is illustrated in Fig. 5.9. The Jahn–Teller effect predominates in the ^2E_g excited state, removing all of the orbital degeneracy and leaving only Kramers doublets, while trigonal distortion and spin–orbit interaction combine to remove the orbital degeneracy of the $^2T_{2g}$ ground state.

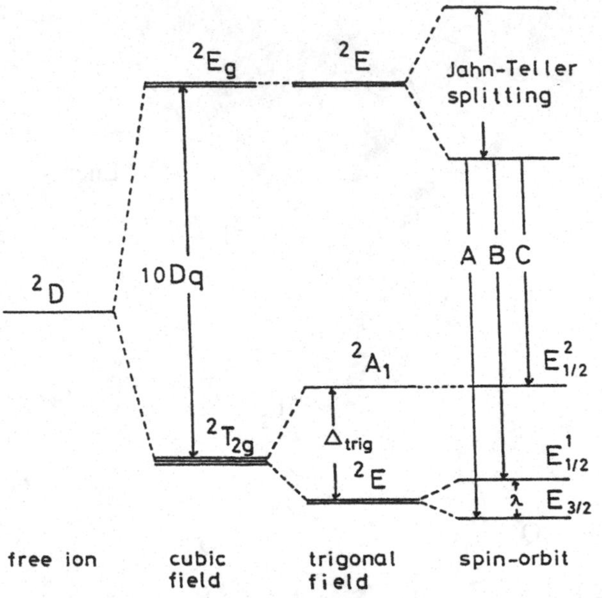

Figure 5.9. Crystal-field, spin-orbit and Jahn–Teller splitting of Ti^{3+} energy levels in trigonally distorted octahedral coordination [after Yamaga *et al.* (1991b)].

5.5 Approximate line-shape functions

5.5.1 Alternative energy units

A general expression for the normalized line-shape function $G(\Omega)$ associated with an optical transition from electronic state a to electronic state b is provided by Eq. (5.88), where the transition energy is defined by Eq. (5.87). However, it is sometimes more convenient to employ normalized line-shape functions of photon frequency or transition energy, related by

$$h^{-1}g(\nu) \equiv I(E) \equiv \hbar^{-1}G(\Omega), \tag{5.140}$$

where

$$h\nu \equiv E \equiv |\hbar(\Omega_0 + \Omega)|. \tag{5.141}$$

For example, in the case of the configuration-coordinate model with linear coupling in the low temperature limit, Eqs. (5.90) and (5.94) imply

$$G^{(0)}(\Omega) = \sum_{p=0}^{\infty} \frac{\exp(-S_0)S_0^p}{p!}\,\delta(p\omega_0 - \Omega). \tag{5.142}$$

This line shape is the same for both absorption and emission, but Ω_0 is positive in absorption and negative in emission. On the other hand, the energy gap of Eq. (5.132), defined by

$$E_0 \equiv |\hbar\Omega_0|, \tag{5.143}$$

is always positive, and the line-shape functions of photon frequency and transition energy are given, respectively, by

$$g(\nu) = \sum_{p=0}^{\infty} \frac{\exp(-S_0)S_0^p}{p!} \delta(E_0/h \pm p\omega_0/2\pi - \nu), \tag{5.144}$$

$$I(E) = \sum_{p=0}^{\infty} \frac{\exp(-S_0)S_0^p}{p!} \delta(E_0 \pm p\hbar\omega_0 - E), \tag{5.145}$$

where the upper sign applies in absorption and the lower sign in emission.

Line-shape functions $G(\Omega)$ for absorption and emission are also identical in the more general case of linear coupling to many symmetrical modes with a range of frequencies at arbitrary temperature. It follows that the absorption and emission bands as functions of photon frequency ν or transition energy E are mirror images of one another with respect to the zero-phonon line, $p=0$; thus, in the low-temperature limit, they have only the zero-phonon line in common.

5.5.2 *Typical Huang–Rhys factors*

The electronic wavefunctions of point imperfections in insulators and semiconductors range from extremely compact to extremely diffuse. This variation in scale is reflected in the strength of electron–lattice coupling, as manifest both in crystal-field splitting and in the Huang–Rhys factor, S_0. Lanthanide ions, in which the valence charge density shields the electrons in open-shell ($4f$) orbitals from the charges of neighbouring ions rather than from the charge of the central nucleus, provide an extreme example of compact wavefunctions. Their Huang–Rhys factors are typically of the order $S_0 \cong 0.01$. It follows that zero-phonon lines account for 99% of the intensity of optical spectra, and single-phonon sidebands account for most of the remaining 1%; thus rare-earth spectra are characterized by very narrow lines. Iron-group transition metals have significantly less compact open-shell ($3d$) orbitals. Their typical Huang–Rhys factors range from $S_0 \cong 0.01$ to $S_0 \cong 10$, depending on whether crystal-field splitting is involved, as is evident from Table 5.2. Their zero-phonon lines are still prominent, but their sidebands are often dominant, as in the example of Fig. 5.3. Impurity-related colour centres, such as the $Tl^0(1)$ centre, have more diffuse wavefunctions with $S_0 \cong 10$. *F*-centre wavefunctions are even more diffuse, comparable with the lattice spacing. Crystal-field splitting is responsible for their entire electronic structure, and they exhibit the largest Huang–Rhys factors, as high as $S_0 \cong 100$. For them, the zero-phonon line is no longer detectable and the line-shape function, smoothed by multiple convolutions, is nearly gaussian. Shallow donor states in semiconductors provide the extreme example of diffuse wavefunctions, which extend over very many lattice sites. Since their excess electrons sample bulk properties, they are relatively insensitive to ionic displacements and their typical Huang–Rhys factors are again $S_0 \cong 0.01$.

Optical transitions in potential laser materials tend to fall into one of two categories. Lasing transitions in fixed frequency lasers, e.g., intra-$4f$ transitions of rare-earth ions and ruby R lines, involve weak coupling, $S_0 \ll 1$, while pump bands and lasing transitions in vibronic tunable lasers, e.g., alexandrite, involve strong coupling, $S_0 \gg 1$.

5.5.3 *Strong-coupling limit*

The line-shape function $G^{(0)}(\Omega)$ for the configuration-coordinate model with linear coupling is given for arbitrary temperature by Eqs. (5.90), and its first three moments are listed in Eqs. (5.95).

It is evident from Eq. (5.95a) that S_0 is the mean number of phonons emitted in the transition. This result suggests that an appropriate value for the Stokes shift is $2S_0\hbar\omega_0$, in agreement with the classical Franck–Condon principle. On the other hand, Eq. (5.94) implies that the intensity ratio of successive lines in the absorption spectrum at low temperature is $R_p/R_{p-1} = S_0/p$. It follows that, for sufficiently strong coupling, the envelope function is maximum near $p \cong S_0 - \frac{1}{2}$. The Stokes shift based on the energy difference between the peaks of the absorption and emission bands is then $(2S_0 - 1)\hbar\omega_0$. The peak energies differ from the first moments because the line shapes are asymmetrical. The former are more convenient to determine experimentally than the latter; however, the Stokes shift formula based on peak energies fails in the limit $S_0 \to 0$, whereas the formula based on moments behaves correctly in this limit and remains applicable at finite temperature.

For sufficiently strong coupling, $S_0 \gg 1$, the third moment, Eq. (5.95c), becomes relatively less important than the second, Eq. (5.95b), and the envelope of the line-shape function can be approximated by a gaussian function with the same first and second moments as the actual function,

$$G^{(0)}(\Omega) \cong \frac{1}{\sqrt{2\pi S_0\omega_0^2(2\bar{n}+1)}}\exp\left[-\frac{(\Omega - S_0)^2}{2S_0\omega_0^2(2\bar{n}+1)}\right], \tag{5.146}$$

It follows from the definition of $\bar{n}$, Eq. (5.91), that

$$2\bar{n} + 1 = \coth(\hbar\omega_0/2k_{\mathrm{B}}T). \tag{5.147}$$

The full width at half maximum (FWHM) is then given by

$$\mathrm{FWHM} = \left[(8\ln 2)S_0\omega_0^2\coth(\hbar\omega_0/2k_{\mathrm{B}}T)\right]^{1/2}. \tag{5.148}$$

In the high temperature limit, $k_{\mathrm{B}}T \gg \hbar\omega_0$, Eq. (5.146) reduces to

$$G^{(0)}(\Omega) \cong \frac{1}{\sqrt{4\pi S_0\omega_0 k_{\mathrm{B}}T/\hbar}}\exp\left[-\frac{(\Omega - S_0)^2}{4S_0\omega_0 k_{\mathrm{B}}T/\hbar}\right]. \tag{5.149}$$

This approximation can be derived alternatively from a purely classical model in which the line shape of the optical transition is interpreted as the projection of the thermal density distribution of the initial-state configuration-coordinate values on the slope of the final-state configuration-coordinate curve; it is thus a manifestation of the correspondence principle. The corresponding FWHM is given by

$$\mathrm{FWHM} \cong \left[(16\ln 2)S_0\omega_0 k_{\mathrm{B}}T/\hbar\right]^{1/2}. \tag{5.150}$$

Eq. (5.150) was adapted to a Jahn–Teller system in Eq. (7.26) by replacing $S_0\hbar\omega_0$ with E_{JT}.

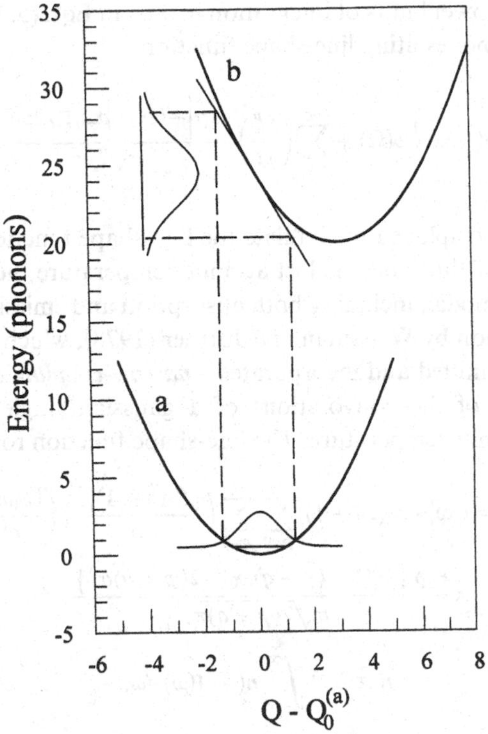

Figure 5.10. Semi-classical approximation for the optical absorption line-shape function for linear coupling to a single mode in the strong-coupling, low-temperature limit.

In the low-temperature limit, Eq. (5.146) reduces to

$$G^{(0)}(\Omega) \cong \frac{1}{\sqrt{2\pi S_0 \omega_0^2}} \exp\left[-\frac{(\Omega - S_0)^2}{2S_0\omega_0^2}\right], \tag{5.151}$$

with FWHM given by

$$\text{FWHM} = \left[(8\ln 2)S_0\omega_0^2\right]^{1/2}. \tag{5.152}$$

This approximation can be derived alternatively from a semi-classical model in which the line shape of the optical transition is interpreted as the projection of the square of the lowest initial-state vibrational wave function, Eq. (5.80b), on the slope of the final-state configuration-coordinate curve, as illustrated in Fig. 5.10.

5.5.4 Approximations for linear coupling to many modes

A model based on the theory of Pryce (1966) for linear coupling to many modes (§5.2.5) can be implemented in the low-temperature limit by adopting an approximate gaussian single-phonon sideband,

$$A(\omega) = \frac{S_0}{\sigma\sqrt{2\pi}} \exp\left[-(\omega - \omega_0)^2/2\sigma^2\right], \tag{5.153}$$

and by extending the lower limits of integration to $-\infty$ in Eqs. (5.96) to accommodate the approximation. The resulting line-shape function is

$$G(\Omega) = \exp(-S_0)\left\{\delta(\Omega) + \sum_{p=1}^{\infty}\left(\frac{S_0^p}{p!}\right)\frac{\exp\left[-(\Omega - p\omega_0)^2/2p\sigma^2\right]}{\sigma\sqrt{2p\pi}}\right\}. \tag{5.154}$$

Equation (5.154) was employed to simulate the line-shape functions in Figs. 5.3 and 9.6b. The simplicity of this model is lost at finite temperature, however, since $B_1(\Omega)$ becomes distinctly bimodal, including both absorption and emission of a phonon.

A related formulation by Weissman and Jortner (1978), which emphasizes the net number of phonons emitted and incorporates a *narrow-coupling approximation*, preserves the simplicity of the convolutions of a gaussian single-phonon sideband, Eq. (5.153), even at finite temperature. The line-shape function for this model is

$$G(\Omega) = \exp[-S_0(2\bar{n}+1)]\sum_{p=0}^{\infty}\sum_{q=0}^{\infty}\left(\frac{[S_0(\bar{n}+1)]^p}{p!}\right)\left(\frac{[S_0\bar{n}]^q}{q!}\right)$$
$$\times\frac{\exp\left\{-[\Omega-(p-q)\omega_0]^2/2(p+q)\sigma^2\right\}}{\sigma\sqrt{2(p+q)\pi}}, \tag{5.155a}$$

$$\bar{n} \equiv S_0^{-1}\int_{-\infty}^{\infty} n(\omega)A(\omega)\,d\omega, \tag{5.155b}$$

where the term in the summation with $p=q=0$ is identified with $\delta(\Omega)$. Note that Eq. (5.155a) reduces to Eq. (5.154) in the low-temperature limit.

The moments of the line-shape function of Eq. (5.155a) are [Bartram *et al.* (1986b)]

$$M_1 = S_0, \tag{5.156a}$$

$$M_2 = S_0(1+\sigma^2/\omega_0^2)(2\bar{n}+1), \tag{5.156b}$$

$$M_3 = S_0(1+3\sigma^2/\omega_0^2); \tag{5.156c}$$

consequently, $G(\Omega)$ can be approximated in the strong-coupling limit by

$$G^{(0)}(\Omega) \cong \frac{1}{\sqrt{2\pi S_0(\omega_0^2+\sigma^2)(2\bar{n}+1)}}\exp\left[-\frac{(\Omega - S_0\omega_0)^2}{2S_0(\omega_0^2+\sigma^2)(2\bar{n}+1)}\right], \tag{5.157}$$

with line width

$$\text{FWHM} = \left[(8\ln 2)S_0(\omega_0^2+\sigma^2)\coth(\hbar\omega_0/2k_BT)\right]^{1/2}. \tag{5.158}$$

5.5.5 *Lattice Green's function method for linear coupling to many modes*

The single-phonon sideband, which was approximated by a single gaussian function in Eq. (5.153), actually exhibits complex structure related to the phonon density of states of the host lattice [DeLeo *et al.* (1981)]. In the harmonic approximation with assumed time dependence $\exp(-i\omega t)$, perfect-lattice ion displacements $\mathbf{u}$, with elements labeled by lattice site ℓ, ion type κ and cartesian component α, satisfy Lagrange's equations in the form [Maradudin *et al.* (1971)]

$$\mathbf{L}^0(\omega^2)\cdot\mathbf{u}=0, \tag{5.159a}$$

$$\mathbf{L}^0(\omega^2)\equiv\mathbf{M}^0\omega^2-\mathbf{\Phi}^0, \tag{5.159b}$$

where $\mathbf{M}^0$ and $\mathbf{\Phi}^0$ are square mass and force-constant matrices in the same space as $\mathbf{u}$. Normal modes of vibration of the perfect lattice are defined by the set of eigenvalues ω_j^2 and eigenvectors χ_j^0, which satisfy orthonormality and closure relations of the form

$$\tilde{\chi}_j^{0*}\cdot\mathbf{M}^0\cdot\chi_{j'}^0=\delta_{jj'}, \tag{5.160a}$$

$$\sum_j\mathbf{M}^0\cdot\chi_j^0\tilde{\chi}_j^{0*}=\mathbf{I}. \tag{5.160b}$$

The corresponding perfect-lattice Green's function matrix and its analytic continuation are given by

$$\mathbf{G}^0(\omega^2)\equiv\mathbf{L}^0(\omega^2)^{-1}=\sum_j\frac{\chi_j^0\tilde{\chi}_j^{0*}}{\omega^2-\omega_j^2}, \tag{5.161a}$$

$$\operatorname{Re}\mathbf{G}^0(\omega^2-i\varepsilon)=\pi^{-1}P\int_0^{\omega(\max)}\operatorname{Im}\mathbf{G}^0(\omega'^2-i\varepsilon)\frac{2\omega'd\omega'}{\omega^2-\omega'^2}, \tag{5.161b}$$

$$\operatorname{Im}\mathbf{G}^0(\omega^2-i\varepsilon)=\frac{\pi}{2\omega}\sum_j\chi_j^0\tilde{\chi}_j^{0*}\delta(\omega-\omega_j), \tag{5.161c}$$

where the limit $\varepsilon\to 0+$ is understood and P denotes the Cauchy principal value.

A point imperfection, such as a colour centre or impurity, modifies both the mass and force-constant matrices and introduces additional forces acting on the ions,

$$F_{l\kappa\alpha}=-\frac{\partial}{\partial u_{\ell\kappa\alpha}}V(\mathbf{u}), \tag{5.162}$$

where $V(\mathbf{u})$ is the adiabatic potential energy associated with a particular electronic state. The resulting single-phonon sideband for a transition between non-degenerate states with linear coupling to many modes is given by

$$A(\omega)=\frac{1}{\pi\hbar\omega^2}\Delta\tilde{\mathbf{F}}^*\cdot\operatorname{Im}\left\{\left[\mathbf{I}-\mathbf{G}^0(\omega^2-i\varepsilon)\cdot\delta\mathbf{L}\right]^{-1}\cdot\mathbf{G}^0(\omega^2-i\varepsilon)\right\}\cdot\Delta\mathbf{F}, \tag{5.163a}$$

$$\delta\mathbf{L}=-\delta\mathbf{M}\omega^2+\delta\mathbf{\Phi}, \tag{5.163b}$$

where $\Delta\mathbf{F}$ is the change in the force vector associated with the transition between electronic states. Equations (5.163) can then be used in conjunction with the Pryce formalism, Eqs. (5.96)–(5.98), to generate a line-shape function.

Rigid ions are assumed in the preceeding development. Polarizable ions can be accommodated by employing the shell model for lattice dynamics [Dick and Overhauser (1958)] or breathing shell model [Schröder (1966)]. The method can also be extended to transitions between degenerate electronic states by exploiting the theory of Jahn–Teller coupling to many modes with a range of frequencies [O'Brien (1972)]. Both of these refinements were incorporated in an application of this method to the R centre in KCl [DeLeo *et al.* (1981)], based on independently determined shell-model parameters [Copley *et al.* (1969)]. In that application, the predicted single-phonon sideband was compared directly with an experimental single-phonon sideband extracted from the transition line shape by a deconvolution procedure [Giesecke *et al.* (1972)].

5.5.6 *Approximations for quadratic and anharmonic coupling*

The line-shape function $G(\Omega)$ for nonlinear coupling to a single mode in the harmonic approximation is obtained by specializing Eq. (5.88),

$$G(\Omega) = [1 - \exp(-\hbar\omega_a/k_B T)] \sum_\alpha \exp(-\alpha\hbar\omega_a/k_B T)$$

$$\times \sum_\beta \left| \int \chi_\beta^{(b)}\left(Q - Q_0^{(b)}\right) \chi_\alpha^{(a)}\left(Q - Q_0^{(a)}\right) dQ \right|^2 \delta(\beta\omega_b - \alpha\omega_a - \Omega). \qquad (5.164)$$

Pure quadratic coupling was considered by Kubo (1952), Kubo and Toyozawa (1955) and Keil (1965); the latter expressed the vibrational overlap integral in terms of an associated Legendre polynomial. Qualitatively, $G(\Omega)$ in the low temperature limit consists of a series of equally spaced delta functions multiplied by an envelope function which exhibits a sharp cut-off on one side and an exponential tail on the other.

In the more general case of combined linear and quadratic coupling to a single mode, Struck and Fonger (1975) appealed to recursion relations due to Manneback (1951) for evaluation of the vibrational overlap integral and calculated the line-shape function $G(\Omega)$ numerically as a histogram.

Woods *et al.* (1994) developed an even more general method for optical line-shape simulation by direct diagonalization of the matrix of the vibrational Hamiltonian within each electronic state in a common harmonic-oscillator basis. The vibrational Hamiltonian for mode a of electronic state n is expressed in terms of ladder operators b_a and b_a^+, defined by Eqs. (5.81), in the form

$$\frac{H_a^{(n)}}{\hbar\omega_a} = b_a b_a^+ + \frac{1}{2} + \sum_k C_{ak}^{(n)} (b_a + b_a^+)^k. \qquad (5.165)$$

Evaluation of matrix elements is facilitated by Eqs. (5.82). Matrices are truncated to finite dimension; typically, convergence is achieved for dimensions less than 100. Vibrational overlap integrals are evaluated as direct products of eigenvectors, and the corresponding energy eigenvalues are employed in evaluation of delta-function factors and thermal weighting factors. Line-shape functions are accumulated as histograms, and the composite line shape for a finite set of discrete modes is calculated by numerical convolution of the component line shapes. This method accommodates arbitrary combinations of linear, quadratic and anharmonic coupling.

5.5.7 *Zero-phonon line*

The zero-phonon line, which predominates in most rare-earth spectra, is extremely narrow compared with the single-phonon sideband. Nevertheless, its width is generally orders of magnitude greater than lifetime broadening, Eq. 5.70, would suggest, and is strongly temperature dependent. For a system in which the energy separation of initial and final electronic states, both from one another and from other electronic states, exceeds the maximum phonon energy, the homogeneous broadening of the zero-phonon line is mediated by Raman scattering which does not affect the radiative lifetime [DiBartolo (1968)]. It is manifest in a one-electron density of states function $\rho(E)$ for each electronic state, given by [Stoneham (1985)]

$$\rho(E) = \frac{\Gamma/2\pi}{(E - E_0 - \Lambda)^2 + (\Gamma/2)^2}. \tag{5.166a}$$

Approximate expressions for Γ and Λ can be derived from the Debye model, in which the crystal is treated as a homogeneous, isotropic elastic continuum with only acoustic modes, but the phonon density of states is truncated at a maximum angular frequency $\omega_{\max} = k_B T_D/\hbar$ to preserve the number of degrees of freedom [DiBartolo (1968)],

$$\Gamma \propto \left(\frac{T}{T_D}\right)^7 \int_0^{T_D/T} \frac{x^6 e^x}{(e^x - 1)^2}\, dx, \tag{5.166b}$$

$$\Lambda \propto \left(\frac{T}{T_D}\right)^4 \int_0^{T_D/T} \frac{x^3}{e^x - 1}\, dx. \tag{5.166c}$$

The energy width Γ is proportional to T^7 at low temperatures, $T \ll T_D$, and to T^2 at high temperatures, $T \gg T_D$, while the energy shift Λ is proportional to T^4 at low temperatures and to T at high temperatures. The absolute magnitudes of these parameters depend on the strength of coupling of the electronic system to the strains associated with acoustic modes. The line-shape function for the zero-phonon line is just the convolution of the functions $\rho(E)$ for the initial and final electronic states. It is evident that the energy shift for a zero-phonon transition between two electronic states is just the difference of their individual energy shifts Λ. It follows from the convolution theorem that the line-shape function is also Lorentzian and its energy width is the sum of the individual energy widths Γ for the two states. In systems which fail to satisfy the assumed restrictions on energy separations, phonon-induced transitions to nearby electronic states make a temperature-dependent contribution to the homogeneous linewidth by enhanced lifetime broadening. Inhomogeneous broadening by random strain distributions [Stoneham (1969)], for which the zero-phonon line shape is gaussian, tends to be the dominant mechanism at very low temperatures in crystals and at all temperatures in glasses. The line shape is intermediate between gaussian and Lorentzian when neither homogeneous nor inhomogeneous broadening is dominant. Since the interplay between homogeneous and inhomogeneous processes as a function of temperature is both ion and host dependent, the preceding observations should be regarded as broad generalizations.

5.6 Nonlinear susceptibilities

Linear susceptibilities of optical materials are related to optical spectra by Kramers–Kronig relations. They are calculated quantum-mechanically by evaluating the expectation value of the dipole-moment operator, $-e\sum_i \mathbf{r}_i$, in first-order time-dependent perturbation theory, with the perturbation of Eq. (5.11). Thus for a system of identical ions of density N, the non-resonant linear susceptibility is given by

$$\chi^{(1)}(\omega) = \frac{Ne^2}{\varepsilon_0 \hbar} \sum_{a,b} P_a \left[\frac{\mathbf{r}_{ab}\mathbf{r}_{ba}}{\omega_b - \omega_a - \omega} + \frac{\mathbf{r}_{ba}\mathbf{r}_{ab}}{\omega_b - \omega_a + \omega} \right], \tag{5.167a}$$

$$P_a = \frac{\exp(-\hbar\omega_a/k_B T)}{\sum_c \exp(-\hbar\omega_c/k_B T)}, \tag{5.167b}$$

where $\mathbf{r}_{ab}$ is concise notation for $\langle a|\sum_i \mathbf{r}_i|b\rangle$.

Nonlinear susceptibilities, discussed in §3.3.5, are manifest only in the presence of the intense electric fields inherent in laser radiation. Their quantum-mechanical calculation proceeds by application of higher orders of time-dependent perturbation theory [Bloembergen (1965)]. A general expression can be derived for the nth-order non-resonant nonlinear susceptibility [Butcher and Cotter (1990)],

$$\chi^{(n)}_{\mu\alpha_1\cdots\alpha_n}(-\omega_\sigma; \omega_1, \ldots, \omega_n)$$
$$= \frac{N e^{n+1}}{\varepsilon_0\, n! \hbar^n} \mathbf{S}_\mathrm{T} \sum_{ab_1\cdots b_n} P_a \left[\frac{r^{\mu}_{ab_1} r^{\alpha_1}_{b_1 b_2} \cdots r^{\alpha_{n-1}}_{b_{n-1}b_n} r^{\alpha_n}_{b_n a}}{(\Omega_{b_1 a} - \omega_1 - \cdots - \omega_n)(\Omega_{b_2 a} - \omega_2 - \cdots - \omega_n)\cdots(\Omega_{b_n a} - \omega_n)} \right], \tag{5.168a}$$

$$\omega_\sigma = \omega_1 + \omega_2 + \cdots + \omega_n, \tag{5.168b}$$

$$\Omega_{ba} = \omega_b - \omega_a, \tag{5.168c}$$

where μ, α denote cartesian components. The total symmetrization operator $\mathbf{S}_\mathrm{T}$ implies that the expression which follows it is to be summed over all $(n+1)!$ permutations of the pairs $(\mu, \omega_\sigma), (\alpha_1, \omega_1), \ldots, (\alpha_n, \omega_n)$.

The second-order susceptibility derived from Eqs. (5.168) is discussed in terms of the structural properties of important nonlinear optical crystals in §9.8. The applications of optical nonlinearities in diode-pumped solidstate lasers are reviewed in §10.8.3.

6

Radiationless transitions

6.1 Physical principles

The present chapter is concerned with transitions between optical levels of point imperfections in solids in which energy is conserved by emission of phonons rather than photons. Such transitions can introduce an undesired loss mechanism in laser materials which competes with stimulated emission, but they also play an essential role in the performance of laser materials, since excited states reached in electric-dipole-allowed pumping transitions subsequently relax to metastable lasing states by a sequence of radiationless processes.

Radiationless transitions are, paradoxically, both ubiquitous and elusive. They are the rule, rather than the exception, following optical excitation; yet, they are inaccessible to direct observation. Radiationless relaxation is generally inferred from its consequences, including such familiar phenomena as radiant heating and the absence of luminescence. In favourable circumstances, the observation of luminescence with diminished intensity and duration provides more detailed information about radiationless processes. Quantitative understanding of some aspects of radiationless relaxation has also proved elusive, in that theory has been more successful in elucidating trends than in the prediction of absolute transition rates. Radiationless transitions generally belong to one of two categories: static processes which are thermally activated from a metastable state, and dynamic processes which occur during rapid relaxation immediately following excitation [Bartram (1990)]. Point imperfections in solids provide examples of both categories.

6.1.1 Prepared state

Radiationless transitions can only occur between non-stationary states of a system; thus the radiationless transition rate depends critically on the sort of non-stationary state which is prepared in a given experiment. The structure of the prepared excited state, the initial state for the radiationless process, depends both on the spectral characteristics of the exciting light and on the absorption spectrum. The projected radiative state [Rhodes (1977)] is a coherent superposition of those exact eigenstates which are accessible by radiative transitions from the ground state in a frequency range $2\pi\tau^{-1}$, where τ is the duration of a minimum uncertainty pulse or, equivalently, the coherence time for continuous excitation. The interaction of the projected radiative state with a forbidden manifold (i.e., inaccessible by radiative transitions from the ground state) mediates radiationless transitions.

The prepared excited state for a dynamic process is identified with the projected radiative state itself. In the static case, on the other hand, the prepared state is a relaxed excited state which evolves from the projected radiative state together with its interacting forbidden manifold in a time which is short compared with the duration of the thermally activated radiationless process of interest. In the latter case, the prepared state is more appropriately described as a statistically mixed state rather than a pure state.

The prepared state is generally assumed to be a Born–Oppenheimer state. This assumption is warranted in the circumstance that the frequency range $2\pi\tau^{-1}$ of the exciting pulse exceeds the frequency width $\Delta\omega$ of the absorption band associated with a single Born–Oppenheimer state, which is broadened by its interaction with the forbidden manifold, but does not encompass more than one such state [Rhodes (1977)]. Excitation with radiation whose coherence time exceeds the radiative lifetime of the system excites a nearly exact eigenstate of the Hamiltonian which can only decay radiatively. At the opposite extreme, a very short pulse excites a coherent superposition of Born–Oppenheimer states [Rhodes (1971)].

6.1.2 *Radiationless transition rate*

It is generally assumed that the system Hamiltonian can be partitioned in the form

$$H = H_0 + H', \tag{6.1}$$

where the prepared state $|p\rangle$ is an eigenstate of H_0 and radiationless transitions are mediated by H'. The choice of H_0 is not uniquely determined by the prepared state. One possible approach to the partitioning of H is based on Feshbach projection operators [Feshbach (1962)] $\mathscr{P}$ and $\mathscr{Q}$, defined by [Englman and Zgierski (1985)]

$$\mathscr{P} = |p\rangle\langle p|, \tag{6.2}$$

$$\mathscr{Q} = 1 - \mathscr{P}. \tag{6.3}$$

The prepared state $|p\rangle$ is clearly an eigenstate of H_0, defined by

$$H_0 = \mathscr{P}H\mathscr{P} + \mathscr{Q}H\mathscr{Q}, \tag{6.4}$$

with eigenvalue $\langle p\,|\,H\,|\,p\rangle$. It follows that H' is given by

$$H' = \mathscr{P}H\mathscr{Q} + \mathscr{Q}H\mathscr{P}. \tag{6.5}$$

The radiationless transition rate in first-order time-dependent perturbation theory is then given by

$$W_{\mathrm{NR}} = (2\pi/\hbar)\sum_q |\langle p|\mathscr{P}H\mathscr{Q}|q\rangle|^2\delta(E_p - E_q). \tag{6.6}$$

It should be noted that H' makes a second-order contribution to the energy of state $|p\rangle$ as well.

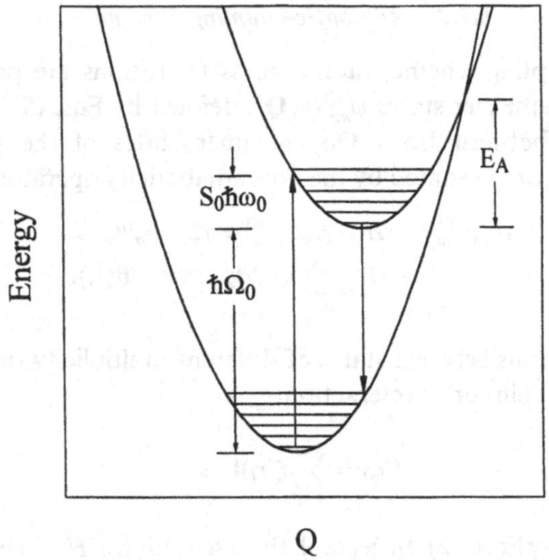

Figure 6.1. Configuration coordinate diagram for linear coupling [after Bartram (1990)].

Equation (6.6) is actually more general than its derivation would suggest, in spite of the fact that the choice of H_0 is not unique [Englman and Zgierski (1985)]. The subsequent development of the prepared state $|p\rangle$ is described by the imaginary part of the projected Green's function $\mathscr{P}G\mathscr{P}$, where G is defined by

$$(E - H)G = 1, \tag{6.7}$$

and where E is assumed to have an infinitesimal positive imaginary part. This approach leads directly to Eq. (6.6) without appeal to the partitioning of H in Eq. (6.1).

6.2 Static processes

6.2.1 Mott theory

An early explanation of thermally activated radiationless transitions [Mott (1938), Gurney and Mott (1939)] can be understood with reference to the configuration-coordinate diagram of Fig. 6.1. The radiationless transition rate is given by the Arrhenius equation

$$W_{\mathrm{NR}} = s \exp(-\Delta E/k_{\mathrm{B}}T), \tag{6.8}$$

where the attempt frequency s is identified with the effective vibration frequency $\omega_0/2\pi$ associated with the configuration coordinate, and the activation energy ΔE with the energy difference E_{A} between the crossing of adiabatic potential energy curves and the minimum of the excited-state curve. The Mott formula, Eq. (6.8), provides insight into the mechanism of radiationless transitions, and is useful in explaining trends, but fails by tens of orders of magnitude to account quantitatively for measured radiationless transition rates [Struck and Fonger (1975)].

6.2.2 Adiabatic-coupling scheme

In the adiabatic-coupling scheme, radiationless transitions are presumed to occur between Born–Oppenheimer states $\psi_{n\nu}^{\mathrm{BO}}(\mathbf{r},\mathbf{Q})$, defined by Eqs. (5.72)–(5.75). Radiationless transitions between Born–Oppenheimer states of the same multiplicity (internal conversion) are mediated by the non-adiabaticity operator, defined by

$$H_{\mathrm{NA}}\psi_{n\nu}^{\mathrm{BO}} = (H - E_{n\nu})\psi_{n\nu}^{\mathrm{BO}} = [T_N, \phi_n]\theta_{n\nu}$$
$$\cong -\hbar^2 \sum_p (\partial\phi_n/\partial Q_p)(\partial\theta_{n\nu}/\partial Q_p). \tag{6.9}$$

Radiationless transitions between states of different multiplicity (inter-system crossing) are mediated by spin–orbit interaction,

$$H_{\mathrm{SO}} = \sum_{i=1}^{N} \xi(r_i)\mathbf{l}_i \cdot \mathbf{s}_i, \tag{6.10}$$

where $\xi(r)$ is defined by Eq. (4.2). In general, the interaction $\mathscr{P}H\mathscr{Q}$ in Eq. (6.6) is given by

$$\mathscr{P}H\mathscr{Q} = \mathscr{P}H_{\mathrm{NA}}\mathscr{Q} + \mathscr{P}H_{\mathrm{SO}}\mathscr{Q}, \tag{6.11}$$

and the operative part of $\mathscr{P}H\mathscr{Q}$ depends on the nature of the forbidden manifold of states with which the prepared state interacts.

Since the prepared state for a static process is a mixed state, the radiationless transition rate W_{NR} between electronic states $|i\rangle$ and $|f\rangle$ is obtained by thermally averaging the right-hand side of Eq. (6.6) over initial vibrational states α and summing over final vibrational states β,

$$W_{\mathrm{NR}} = \left(\frac{2\pi}{\hbar}\right) \sum_\alpha \sum_\beta P_\alpha |\langle\psi_{f\beta}^{\mathrm{BO}}|\mathscr{P}H\mathscr{Q}|\psi_{i\alpha}^{\mathrm{BO}}\rangle|^2 \delta(E_{f\beta} - E_{i\alpha}), \tag{6.12a}$$

$$P_\alpha \equiv \frac{\exp(-E_{i\alpha}/k_{\mathrm{B}}T)}{\sum_{\alpha'}\exp(-E_{i\alpha'}/k_{\mathrm{B}}T)}. \tag{6.12b}$$

In the case of internal conversion, it is convenient to distinguish between promoting coordinates Q_p which mix the initial and final electronic states, and accepting coordinates Q_a which absorb the difference in electronic energy [Lin and Bersohn (1968)]. The matrix element in Eq. (6.12a) for the case of internal conversion has often been approximated by the expression

$$\langle\psi_{f\beta}^{\mathrm{BO}}|H_{\mathrm{NA}}|\psi_{i\alpha}^{\mathrm{BO}}\rangle \cong \sum_p -\hbar^2[U_i(\mathbf{0}) - U_f(\mathbf{0})]^{-1}\langle\phi_f(\mathbf{0})|(\partial H_e/\partial Q_p)_{\mathbf{Q}=\mathbf{0}}|\phi_i(\mathbf{0})\rangle\langle\theta_{f\beta}|\partial/\partial Q_p|\theta_{i\alpha}\rangle, \tag{6.13}$$

proposed by Huang and Rhys (1950) in analogy with the Condon approximation for radiative transitions, Eq. (5.84). More recently, this popular approximation has been shown to be internally inconsistent in the present context, predicting transition rates which are too small by three orders of magnitude [Huang (1981)], and various 'non-Condon' corrections have been proposed [Ridley (1978)]. The difficulty with the Condon approximation arises from evaluation of the energy denominator $U_i(\mathbf{Q}) - U_f(\mathbf{Q})$ at the initial-state equilibrium configuration rather than in the vicinity of the avoided crossing of the adiabatic-potential-energy curves.

6.2.3 Static-coupling scheme

In the static-coupling scheme, radiationless transitions are presumed to occur between crude-adiabatic states $\psi_{n\nu}^{CA}(\mathbf{r},\mathbf{Q})$, defined by Eqs. (5.120). The non-adiabaticity operator H_{NA} is ineffectual in this scheme, since

$$H_{NA}\psi_{n\nu}^{CA}(\mathbf{r},\mathbf{Q}) = 0. \tag{6.14}$$

Instead, radiationless transitions are mediated by the operator $\mathscr{P}[V(\mathbf{r},\mathbf{Q}) - V(\mathbf{r},\mathbf{0})]\mathscr{Q}$.

The approximate equivalence of the adiabatic- and static-coupling schemes has been demonstrated by several authors at a certain level of approximation [Helmis (1965), Pässler (1974, 1982), Gutsche (1982), Burt (1983)], although it has also been demonstrated that they cannot be precisely equivalent [Wagner (1982), Denner and Wagner (1984)]. This approximate equivalence provides a more tractable expression for the required matrix element in the adiabatic-coupling scheme [Bartram and Stoneham (1985)],

$$\langle\psi_{f\beta}^{BO}|H_{NA}|\psi_{i\alpha}^{BO}\rangle \cong \sum_p \langle\phi_f(\mathbf{0})|(\partial H_e/\partial Q_p)_{\mathbf{Q}=\mathbf{0}}|\phi_i(\mathbf{0})\rangle\langle\theta_{f\beta}|Q_p|\theta_{i\alpha}\rangle. \tag{6.15}$$

If one makes the additional assumptions that it is feasible to separate variables in Eq. (5.75), that the harmonic approximation is valid, that the normal coordinates are the same in both initial and final electronic states, and that the symmetry is sufficiently high that promoting and accepting modes are distinct, then the vibrational wavefunctions may be written as

$$\theta_{n\nu}(\mathbf{Q}) = \prod_p \chi_{\nu_p}^{(n)}(Q_p)\prod_a \chi_{\nu_a}^{(n)}(Q_a - Q_{0a}^{(n)}), \tag{6.16}$$

and Eq. (6.12a) becomes

$$W_{NR} = \sum_p \nu_p\omega_p[\bar{n}_p G(-\Omega_0 + \omega_p) + (\bar{n}_p + 1)G(-\Omega_0 - \omega_p)], \tag{6.17}$$

where $G(\Omega)$ is the accepting-mode normalized line shape function of Eq. (5.88) with electronic-state designations a, b replaced by i, f,

$$G(\Omega) = \sum_\alpha\sum_\beta p_\alpha \prod_a \left|\int \chi_{\beta_a}^{(f)}(Q_a - Q_{0a}^{(f)})\chi_{\alpha_a}^{(i)}(Q_a - Q_{0a}^{(i)})dQ_a\right|^2 \delta(\Omega_{f\beta} - \Omega_{i\alpha} - \Omega_0 - \Omega), \tag{6.18}$$

$\bar{n}_p$ is a promoting-mode phonon-occupation number,

$$\bar{n}_p = [\exp(\hbar\omega_p/k_B T) - 1]^{-1}, \tag{6.19}$$

and ν_p is the promoting interaction defined by

$$\nu_p = (\pi/\hbar\omega_p^2)|\langle\phi_f(\mathbf{0})|(\partial H_e/\partial Q_p)_{\mathbf{Q}=\mathbf{0}}|\phi_i(\mathbf{0})\rangle|^2. \tag{6.20}$$

The corresponding expression for inter-system crossing at the same level of approximation is simply

$$W_{NR} = (2\pi/\hbar^2)|\langle\phi_f(\mathbf{0})|H_{SO}|\phi_i(\mathbf{0})\rangle|^2 G(-\Omega_0). \tag{6.21}$$

We have adopted the convention that Ω_0, defined by Eq. (5.86d) with a, b replaced by i, f,

$$\hbar\Omega_0 = U_f(\mathbf{Q}_0^{(f)}) - U_i(\mathbf{Q}_0^{(i)}), \tag{6.22}$$

is positive in absorption and negative in emission, contrary to the convention of Bartram (1990).

The line-shape function $G(\Omega)$, Eq. (6.18), also describes spectral line shapes for radiative transitions, as described in Chapter 5. However, the radiationless transition rate depends on this function for values of its argument which are very far from the range of values accessible to direct observation. Reliable evaluation of this line-shape function is the central problem in the application of the theory of static radiationless transitions, since its exact form depends on the model assumed for accepting-mode interactions. An exploration of alternative models is the agenda of subsequent sections. The vibrational overlap integrals for $\alpha_a \neq \beta_a$ in Eq. (6.18) are non-vanishing only if the equilibrium values of Q_a differ in the initial and final electronic states (linear coupling), if the accepting-mode vibration frequencies differ (quadratic coupling), if there are departures from the harmonic approximation (anharmonic coupling), or any combination of the above. It is convenient to classify models by the nature of their coupling.

6.2.4 *Linear coupling*

In the special case of linear coupling to a single accepting mode (configuration-coordinate model), the radiationless transition rate is obtained by combining Eq. (6.17) with Eqs. (5.90). We will consider several limiting cases of interest, with the additional assumption of a single promoting mode. In weak coupling, $S_0 \ll 1$, typical of rare-earth impurities, the energy-gap law applies [Riseberg and Moos (1968), Moos (1970)],

$$W_{\mathrm{NR}} \cong \nu \exp(-\alpha p)(\bar{n}+1)^p/\sqrt{2\pi p}, \tag{6.23a}$$

$$\alpha = \ln(p/S_0) - 1, \tag{6.23b}$$

where $p(= |\Omega_0|/\omega_0)$ is the number of accepting-mode phonons required to bridge the energy gap between initial and final electronic states. In strong coupling, typical of F centres, thermally activated behaviour is predicted [Englman and Jortner (1970)],

$$W_{\mathrm{NR}} \cong \nu[(2\bar{n}+1)/2\pi S_0]^{1/2} \exp(-E_A/k_{\mathrm{B}}T^*), \tag{6.24a}$$

$$k_{\mathrm{B}}T^* = (\hbar\omega_0/2)\coth(\hbar\omega_0/2k_{\mathrm{B}}T), \tag{6.24b}$$

$$E_{\mathrm{A}} = (\hbar|\Omega_0| - S_0\hbar\omega_0)^2/4S_0\hbar\omega_0. \tag{6.24c}$$

where E_{A} is the curve-crossing energy with respect to the minimum of the upper adiabatic potential energy curve, as indicated in Fig. 6.1. Equations (6.24) reduce to the Mott formula, Eq. (6.8), in the limit $k_{\mathrm{B}}T \gg \hbar\omega_0$, except that the frequency factor s reflects the promoting interaction and can no longer be identified simply with the vibration frequency $\omega_0/2\pi$. However, unlike the Mott formula, they also predict a finite transition rate at $T=0$, due to zero-point vibration, which may be described as 'tunneling'. The effective activation energy at intermediate temperatures is given by

$$\Delta E_{\text{eff}} = -\frac{\partial \ln(W_{\text{NR}})}{\partial (k_{\text{B}}T)^{-1}} = E_{\text{A}}\text{sech}^2(\hbar\omega_0/2k_{\text{B}}T). \tag{6.25}$$

Finally, in the limit $T=0$, the general expression for W_{NR} for arbitrary values of S_0 reduces to

$$W_{\text{NR}} = \nu \exp(-S_0)S_0^{p-1}/(p-1)!. \tag{6.26}$$

A closed form expression for the normalized line-shape function $G(\Omega)$ is no longer available in the case of weak linear coupling to many modes with a range of vibration frequencies. In that case, one can employ the line-shape function of Pryce (1966), Eqs. (5.97) and (5.154), or the refinement of Weissman and Jortner (1978), Eqs. (5.155).

6.2.5 *Quadratic and anharmonic coupling*

Line-shape functions $G(\Omega)$ for nonlinear coupling can be generated numerically and displayed as histograms by the methods described in §5.5.6. The method of Struck and Fonger (1975), which exploits recursion relations due to Manneback (1951), is applicable to combined linear and quadratic coupling to a single mode in the harmonic approximation. A more general method [Woods *et al.* (1994)], which employs direct diagonalization of the matrix of the vibrational Hamiltonian within each electronic state in a common harmonic-oscillator basis, is applicable to combined linear, quadratic and anharmonic coupling to a single mode. In either method, the composite line shape for a finite set of discrete modes can be constructed by numerical convolution of the component line shapes.

6.3 Dynamic processes

6.3.1 *Landau–Zener theory*

A seminal treatment of dynamic radiationless transitions was presented independently by Landau (1932) and Zener (1932, 1933) in the context of molecular dissociation and atomic collisions. In this formulation, the electronic motion is treated quantum-mechanically and the nuclear motion, classically. The contemplated situation is an avoided crossing of adiabatic potential-energy curves $U_k(Q)$, $k=i,f$, at $Q=Q_0$. A parameter of the theory is $v=dQ/dt$ evaluated at $Q=Q_0$. Zener showed that the probability that the system traverses the crossing adiabatically (i.e., that it follows one adiabatic potential-energy curve) is given by

$$P_{if} = 1 - \exp\{-2\pi\varepsilon^2/\hbar v|s_i - s_f|\}, \tag{6.27}$$

$$s_k = (dU_k/dQ)_{Q=Q_0}, \tag{6.28}$$

where 2ε is the minimum separation of adiabatic potential-energy curves.

In the present application, the potential energy curves are parabolas with an energy gap $\hbar|\Omega_0|$ and an avoided intersection at an energy E_{A} above the minimum of the

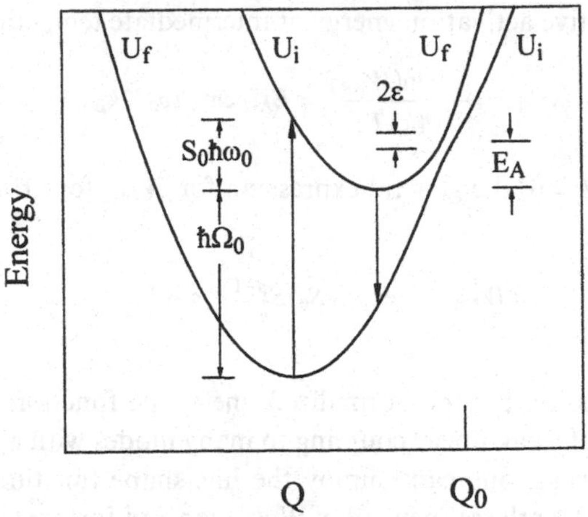

Figure 6.2. Avoided crossing of adiabatic potential energy curves, illustrating the DKR criterion [Dexter *et al.* (1955)] for athermal quenching of luminescence [after Bartram (1990)].

upper curve, as shown in Fig. 6.2. It follows from the Landau–Zener theory that the probability P_{NR} for a radiationless transition in a double traversal of the avoided crossing is given by

$$P_{NR} = 2P_{if}(1 - P_{if}), \tag{6.29}$$

$$v = [2(E - E_A - \hbar|\Omega_0|)]^{1/2}, \tag{6.30}$$

where a mass-weighted configuration coordinate is assumed, and E is measured from the minimum of the lower curve.

The implications of Landau–Zener theory for radiationless transitions of point imperfections in solids depend on the details of the relaxation process following excitation. The configuration coordinate Q which interacts with the electronic system (the 'interaction coordinate') is typically not a true normal coordinate of the crystal. Nevertheless, one can treat it as an approximate normal coordinate by effecting a transformation which introduces coupling of the corresponding localized mode to the remaining lattice modes, thus providing a damping mechanism for vibrational excitation. The rate of damping is proportional to the range of frequencies of normal modes which contribute appreciably to the interaction coordinate. In cases where the interaction coordinate is a normal coordinate for a true normal mode, it is still coupled to lattice modes via anharmonicity. In either case, the strength of this coupling governs relaxation behaviour.

6.3.2 Seitz criterion

An early application of the Landau–Zener approach to point imperfections in solids is embodied in the Seitz criterion for luminescence [Seitz (1939)]. Seitz tacitly assumed sufficiently strong coupling to lattice modes to ensure that the local mode behaves as an over-damped harmonic oscillator. Thus, for the configuration coordinate diagram of Fig. 6.2, the system 'slides' down the upper curve to its minimum following radiative

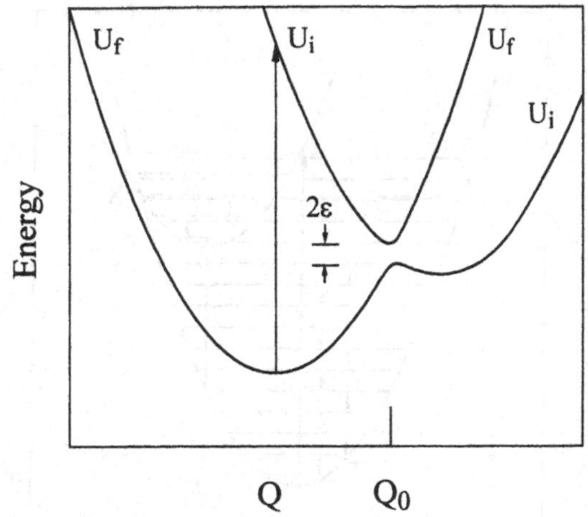

Figure 6.3. Avoided crossing of adiabatic potential energy curves, illustrating the Seitz criterion [Seitz (1939)] for athermal quenching of luminescence [after Bartram (1990)].

excitation, and the avoided crossing never comes into play; consequently, efficient luminescence is expected. A contrasting case is illustrated in Fig. 6.3, in which the minimum of the upper curve lies outside the lower curve. In this case, Seitz assumes that the relaxation proceeds sufficiently rapidly that the avoided crossing is traversed diabatically ($P_{if} \ll 1$), but, by virtue of over-damping, it is traversed just once. In this case also, the system 'slides' to the minimum of the upper curve. By virtue of the Franck–Condon principle, the lower curve is inaccessible from this point by spontaneous emission. Accordingly, the minimum of the lower curve is ultimately reached by a thermally activated process, unaccompanied by luminescence. The Seitz criterion has been designated 'the slide rule' by Englman (1979).

6.3.3 Dexter–Klick–Russell criterion

Dexter, Klick and Russell (DKR) (1955) proceeded from the opposite assumption of a weakly damped harmonic oscillator. The DKR rule predicts radiationless de-excitation of an excited point imperfection in a solid when the energy reached in a Franck–Condon transition from the ground electronic state lies above the avoided crossing of adiabatic potential energy curves, as shown in Fig. 6.2, since the avoided crossing is traversed many times with gradually diminishing velocity as vibrational energy is slowly transferred to the surrounding lattice. The DKR criterion for radiationless de-excitation in the special case of linear coupling, as elaborated by Bartram and Stoneham (1975) can be expressed as $\Lambda > 0.25$, where Λ is defined as the ratio of excited-state relaxation energy to optical absorption energy,

$$\Lambda = \frac{S_0 \hbar\omega_0}{\hbar|\Omega_0| + S_0 \hbar\omega_0}. \tag{6.31}$$

For comparison, the Seitz criterion for radiationless de-excitation in linear coupling is $\Lambda > 0.5$.

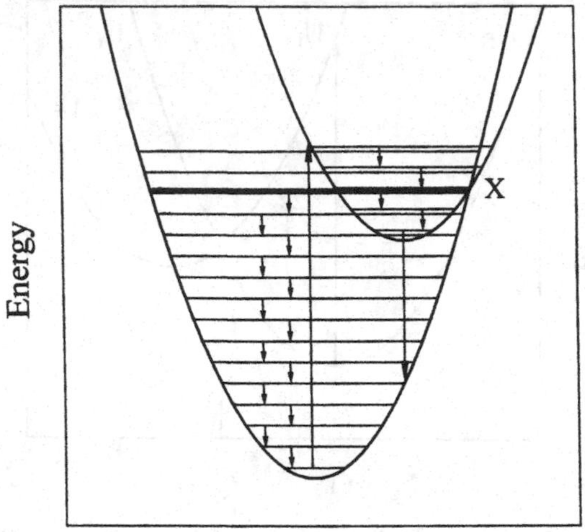

Figure 6.4. Model of Stoneham and Bartram (1978) for the residual quantum efficiency of fluorescence of a system which satisfies the DKR criterion for athermal quenching [after Bartram (1990)]. Branching occurs from a parity-mixed vibronic state at the curve-crossing energy. Subsequent cooling can proceed in either electronic state and fluorescence quantum efficiency is determined by the branching ratio.

Multiple traversals of the avoided crossing in the slow-cooling limit do not ensure an absolute prohibition of luminescence for $0.25 < \Lambda < 0.5$. In order to predict the quantum efficiency of residual luminescence in this range, Stoneham and Bartram (1978) developed a theoretical model for the dynamic process from a different point of view. Their model incorporates a vibronic state of mixed parity involving both accepting and promoting modes.

Essential features of the model are illustrated in Fig. 6.4. Optical absorption is followed by gradual cooling of accepting-mode vibrations until the total energy coincides with the crossing of potential energy curves. This crossing is not avoided in the static-coupling description employed in the theory. At this energy, a strong vibronic mixing of ground and excited accepting-mode crude-adiabatic (CA) states is mediated by the promoting interaction. Subsequent cooling can then proceed in either electronic state until the system either returns directly to the relaxed ground state or reaches the relaxed excited state from which it ultimately decays radiatively on a much longer timescale.

Fluorescence quantum efficiency η is determined by the branching ratio R of cooling from the mixed vibronic state to the excited and ground electronic states, respectively,

$$\eta = \frac{R}{1+R}. \tag{6.32}$$

With the assumption of non-degenerate electronic states, the rate of cooling in each state is approximately proportional to its vibrational quantum number at the crossing, and R is given by

$$R = (1 - 2\Lambda)^2. \tag{6.33}$$

6.3.4 *Extended crossing*

The model of Stoneham and Bartram incorporates the extreme heuristic assumption that the promoting interaction is effective only between a single pair of CA states at the curve-crossing energy. Leung and Song (1980) and Jortner (1979) correctly observed that the promoting interaction is effective between many such states both above and below the curve-crossing energy.

An elaboration of the concept of vibronic states of mixed parity is embodied in the detailed computational model of Englman and Barnett (1970). In their static coupling formalism, the vibronic states are expressed as linear combinations of CA states,

$$\psi_{\gamma}^{\text{mixed}}(\mathbf{r}, Q) = \sum_{\alpha} a_{\gamma\alpha}\psi_{i\alpha}^{\text{CA}}(\mathbf{r}, Q) + \sum_{\beta} b_{\gamma\beta}\psi_{f\beta}^{\text{CA}}(\mathbf{r}, Q), \tag{6.34}$$

where $\psi_{n\nu}^{\text{CA}}(\mathbf{r}, Q)$ is defined by Eqs. (5.120). Coefficients $a_{\gamma\alpha}$ and $b_{\gamma\beta}$ are determined by diagonalizing the matrix of the Hamiltonian

$$H = H_i|i\rangle\langle i| + H_f|f\rangle\langle f| + T(|i\rangle\langle f| + |f\rangle\langle i|), \tag{6.35a}$$

$$H_i = \tfrac{1}{2}\hbar\omega_0[P^2 + (Q-\Delta)^2] - \hbar\Omega_0, \tag{6.35b}$$

$$H_f = \tfrac{1}{2}\hbar\omega_0[P^2 + Q^2], \tag{6.35c}$$

in a representation of CA states. The Frank–Condon offset Δ is defined by Fig. 5.3. This form, in which mixing is mediated by an interaction whose transfer matrix element T between electronic states is treated as an adjustable parameter, correctly represents spin–orbit interaction in inter-system crossing. However, it is only an approximate representation of the promoting interaction in internal conversion and obscures the requirement that a promoting-mode phonon be emitted or absorbed in the radiationless transition. Numerical examples [Barnett and Englman (1970)] reveal not only strong horizontal mixing of CA states near the crossing energy, but also vertical mixing of CA states for which the classical turning point of one potential-energy curve coincides with the minimum of the other.

6.3.5 *Coherent state*

Detailed and rigorous applications of Landau–Zener theory to dynamic radiationless processes of point imperfections in insulators have been presented by several investigators [Kusonoki (1979), Kayanuma and Nasu (1978), Nasu and Kayanuma (1978), Kayanuma (1982), Sumi (1980, 1982a,b)]. A common feature of these applications is the adoption of a constant transfer matrix element T between electronic states, which determines the minimum separation 2ε in Eq. (6.27).

A condition for applicability of the Landau–Zener theory in the low temperature limit is that the prepared state be a coherent vibrational state associated with a single electronic state, of the form

$$\psi_n^{\text{coh}}(\mathbf{r}, Q) = \phi_n(\mathbf{r}, Q)\exp(-\tfrac{1}{2}S_0)\sqrt{S_0^{\nu}/\nu!}\,\chi_{n\nu}(Q)\exp[-i(\nu+\tfrac{1}{2})\omega_0 t], \tag{6.36}$$

where S_0 is the zero-temperature Huang–Rhys factor. Femtosecond pulsed excitation, or an equivalently short coherence time, is required for the preparation of such a state.

The expectation value of the vibrational energy associated with this coherent state is

$$E - E_0 = (S_0 + \tfrac{1}{2})\hbar\omega_0. \tag{6.37}$$

The coherent state is not a stationary state; rather, it is a localized wave packet which retains its shape while oscillating between classical turning points with angular frequency ω_0. Damping of the oscillation in each electronic state appears as a natural consequence of dephasing of the normal modes which are superposed in the interaction coordinate. This dephasing can occur very slowly in the case of a sharp resonance produced by mass or force-constant changes associated with a point imperfection. Since the maximum value of P_{NR}, given by Eq. (6.29), is 0.5, thorough quenching of luminescence can occur only if the cooling rate in the excited state is sufficiently slow to permit many traversals of the avoided crossing, but fast enough in the ground state to suppress the back reaction.

6.4 Manifestations of radiationless transitions

6.4.1 *Thermal activation*

Radiationless relaxation is usually inferred from the observation of luminescence with diminished intensity and duration when radiative and radiationless transition rates are comparable. In a static process, the radiationless transition rate between states of the same multiplicity is given by Eq. (6.17), and that between states of different multiplicity by Eq. (6.21). Since the rate increases very rapidly with increasing temperature, observable effects are confined to a relatively narrow temperature range. Within that narrow range, $\ln W_{NR}$ varies approximately linearly as a function of $1/T$; consequently, W_{NR} can be fitted approximately to an Arrhenius equation, Eq. (6.8). However, the fitting parameters have an uncertain physical significance [Struck and Fonger (1975)]; thus, the activation energy ΔE and frequency factor s should not be identified too literally with specific features of the configuration-coordinate diagram.

The luminescence lifetime as a function of temperature for an electric-dipole allowed transition is then well represented by the expression

$$\frac{1}{\tau(T)} = \frac{1}{\tau_R^S} + \sum_{\Gamma} \frac{1}{\tau_R^{(\Gamma)}} \coth\left(\frac{\hbar\omega^{(\Gamma)}}{2k_B T}\right) + \frac{1}{\tau_{NR}} \exp\left(-\frac{\Delta E}{k_B T}\right), \tag{6.38}$$

which combines the radiative transition rate due to odd components of the static-crystal-field, $1/\tau_R^S$, with the phonon-assisted rate from Eq. (5.122) and the radiationless rate from Eq. (6.8), where the frequency factor s is replaced by $1/\tau_{NR}$. As an example, the measured lifetimes $\tau(T)$ are plotted in Fig. 6.5 for three chromium-doped halide elpasolite crystals [Andrews *et al.* (1986b)]. The solid curves are least-squares fits to the data of Eq. (6.38) with the assumption of a single odd mode and no odd components of the static crystal field, and the optimum parameters are listed in Table 6.1. (A somewhat different fitting function was employed in the original publication.) It can be seen that the onset of thermal quenching occurs rather abruptly. With continuous optical excitation, the emission intensity remains constant below the onset temperature, but diminishes rapidly above that temperature until it is no longer observable. The effects of thermal quenching are apparent only in the narrow temperature range between the

Table 6.1 *Optimum values of parameters for radiative and non-radiative transitions inferred from the temperature dependence of fluorescence life times of chromium-doped halide elpasolites plotted in Fig. 6.5 and chromium-doped scandium borate plotted in Fig. 6.6*

Crystal	$\tau_R^{(\Gamma)}$ (μs)	$\hbar\omega^{(\Gamma)}$ (cm^{-1})	$s = 1/\tau_{NR}$ (sec^{-1})	ΔE (cm^{-1})
$Cs_2NaYCl_6:Cr^{3+}$	130	164	3.8×10^{13}	4250
$K_2NaScF_6:Cr^{3+}$	476	263	1.2×10^{13}	7270
$K_2NaGaF_6:Cr^{3+}$	526	270	5.2×10^{13}	9240
$ScBO_3:Cr^{3+}$	200	255	3.1×10^{9}	3700

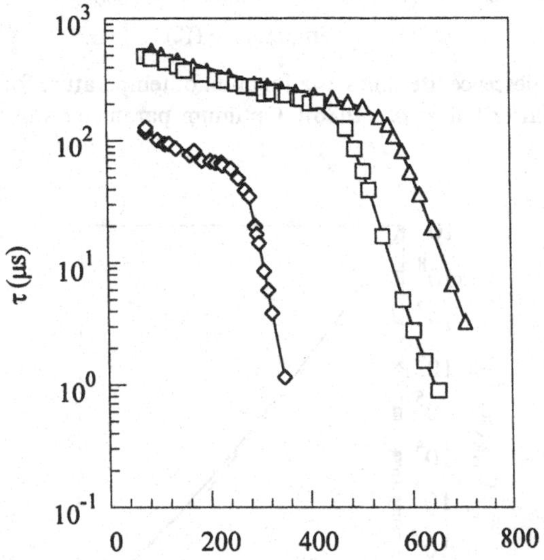

Figure 6.5. Luminescence life times as a function of temperature for substitutional Cr^{3+} in Cs_2NaYCl_6 (◇), K_2NaScF_6 (□) and K_2NaGaF_6 (△) [after Andrews *et al.* (1986b)]. Solid lines are least-squares fits to Eq. (6.38). Optimum parameter values are listed in Table 6.1.

onset temperature and the temperature at which luminescence effectively disappears. A second example with similar features is provided by the laser material $ScBO_3:Cr^{3+}$; measured life times $\tau(T)$ for this crystal [Lai *et al.* (1986)] are plotted in Fig. 6.6, and the optimum fitting parameters are listed in Table 6.1.

Semi-log plots of W_{NR} vs. $1/T$, calculated from Eqs. (6.17)–(6.20), exhibit pronounced curvature over a wide temperature range, contrary to Eq. (6.8). For example, the radiationless transition rate of $K_2NaScF_6:Cr^{3+}$, calculated by the method of Woods *et al.* (1994) for a two-configuration-coordinate, linear- and quadratic-coupling model of Bartram *et al.* (1986a) with parameters optimized to fit experiment over a narrow temperature range, is plotted over a much wider range in Fig. 6.7. Thus the Mott formula (Arrhenius equation) with adjustable parameters should be regarded properly as a convenient device for summarizing experimental data within a limited temperature range, rather than as a fundamental theory.

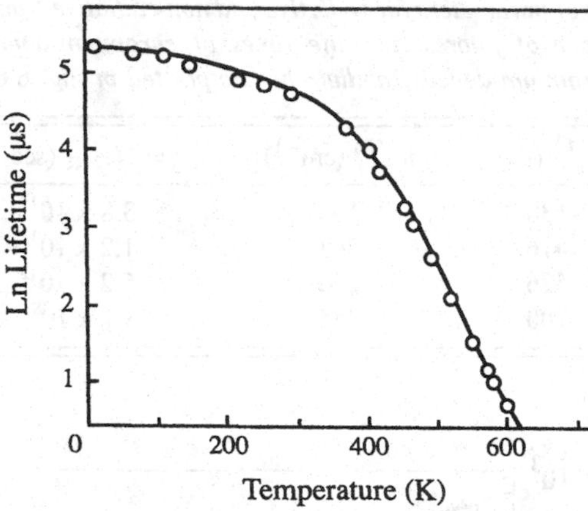

Figure 6.6. Luminescence life times as a function of temperature for substitutional Cr^{3+} in $ScBO_3$ [after Lai *et al.* (1986)]. Optimum parameter values are listed in Table 6.1.

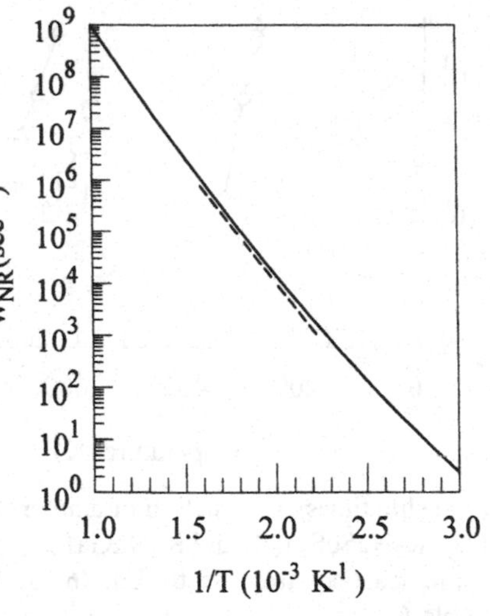

Figure 6.7. Radiationless transition rate as a function of reciprocal temperature for $Cr^{3+}:K_2NaScF_6$, predicted by the two-configuration coordinate, linear- and quadratic-coupling model with optimised parameters [Bartram *et al.* (1986a)]. The dashed line is a plot of the Mott formula, Eq. (6.8), with the parameters of Table 6.1 over the temperature range accessible to direct observation.

6.4.2 Transition-metal and rare-earth impurities

The validity of the energy-gap law, Eqs. (6.23), has been well established experimentally for rare-earth impurities with $S_0 \ll 1$, as evidenced by Fig. 6.8 [Riseberg and Weber (1977)]. The effective accepting-mode vibration frequency ω_0 is that of the highest frequency mode to which there is appreciable coupling, since W_{NR} is an extremely sensitive function of the number p of phonons required to bridge the energy gap. The promoting interaction ν is ordinarily treated as an adjustable parameter.

Transition-metal impurities provide examples of both inter-system crossing and internal conversion. The energy-gap law is not generally applicable to transition-metal

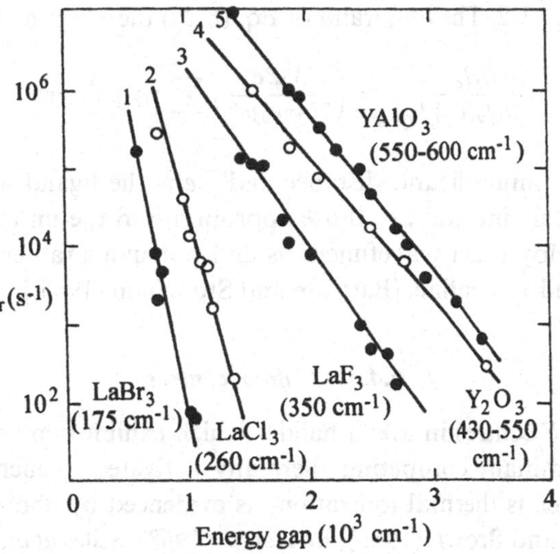

Figure 6.8. Radiationless decay rates from various electronic levels of rare-earth ions as a function of energy gap to the next lower level. The effective phonon energies are given for each host material [after Riseberg and Weber (1977)].

impurities, which are characterized by intermediate electron–lattice coupling ($S_0 \cong 1$). Rigorous calculations for static radiationless processes based on the assumption of linear coupling to a single accepting mode account well for trends, but often fail by many orders of magnitude to explain absolute rates of radiationless de-excitation. These discrepancies have stimulated diverse speculations concerning appropriate modifications of the theoretical model. Sturge (1973) emphasized the extreme sensitivity to anharmonicity of radiationless transition rates, which he illustrated for Co^{2+} in $KMgF_3$ by replacing parabolic potential energy curves with Morse potentials. However, Morse potentials, designed to model the asymmetric potential energy curves of diatomic molecules, are manifestly inappropriate for impurities in solids. Struck and Fonger (1975), on the other hand, have emphasized a comparable sensitivity to quadratic coupling within the harmonic approximation. Bartram *et al.* (1986a) demonstrated that a semi-empirical model which incorporates simultaneous linear coupling to an a_{1g} mode and quadratic coupling to a t_{2g} mode accounts well for ${}^4T_{2g} \rightarrow {}^4A_{2g}$ radiationless de-excitation of Cr^{3+} in host crystals of elpasolite structure, including K_2NaScF_6, K_2NaGaF_6 and Cs_2NaYCl_6, as evidenced by Fig. 6.7. This model, which was designed to explain the ambient pressure life time data plotted in Fig. 6.5 [Andrews *et al.* (1986b)], was also successful in predicting the pronounced pressure dependence of luminescence lifetimes observed in the chloride elpasolite [Rinzler *et al.* (1993)]. Calculations on the fluoride elpasolites were subsequently extended by Woods *et al.* (1994) to include combined linear and quadratic coupling to a_{1g}, e_g and t_{2g} modes, as well as anharmonic coupling to the a_{1g} mode, with parameters determined *ab initio* from an embedded-cluster model, with generally encouraging results.

The promoting mode for ${}^4T_{2g} \rightarrow {}^4A_{2g}$ radiationless de-excitation of Cr^{3+} in octahedral coordination is a t_{1g} mode, corresponding to a rigid rotation of the octahedron, as illustrated in Fig. 3.2. The integrand of Eq. (6.20) then has the form

$$\left(\frac{\partial H_e}{\partial Q_p(t_{1g})}\right)_{Q=0} = \left(\frac{35Ze}{2(4\pi\varepsilon_0)a^6}\right)\sum_i x_i y_i (x_i^2 - y_i^2), \qquad (6.39)$$

where a is the chromium-ligand distance and Ze is the ligand nuclear charge. The wavefunctions in this integral are those appropriate to the un-rotated ligands. It is imperative to employ exact wavefunctions and potentials rather than pseudo-wave functions and pseudopotentials [Bartram and Stoneham (1985)].

6.4.3 *Colour centres*

For the subset of F centres in alkali halides which exhibit luminescence at low temperatures, the dominant competing thermally activated quenching mechanism at higher temperatures is thermal ionization, as evidenced by photoconductivity measurements [Swank and Brown (1963), Bosi *et al.* (1968), Stiles *et al.* (1970), Honda and Tomura (1972)].

F centres in five of the alkali halides (NaI, NaBr, LiCl, LiBr and LiF), as well as in a number of other crystals, exhibit little or no luminescence even at liquid helium temperature. Athermal quenching of F-centre luminescence in these materials was explained as a dynamic process by Bartram and Stoneham (1975) in terms of the DKR rule [Dexter *et al.* (1955)]. Values of Λ inferred from the temperature dependence of absorption spectra on the assumption of linear coupling substantially exceed 0.25 for all of these cases, but not for the cases where efficient luminescence is observed. The absorption bands of F centres in NaCl and KI, with Λ values of 0.260 and 0.231, respectively, span the curve-crossing energy. In these borderline cases, excitation on the low-energy side of the absorption band has been shown to produce luminescence with greater quantum efficiency than excitation on the high-energy side [Wakita *et al.* (1981), Hirai and Wakita (1983)].

The observation of F-centre luminescence at low temperatures in NaI and NaBr with much less than 1% quantum efficiency is consistent with the prediction [Bartram and Stoneham (1983)]

$$R = (1 - 2\Lambda)^2/3, \qquad (6.40)$$

where R is reduced by a factor of three relative to Eq. (6.33) by virtue of the three-fold degeneracy of the excited state. Efficient thermally activated $F \rightarrow F'$ conversion of optically excited F centres at elevated temperatures in these materials does not necessarily imply that an excited F centre passes through its lowest vibrational state; rather, it can be understood as a dynamic process analogous to recombination-enhanced defect reactions, considered below.

The $Tl^0(1)$ centre in alkali halides has proved to be particularly promising as a laser material, as discussed in Chapter 4. The failure of analogous $Ga^0(1)$ and $In^0(1)$ centres [Van Puymbroeck *et al.* (1981)] to exhibit appreciable luminescence even at 10 K was explained successfully in terms of the DKR mechanism [Ahlers *et al.* (1984)].

A unique radiationless process has been observed in F_H centres, consisting of F centres in alkali halides perturbed by molecular anion impurities (CN^- or OH^-) [Yang *et al.* (1985)]. Optical excitation of the F_H centre is followed by efficient radiationless conversion of electronic energy to internal vibrational energy of the molecular anion. Since these vibrations decay radiatively, subsequent time evolution of the system can be followed experimentally. It remains to develop a convincing theoretical model for the radiationless step in the optical cycle.

6.4.4 *Recombination-enhanced defect reactions*

Lang and Henry (1975) have demonstrated that carrier recombination at deep-level impurities in semiconductors occurs by a multiphonon emission process. The relaxation energy released in the capture of an electron at a deep-level impurity is available to promote a defect reaction, e.g. annealing by defect migration [Dean and Choyke (1977), Lang (1982)]. The thermal activation energy of the defect reaction is diminished by the relaxation energy released during this dynamic process. Weeks *et al.* (1975) have speculated on the mechanism for channeling the relaxation energy into the reaction coordinate. Recombination cross sections have been measured by the technique of deep-level transient spectroscopy (DLTS) which utilizes capacitance changes in a *p*–*n* junction.

Photolysis of alkali halides provides a second example of a recombination-enhanced defect reaction. Excitons created by UV radiation are quickly self-trapped and usually decay by radiative recombination. However, in a significant fraction of cases, a competing dynamic radiationless process culminates in the generation of pairs of *F* centres and *H* centres on a picosecond timescale [Bradford *et al.* (1975), Williams *et al.* (1986)].

7

Energy transfer and excited state absorption

Various excited state processes, in addition to radiative decay, are important to laser performance. The quenching of luminescence reduces the excited state lifetime and can cause sample heating, thereby contributing to photothermal effects such as thermal lensing and thermal shock. Luminescence quenching also results in reduced laser slope efficiency, as discussed briefly here. Other excited state processes that require consideration are *excited state absorption* and *energy transfer*. In excited state absorption (ESA) a photon excites an electronic centre from the ground state to an excited state, which then relaxes to some lower lying metastable level. A second photon promotes the centre to an even higher energy state. Energy transfer arises when the optical centres are close enough together to interact, and this occurs when the concentration exceeds some lower bound, which need not be large. Although the energy levels of the interacting ions can be unaffected at such concentrations, the interion interaction is strong enough to enable excitation to be transferred between them. Prior to 1966, energy transfer was understood to involve excited states of donors ($|D^*\rangle$) interacting with the ground states of acceptors ($|A\rangle$). Auzel (1966) pointed out that excited acceptors ($|A^*\rangle$) also receive energy from excited donors ($|D^*\rangle$), and that *energy differences* can be exchanged as well as absolute energies. Energy transfer from excited donors to the *metastable* levels of acceptors can be treated by generalization of the Förster–Dexter theory outlined in §7.1. Three *upconversion luminescence mechanisms* are outlined in §7.4: sequential two-photon excitation pumping (STEP), energy transfer upconversion (ETU) or as it is sometimes called addition of photons by energy transfer (APET) and avalanche absorption pumping (AAP). Excited state absorption leading to STEP is most easily recognized at low concentrations of active ions, since both ground and excited state absorptions are on the same ion. However, ETU and AAP require higher concentrations (> 1 at.%) of donors and acceptors. In principle, energy may be transferred over many lattice sites in ETU and AAP processes before radiative or nonradiative decay returns the system to the ground state. In consequence, ETU and AAP are not restricted to the absorption of two-photons, since the absorption can be repeated on several activator sites: for example, Yb^3-to-Tm^{3+} energy transfer in $Tm^{3+}:Yb^{3+}:BaY_2F_8$ upgrades 970 nm pump radiation to blue upconverted luminescence at 475 nm in a three-photon process.

The consequences of energy transfer are readily studied via the luminescence channel of the optical pumping cycle, including the concentration, excitation and time dependences of the upconverted luminescence spectra. Energy transfer may enable excited state absorption: when an excited donor ion is de-activated and returns to the ground state, the excited acceptor being promoted to an even higher level from which it may decay radiatively or nonradiatively. In general, energy transfer may result in *concentration quenching* of luminescence, *sensitization* of the acceptor luminescence, *parasitic absorption* loss by ESA and upconverted luminescence. There is a growing demand for visible region laser devices for applications in colour displays, high density optical recording, reprographics and underwater optical communications. Upconversion lasers have the potential to provide practical, compact, all-solid-state sources at wavelengths from the ultraviolet to the red. High resolution spectroscopic techniques can distinguish the separate contributions of acceptors and donors at low temperature, including optical hole burning (OHB), fluorescence line narrowing (FLN), transient grating spectroscopy in three- and four-wave mixing geometries and optical free induction decay [Yen and Selzer (1981), Macfarlane (1994a,b)]. However, the electronic structures of the RE^{3+} ions are complex and in experiments at high temperatures (≤ 100 K) many individual lines can overlap due to thermal broadening so that complete assignments of particular inter-multiplet transitions have only occasionally been successful. Irrespective of the details of the upconversion mechanism, the primary function of upconversion is to upgrade infrared photons to blue-green photons.

Microscopic theories of energy transfer started with Förster (1948) and Dexter (1953), who used perturbation calculations to estimate transfer rates in terms of the spectral overlap between the absorption and emission line-shape functions of donor and acceptor ions, respectively. Subsequently Orbach (1967) and Holstein *et al.* (1981) developed a first-principles energy transfer theory for vibronic ions which included the electrostatic and electron–phonon interactions on an equal basis. Strong coupling corrections to these microscopic theories were made subsequently by Kohli and Huang-Liu (1974). In principle, these theories apply to isolated donor–acceptor pairs. Experimental reality relates to a very large number of randomly distributed ions in a material, which must be modelled statistically starting from a set of coupled differential equations. Solutions of these equations are complex and give different time evolution regimes for the *diffusion* and *hopping* models [Huber (1979), (1981), Burshtein (1985)].

7.1 Microscopic theory of donor–acceptor energy transfer

The simplest method of transferring excitation from one ion to another involves the absorption of the photon emitted by the excited donor at the acceptor site. This process of *radiative energy transfer* may be repeated many times before a photon emitted by the donor exits the crystal. Alternatively the ion last to be excited in the excitation chain may decay nonradiatively to the ground state. Radiative energy transfer was first discovered in a ruby crystal that contained only 0.05% Cr^{3+} [Varsanyi *et al.* (1959)]. In single crystals the luminescence lifetime was 15 ms at 77 K. However, the lifetime was shorter in finely dispersed powder increasing with particle size from 4 ms for very fine powders up to the single crystal value of 15 ms. This result indicates that *radiative trapping* within some finite volume of the crystal causes the lifetime to be increased in bulk samples. More generally, energy transfer is nonradiative and the emission–reabsorption mechanism is replaced by site-to-site transfer mediated by some inter-ionic coupling mechanism. The nonradiative energy transfer process, illustrated in Fig. 7.1, involves the simultaneous de-excitation of the donor ion and excitation of the

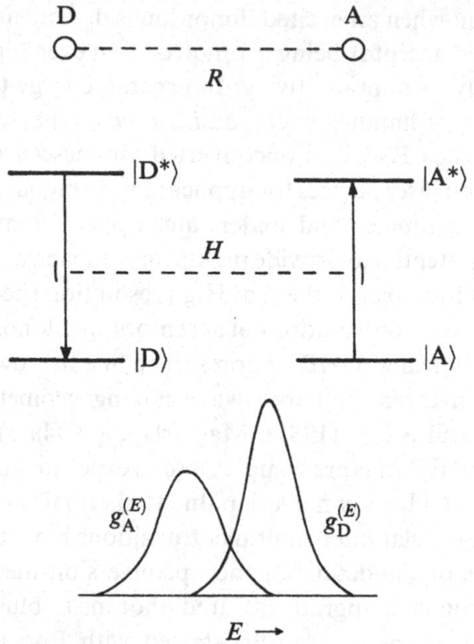

Figure 7.1. Energy transfer between excited donor, D*, and acceptor, A, spatially separated by the vector distance, *R*, over which they are coupled by the interaction, *H'*. The ions have the same electronic energy level structure, with overlap of emission and absorption bands, required by conservation of energy [after Henderson and Imbusch (1989)].

acceptor by an electrostatic, magnetic or exchange coupling between them. The Förster–Dexter theory gives the rate of donor–acceptor energy transfer as

$$W_{DA} = \frac{2\pi}{\hbar} |\langle D, A^* | H' | D^*, A \rangle|^2 \int g_D(E) g_A(E)\, dE. \tag{7.1}$$

The overlap integral in Eq. (7.1) contains normalized shape functions $g_D(E)$ and $g_A(E)$ for homogeneously broadened, radiative emission ($g_D(E)$) and absorption ($g_A(E)$) transitions on donor ($D^* \to D$) and acceptor ($A \to A^*$), respectively. Expanding the electrostatic interaction term between electrons on the donor and acceptor, H', yields an electric dipole–dipole term varying as $(er_D)(er_A)/R^3$, an electric dipole–quadrupole term $(er_D)(er_A^2)/R^4$ and an electric quadrupole–quadrupole term $(er_D^2)(er_A^2)/R^5$; the orbital radii for electrons on the donors and acceptors are, respectively, r_D and r_A, and R is the donor–acceptor separation [Dexter (1953)]. Assuming allowed electric dipole and electric quadrupole transitions for the matrix elements appropriate to radiative transitions in Eq. (7.1) leads to the ratio

$$W_{DA}^{dd} : W_{DA}^{dq} : W_{DA}^{qq} = 1 : (a_0/R)^2 : (a_0/R)^4 \approx 1 : 10^{-2} : 10^{-4},$$

when $R = 10a_0$ and a_0 is the Bohr radius. This estimate shows that energy transfer by electric quadrupole mechanisms can be significant in solids, even though the radiative electric quadrupole transition, which has strength $(a_0/\lambda)^2 \approx 10^{-7}$ relative to the radiative electric dipole transition, is unimportant. Magnetic dipole–dipole processes scale as R^{-6}. Exchange-induced donor–acceptor energy transfer may involve direct exchange between donors and acceptors or super-exchange via overlap with intervening ligand ions. In either case $H'_{ex} \cong J_0 \exp(-R/L)$, in which J_0 is the isotropic exchange constant and L is a characteristic range for the interaction.

For dipole–dipole energy transfer the rate is given by

$$W_{DA}^{dd} = \frac{4\pi}{3\hbar}\left(\frac{1}{4\pi\varepsilon_0 n^2}\right)^2 \frac{|\langle\mu_D\rangle|^2 |\langle\mu_A\rangle|^2}{R^6} \int g_D(E) g_A(E)\, dE, \tag{7.2}$$

where the $|\langle\mu\rangle|^2$ are squared matrix elements for radiative electric dipole transitions on donor and acceptor. Equation (7.2) may be re-written:

$$W_{DA}^{dd} = \frac{4\pi}{3\hbar}\left(\frac{1}{4\pi\varepsilon_0 n^2}\right)^2 \left(\frac{3\hbar e^2}{2m\omega}\right)^2 \frac{f_D f_A}{R^6} \int g_D(E) g_A(E)\, dE, \tag{7.3}$$

using $f_D = (2m\omega/3\hbar e^2)|\langle\mu_D\rangle|^2$ as the oscillator strength of the radiative transition at the central frequency, ω, of the transition [Henderson and Imbusch (1989)]. The oscillator strengths f_D and f_A can be measured from the decay rate of the $D^* \rightarrow D$ radiative transition and the strength of the $A \rightarrow A^*$ absorption transition. A microscopic multipolar D–A transfer rate is defined from Eq. (7.3) as

$$W_{DA} = \alpha_{DA}^{(n)} R^{-n}, \tag{7.4}$$

where $n = 6$, 8 or 10 for dipole–dipole, dipole–quadrupole and quadrupole–quadrupole interactions, successively. Since the luminescence lifetime tends to τ_R as the interaction reduces to zero at some critical acceptor concentration, n_0, it follows from Eq. (7.4) that

$$\tau_R^{-1} = \alpha_{DA}^{(n)} R_0^{-6}, \tag{7.5}$$

where the critical range $R_0 = \sqrt[3]{(4/3\pi n_0)}$ can be determined by measuring the D–A energy transfer rate as a function of acceptor concentration, n_A.

Energy can be transferred also between ions even when there is an energy mismatch, ΔE_{DA}, between the $D^* \rightarrow D$ emission and $A \rightarrow A^*$ absorption transitions, as shown in Fig. 7.2. In this case the perturbation H' includes multipolar interactions between electrons on the donors and electrons on the acceptors in which case energy transfer can involve one-phonon excitations, although interference terms may reduce the strength of such processes. In consequence, two-phonon-assisted energy transfer dominates the one-phonon processes because of a reduction in these interference terms, especially at room temperature. However, when ΔE_{DA} is significantly larger than the bandwidth of the vibrational spectrum multiple-phonon-induced energy transfer is necessary to bridge the energy gap. The effectiveness of quadratic coupling when large numbers of phonons are excited may lead to very efficient nonradiative decay of the excited state [Bartram (1984)]. Phonon-assisted energy transfer, illustrated in Fig. 7.2, requires a matrix element in the transition probability, Eq. (7.1), that takes the form

$$\langle D, A^*, n(\hat{\varepsilon}, q)|H'|D^*, A, n(\hat{\varepsilon}, q)\rangle, \tag{7.6}$$

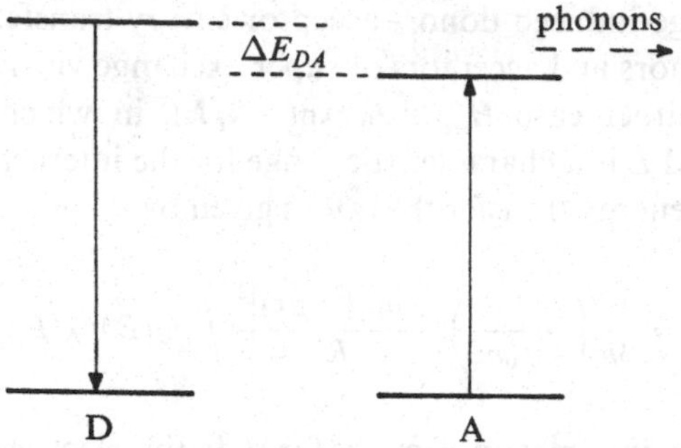

Figure 7.2. Phonon-assisted excitation transfer in which the energy mismatch, ΔE_{DA}, is made up by the absorption or emission of lattice quanta.

in which the vibrational state of the lattice is specified by the phonon occupation numbers $n(\hat{\varepsilon}, q)$ for polarization mode $\hat{\varepsilon}$ and wave vector q. Equation (7.6) shows that the coupling Hamiltonian, H', does not itself change the vibrational state of the lattice. When the energy mismatch between sites is significantly greater than the energy bandwidth of the phonon system it is made good by the excitation of a number of lattice quanta of appropriate energy. The magnitude of the matrix element in Eq. (7.6) depends crucially on the nature of the initial and final states. For states with the same parity at the same site the electric dipole matrix element vanishes and electric quadrupole–quadrupole interactions dominate the energy transfer. However, when states with different parities are involved, the electric dipole–dipole interaction is dominant. Excitation transfer by phonon-assisted processes involves not only site-to-site coupling but also electron–phonon coupling, of the sort described in §5.2, acting independently at each site. The Hamiltonian mediating electron–phonon coupling is proportional to a strain operator linear in the phonon annihilation and creation operators defined by Eqs. (5.81). The electron–phonon coupling strength for the ion at site j (D or A) is denoted by $f(j)$ in the ground state of that ion and $g(j)$ in its excited state. The order in which these perturbations are applied is very important since it can lead to cancellations in the excitation transfer rate that may be complete except that there are differences between the ion–lattice couplings at the D and A sites, and between the ion–lattice couplings in the ground and excited states at a given site (either D or A sites). This latter is the normal situation in phonon-terminated laser transitions. Kohli and Huang-Liu (1974) have shown that all the transition rates for single-phonon-assisted energy transfer processes are reduced by the Debye–Waller factor $\exp[-\phi(T)]$, where

$$\phi(T) = \lambda_s(2n_s + 1), \tag{7.7}$$

and n_s is the Bose occupation factor appropriate to an optical phonon of energy $\hbar\omega_s$ with polarization index s. The explicit dependence on the difference in the ion–lattice coupling in ground and excited state comes from

$$\lambda_s = \frac{[f(j) - g(j)]^2}{(\hbar\omega)^2}, \tag{7.8}$$

where g and f label the ground and excited states respectively, and j indicates either the donor or acceptor site.

The important energy transfer processes were reviewed comprehensively by Holstein *et al.* (1981). They identify one-phonon-assisted transfer, one- or two-phonon-assisted radiative transfer, single-site two-phonon-assisted resonant and non-resonant energy transfer and two-site two-phonon-assisted non-resonant transfer. *One-* or *two-*site specifies the spatial position for the ion-lattice coupling. These processes are expected to be important in vibronic laser transitions. In two-site two-phonon-assisted non-resonant energy transfer, the absorption and emission of phonons takes place at different sites. In consequence, energy transfer cannot be treated by the Förster–Dexter theory involving spectral overlap between transitions. Hamilton *et al.* (1977) showed that the transfer rate varies as $(k_B T)^3$ and is independent of energy mismatch, ΔE_{DA}. In contrast, the single-site two-phonon-assisted resonant and non-resonant transfer processes may be formulated in terms of spectral overlap, even though they have rather different temperature dependences for the transition rates.

7.2 Macroscopic theory of donor–acceptor energy transfer

7.2.1 *No donor–donor transfer*

In principle the Förster–Dexter theory of energy transfer, and its subsequent refinements, relate to isolated D–A pairs within a volume of material large compared with the range of the interaction that couples the donor and acceptor ions. The reality of the experimental situation is rather different, involving the behaviour of a large number of randomly distributed donors and acceptors having a range of discrete values of the donor–acceptor separation, both smaller than and larger than the interaction range. In this case a macroscopic theory of energy transfer must apply statistical techniques to the ensemble of coupled ions and include the possibilities, not only of $D^* \rightarrow A$ energy transfer, but also of $A^* \rightarrow D$ back-transfer. Usually energy back-transfer is ignored because the totality of forward $D^* \rightarrow A$ transfers is much more probable. This affords considerable mathematical simplification. Acceptor–acceptor transfer is also usually omitted. The resulting statistical treatment of energy transfer, including $D^* \rightarrow D$ transfer and $D^* \rightarrow A$ transfer, starts from a set of coupled equations :

$$\frac{dP_n(t)}{dt} = -\left(\frac{1}{\tau_R} + \sum_l X_{nl} + \sum_{n'} W_{nn'}\right) P_n(t) + \sum_{n'} W_{nn'} P_{n'}(t), \tag{7.9}$$

in which τ_R is the intrinsic, excited donor (D^*) lifetime, $W_{nn'}$ is the $D^* \rightarrow D$ transfer rate from the nth to the n'th donor, X_{nl} is the $D^* \rightarrow A$ transfer rate from the nth donor to the lth acceptor and $P_n(t)$ is the probability that the nth donor is excited at time t [Huber (1979)]. If the nth donor is excited then $P_n = 1$ and the summation $X_n = \Sigma X_{nl}$ over all acceptors l depends on the arrangement of ions in the neighbourhood of the nth donor ion. Numerous authors have given solutions to Eq. (7.7); their results have been elegantly reviewed by Huber (1981).

In the static case, when there is no $D^* \rightarrow D$ transfer, Huber (1979) finds an exact solution,

$$\langle P(t)\rangle_c = \exp(-t/\tau_R) \prod_l (1 - c_A(1 - \exp(-X_{0l}t))), \tag{7.10}$$

where the configurational average of $P_n(t)$ over all arrangements of donors and acceptors, $\langle P(t)\rangle_c$ is defined in terms of the total number of donor ions, n_D, by

$$\sum_n P_n(t) = n_D\langle P(t)\rangle_c, \tag{7.11}$$

and c_A is the probability that a site is occupied by an acceptor ion and the summation in Eq. (7.10) is over all sites in the lattice. If a sharp pulse of radiation excites the donors then the donor luminescence signal $I_D(t)$ decays as $\langle P(t)\rangle_c$ in non-exponential fashion with initial rate

$$\frac{1}{I_D(t)}\frac{dI_D(t)}{dt} = \frac{1}{\tau_R} + c_A\sum X_{0l}. \tag{7.12}$$

A continuum approximation used by Inokuti and Hirayama (1965) is physically unrealistic near $t=0$ since there are large contributions to the initial decay from D*–A pairs with vanishingly small separations. The smallest D*–A separations are not permitted by the discreteness of the crystal structure. Watts (1975) solved Eq. (7.9) using 'reasonable approximations' and determined that for D–A transfer by dipole–dipole interaction $\langle P(t)\rangle_c$ decays non-exponentially with time according to

$$\langle P(t)\rangle_c = \exp\left[-\frac{t}{\tau_R} - \frac{4}{3}\pi^{3/2} n_A (\alpha_{DA}^{(6)} t)^{1/2}\right]. \tag{7.13}$$

In the limit that no D*–A transfer occurs then $\langle P(t)\rangle_c$ follows the intrinsic, exponential decay of the excited donors with radiative lifetime τ_R. In contrast, decay at the critical concentration $n_A = n_0$ is much faster, as is illustrated in Fig. 7.3. The physical interpretation of this decay pattern is straightforward. For close D–A pairs the decay is fast and this determines the decay at short times. However, at longer times energy transfer occurs from excited donors, D*, to acceptors at the *average* D–A separation and this occurs at the radiative decay rate. The more distantly separated pairs retain their excitation energy over even longer times, so that the decay rate decreases, without ever reaching the decay rate, τ_R^{-1}, of isolated donor ions.

7.2.2 *Influence of donor–donor energy transfer*

The theory of donor–acceptor energy transfer in the presence of donor–donor transfer is very complex, although the starting point is still Eq. (7.9). An exact solution for $\langle P(t)\rangle_c$ exists when D*–D transfer is exceedingly fast [Huber (1979)]. The donor decay rate is then $(1/\tau_R) + c_A\Sigma X_{01}$, as is the initial decay rate of excited donors in the absence of D*–D transfer. Yokota and Tanimoto (1967) treated the migration between donor ions as a diffusion process, characterized by a *diffusion constant*, *D*, dependant on both the donor concentration and the D*–D transfer rate. At low donor concentrations, assuming dipole–dipole D*–D energy transfer, the diffusion constant is,

$$D = \frac{1}{2}\left(\frac{4}{3}\pi n_D\right)^{4/3} \alpha_{DD}^{(6)}. \tag{7.14}$$

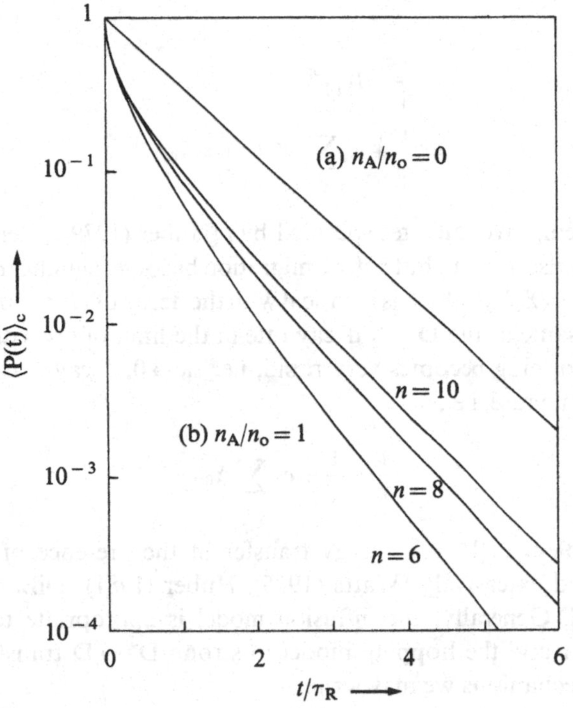

Figure 7.3. The temporal evolution of $\langle P(t)\rangle_c$ after an initial excitation pulse at $t=0$ for D*–A energy transfer in the absence of D*–D transfer. In (a) no D*–A transfer occurs. In (b) D*–A energy transfer occurs by dipole–dipole ($n=6$) interaction at the critical acceptor concentration of $n_A/n_O=1$. In general, the decay is non-exponential, as it is for dipole–quadrupole and quadrupole–quadrupole interactions [after Henderson and Imbusch (1989)].

Generally, the diffusion model is valid when the D*–D transfer is much weaker than the D*–A transfer. The asymptotic decay rate, i.e. when $t\to\infty$, is given by

$$\frac{1}{\tau}=\frac{1}{\tau_R}+W_{DA}^{(\text{diff})}=\frac{1}{\tau_R}+21n_Dn_A\left(\alpha_{DD}^{(6)}\right)^{3/4}\left(\alpha_{DA}^{(6)}\right)^{1/4},$$
$$=\frac{1}{\tau_R}+8.5n_A\left(\alpha_{DA}^{(6)}\right)^{1/4}D^{3/4}, \qquad (7.15)$$

in which D is expected to be temperature dependent.

Energy migration has also been treated as a random walk among donors, in which the excitation is located on an excited donor for an average time τ_0 before *hopping* to another donor [Burshtein (1972)]. According to Watts (1975), the average transfer rate, τ_0^{-1}, for dipole–dipole transfer is

$$\frac{1}{\tau_0}=\left(\frac{2\pi}{3}\right)^2 n_D^2\alpha_{DD}^{(6)}, \qquad (7.16)$$

where n_D is the donor density and the microscopic constant for energy transfer by the dipole–dipole mechanism, $\alpha_{DD}^{(6)}$ is given by Eq. (7.5). The donor decay is also non-exponential, but becomes exponential in the asymptotic limit, where the decay rate is

$$\frac{1}{\tau} = \frac{1}{\tau_R} + W_{DD}^{(hop)}$$
$$= \frac{1}{\tau_R} + c_A \sum_l X_{0l}(1 + \tau_0 X_{0l})^{-1}, \tag{7.17}$$

the summation being over all sites specified by l [Huber (1979)]. Very soon after the sharp excitation pulse at $t = 0$, but before migration becomes significant, the decay rate becomes $(\tau_R)^{-1} + c_A \Sigma_l X_{0l}$, which is identical with the decay of $\langle P(t) \rangle_c$ in Eq. (7.12). This result is also the same as the D*–A decay rate in the limit of exceedingly fast D*–D transfer. When hopping becomes very rapid, i.e. $\tau_0 \to 0$, decay is exponential at all times with this same rate, i.e.,

$$\frac{1}{\tau} = \frac{1}{\tau_R} + c_A \sum X_{0l}. \tag{7.18}$$

The general problem of D*–A energy transfer in the presence of D*–D transfer has been discussed extensively [Watts (1975), Huber (1981), DiBartolo (1984) and Burshtein (1985)]. Generally, the diffusion model is appropriate to weak D* → D excitation transfer and the hopping model to strong D* → D transfer. For dipole–dipole transfer mechanisms we may write:

$$\text{Diffusion model: } \alpha_{DD}^{(6)} \ll \alpha_{DA}^{(6)}, \tag{7.19a}$$

$$\text{Hopping model: } \alpha_{DD}^{(6)} \geq \alpha_{DA}^{(6)}. \tag{7.19b}$$

Henderson and Imbusch (1989) have illustrated these theoretical results schematically in graphs of the donor luminescence intensity as a function of time after the excitation pulse. Figure 7.4(a) is the obvious result that in the absence of any energy transfer the luminescence decays with the radiative lifetime, τ_R. The *static limit*, which is illustrated in Fig. 7.4(b) assuming $n_A = 2n_0$ and $\alpha_{DD}^{(6)} = 0$, is non-exponential at all times. The *diffusion limited case* is illustrated for the same D*–A and D*–D transfer rates (i.e. weak D*–D transfer) by Fig. 7.4(c): the chosen value of $D = \frac{1}{6}(\alpha_{DA}^{(6)} R_0^{-4})$ corresponds to the excitation migrating a distance R_0 through the crystal in a time τ_R. The decay becomes exponential in the asymptotic limit, Eq. (7.15), within a few radiative decay periods.

Recognition of the limiting behaviour in energy transfer processes requires careful measurement of the decay profile with time as a function of the donor concentration. Transfers from Eu^{3+} ions (D*) to Cr^{3+} (A) in Cr^{3+} : Eu phosphate glass serves as an excellent example of diffusion-limited energy transfer (Fig. 7.5). At 77 K and in the absence of Cr^{3+} the 600–700 nm luminescence in the $^5D_0 \to {}^7F_J$ transition of Eu^{3+} decays exponentially with the radiative lifetime of $\tau_R = 2.1$ ms. As the amount of Cr^{3+}-dopant increases both Eu^{3+} and Cr^{3+} luminescence occurs, even though the Eu^{3+} donors are the only ions to be directly excited, indicating that Eu^{3+}–Cr^{3+} transfer is occurring. Initially the Eu^{3+} decay in the presence of Cr^{3+} is non-exponential, becoming eventually exponential, see Fig. 7.5, showing that both $|D^*\rangle \to |D\rangle$ migration and $|D^*\rangle \to |A\rangle$ transfer are occurring [Weber (1971)]. At 77 K the $Eu^{3+} \to Eu^{3+}$ migration is slow relative to $Eu^{3+} \to Cr^{3+}$ transfer. As anticipated from the foregoing

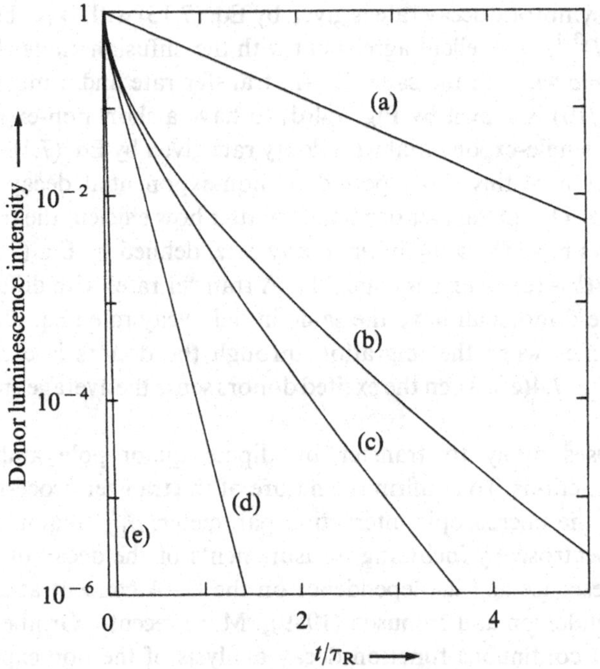

Figure 7.4. The decay of the donor luminescence under various hypothetical conditions [adapted from Henderson and Imbusch (1989)].

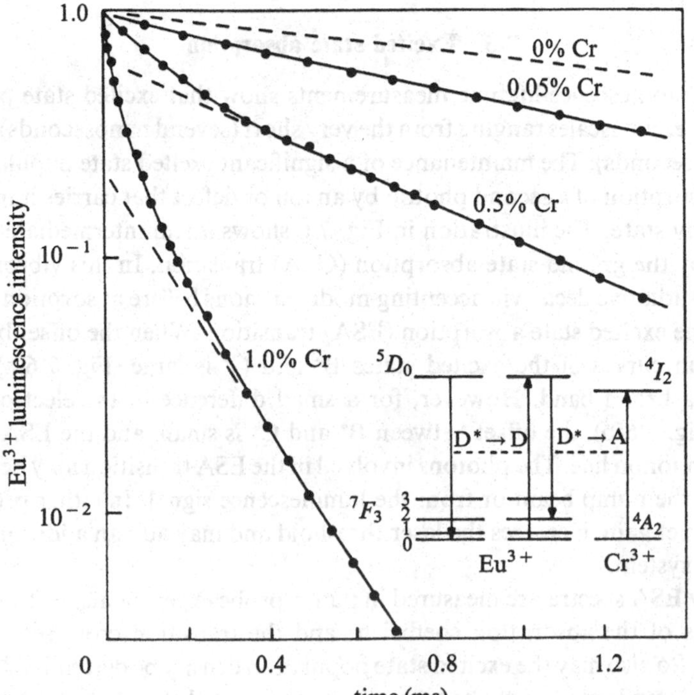

Figure 7.5. The variation in the time dependence of Eu^{3+} luminescence intensity from Cr^{3+}: $Eu(PO_3)_2$ glass with increasing Cr^{3+} concentration [after Weber (1971)].

discussion, the asymptotic decay rate is given by Eq. (7.15) with W_{DA} being accurately proportional to $D^{3/4}$, in excellent agreement with the diffusion model. The prediction for the *hopping model*, with the same D*–A transfer rate and a much faster D*–D transfer ($\tau_0 = \tau_R/10$) is shown by Fig. 7.4(d) to have a short non-exponential decay before becoming single-exponential with decay rate given by Eq. (7.18). As the D*–D transfer rate increases this short period of non-exponential decay becomes even shorter. Whenever D*–D transfer occurs, no matter how efficient the transfer between donors, all decays have the same initial decay rate, defined by Eq. (7.12). The decay curves in Fig. 7.4(b)–(e) assume identical D–A transfer rates with different migration rates through the donors: all have the same initial decay rate (Eq. (7.12)). This rate persists at all times when the migration through the donors is exceedingly rapid (Eq. (7.18) and Fig. 7.4(e)), when the excited donors sense the average environment due to acceptors.

Similar analyses apply to transfer by dipole–quadrupole and quadrupole–quadrupole interactions. To confirm the nature of the transfer process, including the multipolarity of the microscopic interaction parameter, $\alpha_{DA}^{(n)}$, requires very accurate time resolved spectroscopy including measurements of the decay of donor (and/or acceptor) luminescence and its dependence on the D–A concentrations and sample temperature [Henderson and Imbusch (1989)]. More recently Grinberg *et al.* (1998) have shown that continuous function decay analysis of the non-exponential decay profile can determine the distribution of acceptors around donor sites and identify as many as 100 multisites of luminescent ions in glasses and disordered crystals [Yamaga *et al.* (1992), (1996), Grinberg *et al.* (1995)].

7.3 Excited state absorption

In general, luminescence lifetime measurements show that excited state populations can decay over timescales ranging from the very short (several nanoseconds) to the very long (many seconds). The maintenance of a significant excited state population facilitates the absorption of a second photon by an ion or defect that carries it into an even higher energy state. The illustration in Fig. 7.6 shows a real intermediate state to be populated by the ground state absorption (GSA) transition. In this vibronic system, there is nonradiative decay via accepting mode phonons before absorption of another photon in the excited state absorption (ESA) transition. When the offset between the configuration curves of the excited states B* and C* is large, Fig. 7.6(a), the ESA appears as a broad band. However, for a small difference in the electron–phonon coupling, Fig. 7.6(b), the offset between B* and C* is small, and the ESA signal is a sharp zero-phonon line. The photons involved in the ESA transition may be abstracted either from the pump beam or from the luminescence signal. In either process, ESA reduces the net gain, increases the laser threshold and may add an additional thermal load on the system.

Normally ESA spectra are measured in pump-probe experiments. Both the spectral dependences of the absorption coefficient and the transition cross sections can be determined. So also may the excited state population density be determined. Typically, a xenon arc lamp is used to probe transversely the excited state population created by a flashlamp pump beam [Fairbank *et al.* (1975), Andrews *et al.* (1986a), Caird *et al.* (1988)]. Although this technique is useful in identifying centres, it can be difficult to

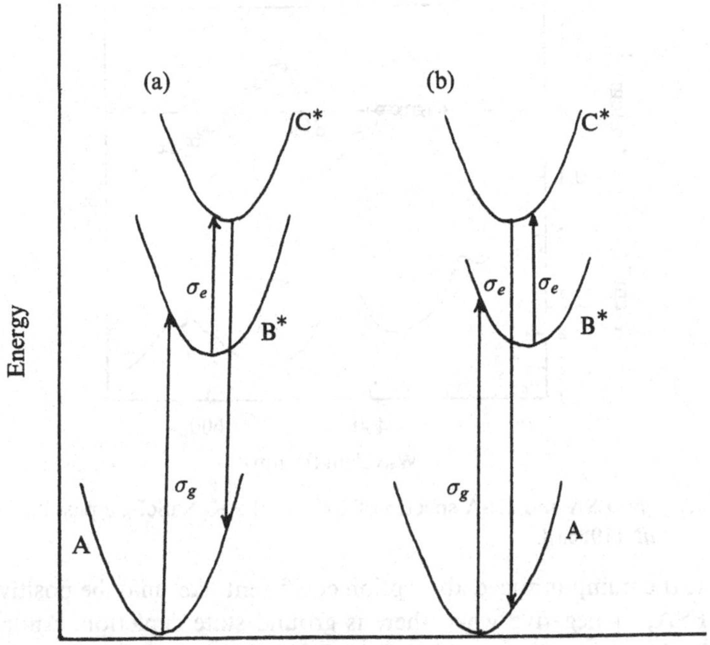

Figure 7.6. Configurational coordinate description of an excited state absorption transition leading to (a) a broad vibronic band and (b) a sharp zero-phonon line.

discriminate between ESA and fluorescence signals because both decay under pulsed excitation with the same emission lifetime. Petermann (1990) has used a CW technique to overcome this difficulty. An excited state population is created by focusing an intense pump beam from an Ar^+ or Kr^+ laser at some point in the sample. A probe beam from a Xe arc lamp is focused through a pinhole to maximize overlap at the pump and probe beam foci. The intensity of broadband light from the arc lamp transmitted by the excited sample is then measured by a photomultiplier tube. The probe beam is chopped using an electromechanical chopper that provides a reference signal for the lock-in detector. The intensities of light passing through the pumped (I_p) and unpumped (I_u) sample are then measured or compared during the ON/OFF periods, after adjustment of the chopping frequency to null out the luminescence signal. From the Lambert–Beer law, Eqs. (1.1) and (5.43), these intensities for a sample of length L are

$$I_u(\nu, \hat{\varepsilon}) = I_0(\nu, \hat{\varepsilon}) \exp(-\alpha(\nu, \hat{\varepsilon})L), \tag{7.20}$$

and

$$I_p(\nu, \hat{\varepsilon}) = I_0(\nu, \hat{\varepsilon}) \exp(-\alpha_p^2(\nu, \hat{\varepsilon})L - \alpha_p^1(\nu, \hat{\varepsilon})L), \tag{7.21}$$

where the subscripts u and p refer to the unpumped and pumped conditions, and the superscripts refer to absorption to first (1) and second (2) excited levels, respectively. Hence the additional absorption, $\Delta\alpha$, associated with the excited state population is

$$\Delta\alpha(\nu, \hat{\varepsilon}) = (1/L) \ln(I_u/I_p), \tag{7.22}$$

in which the $\nu, \hat{\varepsilon}$ in parentheses indicates that the induced absorption coefficient varies with frequency, ν, (or λ) and may be polarization sensitive ($\hat{\varepsilon}$). As the measurements in

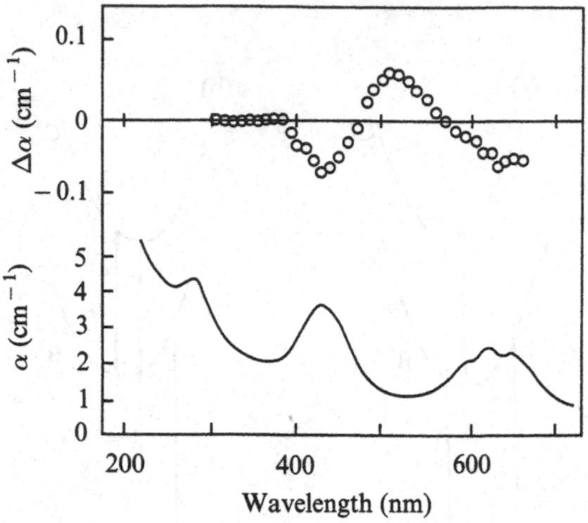

Figure 7.7. The ESA and GSA spectra of Cr^{3+} in the K_2NaScF_6 elpasolite [after Andrews *et al.* (1986a)].

Fig. 7.7 show, the pump-induced absorption coefficient, $\Delta\alpha$, may be positive, corresponding to ESA, or negative when there is ground state depletion. Andrews *et al.* (1986a) made these measurements on $Cr^{3+}:K_2NaScF_6$ over the wavelength range 200–700 nm. The GSA and ESA spectra shown separately in Fig. 7.7 demonstrate that bleaching occurs in the regions of the ${}^4A_2 \rightarrow {}^4T_2$ and ${}^4A_2 \rightarrow {}^4T_{1b}$ GSA bands. In between these bands at $\lambda = 510$ nm there is a strong ${}^4T_2^* \rightarrow {}^4T_{1b}$ ESA transition. (The asterisk indicates that the ESA originates on the relaxed excited state, ${}^4T_2^*$.) There are no other ESA features down to wavelengths as short as 300 nm. Bleaching over the wavelength range of the ${}^4A_2 \rightarrow {}^4T_2$ GSA shows that the observed ESA reduces the efficiency of possible laser action.

Assuming that the total concentration of active ions is $N = N_g^u = N_g^p + N_e^p$, and since the absorption cross-section (σ) and absorption coefficients are related through $\sigma N = \alpha L$, the following equation can be obtained for the cross section for ESA;

$$\sigma_e(\lambda, \hat{\varepsilon}) = \sigma_g(\lambda, \hat{\varepsilon}) + (N_e^p L)^{-1} \ln(I_u/I_p), \tag{7.23}$$

permitting the determination of the cross-section for the ESA transition once N_e^p is determined by rate equation analysis [Petermann (1990)]. Although the absorption coefficient is a convenient experimental quantity to measure, the cross-section, σ, is the more fundamental property of the optical centre. In effect, the cross-section measures the target size presented by the absorber to the incoming photon. Evidently, from Eq. (7.23), the cross-section varies across the shape function of the transition, as is evident in Fig. 7.7.

The first measurements of ESA in Cr^{3+}-doped crystals were reported by Fairbank *et al.* (1975) for Cr^{3+} in Al_2O_3(ruby), $Be_3Al_2(SiO_3)_6$ (emerald) and MgO. In these hosts Cr^{3+} ions occupy strong crystal-field sites and the observed ESA spectra originate on the long-lived ${}^2E$ state. The subsequent recognition that ESA can be an important loss mechanism in numerous transition-metal ion-doped laser crystals has led to rapid developments in this field of spectroscopy [Petermann (1990), Moncorgé and Benyattou (1988), Payne *et al.* (1988b, 1994)]. ESA spectra are of fundamental interest

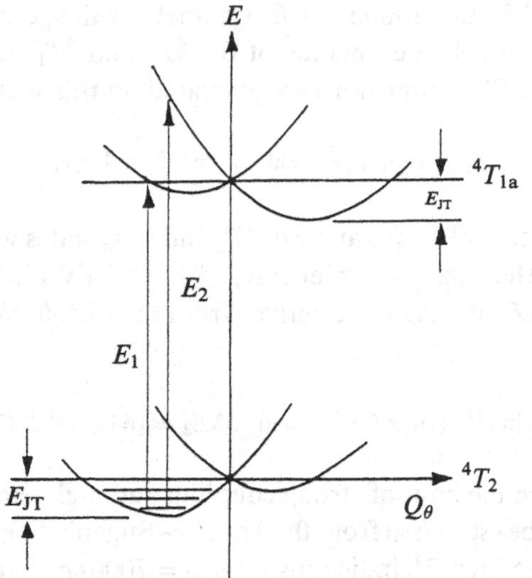

Figure 7.8. The removal of orbital degeneracy in orbital triplet states by the Jahn–Teller effect and the splitting of the $^4T_2 \to {}^4T_{1a}$ ESA band.

not only since they characterize the electronic and vibronic properties of higher lying levels in ways not open to ground state absorption measurements, but also because such parasitic losses may overlap the laser absorption and emission transitions, thereby reducing laser efficiency and increasing the necessary threshold pump power. Partial or complete overlap of an ESA transition with a laser emission reduces the effective emission cross section σ_{eff}, defined by

$$\sigma_{\text{eff}} = \sigma_{\text{em}} - \sigma_{\text{ESA}} \tag{7.24}$$

where σ_{em} and σ_{ESA} are the emission and ESA cross-sections, respectively. To determine σ_{em} accurately requires independent measurements of σ_{eff} and σ_{ESA}. For the $3d^3$ ions such as V^{2+} and Cr^{3+} there may be problems of determining the initial and terminal levels coupled by the radiation field. For these ions the ordering of low-lying excited states depends on the strength of the crystal-field. In strong field sites where $Dq/B > 2.4$ and $\Delta E = E(^4T_2) - E(^2E) > 1000\,\text{cm}^{-1}$ (§5.4), ^2E is lowest and the excited transitions will be spin-allowed to higher lying spin doublets ($^2E, {}^2T_2, {}^2T_1$, etc.). In contrast, for weak field sites (Cr^{3+} : LLGG, V^{2+} : MgF_2) $^4T_2$ is the lowest lying excited state and $^4T_2 \to {}^4T_{1a,b}$ transitions will dominate the ESA spectrum. In the intermediate crystal-field regime thermally-derived populations of both ^2E and $^4T_2$ levels will lead to a more complex spectrum involving ESA transitions originating on both ^2E and $^4T_2$ levels. Determining the spectral characteristics of such transitions, wavelength and bandwidth, polarization strength and dipolar identity may not be straightforward.

Interpretation of ESA spectra of V^{2+} : MgF_2 and V^{2+} : $KMgF_3$ [Payne *et al.* (1988b)] and Cr^{3+} : colquiriites [Payne *et al.* (1989b), Lee *et al.* (1989)] led to a model of ESA transitions in weak field hosts involving the coupling of E_g phonons to orbital triplet states via the Jahn–Teller effect, §5.2. In these cases, since $^4T_2$ is always the emissive level, the most probable initial state for ESA will also be $^4T_2$. A coupling of the orbital triplet states $^4T_2$ and $^4T_1$ to phonons of E_g symmetry will spontaneously distort the CrF_6 octahedron and lift the degeneracy of the $^4T_2$ and $^4T_1$ states, as is illustrated in Fig. 7.8. Two broad ESA transitions are indicated on this sketch with energies

$$E_1 = E_0 + E_{JT} \quad \text{and} \quad E_2 = E_0 + 4\,E_{JT}, \tag{7.25}$$

in which E_0 is the crystal-field separation of $^4T_2$ and $^4T_{1a}$ states without the distortion that is stabilized by the Jahn–Teller energy, E_{JT}. The FWHM bandwidths in this Jahn–Teller model, ΔE_1 and ΔE_2, are derived from Eq. (5.150) by substituting E_{JT} for $S_0\hbar\omega_0$,

$$\Delta E_1 = (16\ln 2)^{1/2}(E_{JT}kT)^{1/2} \quad \text{and} \quad \Delta E_2 = (64\ln 2)^{1/2}(E_{JT}kT)^{1/2}, \tag{7.26}$$

which are valid only in the limit of strong coupling and high temperature ($E_{JT}, kT \gg h\nu$). Typically E_0 can be estimated from the Tanabe–Sugano diagram (~ 1500 nm, i.e., 6670 cm^{-1}) and $E_{JT} \sim 500\ cm^{-1}$, indicating that $E_1 = 7000\ cm^{-1}$ (or $\lambda_1 = 1430$ nm) and $E_2 = 8500\ cm^{-1}$ ($\lambda_2 = 1180$ nm), with room temperature bandwidths $\Delta E_1 = 1075\ cm^{-1}$ and $\Delta E_2 = 2150\ cm^{-1}$. Obviously $^4T_2 \rightarrow {}^4T_{1a}(1,2)$ ESA bands overlap in the near infra-red region. In contrast, because the Jahn–Teller effect in the $^4T_{1b}$ state is negligible, only a single $^4T_2 \rightarrow {}^4T_{1b}$ band is expected close to the predicted crystal-field energy of 20 000 cm^{-1} (500 nm). In strong field hosts where the ^2E state of $3d^2$ ions is lowest it is expected that $^2E \rightarrow {}^2A_1$, $^2T_{1b}$, $^2T_2$ and ^2E_b will be fairly narrow and overlap near 650 nm, with $^2E \rightarrow {}^2T_{1c}$ occurring near 21 500 cm^{-1} (465 nm). It is evident, therefore, for hosts with intermediate crystal-field sites, that ESA transitions may occur not only in the spectral region (~ 1200 nm) of the $^4T_2 \rightarrow {}^4A_2$ emission but also near 650 nm and 465 nm where they may compete with ground state transitions in absorbing pump radiation.

The possible transitions on the excited states of optical centres include stimulated emission, which can lead to laser amplification, and stimulated absorption from the excited state to an even higher energy excited state. When the ESA spectrum overlaps the wavelength range of the pump or emission band it reduces laser efficiency and increases the threshold pump power. In such cases ESA is an unwanted i.e., *parasitic absorption*. Excited state splittings, including those associated with the Jahn–Teller effect, play an important role in determining whether a particular gain medium will support laser operation. A number of examples of parasitic losses in ESA to higher lying states will be presented in Chapters 9 and 10, where laser operation is catastrophically impaired by such ESA processes, including Ti^{3+}-doped YAG and YAP and Ce^{3+} : YAG also. In some other cases, e.g., Cr^{3+} : alexandrite and Cr^{3+} : colquiriites, excited state processes including luminescence quenching have only modestly adverse impact on laser performance. The detailed effects must be evaluated on a case-by-case basis.

7.4 Experimental studies of excited state processes

Although single-photon absorption and radiative decay are fundamental to the optical properties of defects, transition-metal ions and rare-earth ions in crystalline and glassy materials, there are also excited state processes that influence possible laser operation. The excited state occupied following the absorption of a photon of appropriate energy may decay radiatively by emission of a photon or nonradiatively by the emission of phonons bridging the excited state–ground state energy gap. This latter process is often referred to as *thermal quenching of luminescence*, since it results in the shortening of the luminescence lifetime and decrease in luminescence intensity with increasing temperature (§6.4). In ideal circumstances the spectroscopy of a novel optical centre/host crystal should be carried out at low concentration to record unambiguously the spectra of isolated centres. As the concentration is increased various consequences follow from the interaction of the centre with its environment. In general, the ionic misfit between the centre and the host crystal causes strain, which builds up with increasing concentration. Optical spectra are inhomogeneously broadened by such strain [Stoneham (1969)]; the effects of strain broadening are normally observable in experiments on optical zero-phonon lines. Broadening also occurs with increasing concentration by virtue of weak multipolar and/or exchange interactions between centres. These interactions can be studied by high resolution laser spectroscopy [Yen and Selzer (1981)]. Although they will be too weak to change the energies of levels, they may induce excitation to be transferred nonradiatively between neighbouring sites. Under appropriate conditions, the excitation may transfer from site to site through the crystal before eventually reaching traps or sinks, impurities or defects that de-excite nonradiatively. The decrease in the luminescence efficiency of the material with increasing concentration of active centres is referred to as *concentration quenching*. The decay of the excited state population is then, in general, non-exponential with time and indicative of a reduced lifetime of the upper state reached in the absorption transition. Concentration quenching occurs by energy exchange processes between two ions, one excited $|D^*\rangle$ and one in the ground state $|A\rangle$, resulting in the donor returning to the ground state $|D\rangle$ and the acceptor being excited to a higher level $|A^*\rangle$. A sequence of such events is terminated at the surface by the emission of a photon of appropriate energy or when the excitation reaches a nonradiative trap. Concentration quenching by energy transfer involving two (or more) simultaneously excited ions can result in one excited ion relaxing to the ground state as the second ion is excited to an even higher-lying excited state. This process has been termed *energy transfer upconversion* (ETU). *Sequential two-photon excitation pumping* (STEP) involves a single ion. As Fig. 7.6 shows, the absorbed photon takes an ion into an excited state, which relaxes vibronically before the ESA of a second photon by that ion promotes it to an even higher-lying excited state. Both ETU and STEP transitions can convert infrared photons into visible luminescence, and even visible laser action [Macfarlane (1994a,b)].

7.4.1 *Quenching of luminescence and laser efficiency*

Nonradiative decay leading to the Stokes shift between absorption and emission transitions yields a four-level system of advantage in laser pumping (§1.3). However, nonradiative decay into the ground state of the system, which competes with the radiative process, both reduces laser efficiency and causes sample heating. These are important contributants to optothermal and thermomechanical responses of laser materials. The nonradiative decay is usually probed in measurements of the temperature dependence of the luminescence decay rate, $(\tau(T)^{-1})$, Eq. (6.38), which is related to the luminescence *radiant efficiency*, $\eta(T)$, by

$$\eta(T) = \tau(T)/\tau_R(T), \tag{7.27}$$

where $0 < \eta(T) < 1$ and $\tau_R(T)$ contains contributions to the radiative lifetime induced by static (τ_R^S) and dynamic (τ_R^D) odd-parity distortions of the crystal-field (§6.4.1). The fitting of experimental $\tau(T)$ versus T measurements to Eq. (6.38), which includes nonradiative decay across an energy gap, was discussed in §6.4.2 for octahedral Cr^{3+}-doped halide elpasolites and the laser crystal $Cr^{3+}:ScBO_3$. This partitioning of excited state energy between radiative and nonradiative decay also has important consequences for laser performance. In the presence of energy transfer, values of $\tau_R(T)$ are determined by fitting the luminescence decay to theoretical temporal profiles reviewed in §7.1 and §7.2. Whether the fluorescence is quenched by thermal and/or concentration dependent processes there will be a concomitant decrease in laser efficiency and increase in laser threshold.

Laser efficiency is most usually defined by the slope of the graph of laser output power plotted as a function of laser pump power. This is essentially the gradient of Eq. (1.32), written in power units and involving the Einstein coefficient B_{34} or, from Eq. (1.16), $A_{34} = (\tau_R(T))^{-1}$. In consequence, the measured *slope efficiency*, η_s, is written as

$$\eta_s = f\varepsilon\eta_i\eta(T)T/(L+T) \tag{7.28}$$

in which η_i is the *intrinsic slope* efficiency from Eq. (1.32), the factor, f, represents the effective fraction of pump photons absorbed by the gain medium and $T/(L+T)$ reflects the division of stimulated emission into that transmitted by the output coupler and that part dissipated internally. In a co-axially pumped laser with large absorption coefficient $f \rightarrow 1$, but in transverse pumping f is usually small. For lasers where the gain per pass is large (e.g. colour centre and Ce^{3+} lasers) efficient operation is achievable with T large enough to swamp out the internal cavity losses since $T/(L+T)$ may be close to one. In both three- and four-level lasers, (§1.3), the *laser pump efficiency* $\varepsilon = (\lambda_p/\lambda_l)$ is less than one because of the wavelength shift between pump (λ_p) and laser emission (λ_l) bands. The radiant efficiency, $\eta(T)$ can be much smaller than one when thermal and concentration quenching processes are active, whereas ε is usually a significant fraction of one. In ruby pumped at the peak of the ${}^4A_2 \rightarrow {}^4T_2$ band (570 nm) and operated on the R_1-line (693 nm), $\varepsilon \cong 82\%$ and $\eta(T)$ at 300 K is $71.0 \pm 0.5\%$, the luminescence efficiency of the ${}^4T_2 \rightarrow {}^4A_2$ emission of Cr^{3+} : GSGG at room temperature is 84 ± 1.0 % and that of the ${}^2E \rightarrow {}^2T_2$ emission of $Ti^{3+}:Al_2O_3$ is 69.5 ± 1.0 at.% [Li *et al.* (1991), (1992)]. Comparative data for many tunable solid state lasers activated by $3d^3$ configuration ions are given in Table 10.4: values of η_s vary between 84%(Cr^{3+} : LiCAF) and 0.1 % ($V^{2+}:MgF_2$). The evident variations in η_s follow from the cumulative effects of luminescence quenching by nonradiative decay, energy transfer and ESA, of which ESA is often most important. In the Cr^{4+}-doped garnets nonradiative decay is significant even at low temperature ($T = 20$ K) where $\eta(T)$ never exceeds 35% for any of the garnets [Kück *et al.* (1994), (1995)]. Even for the best examples, $Cr^{4+}:La_3Al_5O_{12}$ and Cr^{4+} : YAG, values of η (300 K) are only 33% and 22%, respectively. In RE^{3+} ion lasers there is the further subtlety that a particular radiative level may emit into several groups of crystal-field transitions. For example, in Nd : YAG the ${}^4F_{3/2} \rightarrow {}^4I_{11/2}$ lines near 1.064 μm compete with the ${}^4F_{3/2} \rightarrow {}^4I_{15/2}$ (1.80 μm), ${}^4I_{13/2}$ (1.35 μm) and ${}^4I_{9/2}$ (0.88 μm) groups, as shown in Fig. 1.5, and with nonradiative decay also (§1.3.2).

7.4.2 *High dopant concentrations*

The very sharpest lines are obviously observed in samples containing the lowest dopant concentration. Increasing the concentration of centres can result in the broadening of spectra and in the appearance of additional sharp lines from pairs of dopant ions, so close together that the exchange interaction between them causes measurable level splittings. Exchange-coupled pairs behave as distinct entities with spectroscopic properties characteristic of the pair of ions rather than the single ions from which they are formed. Exchange-coupled pairs of ions have been much studied by magnetic resonance and optical spectroscopic techniques. Pair interactions in ruby are apparent at Cr^{3+} concentrations as low as 0.1% Cr^{3+}, and pairs of Cr^{3+} ions in first, second, third and fourth nearest neighbour sites have been studied. At even higher concentrations additional broad lines are observed that are associated with three coupled ions and even larger groups. In general, the exchange interaction between rare-earth ions is much weaker than between transition-metal ions, and it may be necessary to use high resolution laser spectroscopy to distinguish the pair effects from single ion spectra. The presence of pairs and higher-order collections of ions may have impact on the relaxation dynamics of single ions.

Spectroscopic studies of fully concentrated crystals were stimulated by interests in co-operative magnetism [Loudon (1966), Eremenko and Petrov (1977)], where the interactions may be so strong that all intrinsic luminescence is quenched. Chromic oxide, Cr_2O_3, and neodymium trichloride, $NdCl_3$, are fully concentrated crystals that emit no luminescence signal. In contrast, MnF_2 emits strong luminescence although most of the signal arises from the presence of trace impurities. The strong luminescence signals of Cr^{3+} in $LiCaCrF_6$ and Pr^{3+} in $PrCl_3$ are sufficiently efficient that laser action may be stimulated at 300 K. Both luminescence quenching and efficient laser action are consequential upon excitation transfer between ions such that optical transitions occur not through single ions acting alone but because all the ions in the crystal collectively absorb or emit radiation. It is usual to discuss the states of collective excitation in a crystal in terms of exciton absorption or emission transitions. In the field of magnetic insulators, spectroscopic studies of MnF_2 have provided seminal contributions to the understanding of cooperative magnetism, just as have $Cr^{3+}:Al_2O_3$ and $Pr^{3+}:LaF_3$ in their respective fields of transition-metal ion and rare-earth ion impurity spectroscopy.

The antiferromagnetic crystal MnF_2 is orange coloured as a consequence of broad optical absorption bands in the blue and green spectral regions. The absorption/emission cycle of the spin-forbidden ${}^6A_1 \rightarrow {}^4T_1$ transition in the wavelength range 450–650 nm is shown in Fig. 7.9. There is a wealth of fine structure superposed on rather broad vibronic sidebands, the multiple sharp lines being caused by splittings associated with the combined actions of low symmetry crystal field, spin–orbit coupling and exchange interaction. Antiferromagnetic exchange in MnF_2 ($T_N = 68$ K) leads to rapid energy migration from Mn^{2+} ion to Mn^{2+} ion, enabling the ${}^6A_1 \rightarrow {}^4T_1$ transition to sample any traps present in the crystal. In fact, the spectra shown in Fig. 7.9 hardly reflect the intrinsic spectra of Mn^{2+} ions in MnF_2. Rather are they a consequence of the feeding of excitation energy to traps, in this case alkaline earth ions (Mg^{2+}, Ca^{2+} and Zn^{2+}), that are present at concentrations of only a few parts per million even in the most pure samples [Wilson *et al.* (1979)]. Those Mn^{2+} ions perturbed by nearby traps have their zero-phonon levels displaced to lower energies by about 100 cm^{-1} relative to the

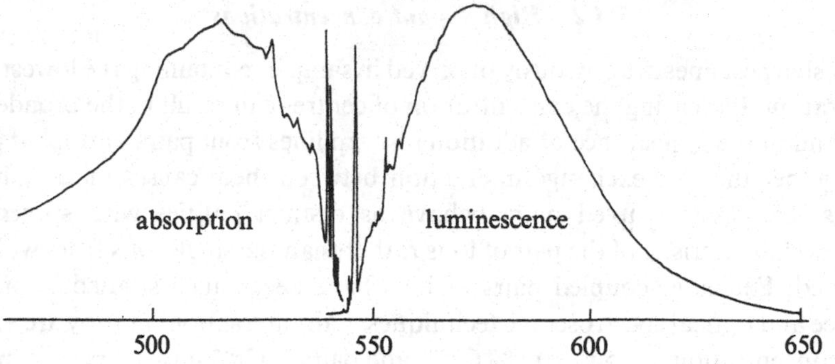

Figure 7.9. The ${}^6A_1 \rightarrow {}^4T_1$ optical absorption and luminescence spectra of moderately pure MnF_2 [after Henderson and Imbusch (1989)] at 2 K.

unperturbed ion levels. In very pure MnF_2 crystals the pure ${}^6A_1 \rightarrow {}^4T_1$ exciton can be sufficiently intense that under resonant laser excitation exciton–exciton interactions are apparent [Wilson *et al.* (1978)].

There have also been experiments on Er^{3+} and Eu^{3+}-doped MnF_2 crystals which enabled measurement of transfer rates to various traps from the intrinsic ${}^6A_1 \rightarrow {}^4T_1$ excitons. In experiments on $Eu^{3+}:MnF_2$ at 4.2 K, the sharp line ${}^5D_0 \rightarrow {}^7F_J$ emission at Eu^{3+} ions is superposed on the ${}^4T_{1g} \rightarrow {}^6A_{1g}$ vibronic sideband of the Mn^{2+} ions. The experiments confirm that Mn^{2+} ions perturbed by alkaline earth ion traps are only $\sim 100\,cm^{-1}$ below the E_1 no-phonon line of the 'pure' Mn^{2+} transition, whereas the Eu^{3+} level is about $1150\,cm^{-1}$ below E_1 with quenching traps more than $8000\,cm^{-1}$ below E_1. At low temperature with resonant excitation into the E_1 zero-phonon line both Mn^{2+} trap and Eu^{3+} trap emissions are observed as a consequence of excitation transfer. The Mn^{2+} luminescence is not observed above about 70 K. The populations of the excited traps can be calculated using rate equations [Hegarty (1976)]. The luminescence intensity $I(T)$ and lifetime $\tau(T)$ dependences on temperature are calculated to be

$$I(T) = [a\exp(-\Delta/kT) + b]^{-1} \tag{7.29}$$

and

$$(\tau(T))^{-1} = (\tau_R(T))^{-1} + c\exp(-\Delta/kT), \tag{7.30}$$

where $\Delta = 1150\,cm^{-1}$ is the energy difference between the E_1 level of Mn^{2+} and the ${}^5D_0$ levels of Eu^{3+}. The data for the Eu^{3+} luminescence intensity plotted as a function of reciprocal temperature, Fig. 7.10, fit equation (7.29) rather well with $\Delta = 1365\,cm^{-1}$. That this activation energy is slightly larger than the spectroscopic value probably reflects the fact that at some temperatures the Eu^{3+} excitation may boil back to both the E_1 level of Mn^{2+} and also the higher lying ${}^4T_1$ levels.

There have been comprehensive studies of concentration quenching of luminescence in Pr^{3-}-doped LaF_3 and $LaCl_3$ and in both PrF_3 and $PrCl_3$. The ${}^3P_0$ state lifetime of

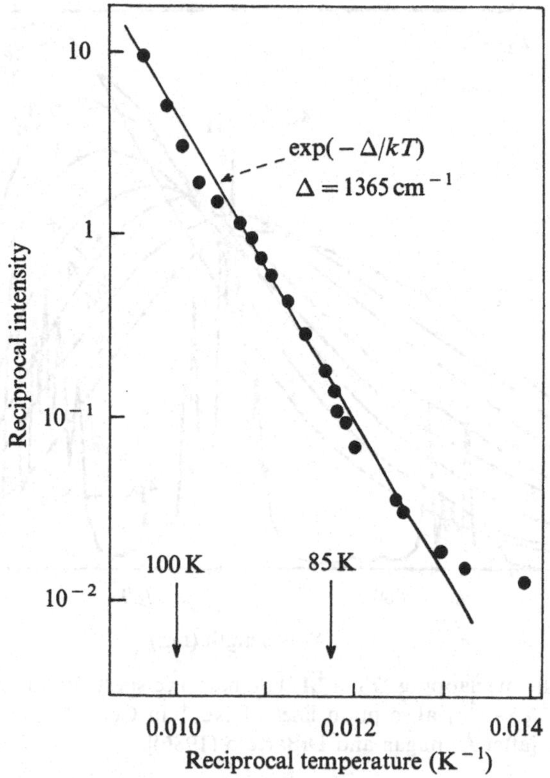

Figure 7.10. The inverse intensity of the excitonic $^5D_0 \rightarrow {}^7F_0$ exciton emission at Eu^{3+} ions in MnF_2 plotted as a function of reciprocal temperature [after Wilson *et al.* (1979)].

Pr^{3+} in the fluoride at low temperature decreases from 40 μs to ~1 μs as the Pr^{3+} concentration increases up to 100%. The lifetime at higher temperatures (~300 K) is completely quenched. In contrast, the same state in $PrCl_3$ shows little evidence of concentration quenching. The behaviour of the Cr^{3+}: doped colquiriites are of particular interest. Even at low concentrations, ~0.5%, fluorescence line narrowing (FLN) techniques resolve first and second nearest neighbour pairs in Cr^{3+}: $LiCaAlF_6$ [Wannemacher and Meltzer (1989)]. The pair interactions are quite weak, < 100 MHz, probably because there are no superexchange paths in $LiCaAlF_6$, and exchange via two intervening F^- ions is expected to be very weak and dominated by magnetic dipole–dipole interaction. In consequence, concentration quenching is not very important even in fully concentrated Cr^{3+}: colquiriites, and is certainly not effective enough to suppress laser operation in $LiSrCrF_6$ and $LiCaCrF_6$.

7.4.3 *Energy transfer and sensitization*

Although energy transfer between centres may result in luminescence quenching, energy may also be transferred to traps which de-excite by radiative emission, emitting their own characteristic luminescence. The acceptor and donor ions are usually

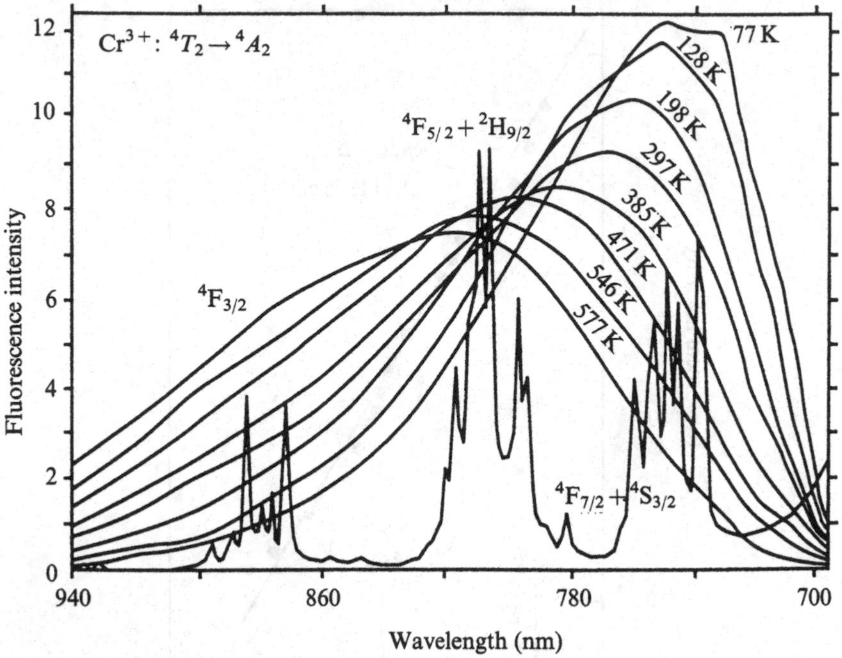

Figure 7.11. The overlapping $^4T_2 \rightarrow {}^4A_2$ luminescence spectrum of Cr^{3+} ions and the low-lying $^4I_{9/2} \rightarrow {}^4F_J$ absorption lines of Nd^{3+} in $Cr^{3+}:Nd^{3+}$: GSGG measured at 300 K [after Armagan and DiBartolo (1986)].

referred to as *activators* and *sensitizers*, respectively, and the emission from the accepting sinks or traps is termed *sensitized luminescence*. The very strong luminescence from trace amounts of Er^{3+} or Eu^{3+} in MnF_2 is an example of this phenomenon. Since the early reviews by Blasse [(1978), (1984)] there have been many other studies stimulated by the potential for improved laser performance in appropriately doped materials [Weber (1981)]. The optical spectra of Cr^{3+}-sensitized RE^{3+}-doped crystals have been of great interest. The absorption spectrum of $Cr^{3+}:Nd^{3+}$: GSGG shows that the $^4A_2 \rightarrow {}^4T_2$, $^4T_1$ bands cover much of the white light spectrum and that the total absorption by the many overlapping Nd^{3+} lines is much weaker. Nonradiative decay from the higher $^4T_1$ level of the Cr^{3+} ions leads to efficient luminescence ($\eta > 80\%$) in the $^4T_2 \rightarrow {}^4A_2$ band, which overlaps the $^4I_{9/2} \rightarrow {}^4F_{3/2}$, $^4F_{5/2}$ and $^4F_{7/2}$ absorption lines of Nd^{3-} ions. As Fig. 7.11 shows, there is also overlap with the $^4I_{9/2} \rightarrow {}^2H_{9/2}$, $^4S_{3/2}$ transitions.

Luminescence in the presence of D → A energy transfer is illustrated by the luminescence decay of Cr^{3+} donors and Nd^{3+} acceptors at 77 K in YAG containing *ca* 5×10^{19} ions cm^{-3} of both dopants. In a simplified model Henderson and Imbusch (1989) assumed that excited donors (D) and acceptors (A) decay radiatively from excited (b) to ground states (a) and that since the donors decay exponentially the D* → A energy transfer can be described by a collective rate W_{DA}. The latter assumption applies even though energy transfer is short-range, occuring between crystallographically close but inequivalent D–A multisites with different W_{DA} values. A* → D back transfer is negligible. The donor and acceptor rate equations following

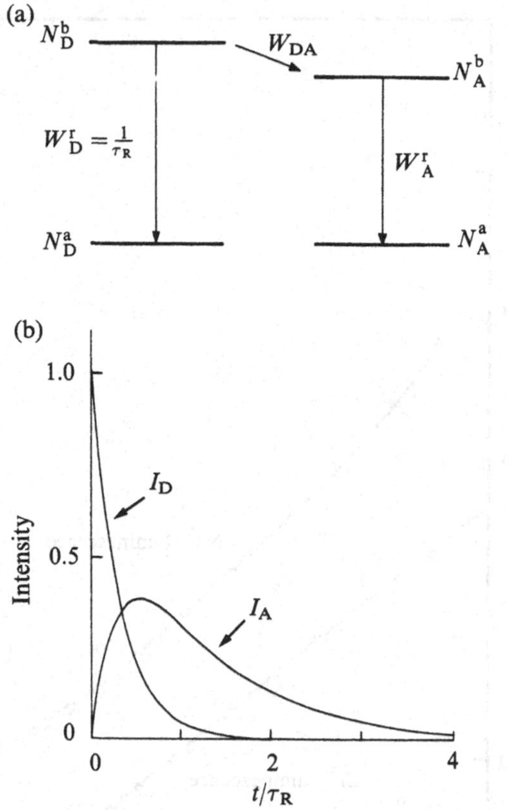

Figure 7.12. (a) Phonon-assisted D*–A energy transfer in the fast migration regime when the D* → A transfer is described by a collective rate W_{DA}. (b) The decay of the donor (I_D) and acceptor (I_A) luminescence intensities following pulsed excitation at $t=0$ is modeled with $W_D^r = W_A^r$, and $W_{DA} = 2W_D^r$ [after Henderson and Imbusch (1989)].

pulsed excitation were represented by

$$\frac{dN_D^b}{dt} = -W_D^r N_D^b - W_{DA} N_D^b \tag{7.31}$$

$$\frac{dN_A^b}{dt} = W_{DA} N_D^b - W_A^r N_A^b, \tag{7.32}$$

where W_D^r and W_A^r are identified with the inverse radiative life times of donors and acceptors, respectively. Solving for $N_D^b(t)$ and $N_A^b(t)$, assuming that $N_D^b(0)$ donors are excited by the initial pulse at $t=0$ without direct excitation of the acceptors, gives

$$N_D^b(t) = N_D^b(0)\exp[-(W_D^r + W_{DA})t] \tag{7.33}$$

and

$$N_A^b(t) = N_D^b(0)\frac{W_{DA}}{(W_D^r + W_{DA}) - W_A^r}\{\exp(-W_A^r t) - \exp[-(W_D^r + W_{DA})]t\} \tag{7.34}$$

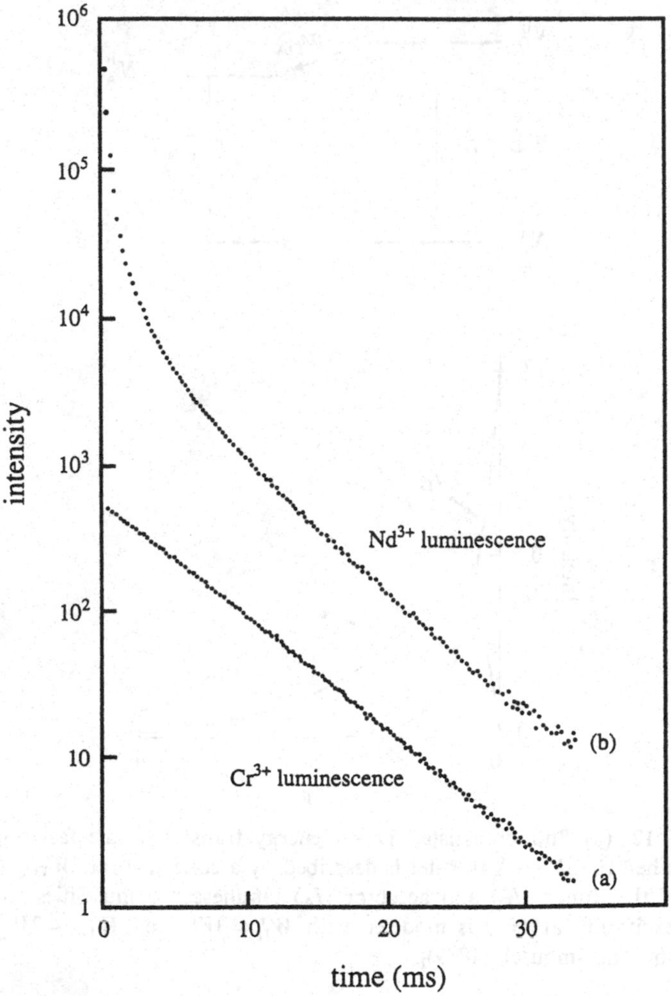

Figure 7.13. Luminescence decay of Cr^{3+} donors and Nd^{3+} acceptors in Cr^{3+} : Nd^{3+} : YAG crystals excited with a white light pulse at 77 K [after Henderson and Imbusch (1989)].

Plots of Eqs. (7.33) and (7.34) in Fig. 7.12 represent the decaying populations of donors and acceptors with the assumption that $W_D^r = W_A^r$ and $W_{DA} = 2W_D^r$. The solution for $N_D^b(t)$ shows the total donor decay rate to be $\tau_D^{-1} = \tau_R^{-1} + \tau_{DA}^{-1}$. The intensity of acceptor luminescence, which follows the temporal evolution of $N_A^b(t)$, rises from zero intensity at $t = 0$ to a peak value, after which it decays exponentially with time constant W_A^r or $W_D^r + W_{DA}$, whichever is the smaller [Henderson and Imbusch (1989)].

In co-doped Cr^{3+} : Nd^{3+} : YAG, $D^* \rightarrow A$ transfer is so fast that the peak intensity in the acceptor decay pattern occurs at times too short to be resolved with the available apparatus. The white light pulse that directly excites both donors and acceptors also excites acceptors by $D^* \rightarrow A$ energy transfer. Since $D^* \rightarrow A$ excitation transfer occurs at a rate much faster than the radiative decay rate of donors, the supply of acceptors excited by $D^* \rightarrow A$ transfer is rapidly exhausted after the excitation pulse. Accordingly, energy transfer is not detected on the decaying donor luminescence, which Fig. 7.13 shows to be exponential with time constant equal to the intrinsic decay rate, $W_D^r = 1.88 \times 10^2\,s^{-1}$, the concentration of long-lived donors greatly exceeding the number of transiently excited D–A pairs. The Nd^{3+} acceptors that are directly excited by the white light pulse are so numerous as to mask the peak in the acceptor decay curve, Eq. 7.34. They decay at the intrinsic rate of $W_A^r \sim ca\ 5 \times 10^3\,s^{-1}$. This process is superposed on the slower, non-exponential acceptor decay at the total donor decay rate of $W_D^r + W_{DA}$, Eq. (7.34). After times long compared with the pulse duration the decay of the acceptor luminescence asymptotically approaches $W_D^r = 1.88 \times 10^2\,s^{-1}$ as the depletion of D–A pairs is complete. The rate W_{DA}, which is initially faster than either of W_D^r and W_A^r, decreases with increasing time after the excitation pulse due to the exhaustion of the $D^* \rightarrow A$ transfer. It is determined as $W_{DA} \sim 10^5\,s^{-1}$, by fitting to Eq. (7.34) the non-exponential acceptor decay some 2–15 ms after excitation. W_{DA} is large because of the excellent overlap of the Cr^{3+}-luminescence and Nd^{3+} absorption spectra. This is exemplified in Fig. 7.11 for $Cr^{3+}:Nd^{3+}$:GSGG, where the microscopic interaction parameter, $\alpha_{DA}^{(n)}$, defined in Eq. (7.5), is $\sim 2\text{–}3 \times 10^{-39}\,cm^6\,s^{-1}$ [Armagan and DiBartolo (1986)].

Similar studies have been reported for Cr^{3+} sensitization of Nd^{3+}, Er^{3+}, Ho^{3+}, Tm^{3+} and Yb^{3+} luminescence by electric dipole–electric dipole transfer in yttrium and gadolinium-based scandium garnets. The strength of Cr^{3+}–Er^{3+} energy transfer in YSGG is defined by $\alpha_{DA}^{(6)} = 2 \times 10^{-39}\,cm^6\,s^{-1}$, whereas for $Cr^{3+}:Tm^{3+}$ and $Cr^{3+}:Ho^{3+}$ transfer in YSAG $\alpha_{DA}^{(6)} = 4.2 \times 10^{-39}\,cm^6\,s^{-1}$ and $\alpha_{DA}^{(6)} = 2.8 \times 10^{-40}\,cm^6\,s^{-1}$, respectively. The strong $Cr^{3+}:Tm^{3+}$ energy migration follows from the efficient resonant Cr^{3+}–Tm^{3+} transfer based on the overlap of the ${}^4T_2 \rightarrow {}^4A_2$ Cr^{3+} emission with the Tm^{3+} absorption in the ${}^3H_6 \rightarrow {}^3F_4$ transition over the spectral region 700–800 nm. The rather small value of $\alpha_{DA}^{(6)}$ in $Cr^{3+}:Ho^{3+}$ is due to the poor spectral overlap between the ${}^4T_2 \rightarrow {}^4A_2$ emission of Cr^{3+} and the ${}^5I_8 \rightarrow {}^5I_7$ absorption of Ho^{3+}. However, the Ho^{3+} laser emission at 2.8 μm can be sensitized by Cr^{3+}–Tm^{3+} double pumping, in which energy is readily transferred from the Tm^{3+} ${}^3H_4$ level at 1.86 μm to the ${}^5I_7$ level of Ho^{3-}. In these triply-doped YSAG and YSGG crystals the Cr^{3+} activator ions are pumped with the 647.1 nm line from the Kr^+-laser. The major part of the fluorescence is channelled into the ${}^5I_7 \rightarrow {}^5I_8$ emission between 1850 and 2100 nm. The overall efficiency for this Cr^{3+}–Tm^{3+}–Ho^{3+} exceeds 50%.

7.4.4 Upconversion processes

Among various processes that result from ESA is de-excitation by radiative transitions to lower energy states, including the ground state. The energies of the emitted photons can then be greater than the energy of the absorbed photon. Such *upconverted luminescence* has been used as the basis of upconverted lasers, including optical fibre lasers [Smart *et al.* (1991)]. However, the first CW upconversion laser used the STEP pumping scheme in 1% $Er^{3+}:YAlO_3$ shown in Fig. 7.14 [Macfarlane (1994a)]. The first absorbed photon ($\lambda = 792.1$ nm) pumps Er^{3+} ions into the ${}^4I_{9/2}$ level. After non-radiative relaxation to the ${}^4I_{11/2}$ level a second absorbed photon ($\lambda = 839.8$ nm) excites Er^{3+} ions from this metastable level into the ${}^4F_{7/2}$. Nonradiative ${}^4F_{7/2} \rightarrow {}^4S_{3/2}$ decay

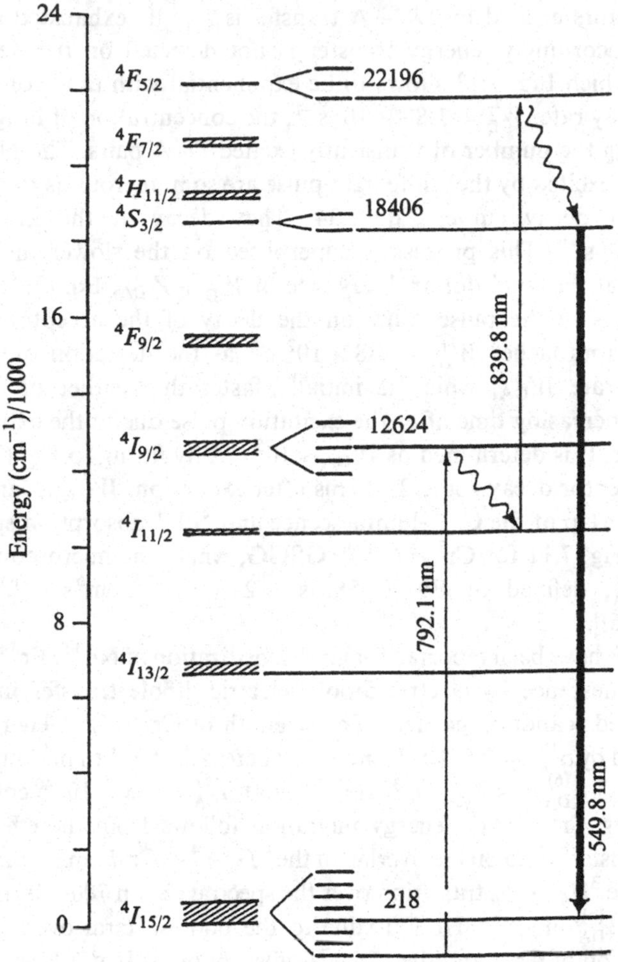

Figure 7.14. STEP and the nonlinear absorption pumping of the green (549.8 nm) $Er^{3+}:YAlO_3$ upconversion laser. The different ground and excited state pump wavelengths (792.1 nm, 839.8 nm) were provided by two tunable dye lasers [after Macfarlane (1994a)].

precedes $^4S_{3/2} \rightarrow {}^4I_{15/2}$ laser emission at $\lambda = 549.8$ nm with typical output of *ca* 1 mW, when pumped using two lasers each with output up to 200–250 mW. When the laser power used to pump the first absorption at 792.1 nm is held constant the output is linear in the pump power used in the second step (Fig. 7.15a). Alternatively, with the pump power in the first step varied and that of the second step fixed the laser output is nonlinear (Fig. 7.15b), showing saturation due to ground state depletion.

Such STEP upconversion schemes as shown in Fig. 7.14 are particularly suited to samples containing low concentrations of dopants where the interaction between optical centres is weak. At higher dopant concentrations, typically near one per cent for rare-earth ions, the nonlinear optical excitation may rely on energy transfer between a pair of excited ions [Auzel (1966), (1978)]. ETU upconversion in the $Er^{3+}:YLF$ laser is

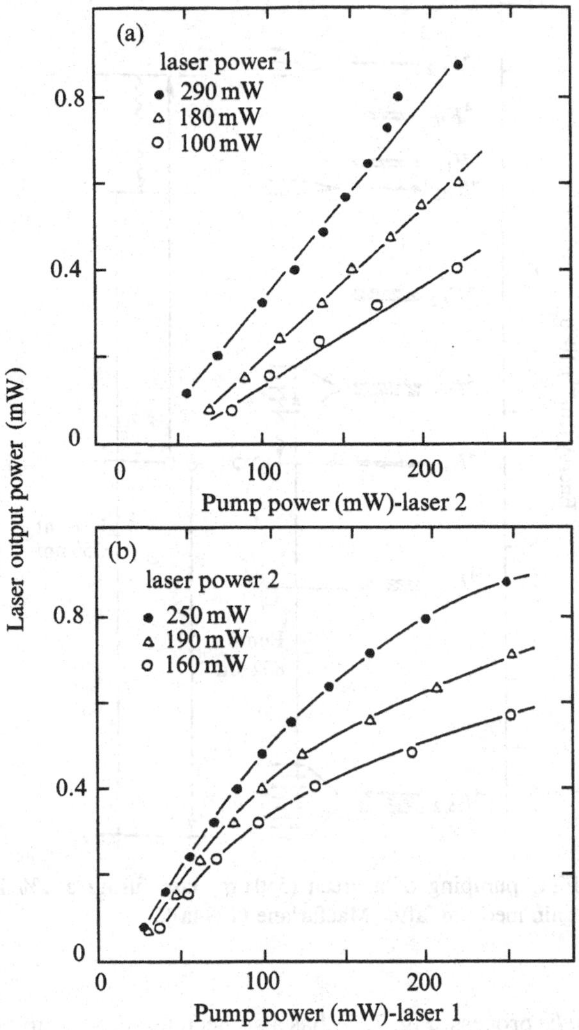

Figure 7.15. Output characteristics of the green (549.8 nm), Er^{3+} : $YAlO_3$ STEP upconversion laser excited at 792.1 nm and 839.8 nm [after Silversmith *et al.* (1987)].

illustrated in Fig. 7.16. Two absorbed photons create two excited ions in the metastable $^4I_{11/2}$ state. These ions are coupled by multipole or exchange interaction which promotes a cross-relaxation step to de-excite one ion into the lower-lying $^4I_{15/2}$ level and promotes the second ion into a higher excited $^4F_{5/2}$ level. Subsequent relaxation into a metastable $^4S_{3/2}$ level results in the laser emission at 550 nm. The cross-relaxation need not be a resonant process, since energy transfer can be phonon-assisted (Fig. 7.2). Nor need the two cross-relaxing ions be the same. This pumping scheme, Fig. 7.16, was used by Lenth *et al.* (1988) in the 1% Er^{3+} : $LiYF_4$ laser operating at 550 nm. A single excitation source was used to produce the two excited ions, the nonlinear pumping being provided by cross-relaxation. The efficiency of this two ion process is higher than that of the STEP upconversion.

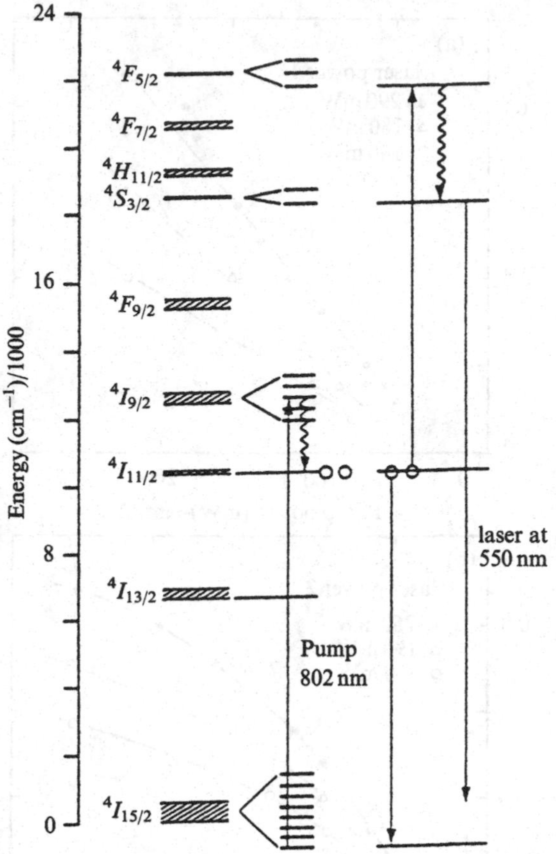

Figure 7.16. ETU pumping of a green (550 nm) laser using a 1% Er^{3+} : YLF crystal as the gain medium [after Macfarlane (1994a)].

A *photon avalanche* process, Fig. 7.17, has also been used to pump an upconversion Nd : $LiYF_4$ laser. A single excitation photon ($\lambda = 603.6$ nm) pumps the weak phonon sideband of the $^4I_{9/2} \rightarrow {}^2H_{11/2}$ transition, seeding a small population in the $^4F_{3/2}$ level. A second photon ($\lambda = 603.6$ nm) resonantly pumps the $^4F_{3/2} \rightarrow {}^4D_{3/2}$ transition. Cross-relaxation from the $^4D_{3/2}$ level effectively pumps a second ground state ion into the ($^4F_{7/2}$, $^4S_{3/2}$) levels which decay nonradiatively to the $^4F_{3/2}$ level on this ion, at the same time relaxing the first ion to its $^4F_{3/2}$ level. Thus a single ion in the $^4D_{3/2}$ level produces two ions in the $^4F_{3/2}$ level, which then absorb pump light to reach the $^4D_{3/2}$ level, and the process is repeated. In Nd : $LiYF_4$ the laser transitions occur at 730 nm and 413 nm. As in the Er : $LiYF_4$ example discussed above, the cross-relaxation need not be resonant; it may be phonon assisted. Furthermore, the excited state transition can be to a level above $^4D_{3/2}$. These relaxations of the strict requirements of the level structure by phonons permit many possible pumping schemes for the different rare-earth ions. Indeed, photon avalanche excitation of laser action has been demonstrated for Pr^{3+} [Kueny *et al.* (1989)], Nd^{3+} [Lenth and Macfarlane (1990)], Sm^{3+} [Krasatskuy (1983)], and Tm^{3+} [Hebert *et al.* (1992)], and for the transition-metal ion Ni^{2+} [Oetliker *et al.* (1992)].

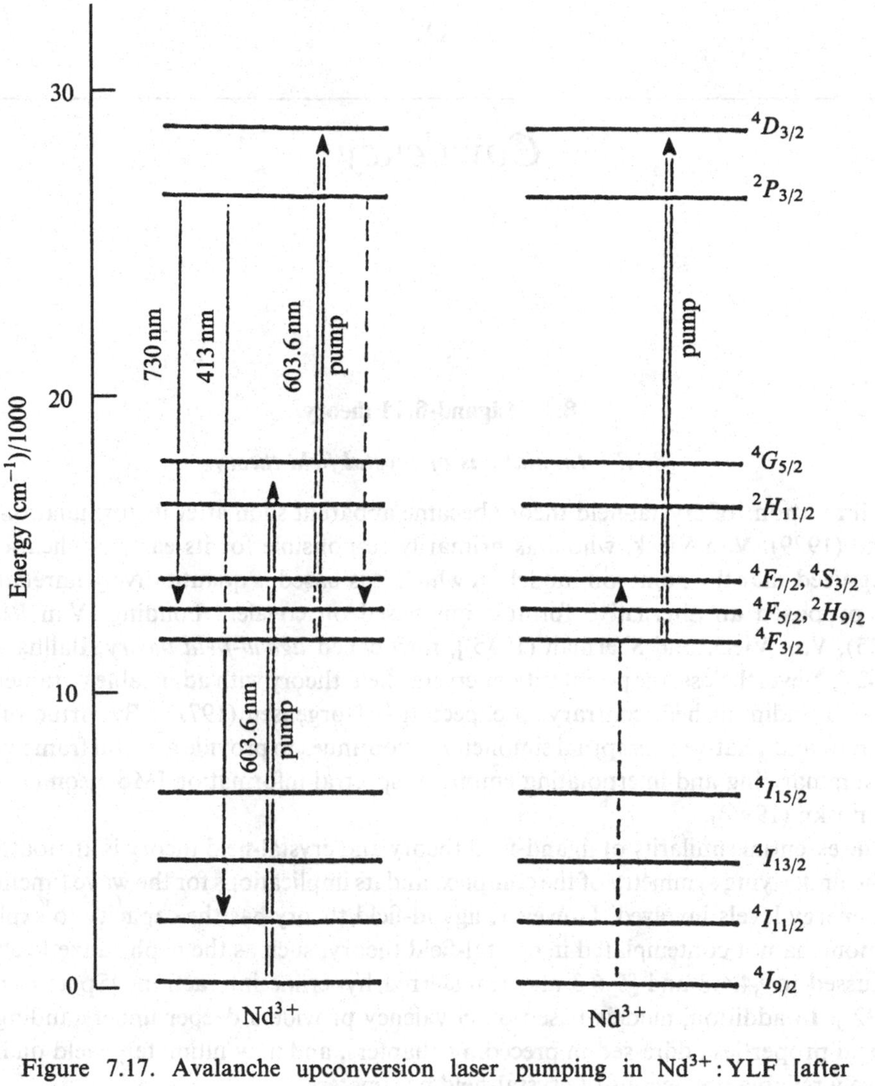

Figure 7.17. Avalanche upconversion laser pumping in Nd^{3+}: YLF [after Macfarlane (1994a)].

8

Covalency

8.1 Ligand-field theory

8.1.1 Limitations of crystal-field theory

The limitations of crystal-field theory became apparent soon after its formulation by Bethe (1929). Van Vleck, who was primarily responsible for its early applications, recognized that the point-ion model on which it is based is quantitatively unreliable, and proposed an alternative formulation based on covalent bonding [Van Vleck (1935), Van Vleck and Sherman (1935)], now called *ligand-field theory* [Ballhausen (1962)]. Nevertheless, the popularity of crystal-field theory with adjustable parameters remains undiminished, contrary to expectation [Jørgensen (1971)]. By virtue of its elegance and relative conceptual simplicity, it continues to provide a useful framework for summarizing and interpolating empirical spectral information [Morrison (1992), Kaminskii (1996)].

The essential similarity of ligand-field theory and crystal-field theory is attributable to the underlying symmetry of the complex, and its implications for the wave functions and energy levels involved. However, ligand-field theory has the capacity to explain phenomena not contemplated in crystal-field theory, such as the nephelauxetic effect discussed in §4.4.8 and §9.6.3 and transferred hyperfine interactions [Spaeth *et al.* (1992)]. In addition, models based on covalency provide a deeper understanding of optical properties addressed in preceding chapters, and may ultimately yield quantitatively reliable predictions of crystal-field parameters.

8.1.2 Molecular orbitals

In the *molecular-orbital theory* of covalency [Hund (1927b), Mulliken (1928)], electrons occupy orbital wave functions which are delocalized over the entire complex consisting, for example, of a transition-metal ion and its immediate ligands. An approximate molecular orbital can be constructed as a linear combination of atomic orbitals (LCAO). If the effective one-electron Hamiltonian is assumed to share the symmetry of the complex, then the molecular orbitals transform as bases for irreducible representations of the crystallographic point group which leaves the complex invariant, and only symmetry-adapted combinations of atomic orbitals contribute to them.

It suffices for a qualitative understanding of ligand-field theory to include only those valence atomic orbitals which dominate covalent bonding. For example, in a complex

Table 8.1. *Character table of reducible representation* Γ *of point group* O_h *spanned by* p_σ*-orbitals*

O_h	E	$8C_3$	$3C_2$	$6C_2'$	$6C_4$	I	$8C_3I$	$3C_2I$	$6C_2'I$	$6C_4I$
Γ	6	0	2	0	2	0	0	4	2	0

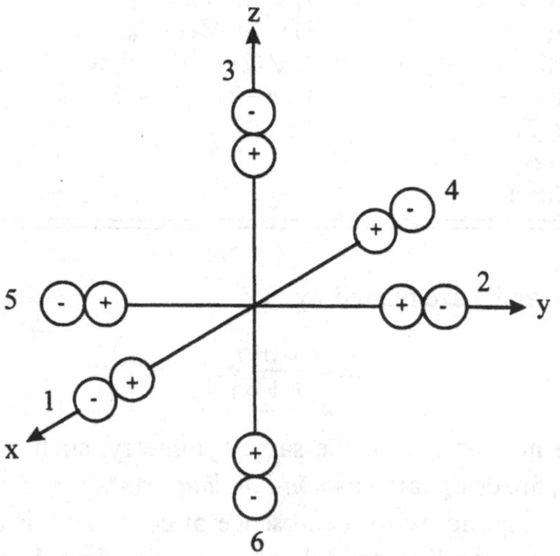

Figure 8.1. Ligand orbitals of p_σ type numbered according to the convention of Van Vleck (1935).

consisting of an iron-group transition-metal ion octahedrally coordinated to F^- or O^- anion ligands, these include 3*d*, 4*s* and 4*p* transition-metal orbitals and the 2*p* ligand orbitals which are directed toward the central metal ion, called p_σ-orbitals, illustrated in Fig. 8.1, where the ligands are numbered according to the convention of Van Vleck (1935). These orbitals are transformed into one another by the symmetry elements of point group O_h, and are thus a basis for a six-dimensional reducible representation Γ. The character of this representation, displayed in Table 8.1, is constructed simply by counting the number of p_σ-orbitals left invariant by a symmetry operation of each class. It is then readily verified by application of Eq. (2.10) and Tables 2.7 and 2.11 that the reducible representation contains irreducible representations A_{1g}, T_{1u} and E_g. Symmetry-adapted linear combinations of p_σ-orbitals are presented in Table 8.2, together with corresponding transition-metal orbitals defined in Table 4.4.

In order to describe covalency effects, *bonding* and *antibonding* molecular orbitals, ψ^b and ψ^a, respectively, are each constructed from a transition-metal orbital ϕ and a symmetry-adapted combination of ligand orbitals χ, as follows [Owen and Thornley (1966) Simánek and Sroubek (1972)]:

$$\psi^a = (1 - 2\lambda S + \lambda^2)^{-1/2}(\phi - \lambda\chi), \tag{8.1a}$$

$$\psi^b = (1 + 2\gamma S + \gamma^2)^{-1/2}(\chi + \gamma\phi), \tag{8.1b}$$

where S is the overlap integral

$$S = \int \phi(\mathbf{r})\chi(\mathbf{r})\, d\tau, \tag{8.1c}$$

Table 8.2. *Symmetry-adapted combinations of transition-metal and ligand p_σ-orbitals*

Symmetry	central-ion orbitals	ligand p_σ-orbitals
a_{1g}	$4s$	$(1/\sqrt{6})(-X_1+X_4-Y_2+Y_5-Z_3+Z_6)$
$t_{1u}\alpha$	$4p(x)$	$(1/\sqrt{2})(-X_1-X_4)$
$t_{1u}\beta$	$4p(y)$	$(1/\sqrt{2})(-Y_2-Y_5)$
$t_{1u}\gamma$	$4p(z)$	$(1/\sqrt{2})(-Z_3-Z_6)$
$e_g\theta$	$3d(2z^2-x^2-y^2)$	$(1/\sqrt{12})(-2Z_3+2Z_6+X_1-X_4+Y_2-Y_5)$
$e_g\varepsilon$	$3d(x^2-y^2)$	$(1/2)(-X_1+X_4+Y_2-Y_5)$
$t_{2g}\xi$	$3d(yz)$	
$t_{2g}\eta$	$3d(zx)$	
$t_{2g}\zeta$	$3d(xy)$	

and the parameters λ and γ are related by

$$\lambda = \frac{S+\gamma}{1+S\gamma}. \tag{8.1d}$$

Orbitals which have no partners of the same symmetry, such as the central-ion t_{2g} orbitals in Table 8.2, are designated as *non-bonding* orbitals, ψ^{n}.

The $\gamma=0$ case, corresponding to the absence of covalency, is designated the *ionic model.* In that case, the bonding orbital, ψ^{b}, is a pure ligand orbital; however, the antibonding orbital, ψ^{a}, is not then a pure transition-metal orbital, but instead retains an admixture of ligand orbital as a consequence of the non-orthogonality of the basis functions ϕ and χ.

8.1.3 *Variational principle*

An approximate orbital energy E_{t} can be determined by application of the variational principle with the assumption of a common one-electron Hamiltonian h,

$$E_{\mathrm{t}} = \frac{\langle\psi_{\mathrm{t}}|h|\psi_{\mathrm{t}}\rangle}{\langle\psi_{\mathrm{t}}|\psi_{\mathrm{t}}\rangle}, \tag{8.2}$$

where ψ_{t} is an approximate wave function, called a *variational trial function.* This expression provides an upper bound on the ground state energy whose error is quadratic in the wavefunction error; consequently, the approximation can be optimized by varying ψ_{t} to minimize E_{t}. If the trial function ψ_{t} is constrained to transform as a basis function for an irreducible representation of the group G of the Hamiltonian h, then the corresponding energy E_{t} is an upper bound on the lowest energy of the same representation, as well.

In the special case in which the trial function is expressed as a linear combination of fixed basis functions ψ_i,

$$\psi_{\mathrm{t}} = \sum_i \psi_i C_i, \tag{8.3}$$

variation of the coefficients C_i to minimize E_{t} leads to the matrix equation

$$\sum_j (h_{ij} - E_{\mathrm{t}} S_{ij}) C_j = 0, \tag{8.4a}$$

$$h_{ij} = \langle \psi_i | h | \psi_j \rangle, \tag{8.4b}$$

$$S_{ij} = \langle \psi_i | \psi_j \rangle, \tag{8.4c}$$

and the corresponding secular equation

$$\det|h_{ij} - E_t S_{ij}| = 0. \tag{8.5}$$

The lowest root of the secular equation is then a least upper bound on the lowest exact energy of the same representation, and elements of the corresponding eigenvector are the coefficients c_i of the fixed basis functions ψ_i in the optimum trial function. The remaining roots are upper bounds on the energies of successive excited states of the same symmetry, and the corresponding eigenvectors define their approximate wave functions, but the quality of the approximation diminishes with increasing energy.

In the σ-bonding model described previously, the basis functions are the symmetry-adapted combinations of atomic orbitals listed in Table 8.2. Since there are two basis functions for each row of each of the irreducible representations A_{1g}, T_{1u} and E_g, there is an independent 2×2 secular equation for each irreducible representation, independent of row. With the fixed basis functions identified as

$$\psi_1 = \chi, \tag{8.6a}$$

$$\psi_2 = \phi, \tag{8.6b}$$

and the assumption that χ is normalized (overlap integrals of p_σ-orbitals neglected), the orbital energies are given by

$$E^{\mathrm{a,b}} = \frac{h_{11} + h_{22} - 2Sh_{12}}{2(1 - S^2)} \pm \frac{\sqrt{(h_{12} + h_{22} - 2Sh_{12})^2 - 4(1 - S^2)(h_{11}h_{22} - h_{12}^2)}}{2(1 - S^2)}, \tag{8.7}$$

and the covalency parameter γ by

$$\gamma = -\frac{h_{11} - E^{\mathrm{b}}}{h_{12} - E^{\mathrm{b}}S} = -\frac{h_{12} - E^{\mathrm{b}}S}{h_{22} - E^{\mathrm{b}}}. \tag{8.8}$$

All of the bonding orbitals are doubly occupied in the ground state. The $3d$ electrons of the free transition-metal ion occupy non-bonding t_{2g}^{n} and antibonding e_g^{a} orbitals of the complex, as shown schematically in Fig. 8.2. The separation of the corresponding energy levels, $\Delta = 10Dq$, is then seen as a measure of the strength of covalent bonding rather than of the crystal field as in Chapter 4. Note that Fig. 8.2 is appropriate to an independent-particle model with a common one-electron Hamiltonian; the multiplet structure arising from mutual electrostatic interaction is obscured at this level of approximation. Optical transitions of the type $t_{2g}^{\mathrm{n}} \rightarrow e_g^{\mathrm{a}}$ are parity forbidden; they appear either as magnetic-dipole transitions, or are weakly enabled by odd phonons or odd components of the crystal field. Transitions of the type $t_{2g}^{\mathrm{n}} \rightarrow t_{1u}^{\mathrm{a}}$ or $e_g^{\mathrm{a}} \rightarrow t_{1u}^{\mathrm{a}}$, which are fully electric-dipole allowed, usually appear as intense charge-transfer bands.

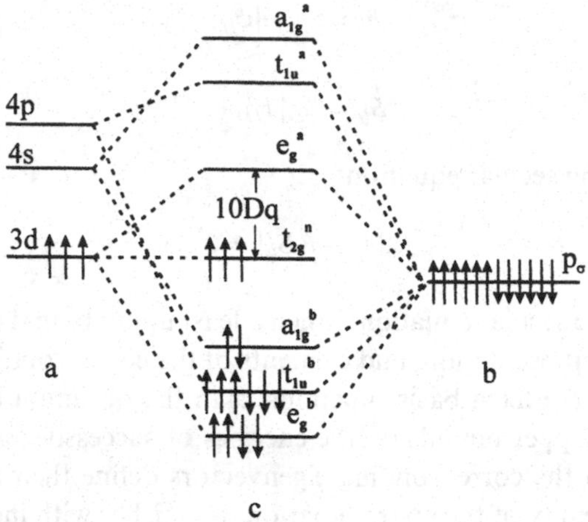

Figure 8.2. Ligand-field theory bonding scheme for an octahedral transition metal complex with a d^3 configuration, such as Cr_6^{3-}.

The basis set can be extended by including 2s ligand orbitals, which transform the same way as p_σ-orbitals, and transverse p_π-orbitals which can combine with central-ion t_{2g} orbitals to form weak π-bonds, manifest in transferred hyperfine interaction. The example of Ti^{3+} discussed in §8.6.3 involves such an extended basis set.

8.1.4 Valence bonds

An alternative description of covalent bonding, *valence-bond theory* [Heitler and London (1927), Pauling (1939)], emphasizes the role of individual ligands. This theory employs the individual p_σ-orbitals together with linear combinations of a_{1g}, t_{1u} and e_g central-ion orbitals which transform as bases for the six-dimensional reducible representation Γ. These combinations of central-ion orbitals, called sp^3d^2 hybrids, are equivalent in the sense that they transform into one another under the symmetry operations of group O_h. Variational trial functions are constructed as linear combinations of a ligand orbital and a central-ion hybrid orbital, and Eqs. (8.5) and (8.6) are assumed to apply to each bond independently, leading to a 2×2 secular equation, the same for each bond. However, in this case the partitioning of the Hamiltonian and overlap matrices into 2×2 blocks is only an approximate consequence of the small mutual overlap of p_σ-orbitals, rather than a rigorous consequence of group theory [Hurley *et al.* (1953)].

Valence bonds conform more closely to chemical intuition than molecular orbitals, and provide a more convincing explanation of the directional stability of chemical bonds. However, they obscure configurations as well as multiplet structure; for example, in the valence-bond model, the three antibonding levels in Fig. 8.2 are merged into a single six-fold degenerate antibonding level. The computational methods employed in modern quantum-chemistry, which are addressed in subsequent sections of this chapter, are based primarily on molecular-orbital theory instead.

8.1.5 *Charge-transfer model*

Since the doubly occupied bonding orbitals are pure ligand orbitals in the ionic model, but contain a substantial admixture of central-ion orbitals in the ligand-field model, it is evident that covalency involves a substantial transfer of electronic charge from the ligands to the transition-metal ion which is only partially offset by delocalization of electrons in non-bonding and antibonding orbitals. This observation is made more explicit in an alternative *charge-transfer model* [Freeman and Watson (1965)].

For example, consider a hypothetical three-electron system whose ground-state wave function can be represented in terms of the orthogonal molecular orbitals by a single Slater determinant,

$$\Psi^{\text{mo}} = \frac{1}{\sqrt{3!}}\det|\psi_\uparrow^{\text{a}}\psi_\uparrow^{\text{b}}\psi_\downarrow^{\text{b}}|. \tag{8.9}$$

Covalency can be described alternatively in terms of configuration mixing. The ground state wave function for the three-electron system in the ionic model can be represented by the single Slater determinant

$$\Psi_{\text{G}} = \frac{1}{\sqrt{3!(1-S^2)}}\det|\phi_\uparrow\chi_\uparrow\chi_\downarrow|. \tag{8.10}$$

An improved ground-state wavefunction can be constructed by admixture of an excited charge-transfer configuration in which one electron is transferred from a ligand orbital to a central-ion orbital,

$$\Psi^{\text{ct}} = \frac{1}{\sqrt{1+2\gamma S+\gamma^2}}(\Psi_{\text{G}}+\gamma\Psi_{\text{E}}), \tag{8.11a}$$

$$\Psi_{\text{E}} = \frac{1}{\sqrt{3!(1-S^2)}}\det|\phi_\uparrow\chi_\uparrow\phi_\downarrow|. \tag{8.11b}$$

It can be shown that the wavefunctions of Eqs. (8.9) and (8.11a) are identical with the same value of γ; accordingly, the charge-transfer and molecular-orbital models are equivalent.

8.2 Hartree–Fock method

8.2.1 *Hamiltonian*

The Hartree–Fock method provides a more realistic description of covalency than the independent particle model, while preserving some of its essential features, including molecular orbitals [Hartree (1957), Fischer (1977), Szabo and Ostlund (1989)]. The ultimate source of the Hamiltonian employed in this method is the relativistic, semi-classical *Dirac equation* for a single electron in external electric and magnetic fields [Dirac, 1958],

$$i\hbar\frac{\partial\psi}{\partial t} = [-\boldsymbol{\alpha}\cdot(c\mathbf{p}+e\mathbf{A}) - \beta mc^2 - e\phi]\psi, \tag{8.12}$$

where ϕ and $\mathbf{A}$ are electric scalar and magnetic vector potentials, respectively, $\mathbf{p}=-i\hbar\nabla$ is the momentum operator, $\psi(\mathbf{r}, t)$ is a four-component wavefunction, and α_x, α_y, α_z and β are 4×4 matrices which anti-commute in pairs.

Two components of ψ become vanishingly small for positive-energy solutions in the non-relativistic limit. One proceeds by eliminating the two small components in favour of the two large ones; by expanding in ν/c, retaining terms out to $(\nu/c)^2$; and by effecting a *Foldy–Wouthuysen* transformation which ensures normalization of the remaining two-component wavefunction to the same order of approximation [Foldy and Wouthuysen (1950), Bjorken and Drell (1964)]. The approximate Hamiltonian obtained by this procedure has the form

$$H=\frac{\mathbf{p}^2}{2m}-e\phi+\frac{e\hbar}{2m^2c^2}\mathbf{s}\cdot\mathbf{E}\times\mathbf{p}+\frac{e}{mc}\mathbf{A}\cdot\mathbf{p}+2\mu_B\mathbf{H}\cdot\mathbf{s}$$
$$+\frac{1}{[1+(e\phi/2mc^2)]^2}\frac{\hbar e^2}{2m^2e^4}\mathbf{s}\cdot\mathbf{E}\times\mathbf{A}-\frac{\mathbf{p}^4}{8m^3c^2}+\frac{e\hbar^2}{8m^2c^2}\nabla\cdot\mathbf{E}, \qquad (8.13)$$

where $\mathbf{s}$ is the spin angular momentum operator, whose components can be represented by 2×2 matrices operating on the two-component wavefunction, and $\mathbf{E}=-\nabla\phi$ is the electric-field intensity. Generalization of this Hamiltonian for a many-electron system is accomplished by summing over electrons and incorporating internal interactions.

Equation (4.1) corresponds to just the first three terms on the right-hand side of Eq. (8.13), adapted to a many-electron system. The second term accommodates electrostatic interactions with a central nucleus, with other electrons and with a crystal field. The third term is the spin–orbit interaction which, for a many-electron system with a central potential, can be written in the form indicated in Eqs. (4.1) and (4.2). The fourth and fifth terms on the right-hand side of Eq. (8.13) accommodate both Zeeman and hyperfine interactions. The sixth term, which is bilinear in electric and magnetic fields, is negligible everywhere except in the immediate vicinity of the nucleus, but it makes a finite integrated contribution to the hyperfine interaction (*contact hyperfine interaction*). The seventh and eighth terms are relativistic corrections which preserve the symmetries and degeneracies of the first two terms but are nonetheless important for precise atomic-structure calculations on heavy atoms.

The Hamiltonian which is relevant to the present discussion of covalency corresponds to just the first two terms on the right-hand side of Eq. (8.13), adapted to several nuclei as well as to many electrons,

$$H=\sum_{i=1}^{N}\left[-\frac{\hbar^2}{2m}\nabla_i^2-\sum_{\alpha=1}^{R}\frac{Z_\alpha e^2}{(4\pi\varepsilon_0)r_{i\alpha}}\right]+\frac{1}{2}\sum_{i\neq j=1}^{N}\frac{e^2}{(4\pi\varepsilon_0)r_{ij}}+\frac{1}{2}\sum_{\alpha\neq\beta=1}^{R}\frac{Z_\alpha Z_\beta e^2}{(4\pi\varepsilon_0)Q_{\alpha\beta}}, \qquad (8.14)$$

where the sums are over the N electrons and R nuclei of the complex. The nuclear kinetic energy operator T_N is suppressed in Eq. (8.14), in conformity with the Born–Oppenheimer approximation, Eq. (5.73). The last term on the right-hand side of Eq. (8.14), the mutual electrostatic interactions of the nuclei, must be included in a geometry search or in an investigation of normal modes of vibration, but since it has no bearing on electronic structure, it will be omitted in the remainder of the present discussion. It is reasonable to treat the remaining six terms on the right-hand side of

Eq. (8.13) as small perturbations for iron-group transition-metal complexes; however, the much larger relativistic corrections for rare-earth ions, including spin–orbit interaction, require special consideration.

8.2.2 Hartree–Fock approximation

In the Hartree–Fock approximation, the variational principle is employed with the Hamiltonian H of Eq. (8.14), together with a many-electron variational trial function $|\Phi_0\rangle$ in the form of a single Slater determinant constructed from molecular spin orbitals $\phi_k(\mathbf{x})$, where $\mathbf{x}$ includes both space coordinates $\mathbf{r}$ and spin coordinates s,

$$E_{\mathrm{HF}} = \frac{\langle\Phi_0|H|\Phi_0\rangle}{\langle\Phi_0|\Phi_0\rangle}, \tag{8.15}$$

$$|\Phi_0\rangle = \frac{1}{\sqrt{N!}}\det|\phi_k|. \tag{8.16}$$

This form of trial function is chosen to satisfy the Pauli principle, in analogy with Eqs. (4.9). Variation of the molecular spin-orbitals to minimize the energy, subject to the constraint that they remain orthonormal, then leads to the Hartree–Fock equations,

$$F\phi_k(\mathbf{x}_1) = \left[-\frac{\hbar^2}{2m}\nabla_1^2 - \sum_{\alpha=1}^{R}\frac{Z_\alpha e^2}{(4\pi\varepsilon_0)r_{1\alpha}} + \frac{e^2}{(4\pi\varepsilon_0)}\sum_l \int \frac{|\phi_l(\mathbf{x}_2)|^2}{r_{12}}\,d\mathbf{x}_2\right]\phi_k(\mathbf{x}_1)$$
$$-\left[\frac{e^2}{(4\pi\varepsilon_0)}\sum_l \int \frac{\phi_l^*(\mathbf{x}_2)\phi_k(\mathbf{x}_2)}{r_{12}}\,d\mathbf{x}_2\right]\phi_l(\mathbf{x}_1) = \varepsilon_k\phi_k(\mathbf{x}_1), \tag{8.17}$$

where the sum over l extends over occupied molecular orbitals. The *Fock operator F* plays a role analogous to a one-electron Hamiltonian, and the diagonal Lagrange multipliers ε_k serve as its energy eigenvalues; off-diagonal Lagrange multipliers are assumed to vanish, since they can be eliminated by a unitary transformation of molecular spin-orbitals. However, the Hartree–Fock energy E_{HF}, given by Eq. (8.15), differs from the sum of diagonal Lagrange multipliers, since the latter counts twice the integrals which embody the Coulomb and exchange interactions with other electrons. In a concise notation,

$$E_{\mathrm{HF}} = \langle\Phi_0|H|\Phi_0\rangle = \sum_a \varepsilon_a - \tfrac{1}{2}\sum_{ab}\langle ab||ab\rangle, \tag{8.18}$$

$$\langle ab||cd\rangle = \langle ab|cd\rangle - \langle ab|dc\rangle, \tag{8.19}$$

$$\langle ab|cd\rangle = \int\!\!\int \phi_a(\mathbf{x}_1)^*\phi_b(\mathbf{x}_2)^*\frac{e^2}{(4\pi\varepsilon_0)r_{12}}\phi_c(\mathbf{x}_1)\phi_d(\mathbf{x}_2)\,d\mathbf{x}_1\,d\mathbf{x}_2, \tag{8.20}$$

where the sums are over occupied molecular spin-orbitals.

Since the Fock operator F depends implicitly on the occupied molecular spin-orbitals, the Hartree–Fock equations are nonlinear and must be solved by a self-consistent-field (SCF) method in which F is evaluated from the orbitals of the previous iteration. For the case of a single nucleus, it is feasible to separate variables in the atomic orbitals as in Eq. (4.7) and to perform the iterative solution of the radial equation by numerical integration. The numerical procedure is impractical for a molecular complex, however, so an LCAO method is employed instead in which each molecular orbital $\psi_k(\mathbf{r})$ is expressed as a linear combination of non-orthogonal basis functions $\chi_i(\mathbf{r})$,

$$\psi_k(\mathbf{r}) = \sum_i \chi_i(\mathbf{r}) C_{ik}. \tag{8.21}$$

In the standard *spin-restricted Hartree–Fock* (RHF) method, the molecular spin-orbitals $\phi_k(\mathbf{r}, s)$ are further constrained by associating each molecular orbital $\psi_k(\mathbf{r})$ with two spin functions. For a configuration of closed molecular shells, the expansion coefficients then satisfy the *Roothaan equations* [Roothaan (1951)]

$$\sum_j (F_{ij} - \varepsilon_k S_{ij}) C_{jk} = 0, \tag{8.22a}$$

$$F_{ij} = \langle \chi_i | F | \chi_j \rangle, \tag{8.22b}$$

$$S_{ij} = \langle \chi_i | \chi_j \rangle. \tag{8.22c}$$

These equations are analogous to Eqs. (8.4), except that they must be solved iteratively since the elements F_{ij} of the Fock matrix depend implicitly on the occupied molecular orbitals,

$$F_{ij} = h_{ij} + \sum_r \sum_s P_{rs}[(ij|rs) - \tfrac{1}{2}(is|rj)], \tag{8.23}$$

where h_{ij} is the matrix of one-electron operators in a concise notation, P_{rs} is the *density matrix* defined by

$$P_{rs} = 2 \sum_l C^*_{rl} C_{sl}, \tag{8.24}$$

where the sum over l is over occupied molecular orbitals, and the two-electron integrals over basis functions are defined by

$$(ab|cd) = \int\int \chi^*_a(\mathbf{r}_1)\chi_b(\mathbf{r}_1) \frac{e^2}{(4\pi\varepsilon_0)r_{12}} \chi^*_c(\mathbf{r}_2)\chi_d(\mathbf{r}_2)\, d\mathbf{r}_1\, d\mathbf{r}_2. \tag{8.25}$$

The Roothaan equations, Eq. (8.22a), can be written more concisely in matrix notation,

$$\mathbf{FC} = \mathbf{SC}\varepsilon. \tag{8.26}$$

The SCF procedure is then initiated by choosing a set of *starting vectors* **C**. At each iteration, the Fock matrix **F** is evaluated from the density matrix **P** of the previous iteration, and convergence is achieved when the change in either the ground-state energy or some measure of the density between successive iterations is smaller than some specified value (*convergence criterion*).

8.2.3 Basis functions

Specification of the basis functions $\chi_i(\mathbf{r})$ is a critical consideration in application of the Roothaan equations. *Slater-type orbitals* (STOs) [Slater (1930)] of the form

$$\chi_{nlm}^{\rm STO}(\mathbf{r}) = N(n,l)r^{n-1}\exp(-\zeta r)Y_l^m(\theta,\phi), \tag{8.27}$$

where N is a normalizing factor, were employed as basis functions in early calculations because of their qualitative resemblance to atomic orbitals, and much effort was expended on evaluation of multi-centre integrals involving these functions [Mulliken *et al.* (1949), Barnett and Coulson (1951), Löwdin (1956), Barnett (1963)]. On the other hand, it was recognized early [Boys (1950)] that the evaluation of multi-centre integrals is greatly facilitated by employing *gaussian* basis functions of the form

$$\chi_{abc}^{\rm GF}(\mathbf{r}) = N(a,b,c)x^a y^b z^c \exp(-\alpha r^2), \tag{8.28}$$

by virtue of the fact that the product of two gaussian functions with different origins is a third gaussian function with an intermediate origin. However, since gaussian functions are poorer representations of atomic orbitals, a larger basis set is required with consequent adverse effect on convergence time, which increases as the fourth power of the number of basis functions.

The conflict is resolved in modern quantum chemistry by employment of *contracted gaussian functions* [Hehre *et al.* (1969), Dunning (1970), Hehre *et al.* (1986)]

$$\chi^{\rm CGF}(\mathbf{r}) = \sum_{p=1}^{L} d_p \chi_p^{\rm GF}(\mathbf{r}), \tag{8.29}$$

which combine the virtues of both alternatives. In the standard minimal basis set, designated STO-3G, each basis function is a linear combination (*contraction*) of three gaussian (*primitive*) functions with coefficients and exponents adjusted for a best fit to a Slater-type orbital. This basis set is capable of replicating molecular properties, but a significant improvement is obtained by employing at least two independent contractions for each valence atomic orbital (*split-valence* or *double-zeta quality* basis set). An example is a 6-31G basis for first row atoms in which a contraction of six primitives is employed for the 1s orbital, and two contractions with three and one primitives, respectively, for each valence orbital (2s, 2p_x, 2p_y and 2p_z). The coefficients and exponents are optimized by *ab initio* atomic structure calculations [Huzinaga (1965)] rather than by fitting Slater-type orbitals. Further refinements are achieved by addition of polarizing functions such as *d*-orbitals for first-row atoms, to allow for distortion of atomic orbitals in a molecular environment, and by addition of diffuse functions.

8.2.4 Open shells

The RHF method is appropriate for configurations of closed shells, since the spin-restricted Slater determinant is an eigenfunction of spin operators $\mathbf{S}^2$ and S_z with quantum numbers $S=0$ and $M_S=0$. Furthermore, it shares the symmetry of the molecular complex if the individual molecular orbitals are symmetry-adapted. However, it is a less than optimum trial function for open-shell configurations. It is still an eigenfunction of S_z with M_S equal to half the difference in the numbers of up and down spins, but it is not necessarily an eigenfunction of $\mathbf{S}^2$ unless all of the unpaired spins are parallel, nor is it symmetry-adapted in general. Consequently, multiplet structure of the molecular configuration is obscured.

In a refinement of the RHF method for open-shell configurations, spin- and symmetry-adapted linear combinations of spin-restricted Slater determinants are employed as trial functions [Roothaan (1960), Roothaan and Bagus (1963), Hurley (1976)]. Separate Roothaan equations for open- and closed-shell orbitals,

$$\mathbf{F}^{\mathrm{O}}\mathbf{C}^{\mathrm{O}} = \mathbf{S}\mathbf{C}^{\mathrm{O}}\varepsilon^{\mathrm{O}}, \tag{8.30a}$$

$$\mathbf{F}^{\mathrm{C}}\mathbf{C}^{\mathrm{C}} = \mathbf{S}\mathbf{C}^{\mathrm{C}}\varepsilon^{\mathrm{C}}, \tag{8.30b}$$

are intrinsically coupled by the dependence of each Fock operator on both density matrices, $\mathbf{P}^{\mathrm{O}}$ and $\mathbf{P}^{\mathrm{C}}$. Equations (8.30) are valid for no more than one open shell per symmetry type. While this RHF approach preserves multiplet structure, the constraint of spin-restricted Slater determinants prevents it from yielding the lowest energy.

An alternative approach, the *unrestricted Hartree–Fock* (UHF) method, employs a single, spin-unrestricted Slater determinant as the variational trial function. The use of different spatial orbitals for different spin projections reflects their different exchange interactions, and results in a lower total energy. The LCAO approximation then leads to the *Pople–Nesbet equations* [Pople and Nesbet (1954)],

$$\mathbf{F}^{\alpha}\mathbf{C}^{\alpha} = \mathbf{S}\mathbf{C}^{\alpha}\varepsilon^{\alpha}, \tag{8.31a}$$

$$\mathbf{F}^{\beta}\mathbf{C}^{\beta} = \mathbf{S}\mathbf{C}^{\beta}\varepsilon^{\beta}, \tag{8.31b}$$

where α and β distinguish spin projections. Again, these equations are coupled by the dependence of each Fock operator on both density matrices, $\mathbf{P}^{\alpha}$ and $\mathbf{P}^{\beta}$. However, the UHF wave functions are not symmetry-adapted, and they fail as eigenfunctions of $\mathbf{S}^2$ even when all of the unpaired spins are parallel; rather, they are contaminated with higher spin values, $S > M_S$. In the *projected* unrestricted Hartree–Fock (PUHF) method, spin- and symmetry-adapted functions are recovered from optimized UHF functions by subsequent application of projection operators [Löwdin (1955)].

8.3 Correlation

8.3.1 Correlation energy

In the Hartree–Fock approximation, each electron is considered to move in an average potential due to all of the electrons, except that electrons of parallel spin tend to avoid one another as a consequence of the Pauli principle, giving rise to an *exchange hole* which accompanies each electron. In reality, the motions of electrons of opposite spin are also correlated by virtue of their repulsive coulomb interactions, and this additional correlation must be incorporated in any theory which aspires to improve on the Hartree–Fock approximation.

The *correlation energy* is defined as the difference between the exact energy eigenvalue $\mathscr{E}_0$ of the non-relativistic Hamiltonian, for either the ground state or the lowest state of a particular symmetry, and the corresponding Hartree–Fock energy E_0 in the

Hartree–Fock limit, the energy which would be obtained with a *complete* infinite basis set,

$$E_{\text{corr}} = \mathscr{E}_0 - E_0. \tag{8.32}$$

An enormous effort has been expended on devising methods to calculate estimates of the correlation energy. The methods discussed in this section are representative, but by no means exhaustive.

8.3.2 *Configuration interaction*

Configuration interaction (CI) provides the most transparent, though by no means the most efficient, approach to the treatment of correlation. An unrestricted Hartree–Fock calculation with K basis functions on a molecule with N electrons yields K orthonormal molecular spin-orbitals, of which N are occupied and $K-N$ are *virtual* (unoccupied). One can construct $N\times N$ Slater determinants by substituting *virtual* spin-orbitals for occupied spin-orbitals in the original Slater determinant. These Slater determinants with fixed molecular spin-orbitals can then be used as an orthogonal basis set for expanding a variational trial function $|\Psi_0\rangle$,

$$|\Psi_0\rangle = |\Phi_0\rangle + \sum_{a,r} c_a^r|\Phi_a^r\rangle + \sum_{\substack{a<b\\r<s}} c_{ab}^{rs}|\Phi_{ab}^{rs}\rangle + \sum_{\substack{a<b<c\\r<s<t}} c_{abc}^{rst}|\Phi_{abs}^{rst}\rangle + \cdots, \tag{8.33}$$

where $|\Phi_a^r\rangle$ differs from $|\Phi_0\rangle$ by substitution of spin-orbital ϕ_r for spin-orbital ϕ_a, etc. Substitution of $|\Psi_0\rangle$ for $|\Phi_0\rangle$ in Eq. (8.15) and variation of the expansion coefficients to minimize the energy then leads to a matrix diagonalization problem. Since this trial function contains the Hartree–Fock solution $|\Phi_0\rangle$ as one term, it must yield a lower energy than the Hartree–Fock energy. In the Hartree–Fock limit, where the molecular spin-orbitals form a complete set of one-electron functions, the $N\times N$ Slater determinants constructed from them form a complete, orthonormal set of N-electron, antisymmetrized basis functions; consequently, employment of the full CI trial function of Eq. (8.33) yields an exact solution of the non-relativistic Schrödinger equation.

In practice, even for a finite basis set of modest size, the number of determinants, $K!/[(K-N)!N!]$, is much too large, and one must employ a truncated CI trial function. An explicit expression for the correlation energy is [Szabo and Ostlund (1989)]

$$E_{\text{corr}} = \langle\Phi_0|H - E_{\text{HF}}|\Psi_0\rangle = \sum_{\substack{a<b\\r<s}} c_{ab}^{rs}\langle\Phi_0|H|\Phi_{ab}^{rs}\rangle, \tag{8.34}$$

since matrix elements of the Hamiltonian between the Hartree–Fock determinant and excited determinant vanish for triple and higher excitations by virtue of the fact that the Hamiltonian contains at most two-electron operators, and for single excitations by Brillouin's theorem,

$$\langle\Phi_i^a|H|\Phi_0\rangle = \langle a|h|i\rangle + \sum_j \langle aj \,\|\, ij\rangle = \langle a|F|i\rangle = \varepsilon_i\langle a|i\rangle = 0. \tag{8.35}$$

These excitations contribute indirectly to the coefficients in Eq. (8.34), but it is plausible from Eq. (8.34) that the double excitations predominate. Accordingly, a popular truncation, singly and doubly excited CI (CISD), retains only the first three terms on the right-hand side of Eq. (8.33). The CI matrix may be partitioned as well by employing spin- and symmetry-adapted combinations of Slater determinants. Its size may be further reduced by a *frozen-core approximation* which excludes excitations from occupied core orbitals.

The CI method has the virtue that it is variational, and thus always yields an upper bound on the true energy. For a given truncation, the approximation can be improved by varying the orbitals, since HF orbitals are not optimum; alternative choices include *natural orbitals* [Löwdin (1955)] and orbitals calculated by the *multi-configuration self-consistent-field* (MCSCF) method [Wahl and Das (1977)] and *generalized-valence-bond* (GVB) method [Hunt *et al.* (1973)]. However, a deficiency of truncated CI is a lack of *size-consistency* (*extensivity*); the quality of the approximation deteriorates rapidly with increasing molecular size, and the correlation energy of a system composed of non-interacting parts differs from the sum of individual correlation energies. For example, in a system of N identical non-interacting molecules, the correlation energy calculated by CISD increases only as $\sqrt{N}$, since simultaneous double excitations of two or more independent molecules are disallowed. Full CI is exact, and therefore size-consistent.

8.3.3 Perturbation theory

An alternative to CI for calculating correlation energies is Rayleigh–Schrödinger perturbation theory (RSPT), also known in this context as *Møller–Plesset perturbation theory* (MPPT) [Møller and Plesset (1934)]. A Slater determinant $|\Phi_0\rangle$ constructed from molecular orbitals which satisfy the Hartree–Fock equations is not only an approximate eigenfunction of the Hamiltonian H, Eq. (8.14), but also an exact eigenfunction of the *Hartree–Fock Hamiltonian* H_0,

$$H_0 = \sum_{i=1}^{N} F(i) = \sum_{i=1}^{N} [h(i) + v^{\mathrm{HF}}(i)], \tag{8.36}$$

where $F(i)$ is the Fock operator for the ith electron, defined by Eq. (8.17), $h(i)$ is the sum of one-electron operators, and $v^{\mathrm{HF}}(i)$ comprises the Coulomb and exchange terms. The energy eigenvalue of H_0 is

$$E_0 = \sum_a \varepsilon_a, \tag{8.37}$$

where the sum is over occupied spin-orbitals. The Hamiltonian H can then be decomposed (*Møller–Plesset separation*) as

$$H = H_0 + H_1, \tag{8.38a}$$

$$H_1 = \sum_{i<j=1}^{N} \frac{e^2}{(4\pi\varepsilon_0)r_{ij}} - \sum_{i=1}^{N} v^{\mathrm{HF}}(i), \tag{8.38b}$$

The Hartree–Fock energy is the sum of the zeroth- and first-order energies,

$$\begin{aligned} E_{HF} &= \langle\Phi_0|H|\Phi_0\rangle = \langle\Phi_0|H_0|\Phi_0\rangle + \langle\Phi_0|H_1|\Phi_0\rangle \\ &= E_0 + E_1 = \sum_a \varepsilon_a - \frac{1}{2}\sum_{ab} \langle ab\|ab\rangle. \end{aligned} \tag{8.39}$$

The second-order energy then provides a first approximation to the correlation energy,

$$E_{\mathrm{corr}} \cong E_2 = \sum_{n \neq 0} \frac{|\langle 0|H_1|n\rangle|^2}{E_0 - E_n} = \sum_{n \neq 0} \frac{|\langle 0|H|n\rangle|^2}{E_0 - E_n}. \tag{8.40}$$

Since only double excitations survive in the sum over n by virtue of Brillouin's theorem and the limitation to one- and two-electron operators, the second-order contribution to the correlation energy is

$$E_{\mathrm{corr}} \cong E_2 = \sum_{\substack{a<b\\ r<s}} \frac{|\langle \Phi_0|H|\Phi_{ab}^{rs}\rangle|^2}{\varepsilon_a + \varepsilon_b - \varepsilon_r - \varepsilon_s} = \sum_{\substack{a<b\\ r<s}} \frac{|\langle ab\|rs\rangle|^2}{\varepsilon_a + \varepsilon_b - \varepsilon_r - \varepsilon_s}. \tag{8.41}$$

Higher orders of perturbation theory make additional contributions to the correlation energy; the general expression for $E_1 + E_{\mathrm{corr}}$ in RSPT is

$$E - E_0 = \left\langle \Phi_0 \middle| H_1 \sum_n [(E_0 - H_0)^{-1} P (H_1 - \Delta E)]^n \middle| \Phi_0 \right\rangle, \tag{8.42a}$$

$$P = 1 - |\Phi_0\rangle\langle\Phi_0|. \tag{8.42b}$$

This expression is size-consistent in all orders, provided the space of P is not restricted; e.g., limited to double excitations. An alternative formulation of the problem, *many-body perturbation theory* (MBPT) [Bartlett and Purvis (1978)] relies on a linked-diagram expansion [Brueckner (1955); Goldstone (1957)],

$$E - E_0 = \left\langle \Phi_0 \middle| H_1 \sum_n [(E_0 - H_0)^{-1} H_1]^n \middle| \Phi_0 \right\rangle_L, \tag{8.43}$$

where the sum is restricted to linked *Goldstone diagrams*. Goldstone diagrams are modified Feynman diagrams used as book-keeping devices to keep track of terms in the perturbation expansion; examples are presented in Fig. 8.3. In MBPT, size consistency is guaranteed by complete exclusion of un-linked diagrams, but this is achieved at the expense of mixing different levels of excitation. These distinctions become apparent in fourth-order perturbation theory, in which single, double, triple and quadruple excitations all contribute. However, perturbation theory is not variational and may actually over-estimate the correlation energy in second order.

Configuration interaction and perturbation theory are complementary approaches to correlation. The *quadratic-configuration-interaction* (QCI) method [Pople *et al.* (1987)] combines the virtues of both approaches by adding terms derived from MBPT to the truncated CI expansion to restore size consistency. Other approaches include the *independent electron-pair approximation* (IEPA) [Sinanoglu (1964); Nesbet (1965)], suggested by Eq. (8.34), and related approximations.

8.3.4 Excited states

The variational principle fails for optically excited states other than those which have the lowest energy of their symmetry type. *Koopmans' theorem*, based on a

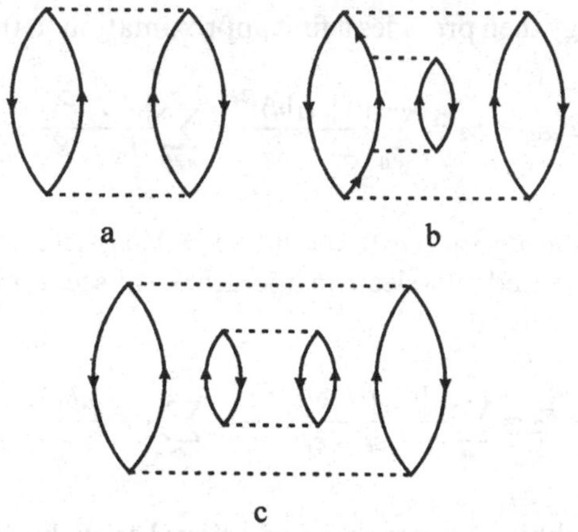

Figure 8.3. Selected examples of antisymmetrized Goldstone diagrams. Horizontal dashed lines represent the perturbation H_1, and their number corresponds to the order of perturbation theory. Solid lines are propagators $(E_0 - H_0)^{-1}$ for holes (down arrows) or particles (up arrows), relative to Hartree–Fock ground state as the vacuum state. Diagram (c) is unlinked, and therefore makes no contribution.

frozen-orbital approximation, states that the diagonal Lagrange multipliers ε_k for the occupied molecular orbitals in the Hartree–Fock approximation provide a first approximation to the corresponding ionization potentials. However, the difference in the values of ε_k between unoccupied and occupied molecular orbitals is generally a poor approximation to the excitation energy, since ε_k is not calculated self-consistently for the unoccupied orbital. In particular, the HOMO–LUMO gap (highest occupied molecular orbital to lowest unoccupied molecular orbital) is a poor approximation to the lowest excitation energy. A much better approximation is provided by the method of configuration interaction with single excitations only (CIS). Although there are no matrix elements of the Hamiltonian connecting singly excited determinants with the Hartree–Fock ground state wave function, by virtue of Brillouin's theorem, there may be matrix elements connecting them with one another, since H contains two-electron operators,

$$\langle \Phi_0 | H | \Phi_a^r \rangle = 0, \tag{8.44a}$$

$$\langle \Phi_a^r | H | \Phi_b^s \rangle \neq 0. \tag{8.44b}$$

Diagonalization of the Hamiltonian matrix for excited configurations then yields a much improved approximation to the excited-state energies and wave functions. The resulting eigenvectors can be used to calculate transition dipole moments which are related to transition probabilities as indicated in Chapter 5. As before, the CI matrix can be reduced in size by a frozen-core approximation and can be partitioned by spin- and symmetry-adapted combinations of determinants. The method is obviously susceptible to further refinements which will not be considered here.

8.4 Additional approximations

8.4.1 Effective core potentials

All-electron *ab initio* molecular-orbital calculations are rather cumbersome for large molecules and for molecules which incorporate heavy elements, since the computation time increases as the fourth power of the number of basis functions. This problem can be addressed by introducing additional approximations to increase the efficiency of the Hartree–Fock method.

Much of the computational effort in an all-electron calculation on a transition-metal or rare-earth complex is expended in reproducing the occupied core orbitals, which are relatively insensitive to the molecular environment. Their ionization potentials, which are sensitive, are of interest in photoelectron spectroscopy, and the core orbitals must be included in a determination of the total energy. However, if the emphasis is on optical spectra, a frozen-core approximation is an appropriate device to expedite the calculation of valence orbitals and of optically excited states. Application of the frozen-core approximation is greatly facilitated by the use of *pseudopotentials*.

A pseudopotential replaces the kinetic energy associated with orthogonalization of a valence wave function ψ_ν to occupied ion-core orbitals ψ_c by an effective potential for use with a smooth pseudo-wave function ϕ to obtain the same energy eigenvalue E_ν [Phillips and Kleinman (1959)]. For clarity, it is assumed that both valence and core orbitals are solutions of the same Schrödinger equation with a one-electron Hamiltonian,

$$H\psi = E\psi, \tag{8.45a}$$

$$H = T + V, \tag{8.45b}$$

where T is the kinetic energy operator and V is the potential. A smooth pseudo-wave function ϕ is postulated, which can be orthogonalized to the occupied core orbitals to obtain a valence orbital ψ_ν,

$$\psi_\nu = (1 - P)\phi, \tag{8.46a}$$

$$P = \sum_c |\psi_c\rangle\langle\psi_c|, \tag{8.46b}$$

where P is a projection operator for the occupied core states. Note that ψ_ν and ϕ cannot both be normalized in Eq. (8.46a). Substitution of Eq. (8.46a) in Eq. (8.45a) then gives

$$[T + V_p^{\text{PK}}]\phi = E_\nu\phi, \tag{8.47}$$

with the same energy eigenvalue E_ν, where V_p^{PK} is a non-local pseudopotential,

$$V_p^{\text{PK}} = V + (E_\nu - H)P = V + \sum_c (E_\nu - E_c)|\psi_c\rangle\langle\psi_c|. \tag{8.48}$$

The pseudo-wave function is not unique, however, since Eq. (8.47) is satisfied by any function of the form

$$\phi = \psi_\nu + \sum_c \alpha_c\psi_c. \tag{8.49}$$

A broader definition of the pseudopotential [Austin *et al.* (1962)],

$$V_p = V + POp, \tag{8.50}$$

where Op is an arbitrary operator, includes V_p^{PK} as a special case, since P commutes with H. Another special case is the pseudopotential which corresponds to the smoothest pseudo-wave function [Cohen and Heine (1961)],

$$V_p^{\mathrm{CH}} = V + P(\bar{V} - V), \tag{8.51a}$$

$$\bar{V} = \frac{\langle\phi|V_p^{\mathrm{CH}}|\phi\rangle}{\langle\phi|\phi\rangle}. \tag{8.51b}$$

The BSG local model pseudopotential for F centres, Eqs. (4.67), is derived from V_p^{CH}. Unfortunately, the pseudopotential of Eq. (8.50) is not generally hermitian, and is therefore incompatible with the variational principle, although local model pseudopotentials derived from it usually are hermitian [Weeks and Rice (1968)].

A popular approach to the generation of a local model pseudopotential is to construct a smooth, nodeless pseudo-wave function ϕ for a chosen reference state with energy E_ν, and then to invert the Schrödinger equation [Melius and Goddard (1974)],

$$V_p = \frac{(E_\nu - T)\phi}{\phi}. \tag{8.52}$$

The reference pseudo-wave function might be constrained to satisfy Eq. (8.49) by requiring it to be a linear combination of valence and core orbitals [Kahn *et al.* (1978)]; however, Eq. (8.46a) then implies that when ϕ and ψ_ν are both normalized, they exhibit different long-range behaviours. By relaxing the constraint, it is possible to construct *norm-conserving pseudopotentials* such that ϕ and ψ_ν exhibit identical long-range behaviours when both are normalized [Christiansen *et al.* (1979), Hamann *et al.* (1979)].

Ab initio, norm-conserving, angular-momentum-dependent pseudopotentials are now available for the entire periodic table [Hay and Wadt (1985), Wadt and Hay (1985)]. Reference pseudo-wave functions $\phi_l(r)$, chosen to be identical to Hartree–Fock atomic orbitals beyond a cut-off radius r_c, were joined smoothly for $r < r_c$ to a nodeless function chosen to preserve normalization. *Effective core potentials* (ECPs) $U(r)$ derived from them are given, in Hartree atomic units, by

$$U(r) = U_L(r) + \sum_{l=0}^{L-1} \sum_{m=-l}^{l} [U_l(r) - U_L(r)]|lm\rangle\langle lm|, \tag{8.53a}$$

$$U_l(r) = \varepsilon_l - \frac{l(l+1)}{2r^2} + \frac{Z}{r} + \frac{\phi_l''(r)}{2\phi_l(r)} - \frac{V_{\mathrm{val}}\phi_l(r)}{\phi_l(r)}, \tag{8.53b}$$

where the operator V_{val} includes Coulomb and exchange interactions among occupied valence pseudo-wave functions. *Relativistic effective potentials* (REPs) which depend on angular-momentum quantum numbers l, j, m can be derived from *Dirac–Fock equations*, analogous to Hartree–Fock equations [Ermler *et al.* (1988)]. Weighted averages of these over fine-structure levels yield *average relativistic effective potentials*

(AREPs) which depend only on l, exactly as in Eq. (8.53a); thus relativistic effects can be incorporated in the ECPs for heavy elements in a non-relativistic formalism for the valence orbitals [Kahn *et al.* (1978)]. Finally, REPs and AREPs can be combined to serve as effective *spin–orbit operators* for valence electrons.

8.4.2 *Local exchange approximation*

Most of the computational effort involved in Hartree–Fock calculations on large molecules is attributable to multi-centre integrals involved in the exchange interaction; significant economies therefore accrue from an approximate treatment of exchange. Two distinct methods which share the same exchange approximation are the topics of this section.

The energy of a uniform electron gas with a stabilizing positive background charge can be calculated by application of perturbation theory to the Hamiltonian

$$H = H_0 + H_1, \tag{8.54a}$$

$$H_0 = T, \tag{8.54b}$$

$$H_1 = U + V, \tag{8.54c}$$

where T is the kinetic energy operator, U is the potential arising from the uniform positive background charge, and V is the mutual electrostatic interaction of the electrons. The first-order contribution to the ground-state energy is just the exchange interaction, since the coulomb interaction is canceled by the background charge. The exchange energy per electron E_X is then given in terms of the electron density ρ by

$$E_X = -\frac{3}{2}\left(\frac{e^2}{4\pi\varepsilon_0}\right)\left(\frac{3}{8\pi}\right)^{1/3}\rho^{1/3}. \tag{8.55}$$

Slater (1951) substituted twice this quantity, evaluated with the local electron density, for the exchange term in the Hartree–Fock equations, Eq. (8.17), to obtain the *Hartree–Fock–Slater approximation*,

$$F\phi_k(\mathbf{x}_1) = \left[-\frac{\hbar^2}{2m}\nabla_1^2 - \sum_{\alpha=1}^{R}\frac{Z_\alpha e^2}{(4\pi\varepsilon_0)r_{1\alpha}} + \frac{e^2}{(4\pi\varepsilon_0)}\sum_l \int \frac{|\phi_l(\mathbf{x}_2)|^2}{r_{12}}\,d\mathbf{x}_2 \right.$$
$$\left. - \frac{3e^2}{(4\pi\varepsilon_0)}\left(\frac{3}{8\pi}\right)^{1/3}\left[\sum_{l(\text{spins}\|)}\phi_l^*(\mathbf{x}_1)\phi_l(\mathbf{x}_1)\right]^{1/3}\right]\phi_k(\mathbf{x}_1) = \varepsilon_k\phi_k(\mathbf{x}_1), \tag{8.56}$$

where $\phi_k(\mathbf{x})$ is a molecular spin-orbital. This approximation was applied by Herman and Skillman (1963) to atomic-structure calculations for the entire Periodic Table.

Essentially the same exchange approximation was employed in the *Thomas–Fermi–Dirac statistical model* [Dirac (1930)], which was subsequently extended as the *density functional method* [Hohenberg and Kohn (1964), Kohn and Sham (1965)] to include density-dependent contributions to the correlation energy [Wigner (1934), Gell-Mann and Brueckner (1957)] as well as density-gradient corrections. In this local density approximation, the energy is expressed as a functional of the electron density, $E[\rho(\mathbf{r})]$, which is varied to minimize the energy subject to the constraint that the total number of electrons be fixed,

$$\frac{\delta\{E[\rho] - \mu(N[\rho] - N_0)\}}{\delta\rho} = 0. \tag{8.57}$$

The resulting chemical potential μ, which can be incorporated in a modified Hartree equation analogous to Eq. (8.56), contains a local exchange term which is just 2/3 of that in the Hartree–Fock–Slater approximation. The density functional method is essentially confined to ground-state properties [Callaway and March (1984)].

In response to the density-functional result, Slater subsequently incorporated an adjustable factor α in his local exchange term which satisfies the inequality

$$1 \geq \alpha \geq \tfrac{2}{3}; \tag{8.58}$$

hence the designation *Xα method*, where X stands for exchange. The Xα method as applied to molecules was extended to include a spherically symmetrical potential within a spherical shell enclosing each atom plus a uniform potential between spheres (*muffin-tin potential*). In the case of a negatively charged molecule ion, the entire cluster is surrounded by a stabilizing sphere of positive charge (*Watson well*). Interior solutions obtained by integration of the Hartree–Fock–Slater equations are matched to a plane-wave expansion at the spherical boundaries, and energies and potentials are iterated to self-consistency. The contemporary designation for this extended approach is the *SCF-Xα-SW method* [Johnson (1973)]. The SCF-Xα-SW method has found extensive applications in molecular physics [Slater (1974)]. Its principal virtue is rapidity of computation; elimination of two-electron, multi-centre integrals reduces CPU times by orders of magnitude. In the Xα method, optical transitions are investigated by application of Koopmans' theorem to *transition states* with fractional occupancy of molecular spin orbitals averaged between initial and final states. Multiplet structure, which is essential to the description of crystal-field transitions, is obscured in this method, but may be recovered in favourable cases by employing a different local exchange potential for each distinct Slater determinant, in conjunction with a sum rule for term energies [Ziegler *et al.* (1976)]. A second limitation of the method is its failure to yield reliable total energies [Ziegler *et al.* (1977)], which precludes accurate determination of lattice relaxation.

8.4.3 Approximate SCF semi-empirical methods

Semi-empirical methods which incorporate simplifying approximations in the Hartree–Fock equations and compensate for the resulting loss of information by substituting empirical information include complete neglect of differential overlap (CNDO) and intermediate neglect of differential overlap (INDO) [Pople and Beveridge (1970)]. In these methods, the basis set is limited to valence atomic orbitals, represented by Slater-type orbitals. Under the zero-differential overlap approximation, integrals of the type defined by Eq. (8.25) are assumed to satisfy the restriction

$$(ac|bd) = (aa|bb)\delta_{ac}\delta_{bd}, \tag{8.59}$$

and at the same time off-diagonal elements of the overlap matrix, S_{ij}, are neglected in the Roothaan equations, Eqs. (8.22). However, in order to insure invariance of the approximation under unitary transformations of the basis functions on each atom, it is necessary to assume that the surviving integrals depend only on the atoms and not on individual orbitals,

$$(aa|bb) = \gamma_{AB} \begin{cases} \text{all } \chi_a \text{ on atom } A \\ \text{all } \chi_b \text{ on atom } B. \end{cases} \tag{8.60}$$

After additional approximations are introduced in a similar spirit, the final result for the Fock matrix in CNDO/2 parameterization is

$$F_{aa} = -\tfrac{1}{2}(I_a + A_a) + [(P_{AA} - Z_A) - \tfrac{1}{2}(P_{aa} - 1)]\gamma_{AA} + \sum_{B(\neq A)} (P_{BB} - Z_B)\gamma_{AB}, \tag{8.61a}$$

$$F_{ab} = \beta^0_{AB} S_{ab} - \tfrac{1}{2} P_{ab}\gamma_{AB}, \tag{8.61b}$$

$$F_{aa'} = -\tfrac{1}{2} P_{aa'}\gamma_{AA}, \tag{8.61c}$$

where I_a is an ionization potential, A_a is an electron affinity, Z_A is the core charge of atom A, β^0_{AB} relates the *resonance integral* to the overlap integral, and P_{AA} is the sum of diagonal density matrix elements for orbitals on atom A,

$$P_{AA} = \sum_a P_{aa}. \tag{8.62}$$

All three- and four-centre integrals are eliminated in this approximation. The INDO method is similar, but the approximations are somewhat less drastic. In particular, INDO retains monatomic differential overlap in one-centre integrals, which is essential to proper treatment of exchange interaction.

Although these approximations greatly simplify the Hartree–Fock procedure, they appear to offer little advantage for transition-metal complexes, since it is difficult to establish invariant values of the empirical parameters which would ensure predictive capability.

8.4.4 *Extreme semi-empirical methods*

Extreme semi-empirical methods proceed from the approximation of a common one-electron Hamiltonian invoked in §8.1.3. In the Hückel method [Hückel (1931)], matrix elements of the one-electron Hamiltonian are treated as adjustable parameters, coulomb integrals α_i for diagonal elements and resonance integrals β_{ij} for off-diagonal elements. Overlap integrals S_{ij} are neglected, although they are retained in the extended Hückel method [Hoffman (1963)]. The Hückel and extended Hückel methods were devised primarily for π electrons of planar organic molecules.

The Wolfsberg–Helmholz (1952) model of a transition-metal complex provides another example of an extreme semi-empirical method. In this method, diagonal matrix elements h_{ii} are derived from empirical ionization potentials, overlap integrals S_{ij} are calculated from Slater orbitals [Slater (1930)], and off-diagonal matrix elements

h_{ij} are given by

$$h_{ij} \cong S_{ij}(h_{ii} + h_{jj}). \tag{8.63}$$

The angular overlap (AOM) method [Schäffer and Jørgensen (1965), Jørgensen (1971), Gerloch and Slade (1973)] is an even more approximate variant of the Wolfsberg–Helmholz method. This approximation is predicated on the assumption that off-diagonal elements of the Hamiltonian are small compared with the difference of diagonal elements, and that overlap integrals are small compared with unity. The secular equation, Eq. (8.5), is then linearized by replacing the ith eigenvalue E_i by H_{ii} in every element of the determinant except for the element $H_{ii} - E_i$. For example, consider the basis set defined by Eqs. (8.6). Then H_{11} and H_{22} are the energies associated, respectively, with the symmetry-adapted combination of ligand orbitals χ and the transition-metal orbital ϕ in the absence of covalency. The inequality $H_{11} < H_{22}$ is also assumed, so that χ and ϕ dominate the bonding and antibonding molecular orbitals, respectively. Finally, the overlap integral S is written as a product of two factors, a radial part S^* which is a function of the radial properties of the orbitals and the bond length only, and an angular part F_λ^l which is a property of the number and orientations of overlapping orbitals. With these approximations and assumptions, the energy of the antibonding orbital is given by

$$E^a = h_{22} + e_\lambda (F_\lambda^l)^2, \tag{8.64a}$$

$$e_\lambda = \frac{h_{11}^2}{(h_{22} - h_{11})} S^{*^2}, \tag{8.64b}$$

for a complex in which all of the ligands are of the same type at the same distance. The factors e_λ are treated as adjustable parameters, and the numerical angular factors F_λ^l are calculated from geometrical considerations. For example, the energy difference Δ between antibonding levels e_g and t_{2g} in octahedral symmetry, equal to $10Dq$ in crystal-field theory, is given by

$$\Delta = 3e_\sigma - 4e_\pi. \tag{8.65}$$

The AOM model remains a popular alternative to crystal-field theory, especially for angular distortions of transition-metal complexes. An example of its application is presented in Chapter 9 [Riley *et al.* (1999)].

8.5 Embedded clusters

8.5.1 Embedding potentials

Thus far in this chapter the complex associated with an impurity or defect has been treated as an isolated molecule ion of fixed geometry. However, important effects which arise from the fact that the complex is a cluster of ions embedded in a larger crystal must be taken into account. Purely electronic effects arise not only from the charge distribution and polarizability of the surrounding medium, but also from the mobility of electrons across the boundary of the cluster [Fisher (1991)].

Several naive approaches have proved successful. The simplest approach in an ionic crystal is to embed the cluster in an array of point ions; one can then perform the conditionally convergent infinite lattice sums [Nijboer and de Wette (1957)] to determine their potential distribution within the cluster, including the Madelung potential, or one can incorporate a finite point-ion array in the SCF procedure. In covalent crystals, a popular approach is to add pacifying atoms, typically hydrogen atoms, to saturate dangling bonds at the boundary of the cluster.

More formal approaches introduce *embedding potentials* which are designed to confine the electrons to the cluster. One type of non-local embedding potential [Adams (1961), Kunz and Klein (1978), Kunz and Vail (1978)] is of the form $\rho W \rho$, where W is an arbitrary operator and ρ is a density operator which projects on a sub-set of occupied orbitals. This potential affects the individual orbitals, but not the many-electron Hartree–Fock wave function. Another type of non-local embedding potential, developed for defects in semiconductors [Baraff and Schlüter (1986)], is of the form

$$\Sigma = (G^{-1})_{11} - G_{11}^{-1}, \tag{8.66}$$

where G is the perfect-lattice electronic Green's function,

$$G = (E - H)^{-1}, \tag{8.67}$$

and the subscript 1 refers to the sub-space of Hilbert space spanned by the basis functions of the cluster.

8.5.2 *Lattice relaxation*

A common feature of luminescent centres in solids is a pronounced Stokes shift between peak energies for absorption and emission. In some cases, this shift is attributable in part to inter-system crossing from the final state in absorption to a lower emitting state of different multiplicity. Lattice relaxation also makes a major contribution to the Stokes shift when the geometrical configuration of the relaxed emitting state differs significantly from that of the ground state; this phenomenon was interpreted in terms of a configuration-coordinate diagram in §5.2.4. To be successful, a geometry search for the equilibrium configuration in each electronic state must be performed on the embedded cluster and must include relaxation of the surrounding lattice.

Molecular-dynamics simulations [Sangster and Dixon (1976)] provide one approach to the determination of equilibrium geometries. In this method, a superlattice of defects or impurities is constructed by imposition of periodic boundary conditions. Phenomenological pair potentials are adopted, random initial conditions are imposed and the classical equations of motion are integrated over a large number of time steps as the system is artificially 'cooled'. An extension of this method to electronic structure determination [Car and Parinello (1985), De Vita *et al.* (1992)] may ultimately be applicable to transition-metal complexes.

An alternative approach is based on the lattice-statics program HADES (Harwell Automatic Defect Evaluation System) [Norgett (1974)], designed to calculate both the defect energy and the static lattice distortion associated with a defect or impurity in an ionic crystal. Ions are represented either by a rigid-ion model or by a *shell model* in which the ionic charge is distributed between a massive core and a massless shell connected by a harmonic potential [Dick and Overhauser (1958)]. Ions or shells within a central Region I interact via pair potentials such as a *Buckingham potential* which includes short-range Born–Mayer repulsive and Van der Waals attractive terms as well as a long-range coulomb interaction,

$$V(r) = \frac{Z_1 Z_2 e^2}{(4\pi\varepsilon_0)r} + A_{12}\exp\left(-\frac{r}{\rho_{12}}\right) - \frac{C_{12}}{r^6}. \qquad (8.68)$$

The interactions of ions in Region I with those in the outer Region II are treated more approximately by the *Mott–Littleton method* [Mott and Littleton (1938)] in which ion displacements are related to the polarization of a dielectric continuum. Ion or core and shell positions, specified as components of a vector $\mathbf{u}$, are relaxed iteratively according to the algorithm

$$\mathbf{u}^{m+1} = \mathbf{u}^m - (\mathbf{W}^{-1})^m \mathbf{g}^m. \qquad (8.69)$$

The total energy E, forces $\mathbf{g} = \partial E/\partial \mathbf{u}$ and force-constant matrix $\mathbf{W} = \partial^2 E/\partial \mathbf{u}\partial \mathbf{u}$ are updated at each iteration until the forces are smaller than some predetermined convergence limit. HADES is commercially available as part of the CASCADE system of programs [Leslie (1982)].

Two implementations of the lattice-statics approach are of particular interest in the present context. The first of these is a commercial program called ICECAP (Ionic Crystal and Electronic Cluster: Automated Program) [Harding *et al.* (1985), Vail (1990)]. ICECAP combines HADES with an *ab initio* electronic structure program called UHF (Unrestricted Hartree–Fock) in a single, user-friendly package. At each step in the optimization of the cluster geometry, which requires a complete SCF-UHF-LCAO calculation, the remainder of the lattice is relaxed via HADES and the corresponding defect energy is included in the convergence criterion together with the electronic energy of the cluster. The package also contains several optional features such as effective core potentials, a second-order MPPT treatment of correlation, and a Kunz–Klein embedding potential with W chosen to cancel ion-size corrections to the point-ion potential of the surrounding lattice.

The second implementation of the lattice-statics approach is based on a modification of HADES, called HADESR [Woods *et al.* (1993), Donnerberg and Bartram (1996)]. Although similar in concept to ICECAP, it employs a different algorithm. In HADESR, pair potentials are replaced by an *ab initio* potential energy function within a cluster comprising a small portion of Region I. Interactions of ions within the cluster with ions of Region I lying outside the cluster are still mediated by pair potentials. In order to accomplish simultaneous relaxation of the cluster and its surrounding lattice, the total energy of the cluster is calculated on a mesh of symmetry-adapted displacements within each electronic state. The difference between *ab initio* and pair-potential calculations of the total cluster energy is then fitted to a polynomial in symmetry-adapted displacements. This polynomial is consulted in each iteration of HADESR to update both the total energy and the forces acting on each ion. HADESR has the virtue that it can be employed in conjunction with any electronic-structure program, such as the commercially available GAUSSIAN94 system of programs [Frisch *et al.* (1995)] which includes as options most of the methods described in this chapter, or MELD

(Many ELectron Description) [Davidson (1991)], available through Quantum Chemistry Program Exchange. Furthermore, it provides information beyond the energies and geometries of relaxed configurations, since the energy polynomial for each electronic state describes an adiabatic potential energy surface which is useful for investigating force-constant changes, the frequencies of localized vibrations and anharmonicity. The principal limitation of HADESR is that its practical application is restricted to highly symmetrical systems with only a few relevant symmetry-adapted displacements. Examples of the application of HADESR are presented in the next section.

8.6 Applications

The principles and methods developed in this chapter are illustrated in the present section by selected examples. *Ab initio* electronic-structure and lattice-statics calculations on chromium-doped halide elpasolites, described in §8.6.1, serve to explain both pressure dependence and thermal quenching of optical spectra. Similar calculations on the $Tl^0(1)$ centre and its analogues, presented in §8.6.2, are concerned with athermal quenching of optical spectra. Odd-parity distortions of octahedrally coordinated transition-metal ions that enable electric-dipole transitions between *3d* levels are addressed in §8.6.3 for Ti^{3+} and in §8.6.4 for Cr^{3+}.

8.6.1 Cr^{3+} in halide elpasolites

Ab initio calculations on the chromium-doped halide elpasolites K_2NaGaF_6, K_2NaScF_6 and Cs_2NaYCl_6 [Woods *et al.* (1993)] exemplify the methods of this chapter. This class of materials possesses the virtue that trivalent cation substitutional impurities can be accommodated in well-separated, rigorously octahedral sites without charge compensation. Optical properties were measured both at ambient pressure [Andrews *et al.* (1986b)] and at elevated pressures in a diamond-anvil cell [Dolan *et al.* (1992), Rinzler *et al.* (1993)]. Optical absorption and emission spectra of octahedrally coordinated Cr^{3+} were described in some detail in §5.4.1, and thermal luminescence quenching in §6.4.1 and 6.4.2. Embedded-cluster RHF-SCF-LCAO molecular-orbital calculations were performed on the twenty-one atom cluster $A_8B_6CrX_6^{11+}$ shown in Fig. 8.4, for each compound $A_2BMX_6 : Cr^{3+}$, by means of the MELD program [Davidson (1991)]. The 1*s*, 2*s* and 2*p* shells of chromium and its nearest chlorine neighbours were replaced by effective core potentials, while double-zeta quality basis sets for valence orbitals were explicitly included. All nearest-neighbour fluorine orbitals were included, but the nearest sodium and potassium ions were represented by bare effective core potentials which, incidentally, ensure that the valence-orbital contribution to an embedding potential of the form of Eq. (8.66) is negligible.

For each electronic state of each compound, the total energy of the molecular cluster was calculated on a mesh of 125 combinations of symmetrical (breathing) displacements of the three shells of ions, X^-, A^+ and B^+, surrounding the Cr^{3+} impurity, and was fitted to a fourth-degree polynomial in the three displacement coordinates. The symmetrical equilibrium geometry of the embedded cluster together with its surrounding lattice was then calculated by HADESR as a function of lattice parameter, with rigid-ion Buckingham pair potentials, Eq. (8.68), in order to simulate the effects of

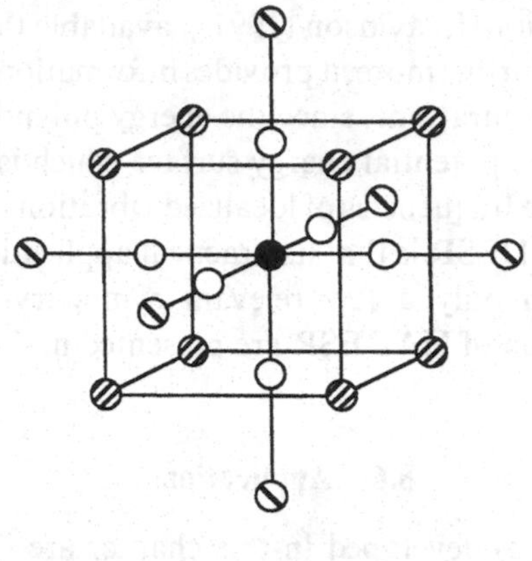

Figure 8.4. Twenty-one atom molecular ion cluster $A_8B_6CrX_6^{11+}$ of the chromium-doped elpasolite $A_2BMX_6:Cr^{3+}$ [after Woods *et al.* (1993)].

hydrostatic pressure. The relation between pressure and lattice parameter was determined by an independent HADES calculation. By virtue of its orbital degeneracy, the $^4T_{2g}$ state can also couple linearly to e_g and t_{2g} displacements of the ligands illustrated in Fig. 3.2 (Jahn–Teller effect). The Jahn–Teller stabilization energies were calculated by HADESR for a seven-atom cluster at ambient pressure only. The pressure dependence of the Jahn–Teller stabilization energy was calculated more approximately from ligand displacements alone. Jahn–Teller coupling to the e_g displacement was found to be an order of magnitude greater than that to the t_{2g} displacement, as expected, and to be comparable with linear coupling to the a_{1g} displacement.

The total energy of each state was calculated in the relaxed configuration of the other state to determine optical transition energies in accordance with the Franck–Condon principle. The predicted pressure dependence of the photoluminescence peak energy of $K_2NaScF_6:Cr^{3+}$ agrees well with experiment, as shown in Fig. 8.5. The predicted Stokes shift at ambient pressure, $2083\,cm^{-1}$, is in reasonable agreement with the experimental value of $2748\,cm^{-1}$. Vibration frequencies of defect normal modes were determined approximately by calculating second derivatives of the total energy of the cluster as a function of small a_{1g}, e_g and t_{2g} displacements of the six nearest halogen neighbours of the chromium impurity. Approximate vibration frequencies of $K_2NaScF_6:Cr^{3+}$, calculated as functions of pressure, compare favourably with experimental values inferred from the vibronic structure of emission spectra. Radiationless transition rates were predicted *ab initio* from adiabatic potential energy functions of a_{1g}, e_g and t_{2g} displacements of the six nearest halogen neighbours in both $^4A_{2g}$ and $^4T_{2g}$ states [Woods *et al.* (1994)], as described in §6.4.2, with encouraging results. Matrix elements of the t_{1g} promoting-mode interaction, Eq. (6.39), were evaluated approximately with free-ion chromium orbitals [Bartram and Stoneham (1985)]. These favourable results, which confirm the predictive capability of the model, can also serve to parameterize the crystal-field model.

Similar calculations performed on Mn^{4+} in $BaTiO_3$ [Donnerberg and Bartram (1996)] incorporated additional refinements, including correlation corrections at the level of CISD and shell-model pair potentials for lattice relaxation.

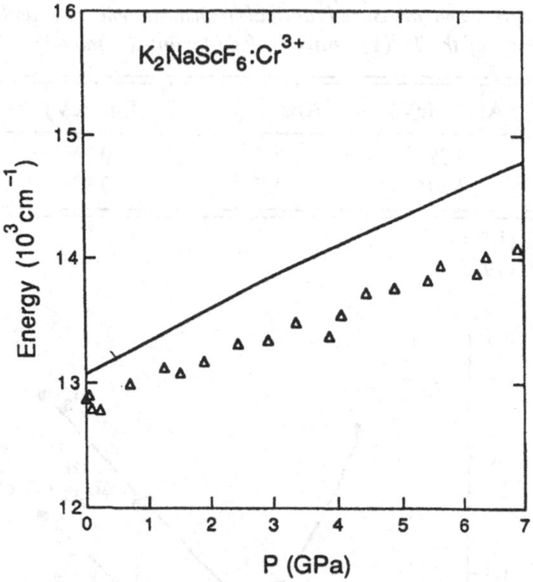

Figure 8.5. Pressure dependence of the photoluminescence peak energy for Cr^{3+} : K_2NaScF_6. The continuous curve is the theoretical prediction [Woods *et al.* (1993)] and the symbols are experimental results [Dolan *et al.* (1992)].

8.6.2 The $Tl^0(1)$ centre and its analogues

The $Tl^0(1)$ centre in KCl, described in §4.6.5, is the prototypical impurity-related laser-active colour centre. *Ab initio* embedded-cluster RHF-SCF-LCAO molecular-orbital calculations were performed on both ground and excited states of this centre [Gryk and Bartram (1995)]. Effective core potentials and valence orbitals were employed on the thallium atom and its nearest-neighbour chlorine ions. The remaining ions in the cluster were represented either by bare effective core potentials or by point ions. Spin–orbit effects were calculated perturbatively in the intermediate coupling regime, Eqs. (4.75) and (4.76). External interactions of the molecular cluster were represented by rigid-ion pair potentials, and simultaneous relaxation of the cluster and surrounding lattice was accomplished with the HADESR program. Optical absorption energies for transitions to the first two excited states, the optical emission energy and a ground-state vibration frequency associated with thallium displacement were determined successfully.

A simplified point-ion model, based on a single-centre expansion with polarizing functions and a fixed point-ion lattice, validated for the $Tl^0(1)$ centre in KCl, was also employed to determine absorption energies of the analogous $In^0(1)$ and $Ga^0(1)$ centres in KCl and to explain the nonradiative de-excitation of these centres in terms of the DKR mechanism described in §6.3.3. In this model, the configuration coordinate corresponds to the displacement of the thallium atom away from the anion vacancy. The results of point-ion calculations for the $Tl^0(1)$ centre are summarized in Table 8.3 and Fig. 8.6. The success of the point-ion model suggests that covalent bonding of the thallium atom with its halogen ligands may not be very important in this type of colour centre.

Table 8.3. *Predicted and measured optical transition energies and ground-state vibration frequency of the $Tl^0(1)$ centre in KCl (point-ion model)*

	Abs. 1 (eV)	Abs. 2 (eV)	Em. (eV)	ω (10^{13} s^{-1})
Theory	1.29	1.77	0.75	0.56
Experiment	1.20*	1.72*	0.83*	0.57**

* [Fockele *et al.* (1985)]
** [Joosen *et al.* (1985)]

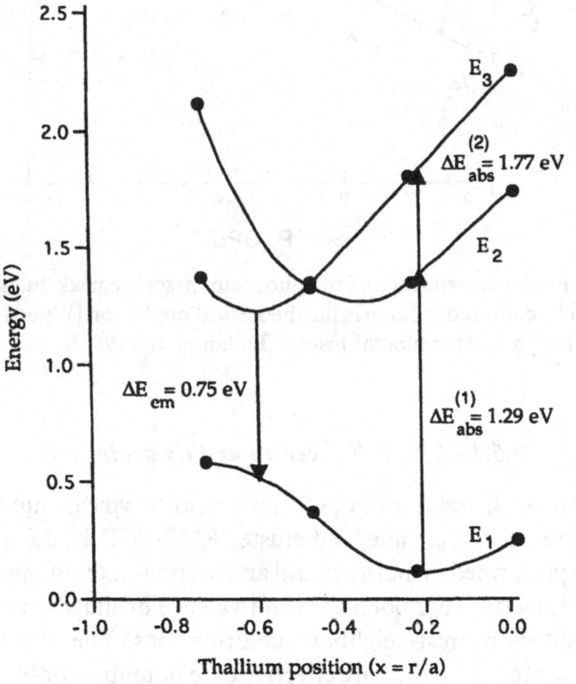

Figure 8.6. The configuration coordinate diagram for the $Tl^0(1)$ centre in KCl [after Gryk and Bartram (1995)]. The total energy of each state is plotted as a function of thallium displacement, negative toward the vacancy, with the remaining ions as point ions fixed in their perfect-lattice positions.

8.6.3 Ti^{3+} in distorted octahedral coordination

The theory of electric-dipole transitions between 3*d* levels of transition-metal ions developed in §5.3 assumes parity mixing of transition-metal orbitals by static or dynamic odd-parity components of the crystal field. Although the closure approximation is invoked in Judd–Ofelt theory, §5.3.4, it is generally supposed that transition-metal 4*p* orbitals are the dominant intermediate states. However, a molecular-orbital model developed by Yamaga *et al.* (1991a) for calculating polarized intensities of zero-phonon lines and broadband luminescence spectra of Ti^{3+} in Al_2O_3, $YAlO_3$ and YAG proceeds from a very different assumption. In this model, $d \rightarrow d$ transitions are explained instead by admixture of odd-parity combinations of ligand orbitals by T_{1u} and T_{2u} distortions of the octahedral complex, illustrated in Fig. 3.2.

The molecular orbitals of interest are e_g and t_{2g} antibonding orbitals constructed from central-ion $3d$ orbitals and ligand s, p_σ and p_π orbitals, for example,

$$|\varepsilon\rangle = N_\sigma\left[|x^2 - y^2\rangle - \frac{\lambda_\sigma}{2}(S_1 + S_4 - S_2 - S_5) - \frac{\lambda'_\sigma}{2}(-X_1 + X_4 + Y_2 - Y_5)\right], \tag{8.70a}$$

$$|\theta\rangle = N_\sigma\left[|2z^2 - x^2 - y^2\rangle - \frac{\lambda_\sigma}{\sqrt{12}}(2S_3 + 2S_6 - S_1 - S_4 - S_2 - S_5) - \frac{\lambda'_\sigma}{\sqrt{12}}(-2Z_3 + 2Z_6 + X_1 - X_4 + Y_2 - Y_5)\right], \tag{8.70b}$$

$$|\xi\rangle = N_\pi\left[|yz\rangle - \frac{\lambda_\pi}{2}(Z_2 - Z_5 + Y_3 - Y_6)\right], \tag{8.70c}$$

$$|\eta\rangle = N_\pi\left[|zx\rangle - \frac{\lambda_\pi}{2}(X_3 - X_6 + Z_1 - Z_4)\right], \tag{8.70d}$$

$$|\zeta\rangle = N_\pi\left[|xy\rangle - \frac{\lambda_\pi}{2}(Y_1 - Y_4 + X_2 - X_5)\right], \tag{8.70e}$$

where N_σ and N_π are normalizing constants. The molecular orbital $|\xi\rangle$ is illustrated in Fig. 8.7, together with $T_{1u}(\pi)$ and $T_{2u}(\pi)$ distortions. These distortions admix odd-parity combinations of ligand orbitals. For example, the $T^z_{2u}(\pi)$ distortion mixes $|\xi\rangle$ with $(Y_2 + Y_5)/\sqrt{2}$. Relative transition probabilities for polarized light can then be calculated by application of Eq. (5.118). Note that the $T_{1u}(\sigma)$ and $T_{1u}(\pi)$ distortions employed in the present section are linear combinations of the T_{1u} distortions depicted in Fig. 3.2.

In the crystals of interest, the orbital degeneracy of the levels is removed by a combination of the Jahn–Teller effect, low-symmetry even-parity components of the crystal field, and spin–orbit coupling, leaving only Kramers doublets. The splitting in a trigonal field, appropriate to Al_2O_3, is illustrated in Fig. 5.9. The wave functions associated with the Jahn–Teller split ^2E_g state are vibronic admixtures of molecular spin-orbitals $|\theta, \pm\frac{1}{2}\rangle$ and $|\varepsilon, \pm\frac{1}{2}\rangle$. Wave functions associated with the $^2T_{2g}$ state are as follows:

$$\left.\begin{array}{l} E_{3/2} : |\chi_\pm, \pm\frac{1}{2}\rangle \\ E^1_{1/2} : |\chi_\mp, \pm\frac{1}{2}\rangle \\ E^2_{1/2} : |\chi_0, \pm\frac{1}{2}\rangle \end{array}\right\} {}^2T_{2g}, \tag{8.71}$$

$$|\chi_\pm\rangle = \mp\frac{1}{\sqrt{2}}(|\eta_X\rangle \pm i|\eta_Y\rangle), \tag{8.72a}$$

$$|\chi_0\rangle = |\eta_Z\rangle, \tag{8.72b}$$

$$|\eta_X\rangle = \frac{1}{\sqrt{6}}(2|\zeta\rangle - |\xi\rangle - |\eta\rangle), \tag{8.73a}$$

$$|\eta_Y\rangle = \frac{1}{\sqrt{2}}(|\xi\rangle - |\eta\rangle), \tag{8.73b}$$

$$|\eta_Z\rangle = \frac{1}{\sqrt{3}}(|\xi\rangle + |\eta\rangle + |\zeta\rangle), \tag{8.73c}$$

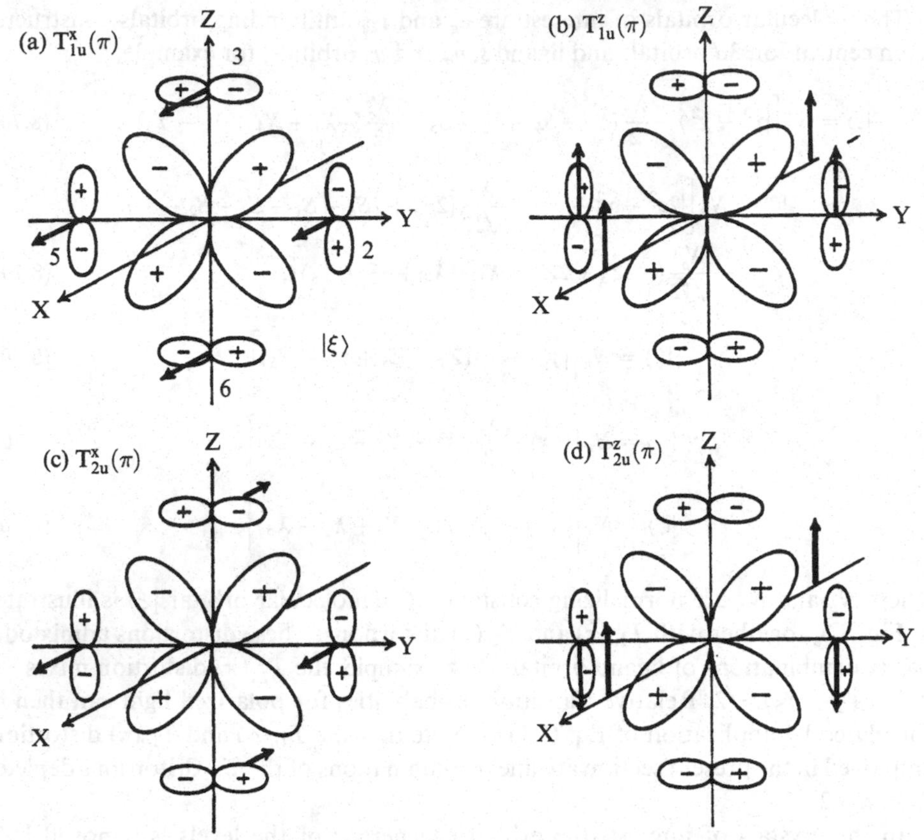

Figure 8.7. Molecular orbital $|\xi\rangle$, Eq. (8.70c), together with $T_{1u}(\pi)$ and $T_{2u}(\pi)$ distortions [after Yamaga *et al.* (1991a)].

where subscripts X, Y and Z refer to a coordinate system rotated with respect to x, y, z. The new axes are chosen to be $X\|\langle\bar{1}\,\bar{1}2\rangle$, $Y\|\langle 1\bar{1}0\rangle$ and $Z\|\langle 111\rangle$.

Relative probabilites for polarized optical transitions in trigonal symmetry, enabled by a $T_{2u}(\pi)$ distortion, are displayed in Fig. 8.8. Similar diagrams were calculated for tetragonal symmetry and for $T_{1u}(\pi)$ and $T_{1u}(\sigma)$ distortions. The squares of the matrix elements should be averaged over ε and θ to take account of vibronic mixing for comparison with experimental intensities. It is evident from Fig. 8.8 that the relative intensities of the A, B and C transitions in Fig. 5.9, predicted for a $T^Z_{2u}(\pi)$ distortion, are in the ratio $I_A:I_B:I_C=1:1:0$, in good agreement with the experimental ratio $I_A:I_B:I_C=1:0.8:0.1$. In addition, the polarization ratio for the allowed transitions, I_π/I_σ, where π and σ correspond, respectively, to Z and to X or Y polarizations, is predicted to be 2:1, which agrees well with experiment for both the zero-phonon line and the broadband. Since no other displacement would be consistent with experiment, it is concluded that a $T^Z_{2u}(\pi)$ distortion is the primary agent in exciting radiative decay via both zero-phonon lines and broadbands. This distortion can be identified with the counter-rotation of the two oxygen triangles about the $\langle 111\rangle$ direction, discussed in §3.3.1.

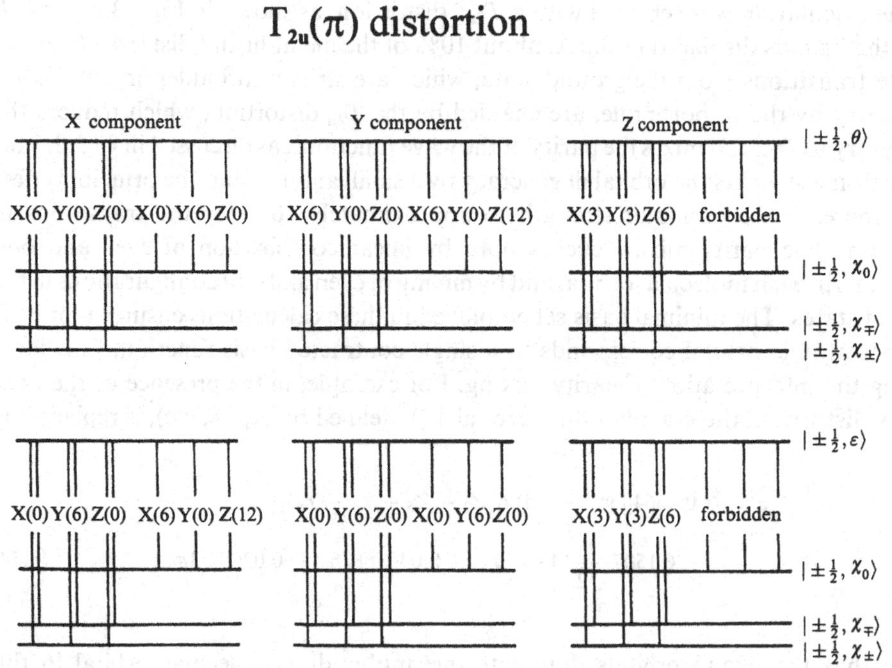

Figure 8.8. Polarizations and relative intensities of the emission induced by a $T_{2u}(\pi)$ distortion in trigonal symmetry. Column headings X, Y and Z denote components of the distortion with respect to trigonal axes. The letters in the vertical lines represent components of polarization with respect to trigonal axes and the numbers in parentheses represent relative intensities [after Yamaga *et al.* (1991a)].

8.6.4 Odd-parity distortions of $(CrF_6)^{3-}$

In further consideration of the effects of odd-parity distortions, *ab initio* molecular-orbital calculations were performed on a $(CrF_6)^{3-}$ cluster, both as a perfect octahedron and with a T_{2u} distortion [Bartram and Henderson (1998)]. The calculations were performed by means of the Gaussian 94W program [Frisch *et al.* (1995)] with a chromium effective core potential and minimal basis set (LanL2MB, equivalent to STO-3G). The metal-ligand distance in octahedral symmetry was taken as the sum of ionic radii, 1.95 Å. The $^4A_{2g}$ ground state was calculated self-consistently with the addition of an MP2 (second-order perturbation theory) correlation correction. The two lowest quartet excited states, $^4T_{2g}$ and $^4T_{1g}$, each with triple orbital degeneracy, were calculated by configuration interaction with single excitations only (CIS), as discussed in §8.3.4. Their energies with respect to the ground state are, respectively, 2.37 eV and 3.47 eV, corresponding to $Dq = 1910\,\text{cm}^{-1}$ and $B = 1180\,\text{cm}^{-1}$. These values are in adequate agreement both with experiment and with the results of more extensive calculations [Woods *et al.* (1993), Aramburu *et al.* (1999)].

The calculation was repeated with a $T_{2u}\zeta$ distortion, as shown in Figs. 3.2 and 8.7, with the ligands displaced by 0.2 Å, about 10% of the metal-ligand distance. Electric-dipole transitions from the ground state, which are strictly forbidden in octahedral symmetry by the Laporte rule, are enabled by the T_{2u} distortion, which reduces the symmetry to D_{2d} and mixes the parity of the wave functions, as discussed in §5.3.2. This distortion also splits the orbital degeneracy by a small amount, but the principal effect is to increase the oscillator strength to approximately 10^{-4} for the transitions in question. The parity mixing occurs both by linear combination of even and odd functions in each molecular orbital and by mixing of even and odd configurations in the excited states. The minimal basis set employed in these calculations ensures that each valence atomic orbital corresponds to a single contracted basis function, thus facilitating the interpretation of parity mixing. For example, in the presence of the odd-parity distortion, the π-antibonding orbital $|\xi\rangle$, defined by Eq. (8.70c), is replaced by

$$\begin{aligned}|\xi'\rangle^a \cong 0.4364[|yz\rangle - 1.106(Z_2 - Z_5 + Y_3 - Y_6)]\\ - 0.1507\frac{1}{\sqrt{2}}(Y_2 + Y_5) + 0.04026|4p_y\rangle - 0.02094|3p_y\rangle.\end{aligned} \tag{8.74}$$

Note that the ligand orbitals dominate the antibonding molecular orbital in this example, contrary to the suppositions of extreme semi-empirical models described in §8.4.4. Configuration mixing provides a second example; the dominant single substitutions in one component of the $^4T_2$ excited state are

$$\begin{aligned}&0.6446(-|\xi'\rangle^a \to |\theta'\rangle) + 0.4106(-|\xi'\rangle^a \to |\varepsilon'\rangle)\\ &- 0.4149(|\xi'\rangle^b \to |\theta'\rangle) - 0.2667(|\xi'\rangle^b \to |\varepsilon'\rangle)\\ &- 0.28962\left[\frac{1}{\sqrt{2}}(Y_2 + Y_5) \to |\theta'\rangle\right] - 0.1974\left[\frac{1}{\sqrt{2}}(Y_2 + Y_5) \to |\varepsilon'\rangle\right],\end{aligned} \tag{8.75}$$

where $|\xi\rangle^b$ is a π bonding orbital.

It is evident from these examples that parity mixing is dominated by odd-parity combinations of ligand orbitals rather than by metal $4p$ and $3p$ orbitals. Which of these orbitals contributes the most to transition dipole moments is less certain, however, since only the totals are given, and the contribution of ligand orbitals is diminished by their reduced overlap with metal orbitals. Nevertheless, one can safely conclude that the effects of transition-metal $4p$ and $3p$ orbitals and odd-parity combinations of ligand orbitals are at least comparable. Thus this model calculation serves to confirm both the traditional crystal-field approach as well as the alternative treatment of Yamaga *et al.* (1991a).

9

Engineering the crystal field

9.1 Principles and objectives

Crystal-field engineering seeks to use present knowledge to establish appropriate design principles for the development of new laser and nonlinear optical materials. First, the wavelength range of the optical device and its possible application (e.g. CW, ultrashort pulse, single frequency or tunable) are specified. This determines the chemical nature of the optical centre. The host environment is then selected, guided by historical knowledge of gain media or intuition of novel hosts with potentially beneficial properties, and then the theoretical and experimental techniques outlined in earlier chapters are invoked. The numerous objectives of crystal-field engineering include shifting the wavelength ranges of optical transitions, increasing the rates of radiative transitions and minimizing loss by nonradiative decay and excited state absorption. In addition, there may be reason to minimize or maximize energy transfer between centres, to avoid concentration quenching and to enhance laser efficiency respectively. Such objectives may be achieved by manipulating the unit cell containing the optical centre using such external perturbations as hydrostatic pressure, uniaxial stress or electric field. More usually, however, manipulating the unit cell is accomplished by changing its chemical composition.

9.1.1 Manipulating the unit cell

Hydrostatic pressure shortens bond lengths, reducing the unit cell dimensions without changing its symmetry. Such hydrostatic pressures will enhance the crystal field and, in consequence, shift spectra to shorter wavelengths. Studies of the Cr^{3+}-doped elpasolites and garnets under pressure demonstrate the continuous tuning of the crystal field and of the coupling of the ^2E and $^4T_2$ states of the Cr^{3+} ion [Dolan *et al.* (1986), Hömmerich and Bray (1995)]. At low pressure the luminescence spectrum of Cr^{3+} : GSGG is a mixture of the $^2E \rightarrow {}^4A_2$ R-line and $^4T_2 \rightarrow {}^4A_2$ broad band [Struve and Huber (1985), Donnelly *et al.* (1988), Yamaga *et al.* (1989a), Wojtowicz *et al.* (1991)] whereas at high pressure (>120 kbar) the $^4T_2 \rightarrow {}^4A_2$ component of the luminescence in Cr^{3+} : GSGG is eliminated. The luminescence spectrum then consists of the R-line and its associated sideband alone with decay time at 300 K lengthening from 110 μs at zero pressure to 4.4 ms at 125 kbar. In contrast, tuning the crystal-field via a series of different garnets will provide a number of discrete values of the crystal-field strength.

The unit cell symmetry can be modified by a *uniaxial stress*. In a definitive experiment, Schawlow *et al.* (1961) studied the R-line of Cr^{3+} in MgO under uniaxial stress along the [100], [110] and [111] directions, to reduce the symmetry at the Cr^{3+} site from octahedral to tetragonal, orthorhombic and trigonal, respectively. The energy shifts of the R-lines were linear in the applied stress, and the splitting patterns and polarizations confirmed the octahedral symmetry. Group theoretical analyses of the effects of uniaxial stress on optical zero-phonon lines [Kaplyanskii (1959, 1964), Hughes and Runciman (1965)], led to the identification of vacancy aggregate centres in a variety of ionic crystals. As discussed in §9.6, applied electric fields have been used to probe the mechanisms of frequency conversion in MO_6-octahedral complexes [Fujii and Sakudo (1976)]. Although a useful tool in the spectroscopist's armoury, piezo-spectroscopy does not lend itself to practical device engineering.

9.1.2 Composition of the unit cell

The most practical means of manipulating the size and symmetry of the unit cell is through its chemical composition. Consider the example of the *F*-centre in the alkali halides. This defect has octahedral symmetry and the lowest energy electronic states transform as the irreducible representations A_{1g} and T_{1u} of the O_h group. Theoretical models of the $A_{1g} \rightarrow T_{1u}$ *F*-band absorption transition discussed in §4.6.1 focus on the finite potential well of the vacancy and predict that the *F*-band energy scales as some function of the dimensions of the well, such as the Mollwo–Ivey relation in Eq. (4.66) [Fowler (1968b), Stoneham (1985), Henderson and Imbusch (1989)]. All the *F*-centres have broad absorption and emission bands, reflecting the strong electron–phonon coupling characterized by Huang–Rhys factors in the range $60 > S > 25$. The luminescence bands of *F*-centres are vibronically shifted into the infrared region, where they cover a wavelength range from 750 nm to *ca* 1450 nm. The luminescence yield at 300 K is very small because nonradiative decay is dominant.

The invention of the broadband tunable alexandrite laser ($Cr^{3+}:BeAl_2O_4$) in the late 1970s catalyzed research on Cr^{3+}-doped insulators. The oxide garnets, with general formula $A_3B_2C_3O_{12}$, were much studied. Changing the chemical content of the unit cell modifies the crystal-fields at A, B and C sites in a systematic way. The octahedral B-site is the preferred substitutional site for Cr^{3+}-dopant ions, at which the strength of the octahedral field may be adjusted so that either of the ^2E or $^4T_2$ excited states is lower in energy. Most Cr^{3+}-doped garnets are laser active. Cr^{3+} : YAG and YGG, like ruby, are three-level lasers. In the strong field B-sites of YAG or YGG, ^2E is lower than $^4T_2$ and their luminescence spectra, for example Fig. 9.1, show the long-lived R-lines near 680–700 nm with their vibronic sidebands. In contrast, four-level vibronic structure of Cr^{3+} : LLGG gives rise to broadband luminescence in the $^4T_2 \rightarrow {}^4A_2$ transition from Cr^{3+} ions in weak crystal-field sites, Fig. 9.2, spanning the range 740–1000 nm and decaying with radiative lifetime $\tau_R \approx 35\ \mu s$ at low temperature [Struve and Huber (1985)]. In Cr^{3+} : GSGG, the ^2E and $^4T_2$ levels are almost degenerate and the luminescence spectrum is a mélange of ^2E, $^4T_2 \rightarrow {}^4A_2$ transitions. Both R-lines and broadband emissions then contribute to laser operation. The semblances of the $^4T_2 \rightarrow {}^4A_2$ band evident in Fig. 9.1 arise from thermal population of the $^4T_2$ level in Cr^{3+} : YGG above *ca* 50 K. The distortions of the CrO_6^{9-} octahedron in the garnets by the odd- and even-parity displacements of neighbouring ions, Fig. 3.5, have important consequences for

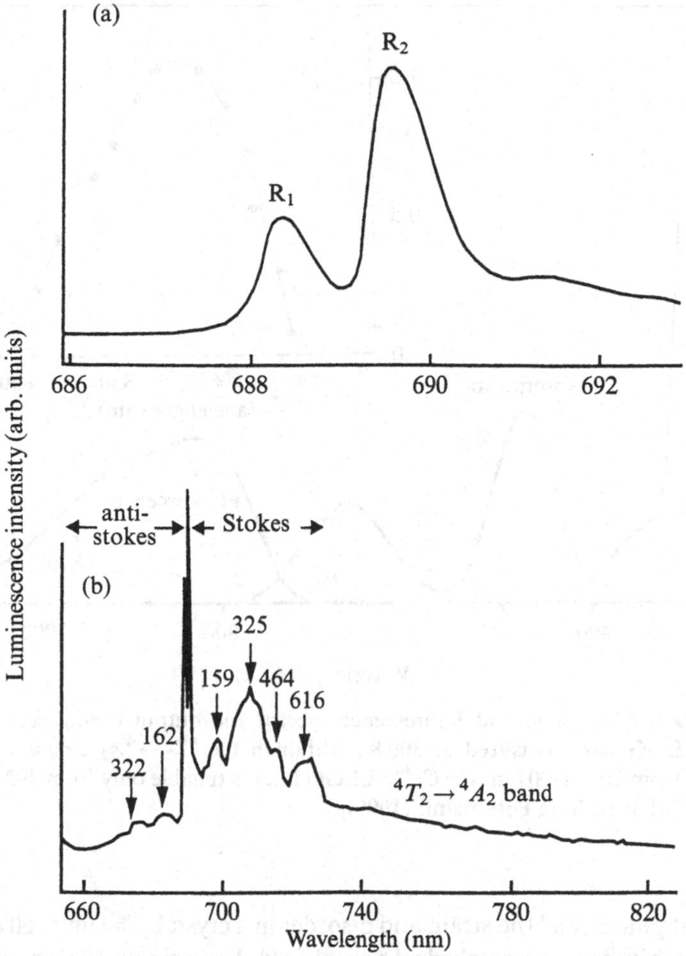

Figure 9.1. Luminescence spectra of Cr^{3+} : YGG at $T = 50$ K showing in (a) the R_1 and R_2 lines and in (b) their Stokes and anti-Stokes vibronic sidebands with a weak underlying $^4T_2 \rightarrow {}^4A_2$ band [after Henderson *et al.* (1990)].

the optical spectra of the Cr^{3+}-doped garnets. The T_{2g} distortion in concert with spin–orbit coupling will split the octahedral energy levels, as is evident from the R-line spectra of Cr^{3+}-doped YGG shown in Fig. 9.1. Odd-parity distortions (T_{1u} and T_{2u}) further reduce the D_{3d} symmetry to C_{3i} and mix wavefunctions of opposite parity into the 3*d*-states, raising the selection rules against electric dipole transitions and determining the polarization intensities and radiative life times of the Cr^{3+} : garnet spectra [Yamaga *et al.* (1989a,b), (1990a)]. Similar static and/or dynamic distortions occur in other crystals with six-fold ligand coordination, including MgO, the fluoride garnets, elpasolites and colquiriites and the ferroelectric ABO_3 compounds such as $LiNbO_3$.

9.2 The positions and shapes of optical transitions

Examples of the changes in spectroscopic properties that are consequent upon variations of chemical composition are now discussed. X-ray diffraction measurements determine the structure and dimensions of the unit cell. They also measure the number

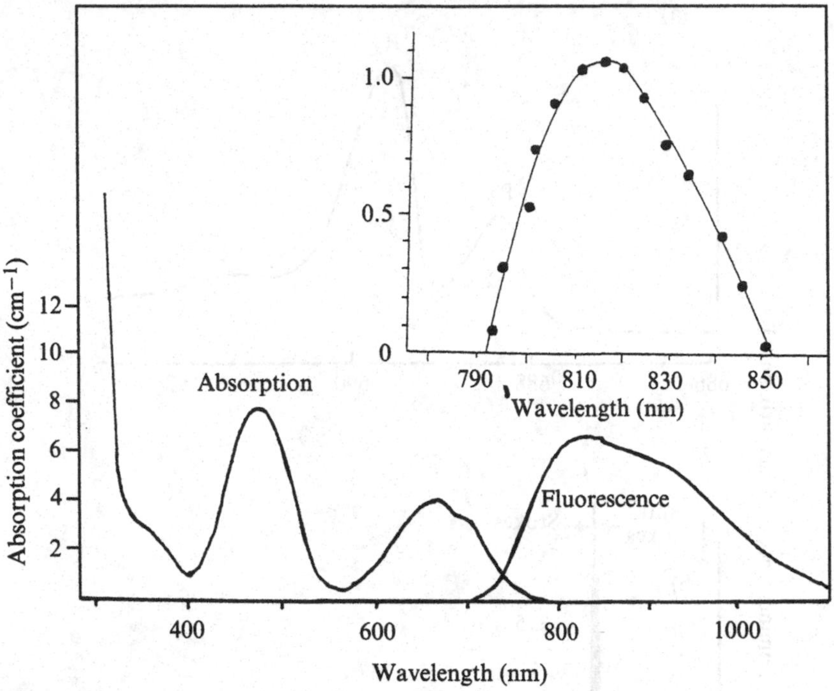

Figure 9.2. Absorption and fluorescence spectra and output tuning curve of a Cr^{3+} : LLGG laser measured at 300 K. Although the ${}^4T_2 \rightarrow {}^4A_2$ emission band extends from 700–1100 nm, the Cr^{3+} : LLGG laser is tunable only from 792 nm to 852 nm [adapted from Petermann (1990)].

of constituent phases, and the strain and disorder in a crystal. The unit cell dimensions define the magnitude of the octahedral crystal-field. For a single 3*d* electron in perfect octahedral symmetry, the energy splitting, $10Dq$, of the e_g and t_{2g} orbitals is derived in the point ion approximation from Eq. (4.50a) as

$$10Dq = \frac{5Ze^2}{12\pi\varepsilon_0}\left[\frac{\langle r^4\rangle}{a^5}\right] \tag{9.1}$$

in which $-Ze$ is the ligand charge at a distance a from the central ion. Assuming harmonic vibrations and the configurational coordinate model, §5.2.4, the peak of the absorption band, λ_0^a, corresponds to a 'vertical' Frank–Condon transition from the mean lattice configuration of the ground state to the upper state configurational coordinate curve, Fig. 5.2. A measurement of the band peak, λ_0^a, determines Dq, since

$$\frac{hc}{\lambda_0^a} = 10Dq. \tag{9.2}$$

The absorbed photon probes the configurational coordinate curve of the upper state, mapping out the absorption band at higher or lower photon energies than the peak as it 'catches' the centre at coordinates other than at the mean lattice configuration. After nonradiative relaxation to the lowest vibrational level in the excited state, radiative

Table 9.1. *Spectral characteristics for Ti^{3+} ions in different ionic crystals, arranged in order of increasing Ti^{3+}–O^{2-} separation*

	YAP	$MgAl_2O_4^+$	Al_2O_3	$BeAl_2O_4$	YAG	GSAG
λ_+^a (nm)	425	490	490	505	500	515
λ_-^a (nm)	490	490	550	565	599	600
λ^e (nm)	610	805	740	750^{++}	752	850
$10Dq$ (cm^{-1})	21 950	20 400	19 300	18 690	18 350	17 940
E_{JT} (cm^{-1})	3420	—	3070	2100	2970	2750

$^+$The Al^{3+} site occupied by Ti^{3+} in this spinel is O_h
$^{++}$Measured with $E \parallel b$: with $E \perp b$ there are two peaks at 710 nm and 875 nm.

decay occurs into the emission band with a peak wavelength, λ_0^e, given by

$$\frac{hc}{\lambda_0^e} = 10Dq - (2S - 1)\hbar\omega_0, \tag{9.3}$$

where S is the Huang–Rhys factor and $\hbar\omega_0$ is the mean energy of the even-parity phonons coupled to the electronic levels.

The $3d^1$ configuration ion Ti^{3+} occupies distorted octahedral complexes. As discussed in §5.4.3, the TiO_6^{9-} octahedron in Al_2O_3 undergoes symmetry-lowering distortions that split the optical absorption spectrum into two overlapping bands separated by *ca* 2500 cm^{-1} and centred near 20 000 cm^{-1} (i.e. 500 nm). The band centroid, according to Eq. (9.2), defines the octahedral splitting, $10Dq$. This equation may be used to compare the $^2T_2 \rightarrow {}^2E$ splitting of Ti^{3+} configuration ions in different crystals: comparative data are given in Table 9.1 for Al_2O_3, $YAlO_3$(YAP), $BeAl_2O_4$, $MgAl_2O_4$, YAG and GSAG. Substituting the mean value of a for each compound in Eq. (9.2) predicts the position of the absorption band centroid, if an appropriate value of $\langle r^4 \rangle_{3d}$ is known. Although Eq. (9.2) is not very precise, it shows the trends to be expected of the $^2T_2 \rightarrow {}^2E$ band centroids. The lattice constant, a, is shortest in $YAlO_3$, which has the largest crystal-field splitting, $10Dq$. Consequently, the mean position of the overlapping absorption bands is shifted to shorter wavelengths relative to Ti-sapphire. In the other crystals the bands are red-shifted by amounts consonant with the appropriate values of a. Assuming the experimental value for $10Dq$ leads to values of $\langle r^4 \rangle_{3d}$ for Ti^{3+} which vary from compound to compound because of their different bonding mechanisms, as noted in Chapter 8.

9.2.1 *F-type centres in the alkali halides*

The alkali halide F centre, an electron trapped at an anion vacancy, is the archetypal colour centre. Optical absorption by F centres occurs via broad bands involving allowed $^2A_{1g} \rightarrow {}^2T_{1u}$ transitions at centres having octahedral, O_h, symmetry, as outlined in §4.6.1. F-band absorption peaks shift to longer wavelength with increasing unit cell length, a, as is shown in Fig. 4.11; the best, straight line fit to the experimental data being given by Eq. (4.66) [Dawson and Pooley (1969)]. After the absorption of an F-band photon, nearby ions relax to new equilibrium positions in response to the spatial diffuseness of the charge distribution in the excited $^2T_{1u}$ state. In consequence, the bottom of the square-well potential is raised to within 0.1 eV of the bottom of the

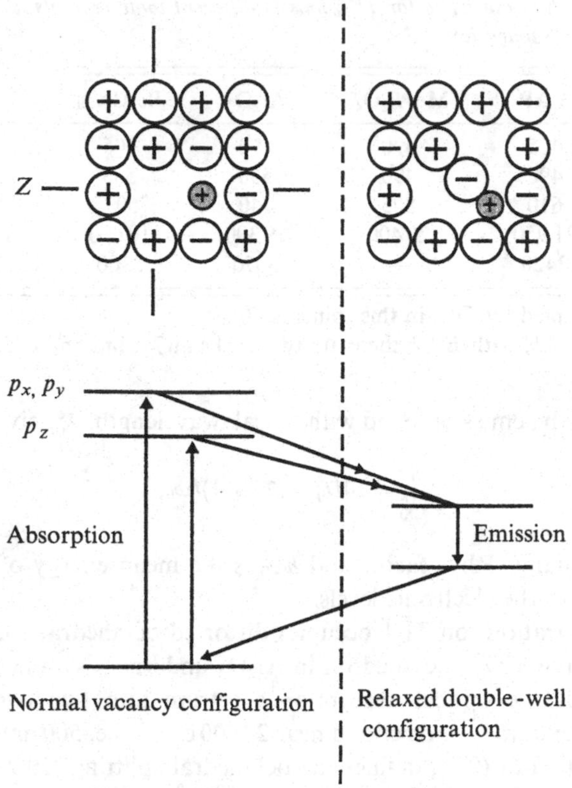

Figure 9.3. Schematic representation of the ionic and the energy level structures of F_A centres, including ionic relaxations in the excited electronic states and their effects on the absorption–luminescence cycle appropriate to F_A (Li) centres in KCl.

conduction band [Fowler (1968b)]. The decay of the excited electron can involve radiative, nonradiative and thermal ionization processes, as a result of which the life times and luminescence yields are strongly temperature dependent [Swank and Brown (1963)]. Photoluminescence is not observed from de-exciting F centres in Li-halides, NaBr and NaI even at low temperature [Dexter *et al.* (1955), Bartram and Stoneham (1975), (1983)]. Vibronic interaction red-shifts the F-centre luminescence in the other alkali halides by ~ 1–2 eV relative to the corresponding absorption band. Variations in luminescence peaks with unit cell size are not nearly as well described by an empirical Mollwo–Ivey law as are the absorption peaks because the finite sizes of ions play an important role in the electronic structure [§4.6, Bartram *et al.* (1968)]. The radiative life times at low temperatures are long (10^{-6} s) compared with allowed $2p \rightarrow 1s$-like transitions ($\sim 10^{-9}$ s). Because of nonradiative processes, F-centre luminescence at 300 K is too weak to support laser action.

The F_A centre has one nearest neighbour host cation replaced by a smaller alkali impurity ion, Fig. 9.3, reducing the local symmetry from octahedral, O_h, to tetragonal, C_{4v}. This perturbed F-centre model was originally deduced from optical and photochemical experiments [Lüty (1968)] and vindicated by magnetic resonance measurements [Spaeth *et al.* (1992)]. For an impurity ion along the z-direction the

Table 9.2. *Absorption and emission properties of F and F_A centres in alkali halides [adapted from Lüty (1968)]*

Crystal	Centre	Absorption peak (eV)			Emission peak	τ_R
		F_{A1}		F_{A2}	eV	(10^{-8} s)
KCl	F		2.31		1.24	58
	F_A (Na)	2.35		2.12	1.12	53
	F_A (Li)	2.25		1.98	0.46	8.5
KBr	F		2.06		0.92	111
	F_A (Na)	2.07		1.90	0.74	100
	F_A (Li)	2.00		1.82	0.45	10
RbCl	F		2.05		1.09	60
	F_A (Na)	2.09		1.85	0.93	60
	F_A (Li)	1.95		1.72	0.45	9

excited state, $^2T_{1u}$ splits into a low-lying singlet state, $^2A_1$, and a higher lying doublet ^2E. Consequently, and as described in §4.6.3, there are two absorption bands labeled F_{A1} and F_{A2}. The F_{A1} absorption band is polarized along a crystal [001] direction, corresponds to $^2A_1 \rightarrow {}^2A_1$ transition and is shifted to lower energies from the F-band. The F_{A2}-band is almost degenerate with the F-band and polarized along the [010] and [100] directions, corresponding to the $^2A_1 \rightarrow {}^2E$ transition at a centre with C_{4v} symmetry. There are two ways that the optical properties of the F_A centres may be engineered. The unit cell dimensions and the size differences between host and impurity cations define the optimal characteristics of F_A centres. For example, in RbCl the Rb^+ ion may be substituted by Li^+, Na^+ and K^+. The data in Table 9.2 for F and F_A centres in KCl and RbCl reveal two distinct categories of emission. The F, F_{A1} and F_{A2} absorption bands scale similarly with lattice parameter, as do the F and F_A (I) luminescence bands, the different sizes of the impurity ions causing only small shifts between the F_A absorption bands. The single photoluminescence bands of the F_A (Na) centres are Stokes-shifted by *ca* 1 eV, are unpolarized and comparatively long-lived ($\tau_R \sim 10^{-6}$ s) because the relaxed excited state has a large orbital radius which is insensitive to perturbations introduced by the impurity ion. In consequence, the excited state splitting is eliminated after relaxation, giving rise to the single, unpolarized emission band. Dramatically different luminescence behaviour occurs when there is a large size difference between the impurity ion and the host cation that it replaces. The F_A (Li) centres have emission bands that are strongly polarized and even more strongly Stokes-shifted ($\sim$ 1.5 eV) than the F_A (Na) centres, with radiative life times shorter by a factor of *ca* ten. These differences are due to the excited state relaxation in which one of the host anions next to the F_A centre moves between its nearest neighbour ions in the cation shell, so producing the double-well configuration shown in Fig. 9.3. After emission this anion may return to its former site or jump into the vacancy. There are four equally probable saddle-point configurations, in which the charge is distributed between the two anion 'vacancies'. The superposition of four $\langle 110 \rangle$-dipole emitters should produce an emission polarization $I_{\parallel}/I_{\perp} = 2.0$ defined relative to the F_A centre axis, a result in excellent accord with experiment. In contrast to the F_A (I) centres, Table 9.2, the emission bands of F_A (II) centres hardly shift from one crystal to another.

The F_B centres are first cousins of the F_A centres, created when the alkali impurity content exceeds several atomic per cent. In these centres two nearest-neighbour cations are substituted by alkali impurity ions. F_B (I) centres retain the anion vacancy configuration during emission, whereas F_B (II) centres emit from the double-well configuration similar to that shown for F_A (II) centres in Fig. 9.3 [Litfin *et al.* (1977), German (1986)]. The trapping of an electron in a double-well potential is reminiscent of the divacancy structure of the F_2^+ centre, Fig. 4.12, which can be treated theoretically as an H_2^+ molecular ion embedded in a dielectric continuum (§4.6.4). The energy levels of the F_2^+ centre are then derived as described in Chapter 4 by scaling the $E(\Gamma)$ versus R relationship for the H_2^+ ion using the effective mass m^* and dielectric constant K as adjustable parameters. These energy levels are classified in Table 4.11 both as irreducible representations of the $D_{\infty h}$ group of the H_2^+ ion or of the exact point group D_{2h} of the F_2^+ centre (Fig. 4.13). The broad band optical spectra of the F_2^+ centres mask the splitting of the $2p\pi_u$ molecular level into B_{2u}, B_{3u} crystal-field levels implied in Table 4.11, which is not resolved. The pumping and lasing transitions A_{1g} $(1s\sigma_g) \leftrightarrow B_{1u}\,(2p\sigma_u)$ involve the same lowest energy states and are appropriately Stokes shifted, thereby eliminating possibilities of self-absorption. The large oscillator strength ($f \cong 0.2$), 100% quantum efficiency and small Stokes shift makes F_2^+ centres almost ideal for four-level lasers. The range of values for K and R_{12} for the family of alkali halides provide a large net tuning range, 800–2600 nm for F_2^+ centre lasers. Unfortunately, these centres are unstable against optical pumping with polarized light, so that the laser intensity fades during operation [Mollenauer (1985a, 1987)]. The mechanism by which F_2^+ centres fade involves centre reorientation during pumping, subsequent random walk through the crystal and further stages of aggregation and deionization. As Mollenauer (1987) observes, the *only truly satisfactory cure to the orientational bleaching problem of* F_2^+ *centres lies in anchoring them to a definite place in the crystal.* The variety of anchoring points include other radiation products and impurities.

The first stabilized F_2^+ centre, designated the $(F_2^+)^*$ centre, was produced in NaF crystals by high dose e-irradiation to create a high density of F_2^+ centres and other concomitant defects [Mollenauer (1980)]. After storing in the dark at room temperature for 24 hours the F_2^+ centre absorption band at 750 nm was usurped by an equally intense absorption band with peak at 870 nm (Fig. 9.4). The $(F_2^+)^*$ centre was shown to have a Stokes-shifted emission band, with band shape, quantum efficiency, polarization and luminescence lifetime in every respect F_2^+ centre-like. Subsequently, Hoffman *et al.* (1985) identified the $(F_2^+)^*$ centre as an F_2^+-centre perturbed by a nearby cation vacancy. Yet another stabilized F_2^+ centre, the $(F_2^+)^{**}$ centre, is produced in crystals doped with 10–70 ppm of OH^- or O^{2-}. As Fig. 9.4 shows, the absorption and emission bands of the $(F_2^+)^{**}$ centres in NaF are shifted even further into the infrared than are the spectra of the $(F_2^+)^*$ centres [Mollenauer (1981)]. The stabilising entity in the $(F_2^+)^{**}$ centres was subsequently identified as O^{2-} impurities [Wandt *et al.* (1986), Pinto *et al.* (1986)]. Given that each species of defect can be created by bombardment with 2 MeV electrons, it is not surprising that F_2^+, $(F_2^+)^*$ and $(F_2^+)^{**}$ centres are created coincidentally. However, there is an unsaturating production of $(F_2^+)^*$ centres with almost 100% conversion of F_2^+ centres, whereas the concentration of $(F_2^+)^{**}$ centres saturates at levels determined by the OH^- or O^{2-} content in the starting crystal. Stabilized F_2^+ centres can also be produced in crystals doped with impurity ions comprising F_2^+

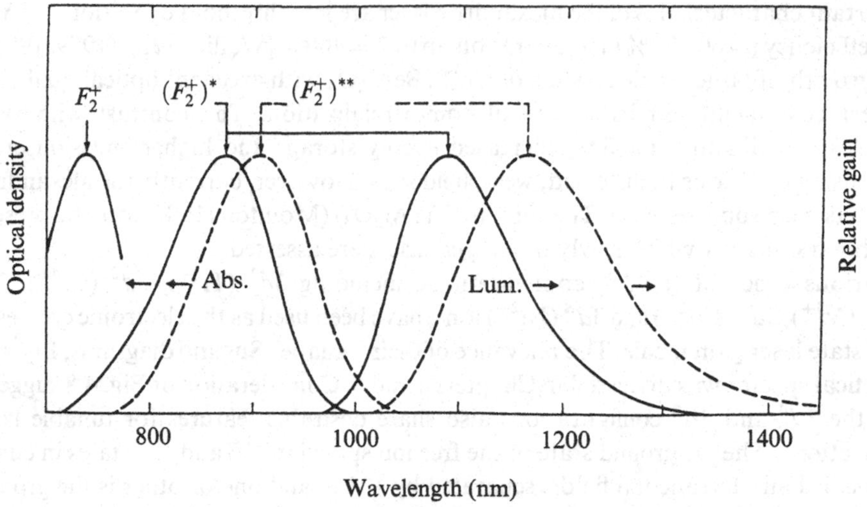

Figure 9.4. The Stokes-shifted absorption and emission bands of the lowest energy transition of $(F_2^+)^*$ and $(F_2^+)^{**}$ centres in NaF relative to the absorption band peak of the F_2^+, measured at 80 K [after Mollenauer (1987)].

centres in KCl, KBr, KI and RbI with a nearest-neighbour Li^+ ion. The fundamental optical transitions of these F_{2A}^+ centres feature luminescence bands at wavelengths near 4.0 μm, as detailed in Chapter 10 [Schneider and Marrone (1979), Schneider and Moss (1983), Schneider and Pollock (1991)].

9.2.2 *Tl⁰(1) centre in the alkali halides*

The characteristic structure of the laser-active $Tl^0(1)$ centre (§4.6.5), a neutral Tl atom next to a halide ion vacancy, was fully established by electron spin resonance and magnetic circular dichroism studies [Goovaerts *et al.* (1981), Ahlers *et al.* (1983)]. Subsequent optically detected magnetic resonance studies showed that the optically active electron, firmly pinned to the Tl-entity in the ground state became much more delocalized in the excited state [Fockele *et al.* (1985)]. The theoretical evaluation of the energy levels of $Tl^0(1)$ centres, fully discussed in §4.6.5, shows why the fundamental optical properties are little dependent upon the host crystal such that the broadband emission in different alkali halides peaks near 1.5 μm [Gellermann *et al.* (1981a,b), Mollenauer *et al.* (1983b)]. The low temperature luminescence decay time for $Tl^0(1)$ in KCl is $\tau_R = 1.6\,\mu s$, corresponding to an oscillator strength of *ca* $7–8 \times 10^{-3}$. When combined with the emission bandwidth, the gain cross section at the band peak is only $\sigma_p \sim 10^{-17}\,cm^2$, a factor of ten or more less than σ_p for F_2^+ centres. Fortunately, the large $Tl^0(1)$ centre densities in the alkali halides compensate for the smaller σ_p.

9.2.3 *Transition-metal ions*

The renewed interest in transition metal ion lasers during the 1980s began with alexandrite ($Cr^{3+}:BeAl_2O_4$), which lased on the vibronic $^4T_2 \rightarrow {}^4A_2$ band. Among important characteristics of the alexandrite laser are low threshold operation (0.5 W), high efficiency (~ 60–70%) and operation up to $T = 400$ K [Walling *et al.* (1979, 1980)]. The growth of large single crystals of $Cr^{3+}:BeAl_2O_4$ with excellent optical quality at modest cost facilitated its successful commercialization. The contrast with ruby includes broadband tunability, increased energy storage and higher emission cross sections by Cr^{3+} ions in distorted, weak field sites. However, currently the alexandrite laser has been superseded by Ti-sapphire ($Ti:Al_2O_3$) (Moulton (1982a,b)], the present benchmark against which newly developed lasers are assessed.

Various other of the $3d^n$ configurations, including $3d^1$ (Ti^{3+}), $3d^2$ (Cr^{4+}), $3d^3$ (V^{2+}, Cr^{3+}), $3d^7$ (Co^{2+}) and $3d^8$ (Ni^{2+}) ions, have been used as the electronic centres in solid state laser gain media. The relevance of their Tanabe–Sugano diagrams, Fig. 4.8, to optical spectra was discussed in Chapters 4 and 5. Consideration of Fig. 4.8 suggests that the $3d^4$ and $3d^6$ configurations also share desirable features for tunable laser applications. The ^5D ground state of the free ion splits into ^5E and $^5T_2$ states in cubic, octahedral and tetrahedral fields, separated by $10Dq$, and one or other is the ground state in the weak field (high-spin) regime. Although there are myriad excited states, all are of lower multiplicity so that ESA of pump and emission beams are largely suppressed by the multiplcity selection rule. The only example of laser action by these configuration ions was reported for Cr^{2+}-doped chalcogenides (DeLoach *et al.* (1996), §10.3.7).

The spectroscopy of $Ti^{3+}:Al_2O_3$ was well-known [Nelson *et al.* (1967), Gächter and Koningstein (1974)] and the symmetry distortions of Al_2O_3 completely understood [Shinada *et al.* (1966)] long before Moulton (1982a,b) invented the Ti : sapphire laser. Several distortions of the $(AlO_6)^{9-}$ octahedron that reduce the symmetry from O_h to C_3 were described in Chapters 2 and 3 in deriving the distorted octahedral structure characteristic of corundum from the perfectly symmetrical unit (Fig. 3.1a). The T_{2u} distortion is often neglected and the symmetry assumed to be C_{3v} in discussions of the optical spectra of Ti^{3+} : sapphire and Cr^{3+} : ruby [Henderson and Imbusch (1989)]. The distorted, six-fold deployment of ligand ions around the central optically-active ion is the heart not only of the $Cr^{3+}:Al_2O_3$ and $Ti^{3+}:Al_2O_3$ lasers but also the Cr^{3+} : garnet lasers, the alexandrite laser, the Cr^{3+} : colquiriite lasers and various Ni^{2+} and Co^{2+} lasers. Indeed, the distorted octahedral unit is also the locus for optical activity in the rare earth alumino- and scandioborates (e.g. $NdAl_3(BO_3)$), rare earth titanyl niobates ($RETiNbO_6$) as well as the SHG crystals $LiNbO_3$, $KNbO_3$ and $KTiOPO_4$.

The garnets $A_3B_xC_{5-x}O_{12}$ with $0 < x < 2$ are an important family of laser hosts for trivalent rare-earth ions (Pr^{3+}, Nd^{3+}, etc.) on the A-site, trivalent transition metal ions (Ti^{3+}, Cr^{3+}, Fe^{3+}) on the B-site and Cr^{4+} ions on the C-site. The concept of an effective radius (§3.3.3) allows variations of crystal-field terms of transition metal ions in the binary ($x = 0$), tertiary ($x = 2$) and non-stoichiometric garnets to be represented on a single graph. Since the octahedral crystal field, Dq, scales approximately as a^{-5} (Eqs. (4.50) and (9.1)) it decreases monotonically along the series $Y_3Al_5O_{12}$ (YAG), $Y_3Ga_5O_{12}$ (YGG), $Gd_3Ga_5O_{12}$ (GGG), $Gd_3Sc_2Al_3O_{12}$ (GSAG), $Gd_3Sc_2Ga_3O_{12}$ (GSGG) and $Lu_3La_2Ga_3O_{12}$ (LLGG). In terms of the Tanabe–Sugano diagram for $3d^3$ ions, then $Dq/B \cong 2.6$ for YAG, 2.3 for GSGG and 2.1 for LLGG. Hence for Cr^{3+} : YAG the first excited state, ^2E, is below $^4T_2$ in energy (strong field), whereas in Cr^{3+} : LLGG (weak field) $^4T_2$ is the lowest excited state. Cr^{3+} : GSGG is an example of intermediate crystal-field strength in which ^2E and $^4T_2$ are almost degenerate [Struve and Huber (1985), Donnelly *et al.* (1988), Marshall *et al.* (1990)]. The consequence of this trend in Dq for $3d^3$ ions in the garnets is clearly demonstrated in their photoluminescence behaviour. Cr^{3+} : YAG and other strong field hosts at low temperature ($T < 200$ K) emit into sharp R-lines with vibronic structure, as was shown in Fig. 9.1. Weak field systems such as Cr^{3+} : LLGG have broad luminescence bands, Fig. 9.2, due to the vibronically broadened $^4T_2 \rightarrow {}^4A_2$ transitions. In contrast, the photoluminescence of Cr^{3+} ions in intermediate field hosts contains an admixture of both R-line and broadband spectra. At ~ 300 K the broad $^4T_2 \rightarrow {}^4A_2$ band is observed for all mixed oxide garnets: there are also semblances of the broadband in Cr^{3+} : YAG or YGG at this temperature. The growth of Cr^{3+}-doped garnets from congruent melts at off-stoichiometric compositions can lead to disorder-induced optical line shapes that can be probed by site-selective laser spectroscopy [Nie *et al.* (1989), Marshall *et al.* (1990)]. The mixed R-line/$^4T_2 \rightarrow {}^4A_2$ bandshape observed for some Cr^{3+}: garnets result partially from disorder, but also from vibronic mixing of ^2E and $^4T_2$ states (§5.4.1) as discussed below (§9.3.1). The impact of static and dynamic distortions of the unit cell are crucial to the tenets of crystal-field engineering. In perfectly octahedral symmetry the R-lines are magnetic dipole transitions, whereas the $^4T_2 \rightarrow {}^4A_2$ band occurs by weak electric dipole transitions. In Cr^{3+}-doped garnets even parity distortions at the B-site split the R-line. However, these $^2E \rightarrow {}^4A_2$ zero-phonon lines still occur by the magnetic dipole interaction, showing that inversion symmetry is preserved with the vibronic sidebands being promoted by T_{1u} and T_{2u} vibrations [Hehir *et al.* (1974)].

The $^4T_2 \rightarrow {}^4A_2$ band supports tunable laser action with varying efficiencies in different garnets. Experiments with non-stoichiometric GSAG, i.e. $Gd_3Sc_{2+x}Al_{3-x}O_{12}$ with ($0.15 < x < 0.30$) illustrate the effects of engineering the strength of the crystal-field; the stoichiometric excess of Sc^{3+} in GSAG reduces Dq and the $^4A_2 \rightarrow {}^4T_2$ splitting (Eq. 9.2), and displaces the broadband emission to longer wavelength because r_{eff} (Eq. 3.1) is increased [Zharikov (1986)]. Since the peak in the vibronically-broadened $^4T_2 \rightarrow {}^4A_2$ band, which is given approximately by Eq. (9.3), occurs at 740 nm in pure GSAG and the shifts are approximately linear in x, the band peak occurs at 752 nm when $x = 0.3$. Non-stoichiometry affords a modest variation in the tuning range of the Cr^{3+} : GSAG laser [Marshall *et al.* (1990)]. Some centrosymmetric fluoride garnets, perovskites and elpasolites provide weak field sites for Cr^{3+} impurities, which emit into the $^4T_2 \rightarrow {}^4A_2$ band induced by electron-phonon coupling to odd-parity phonons (§5.4.1). This contrasts with the Cr^{3+}-luminescence in ruby, alexandrite and the colquiriites, in which static odd-parity distortions of the MX_6-octahedron destroy the inversion symmetry (§2.3.3, §4.4.7). It is then necessary to take into account both static and dynamic odd-parity distortions in calculating the strengths of polarized $^4A_2 \leftrightarrow {}^2E$, $^4T_2$, $^4T_1$ transitions [§5.4, Yamaga *et al.* (1989a,b), Lee *et al.* (1989)].

Another example of crystal-field engineering in practice involves the Cr^{4+} ion in tetrahedral C-sites in the garnets. The Cr^{4+} ion is formed by co-doping the garnet melts with Cr_2O_3 and CaO and growing single crystals in mildly oxidising conditions using a Czochralski furnace. The absence of a centre of inversion at the C-site raises the selection rule against d–d transitions on the Cr^{4+} ion, as discussed in detail in §5.3 and §9.3.2. It is noted that the $^3B_2 \rightarrow {}^3B_1$ laser transition, magnetic dipole allowed in the D_{2d}-distorted tetrahedral sites, results in a weak optical absorption band near 900–1200 nm. The corresponding $^3B_2 \rightarrow {}^3B_1$ luminescence is Stokes shifted to 1300–

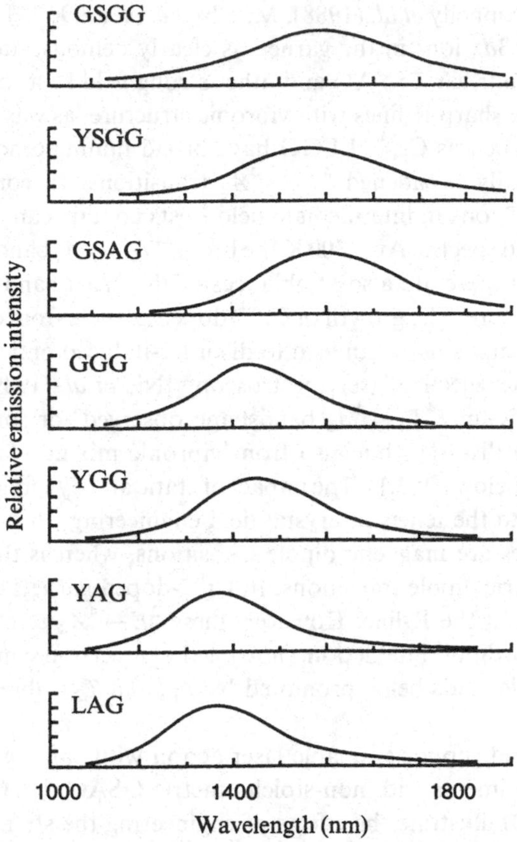

Figure 9.5. The room temperature luminescence spectra of Cr^{4+} ions in several oxide garnet crystals [after Kück *et al.* (1995)].

1600 nm decaying at 300 K with a luminescence lifetime of 2–5 μs. The luminescence band peak, plotted as a function of wavelength in Fig. 9.5 for a number of Cr^{4+}-doped garnets, shifts from $\lambda_p = 1370$ nm in $La_3Al_5O_{12}$ ($a = 1.19$ nm) to 1.580 nm in GSGG ($a = 1.255$ nm). However, no simple relationship exists between the peak shift and lattice constant. Nevertheless, the results in Fig. 9.5, and others for the mixed garnets, demonstrate that the chemical composition of the unit cell is important, as would be expected from the models in Chapter 8 that include finite ion sizes. Research on tunable, room temperature solid state lasers based on tetrahedrally co-ordinated Cr^{4+} ions started with $Cr^{4+}:MgSiO_4$ [Petricevic *et al.* (1988, 1989)], to be followed by reports of laser action in $Cr^{4+}:Ca^{2+}:YAG$ [Sverev and Shestakov (1989)], $Cr^{4+}:Y_2SiO_5$ [Koetke *et al.* (1993a,b)] and $Cr^{4+}:Y_3Sc_xAl_{5-x}O_{12}$ [Kück *et al.* (1994)]. In both forsterite and YAG initial misapprehensions caused by charge state and site multiplicities of the Cr impurity were resolved by careful electron spin resonance and site selective laser spectroscopy [Hoffmann *et al.* (1991), Moncorgé *et al.* (1991), Eilers *et al.* (1993, 1994)]. Extensive searches for other Cr^{4+} hosts with near tetrahedral sites have included the mellilites (e.g., $SrGdGa_3O_7$) and fluoroapatites (e.g., $Ca_5(PO_4)_3F$) [Scott *et al.* (1995, 1997), DeShazer (1994), Mazelsky *et al.* (1968)]. In all these hosts the symmetry is reduced from T_d, with consequences for the number and intensity of optical transitions. The Cr^{4+} photoluminescence is broadened by electron–phonon coupling, which also influences the temperature dependences of radiative and nonradiative decay rates. The interpretation of $3d^2$ ion spectra in silicate, apatite and garnet hosts give an excellent overview of the effects of the dominant (§4.3.4) and low-symmetry (§4.4.7) crystal fields, as discussed in subsequent sections (§9.3.2, §10.3.6).

9.2.4 Rare-earth ions

Rare-earth ions RE^{3+}, where RE = Ce, Pr, Nd, ..., Yb, are the optical centres in numerous prototype solid state lasers [Kaminski (1989)]. Although commercial development has focused particularly on Nd^{3+}: YAG and Nd^{3+}: YLF, there is much current interest in Er^{3+}-doped materials for applications in telecommunications. Most RE^{3+} ions in appropriate hosts support laser action. Since RE^{3+} ions, Ce^{3+} excepted, absorb and emit in sharp line spectra, they are poor abstractors of excitation energy from white light sources. This detracts from their performance in flashlamp-pumped lasers. Sharp line spectra of the RE^{3+} ions result from the spatially compact $4f^n$ orbitals, being weakly coupled to the vibrations of the lattice. Weak coupling of the $4f$-electrons to static even-parity displacements of ligand ions results in a splitting of each ${}^{2S+1}L_J$ fine-structure level into its component crystal-field levels, which may be spread over several hundreds of wavenumbers in energy. Measurement of the positions of the multiplet lines determines the magnitude of the crystal-field splittings of the different $|LSJ\rangle$ fine-structure levels.

The sharpness of line spectra at very low temperatures ($T \sim 4$ K) reflects the weak electron–phonon coupling (§5.2). The small splittings between the crystal-field lines of the fine-structure levels vary from crystal to crystal in response to different symmetries and strengths of the electrostatic crystal field. To elucidate those effects that arise solely from the strength of the electrostatic crystal-field it is important to maintain the symmetry constant. This is done by working within families of compounds, such as the stoichiometric rare-earth fluorides or the ternary oxide garnets. For example, in KYF_4 and $KGdF_4$ crystals the symmetry at Y^{3+} and Gd^{3+} sites, respectively, are identical (D_{2d}). Changing the chemical make-up through the series Cs_2NaYF_6, KYF_4, K_2YF_5 and $LiKYF_5$ (and the La and Gd-isomorphs) reduces the local symmetry at the Y^{3+} site from octahedral to tetragonal to orthorhombic to monoclinic, *et seq.* The symmetry changes are accompanied by different Y–F bond lengths and bond angles. In these crystals the Y^{3+}, La^{3+} or Gd^{3+} host cations may be fully or partially replaced by optically active lanthanides such as Pr^{3+}, Nd^{3+} or Er^{3+}. In consequence rare-earth ion dopants may display extremely high absorption coefficients due to their high concentrations and to odd-parity distortions that are present to a greater or lesser extent in these structures. Such enhanced transition rates and high concentrations of active centres can be exploited in the searches for ideal gain media in diode-pumped lasers.

Zharikov *et al.* (1991) have grown Nd^{3+}-doped crystals from the $(Y_xGd_{1-x})_3Sc_2(Ga_yAl_{1-y})_3O_{12}$ system, where $x = 0$ to 1 and $y = 0$ to 1: the measurable shifts of the Stark-split multiplets are caused by the average crystalline electric field at the A-site in each compound. A significant line broadening is induced by the statistical distributions of Sc^{3+} ions on octahedral and dodecahedral sites in these disordered garnets which result in different environments for the Nd^{3+} ions. Apart from the purely spectral effects there are also modifications in laser performance: in $GdScAl_{0.33}Ga_{0.61}O_{12}$

crystals the ultrashort pulse performance (pulse duration ~260 fs) also is related to disorder in these mixed crystals [Ober *et al.* (1992)]. To counteract the poor absorption of white light by the line spectra of RE^{3+} ions used in flashlamp-pumped solid state lasers, enhanced performance may be effected by additionally doping laser crystals with impurities that have broad absorption bands in the visible region. For example, by co-doping GSGG with Nd^{3+} and Cr^{3+} ions gives much more efficient excitation of Nd^{3+} ions by flashlamps by virtue of absorption in the vibronically broadened ${}^4A_2 \rightarrow {}^4T_2$ and ${}^4T_1$ transitions of the Cr^{3+} ions. The ensuing emission in the ${}^4T_2 \rightarrow {}^4A_2$ band overlaps many crystal-field levels of the Nd^{3+} ions, (Fig. 7.11). There is then very efficient energy transfer from excited Cr^{3+} ions to Nd^{3+} ions, resulting in more efficient Nd^{3+} luminescence at the laser wavelengths of 0.94 μm, 1.06 μm and 1.34 μm (see §7.4 and §10.6).

Normally, RE^{3+} ion lasers based on transitions between the Stark levels of the $|LSJ\rangle$-multiplets are fixed-frequency lasers. In contrast, Ce^{3+} ions support broadly tunable laser action based on crystal-field transitions between $4f^1$ and $5d^0$ levels. The potential applications of Ce^{3+}-doped crystals as phosphors, scintillators and tunable, blue lasers has led to much interest in Ce^{3+}-doped halide crystals LaF_3, $LiYF_4$, $LiCaAlF_6$ and $LiSrAlF_6$. The absorption peaks occur in the wavelength range 250–350 nm, split into two overlapping bands corresponding to transitions into the e_g and t_{1g} levels of the 5*d* orbitals. However, in some phosphors with sufficiently low symmetry sites, all five crystal-field components of the 5*d* levels are resolved in absorption [Kodama *et al.* (1998)]. The 5*d*-levels are similar to the 3*d*-levels in that they are strongly coupled to the lattice vibrations. In consequence, the 4*f*–5*d* transitions are readily shifted to different wavelength regimes by the static crystal-field and broadened by vibronic coupling. These spectra occur by allowed electric dipole transitions in the blue–near-ultraviolet regions of the optical spectrum. The broad emission bands are Stokes shifted into the wavelength range 300–450 nm. The spin–orbit splitting of the ground state into two fine structure levels is often resolved in emission. In LaF_3 and $LiYF_4$ the Ce^{3+} dopant substitutes for a trivalent La^{3+} or Y^{3+} ion. However, in LiCAF and LiSAF Ce^{3+} replaces Ca^{2+} and Sr^{2+}, respectively, requiring charge compensation to maintain electrostatic neutrality. Yamaga *et al.* (1998b) have shown that there are three Ce^{3+} sites in $Ce^{3+}:LiCaAlF_6$: one unperturbed site and two sites in which Ce^{3+} is perturbed by a nearby Li^+ vacancy.

9.2.5 *Optical line shape and laser tuning*

According to Eq. (1.22), the threshold for amplified stimulated emission occurs when the single pass gain coefficient of the laser medium, $\gamma(\nu)$, exceeds the losses per unit length of the cavity, i.e.,

$$\gamma(\nu) = \frac{\lambda^2 \Delta N}{8\pi n^2 \tau_R} g(\nu) \geq \frac{\Delta}{l}, \tag{9.4}$$

implying that the gain spectrum is related to the shape function of the emission, $g(\nu)$. The first solid state laser, i.e., the ruby laser, was a single frequency device. In good quality rubies the half-width of the laser-active R-line is only 0.1 cm^{-1} at low temperature ($T < 30$ K), this width resulting from inhomogeneous broadening by defect-related strains. As discussed in §5.5.7, the temperature dependence of the linewidth can be written as

$$\Delta E(T) = \Delta E_{\text{inh}} + \Delta E_{\text{hom}}(T) \tag{9.5}$$

in which the inhomogeneously broadened width, ΔE_{inh}, is temperature independent and dominant at low temperature (§5.5.7). For a random strain distribution the low-temperature line-shape is gaussian. Lifetime broadening (i.e. $\Delta E_{\text{hom}}(T)$) results in a Lorentzian line-shape, and is dominant at higher temperatures. When neither process is dominant the observed line-shape is a convolution of Lorentzian and gaussian line-shapes.

For the R_1 line in ruby the inhomogeneous width is many times the natural linewidth at low temperatures. Special optical techniques are used to measure the homogeneous linewidth and its temperature dependence. The radiative lifetime at 4.2 K, $\tau_R = 3.4$ ms, corresponds to a homogeneous width of 0.3 kHz. The homogeneous linewidth at 4.2 K, measured using fluorescence line narrowing (FLN) to eliminate strain broadening, is only *ca* 9.3 MHz. The excess linewidth is due mainly to phonon assisted direct relaxation between the $2\bar{A}$ and $\bar{E}$ levels of the ^2E state. However, the temperature dependence of the homogeneous width is determined by a two-phonon Raman process. The homogeneous linewidth at 300 K is *ca* 3×10^4 MHz, whereas the strain broadened width is only *ca* 3×10^3 MHz (i.e. $\sim 0.1\ \text{cm}^{-1}$). Such a room temperature linewidth affords little capability for tunable laser action and the ruby laser operates on the R_1 line at $\lambda = 694.3$ nm with a spectral bandwidth of *ca* 1.2 nm. Other ruby laser wavelengths are based on the R_2-line at 692.9 nm and exchange-coupled $Cr^{3+}-Cr^{3+}$ pairs lines at 700.9 nm and 704.1 nm [Schawlow and Devlin (1961)]. The trivalent rare earth ions, Ce^{3+} and Yb^{3+} excepted, are like ruby in that they emit into very narrow line spectra. The luminescence lines of RE^{3+}-doped crystals are also strain broadened at low temperature ($\sim$4.2 K). Typical homogeneous widths at low temperature, *ca* 10^3 MHz, are much narrower than the strain broadened widths of these gaussian lines (3×10^{10} MHz). However, at 300 K homogeneous broadening is again dominant; the lines are then of Lorentzian shape with half-widths of $2-3 \times 10^4$ MHz. Several of the rare-earths offer multi-line laser action. For Re^{3+}-doped glasses, where the emission linewidths at room temperature are 50–100 times larger than in crystals, a limited degree of tuning is possible, although it has rarely been utilized.

The tunabilities of the vibronically broadened lasers based on Ce^{3+}-ions, transition-metal ions and colour centres in ionic crystals are much more significant. The shape function in Eq. (9.4) then arises out of the electron–phonon interaction. According to §5.5 the single breathing mode, configurational coordinate model gives the low temperature shape-function $g(\nu)$ derived from Eq. (5.94) as proportional to the normalized Poissonian of Eq. (5.144), i.e.,

$$g(\nu) = \sum_{p=0}^{\infty} \frac{\exp(-S)S_0^p}{p!} \delta(E_0/h \pm p\omega/2\pi - \nu), \tag{9.6}$$

where the upper sign applies to absorption ($p = m$) and the lower sign to emission ($p = n$), S_0 is the Huang–Rhys factor at low temperature and p is the effective number of phonons of energy $\hbar\omega$ excited in the transition. The shape function determined using

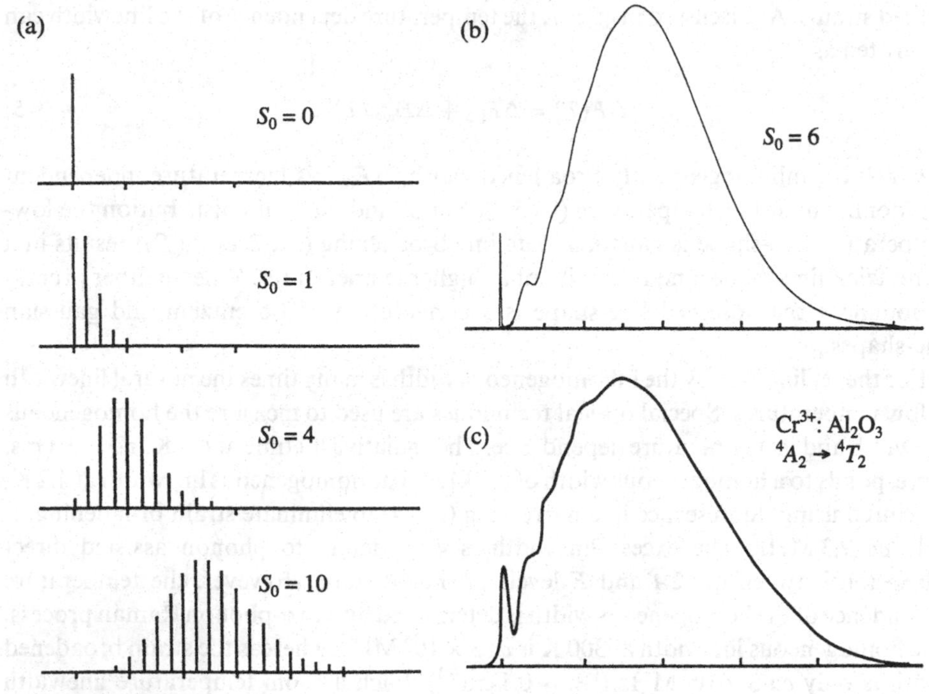

Figure 9.6. (a) Theoretical optical lineshape for different S-values based on the single configurational coordinate and linear coupling model, (b) the predicted absorption bandshape for $S=6$ and mode energy $250\,cm^{-1}$ with FWHM of $250\,cm^{-1}$, and (c) $^4A_2 \rightarrow {}^4T_2$ the absorption band in $Cr^{3+}: Al_2O_3$ at 77 K [adapted from Henderson and Imbusch (1989)].

Eq. (9.6) is a set of δ-functions with different strengths at the different values of p, each separated by $\omega/2\pi$. The predicted bandshapes for different values of S given in Fig. 9.6a show that as S increases the strength of the zero-phonon transition decreases and the vibrational sidebands at $m\hbar\omega$ above the zero-phonon absorption line are enhanced. In the strong coupling limit ($S_0 \gg 1$), the peak of the band is displaced from the zero-phonon line at $p=0$ by $\pm(S_0 - 1/2)(\omega/2\pi)$. The theoretical shape-function in Fig. 9.6a for $S_0 = 4$ is reminiscent of that in Fig. 5.3 except that the δ-function is the preserved shape of the zero-phonon line and of the phonon sideband peaks. An n-fold convolution of the single-phonon sideband increasingly broadens the sideband peaks with increasing m-value, leaving the zero-phonon line very narrow [§5.2.5]. The theoretical shape-function for $S_0 = 4$ in Fig. 9.6a shows the δ-function to be the preserved shape of the zero-phonon line and each phonon sideband.

For the broadband spectra of colour centres and transition-metal ions, the single-mode picture must be modified to take into account the finite width to the spectrum of modes corresponding to the single-phonon sideband. As discussed in §5.2.5 under the assumption of linear coupling, a p-fold convolution of the single-phonon sideband increasingly broadens the sideband peaks with increasing p-value, leaving the zero-phonon line very narrow. A relatively narrow single-phonon sideband was assumed in the example of Fig. 5.3: such a structured sideband is typical of those of molecular ions such as O_2^- and CN^- embedded in crystals. However, in tunable laser materials the single phonon sideband can be rather broad. The theoretical bandshape in Fig. 9.6b was calculated assuming strong ($S_0 = 6$) linear coupling with each sideband within the spectrum of modes having a FWHM of $\sqrt{m}\hbar\omega$, being separated from its neighbour by 250 cm^{-1}. The theoretical bandshape gives peak separations in the predicted sidebands that are similar to those in Fig. 9.6c for the $^4A_2 \rightarrow {}^4T_2$ absorption band of Cr^{3+} in Al_2O_3. However, the intensity distributions in the theoretical and experimental sidebands are quite different. This suggests that the experimental line-shape for this absorption band in ruby is not typical of linear coupling. Indeed, in this linear coupling regime, the intensity in the zero-phonon line relative to that in the entire spectrum is $\exp(-S_0)$, from which one infers that $S_0 \approx 5$. Furthermore, the position of the band peak suggests that $S_0 \approx 6$ and the overall width of the spectrum corresponds to $S_0 \approx 7$–8. These observations suggest strong quadratic coupling. At room temperature broadening of zero-phonon lines and phonon-assisted structure usually results in both $^4A_2 \rightarrow {}^4T_2$ absorption and $^4T_2 \rightarrow {}^4A_2$ emission bands of $3d^3$ ions being almost devoid of structure, as is illustrated by the spectra for Cr^{3+} : LLGG in Fig. 9.2.

A luminescence bandshape similar to that in Fig. 9.6c would support broadband tuning of a solid state laser. In the Ti^{3+}-sapphire laser the absorption spectrum covers the blue-green region and the laser tunability covers *ca* 330 nm about the Stokes-shifted emission peak at 745–750 nm. Almost as impressive is the wavelength tunability of the Cr^{3+} : colquiriite lasers, which are pumped via overlapping $^4A_2 \rightarrow {}^4T_2$, $^4T_1$ visible absorption bands. The Cr^{3+}-dopant in LiCAF has an average cross section for pumping in the $^4A_2 \rightarrow {}^4T_2$, $^4T_1$ absorption bands of *ca* 2.5×10^{-20} cm^2. The emission band peaks at 770 nm and has a FWHM of *ca* 120 nm. For Cr^{3+} : LiSAF and LiSGaF the peak shifts to 850 nm and the bandwidth increases to *ca* 240 nm due to the stronger electron–phonon interaction in these two compounds relative to LiCAF. In the configurational coordinate model, the normalized shape of the absorption band as a function of photon energy E at zero-temperature is given by Eq. 5.145 as

$$I_{ab}(E) = \sum_m \frac{\exp(-S)S^m}{m!} \delta(E_0 + m\hbar\omega - E), \tag{9.7}$$

where E_0 is the energy of the zero-phonon transition and $p = m$ for absorption transitions that sample the configurational coordinate space of the excited electronic state. Since there is a spectrum of phonons rather than a single breathing-mode the vibrational structure is broadened. The higher order sidebands in Fig. 9.6b merge into a smooth band on the high-energy side of this band, in accord with the observed $^4A_2 \rightarrow {}^4T_2$ bandshape of Cr^{3+} in ruby shown in Fig. 9.6c. The same concepts apply to the emission, the normalized shape of which from Eq. (5.145) is given at $T = 0$ by

$$I_{em}(E) = \sum_n \frac{\exp(-S)S^n}{n!} \delta(E_0 - n\hbar\omega - E), \tag{9.8}$$

where $p = n$ is the number of vibrational quanta excited as the emitted photon probes the relaxed electronic ground state. Assuming that the ground-state and excited-state parabolae are identical, the absorption (Eq. (9.7)) and emission (Eq. (9.8)) bandshapes are mirror images of one another and their peaks are shifted by $(2S - 1)\hbar\omega$. Alternatively the Stokes shift, defined as the difference between the first moments of the absorption and emission spectra, is given by $2S\hbar\omega$, according to equation (5.95). On this model the half-width of the band at temperature T is given by

$$\Gamma(T)^2 = \Gamma(0)^2 \coth\left(\frac{\hbar\omega}{kT}\right), \tag{9.9}$$

where the zero-temperature half-width is $\Gamma(0) \cong 2.36\,\hbar\omega\sqrt{S}$. In the absence of other factors the tuning curve of a laser based on the emission spectrum from a vibronic laser gain medium should follow the luminescence profile. The luminescence spectrum of the Cr^{3+} : LLGG laser reproduced in Fig. 9.2 is very much broader than the laser output tuning curve. Such narrowing of the laser tuning curve can result from disorder-induced inhomogeneous broadening and the concomitant nonradiative decay in weaker crystal-field sites and/or competitive absorption from excited state processes [Grinberg *et al.* (1997)].

The laser beam does not contain all wavelengths implied by $g(\nu)$. As discussed in §1.5.2, the laser has a highly monochromatic output imposed by the mode structure of the laser cavity, amplification occuring only for those modes that satisfy $G = G_M G_C > 1$, where $G_M > 1$ is the gain associated with the medium and G_C is the loss (gain) due to the cavity. For other frequencies G_C is small because of destructive interference: when Eq. (1.24) is satisfied G_C is given by $\exp(-\Delta)$. Although the mode number is typically large, $\sim 10^6$, only a small number of modes fall within the luminescence bandwidth. Below threshold all modes within the emission bandwidth are occupied. As the population increases with increasing pump power one mode crosses threshold before any other. This mode is amplified and the output beam is intense, monochromatic and highly directional. By introducing a suitable dispersing element into the cavity then the amplified mode may be tuned to any wavelength within the shape function of the transition.

9.3 Other aspects of transition-metal ion spectroscopy

Two aspects of transition-metal ion spectra distinguish them from rare-earth ion spectra and follow from the sensitivity of the $3d^n$ wavefunctions to the strength and symmetry of the crystal field.

9.3.1 *Mixed vibronic states and avoided level crossings*

The luminescence line shapes of Cr^{3+}-doped crystals, §5.4, are determined by the energy gaps separating the low-lying excited states ($^2E, {}^4T_2$) from the ground state ($^4A_2$). Cr^{3+}-doped GSGG and Cr^{3+} : K_2NaGaF_6, high-field in absorption and low-field in emission, can be converted to high-field in emission by an applied hydrostatic pressure (§5.4.1). In general, however, the Cr^{3+} : elpasolites are weak field emitters whereas the Cr^{3+} : oxide garnets span the range from low field (LLGG) to high field (YAG) emitters. The Tanabe–Sugano diagrams of the $3d^q$ ions in Fig. 4.8 show that level crossings are also a general feature of the energy level structures of $3d^2$ ions in tetrahedral symmetry and $3d^8$ ions in octahedral symmetry. For example, the iso-electronic $3d^2$ ions: V^{3+}, Cr^{4+}, Mn^{5+} and Fe^{6+} change from low field emitters to high field emitters, with Cr^{4+} being intermediate between the two extremes. Furthermore, the $^1E \rightarrow {}^3T_{1a}$ and $^1A_1 \rightarrow {}^3T_{1b}$ level crossings of $3d^8$ ions determine the bandshapes of both green and red emissions of Ni^{2+} ions doped into oxide and fluoride hosts. The

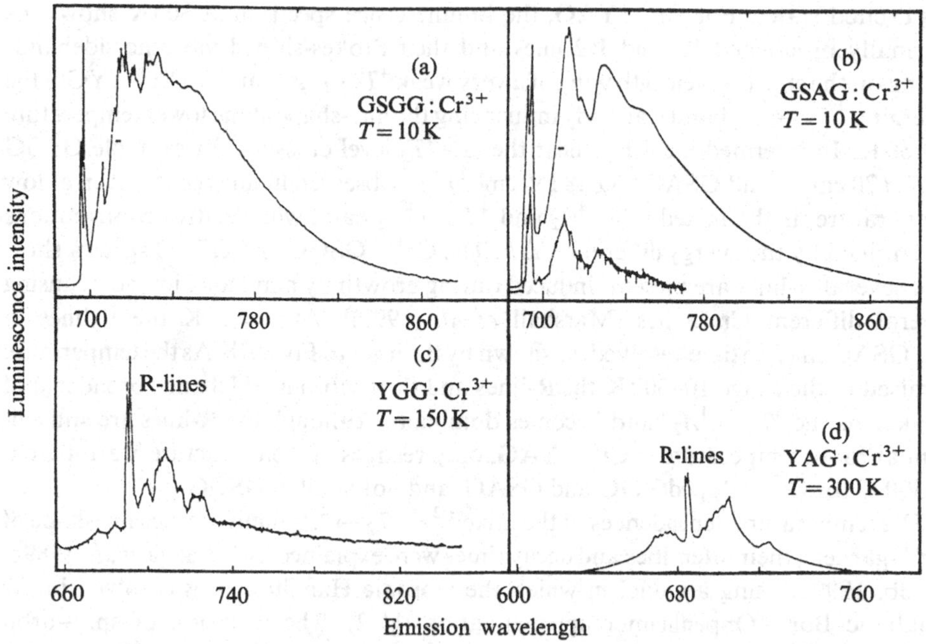

Figure 9.7. The luminescence spectra of some Cr^{3+}-doped garnets showing both the ${}^2E \rightarrow {}^4A_2$ R-lines and their vibronic sidebands atop the broad ${}^4A_2 \rightarrow {}^4T_2$ emission band [after O'Donnell *et al.* (1989)].

consequences of level crossings as determinants of the optical line shapes has only been explored in detail for the Cr^{3+} : garnets. In these systems, the common starting point has been the near degeneracy of ${}^2E$ and ${}^4T_2$ excited states of Cr^{3+} ions in GSGG, which are strongly admixed by spin–orbit and electron–phonon interactions acting in concert [§5.4.1, Donnelly *et al.* (1988), Henderson *et al.* (1988), Yamaga *et al.* (1989b, 1990b,c) and Grinberg (1993)].

The Y- and Gd-garnets permit wide variations of composition by cation substitutions on the A-, B- and C-sublattices, as discussed in §3.3.3 and §9.2.3 [Lutts *et al.* (1990), Zharikov *et al.* (1991)]. Such chemical flexibility can be used to compositionally-tune the value of Dq for Cr^{3+} ion substituents on the octahedral B-sites. Since the equilibrium lattice configuration, Q, of the $A_3B_xC_{5-x}O_{12}$ garnets varies linearly with the effective ionic radius, r_{eff} (Eq. 3.1), it is straightforward to show from Eq. (5.129) when $\Delta Q \ll Q_o$ that the energy separation $\Delta E = E({}^4T_2) - E({}^2E)$ varies linearly with $\Delta Q/Q_o$ around the ${}^2E$, ${}^4T_2$ level crossing. With ΔE positive and large ($\geq 1000\,\mathrm{cm}^{-1}$) luminescence from the lowest excited state, ${}^2E$, occurs via the R-lines and their vibronic sidebands. When $\Delta E < 0$ (GFG, LLGG) only the broad ${}^4T_2 \rightarrow {}^4A_2$ band is observed. A mixture of R-line and broadband is observed when $\Delta E \sim 0$, the observed line-shape being strongly temperature dependent. Figure 9.7 compares the luminescence spectra of Cr^{3+}-doped GSGG (a), GSAG (b), YGG (c) and YAG (d). In Cr^{3+} : YAG ($\Delta E \cong 1000\,\mathrm{cm}^{-1}$) and YGG, $\Delta E = 450\,\mathrm{cm}^{-1}$ [Henderson *et al.* (1988), Yamaga *et al.* (1990b)] the low temperature luminescence is almost exclusively the ${}^2E \rightarrow {}^4A_2$ R-line transitions. At 300 K both R-line and broadband are present due to thermal occupancy of ${}^2E$ and

$^4T_2$ excited states. For Cr^{3+} : YAG, the luminescence spectrum at 300 K shows the thermally broadened R1 and R2 lines and their Stokes-shifted vibronic sidebands sitting on the short wavelength wing of a very weak $^4T_2 \rightarrow {}^4A_2$ band. In Cr^{3+} : YGG the emerging $^4T_2 \rightarrow {}^4A_2$ band is already influencing the line-shape at the lower temperature of 150 K. In intermediate fields, near the ^2E–$^4T_2$ level crossing, for example GSGG ($\Delta E \cong 70\,\text{cm}^{-1}$) and GSAG ($\Delta E \cong 150\,\text{cm}^{-1}$), the observed luminescence, even at low temperature, is the mixed $^2E \rightarrow {}^4A_2$ and $^4T_2 \rightarrow {}^4A_2$ band with relative contributions determined by the energy difference, ΔE. The Cr^{3+} : GSGG and GSAG spectra show that several R-lines are present, induced during growth by non-stoichiometry causing several different Cr^{3+} sites [Marshall *et al.* (1990)]. At $T = 10$ K the R-lines in Cr : GSAG may be time-resolved, as shown by the inset to Fig. 9.7b. As the temperature is raised in the range 10–300 K the R-lines and their vibronic sidebands broaden and weaken as the $^4T_2 \rightarrow {}^4A_2$ band becomes dominant. Although the R-lines are still evident at room temperature in Cr^{3+} : YAG, only vestiges of their structure are apparent at 300 K for Cr^{3+} : doped YGG and GSAG, and not at all in GSGG.

The temperature dependences of the mixed $^2E, {}^4T_2 \rightarrow {}^4A_2$ luminescence lineshape of Cr^{3+}-garnets, their intensities and decay times were explained by Yamaga *et al.* (1989b, 1990b, 1990c) using a model in which the vibronic Hamiltonian was solved in the adiabatic Born–Oppenheimer approximation (§5.2). The inclusion of spin–orbit coupling as a perturbation on the electronic terms mixes the ^2E and $^4T_2$ vibronic wavefunctions. When the energy difference between the excited ^2E and $^4T_2$ states is comparable with the phonon energy tunnelling occurs between the two configurations. The model predicts the luminescence intensities, life times and bandshapes as a function of temperature and their variation from garnet-to-garnet [Yamaga *et al.* (1992)]. The single configurational coordinate model with harmonic coupling to $^4A_2$, ^2E and $^4T_2$ levels depicted in Fig. 5.6 shows that the excited vibronic states, ^2E and $^4T_2$, are separated by a potential barrier, ΔE, the height of which is determined by the strength of the octahedral crystal-field, Dq_o. When the vibronic ^2E and $^4T_2$ states are close to degenerate the adiabatic approximation breaks down and they are mixed by the nuclear kinetic energy and spin–orbit coupling operators. The mixed vibronic wavefunctions are given by

$$\Psi'_E(r,R) = \alpha\Psi_E(r,R) + \beta\Psi_T(r,R) \tag{9.10a}$$

$$\text{and} \quad \Psi'_T(r,R) = \beta\Psi_E(r,R) - \alpha\Psi_T(r,R), \tag{9.10b}$$

where $\Psi_E(r,R)$ and $\Psi_T(r,R)$ are pure vibronic wavefunctions of the ^2E and $^4T_2$ states, respectively, in the absence of tunnelling, α, β are admixture coefficients determined by the energy separation of ^2E and $^4T_2$ and δ is the tunnelling splitting (Eq. 5.136). Equation (9.10) shows that the R-line and its vibronic sideband from ^2E and the broadband emission from $^4T_2$ coexist at low temperature with an intensity ratio proportional to α^2/β^2. This is confirmed by calculations of the Cr^{3+} luminescence lineshapes in the vibronic tunnelling model but including a non-adiabatic term in the wavefunction mixing [Yamaga *et al.* (1992)].

The inverse life times of levels 1 and 2, corresponding to the orthogonal wavefunctions $\Psi'_E(r,R)$ and $\Psi'_T(r,Q)$, respectively, are given by

$$\frac{1}{\tau_1} = \frac{\alpha^2}{\tau_E} + \frac{\beta^2}{\tau_T} \quad \text{and} \quad \frac{1}{\tau_2} = \frac{\beta^2}{\tau_E} + \frac{\alpha^2}{\tau_T}, \tag{9.11}$$

Table 9.3. *Data used to fit Eqs.* 9.16 *and* 9.17 *to the experimental results in Fig.* 9.8. [*after Yamaga et al.* (1990*a,b*)]

Crystal	Lifetime of R-line		Lifetime of broadband		ΔE(o)		
	10 K	300 K	10 K	300 K	cm^{-1}	α^2	β^2
YAG	8.3 ms	1.3 ms		400 μs	1000	1	0
YGG	2.35 ms	—	—	240 μs	450	0.98	0.02
GGG	1.44 ms	—	—	160 μs	350	0.96	0.04
YSGG	1.26 ms	—	—	140 μs	300	0.94	0.06
GSAG	520 μs	—	520 μs	150 μs	150	0.91	0.09
GSGG	260 μs	—	240 μs	120 μs	93	0.82	0.18
LLGG	—	—	85 μs	68 μs	−900	0	1

where τ_E and τ_T are the life times of pure, vibronic ^2E and $^4T_2$ states in the absence of tunnelling. The ratio of the total intensity of the $^4T_2 \rightarrow {}^4A_2$ band to that of the R-line and its sideband is

$$\frac{I_T}{I_E} = \frac{g_T \tau_E}{g_E \tau_T} \cdot \frac{\beta^2 + \alpha^2 \exp(-\Delta E(T)/k_B T)}{\alpha^2 + \beta^2 \exp(-\Delta E(T)/k_B T)}, \tag{9.12}$$

where g_T and g_E are the degeneracies of $^4T_2$ and ^2E states, respectively, including spin. From Eq. (9.12) it follows that at low temperature $I_T/I_E \rightarrow g_T\tau_E/g_E\tau_T$, whereas at high temperature $I_T/I_E \rightarrow (g_T\tau_E/g_E\tau_T)(\beta^2/\alpha^2)$. A rate equation approach to the population changes that occur during emission gives the reciprocal of the lifetime, τ, of this three-level ($^4T_2$, ^2E and $^4A_2$) system, $1/\tau$, as

$$\frac{1}{\tau} = \frac{(1/\tau_1) + (1/\tau_2)\exp(-\Delta E(T)/k_B T)}{1 + \exp(-\Delta E(T)/k_B T)} \tag{9.13}$$

in the absence of nonradiative decay. Equations (9.11)–(9.13) can be modified to include the effects of disorder in garnets, nonradiative decay and oscillator strength enhancement by odd-mode lattice vibrations [Yamaga *et al.* (1990b)].

This model was tested by measurements of I_T/I_E and $1/\tau$ versus temperature for Cr^{3+}-doped GSGG, GSAG and YGG, and by comparison with data in the literature for Cr^{3+}-doped GGG, YSGG and LLGG, assuming three parameters $\tau_T = 60$ μs, $\tau_E =$ 4 ms and $2\delta = 85\ cm^{-1}$. Estimated values of $\Delta E(0)$, α^2 and β^2 presented in Table 9.3 along with the experimental life times at $T = 10$ K and 300 K for Cr^{3+} : YGG, GSAG and GSGG led to close fits between the theoretical I_T/I_E versus T curve and experimental points in Fig. 9.8. The good agreement strongly supports the model, even for the two YGG samples that differed in their compositions, $Y_{3.1}Ga_{4.9}O_{12}$ and $Y_{3.8}Ga_{4.2}O_{12}$, because of disorder introduced during growth by RF sputtering. The trends in the luminescence life times for Cr^{3+} : GSGG were also in accord with the model [Yamaga *et al.* (1990b)]. Evidently, when $\Delta E = (E(^2E) - E(^4T_2))$ is large and positive, e.g. Cr^{3+} : YAG, only the R-line and its vibronic sideband are observed at low temperature, and they should decay with lifetime $\tau_E \cong 4$ ms. In contrast, when ΔE is large and negative (e.g. Cr^{3+} : LLGG) the $^4T_2 \rightarrow {}^4A_2$ luminescence band decays with lifetime $\tau_T \cong 100$ μs. At the level crossing, where $\Delta E \cong 0$ (e.g., Cr^{3+} : GSGG and GSAG), the emission

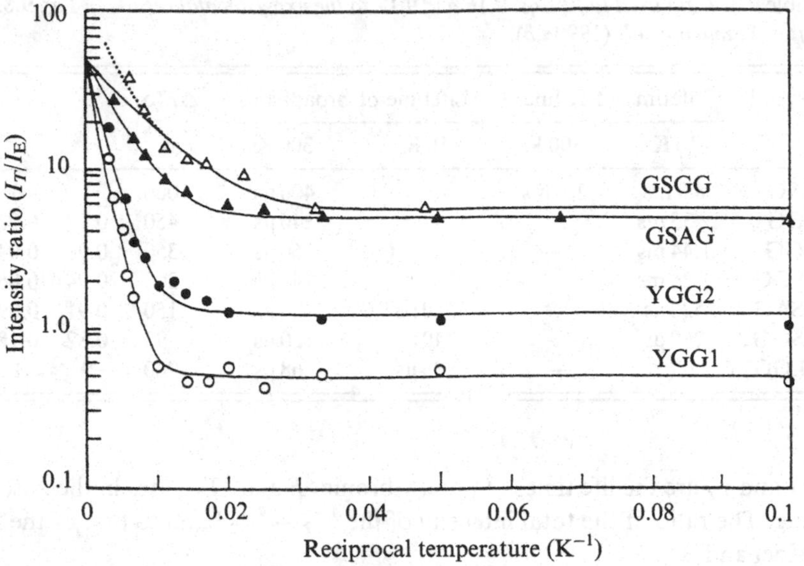

Figure 9.8. The temperature dependences of the intensity ratio I_T/I_E for Cr^{3+}-doped $Y_{3.1}Ga_{4.9}O_{12}$ (YGG1), $Y_{3.8}Ga_{4.2}O_{12}$ (YGG2), $Gd_3Sc_{2.2}Al_{2.8}O_{12}$ (GSAG) and $Gd_3Sc_{1.9}Ga_{3.1}O_{12}$ (GSGG) in the temperature range 10–200 K [after Yamaga *et al.* (1989a)].

lineshape is an intimate mélange of R-lines and broadband for which a single lifetime $\tau = \frac{1}{2}(\tau_E + \tau_T)$ is observed. At higher temperatures, as electrons are excited into the $\Psi'_T(r, Q)$ state by phonon-assisted tunnelling, the lifetime, τ, is dependent on α^2, β^2, ΔE, and the life times τ_E and τ_T for the pure vibronic states $\Psi_E(r, Q)$ and $\Psi_T(r, Q)$. For YAG and LLGG where $\Delta E = 1000\,\text{cm}^{-1}$ and $-900\,\text{cm}^{-1}$, respectively, the observed life times are close to the τ_E and τ_T from the pure vibronic wavefunctions $\Psi_E(r, Q)$ and $\Psi_T(r, Q)$, and measured to be $\tau_E = 8.3$ ms for the R-line in YAG and $\tau_T = 85$ µs for the broadband in LLGG. For other garnets luminescence life times involve the mixed vibronic levels $\Psi'_E(r, Q)$ and $\Psi'_T(r, Q)$ to a greater or lesser extent depending on ΔE. The values of ΔE in Table 9.3 are in reasonable accord with expectation, given the increasing value of the lattice constant, *a*, across the series from YAG to LLGG. The variation of the mixing coefficients α^2 and β^2 in Table 9.3 are physically reasonable, given the trends in ΔE. The data fit for GSGG is sensitive to assumptions made about the several Cr^{3+} multi-sites which have different values of ΔE ($\cong 38\,\text{cm}^{-1}$, $56\,\text{cm}^{-1}$ and $62\,\text{cm}^{-1}$) [Monteil *et al.* (1990)], and hence of α^2 and β^2. Furthermore, the theoretical line-shapes for Cr^{3+} : YGG, GSAG and GSGG agree well with the experimental spectra, as does the temperature dependence of the observed line-shape of Cr^{3+} : GSAG [Yamaga *et al.* (1992)].

A single configurational coordinate model was applied to the ^2E and $^4T_2$ states of Cr^{3+} by Englman and Barnett (1970) and by Struck and Fonger (1975). The former authors distinguish between 'horizontal' and 'vertical' tunnelling. Horizontal tunnelling occurs between the $\Psi_E(r, Q)$ and $\Psi_T(r, Q)$ potential wells separated by a potential barrier and assisted by zero-point vibrations and optical phonons, whereas vertical tunnelling is associated with the mixing of $^4T_2$ and ^2E states at the same configurational coordinate, corresponding to the Frank-Condon principle. The Englman and Barnett

(1970) treatment of vertical tunnelling leads to the same result as Eq. (9.12) with similar consequences for the optical line-shape.

9.3.2 *Dominant symmetry and low symmetry distortions*

The utility of dividing the crystalline potential into dominant and lower symmetry parts was emphasized in the discussion of crystal-field theory (§4.3.4 and §4.4.7). This chapter, and Chapter 10 also, consider several specific examples to show how spectroscopic properties of optical centres respond to the different, low symmetry distortions of the crystal-field. Laser-active colour centres are a case in point. F centres are optically isotropic, with strong optical absorption bands derived from allowed electric dipole transitions ($^2A_{1g} \rightarrow {}^2T_{1u}$) (§4.6.1). A reduction in symmetry from $O_h \rightarrow C_{4v}$ appropriate to F_A centres results in two polarized absorption bands ($^2A_1 \rightarrow {}^2A_2, {}^2E$), split by about 0.2 eV (§4.6.3). Lattice relaxation in the excited state can result in unpolarized luminescence of F_A (I) centres or strongly polarized luminescence from F_A (II) centres. That the polarization axis of F_A (II) centres is a $\langle 110 \rangle$-crystal direction arises out of the potential double-well from which these centres emit. The dominant symmetry is O_h and lower symmetry distortions reduce the symmetry to C_{4v} for F_A (I) and F_A (II) centres in absorption and D_{2h} for F_A (II) centres in emission. The energy levels of F_2^+-like centres transform as the irreducible representations of the D_{2h} point group, their pumping and laser transitions, $^2A_{1g} \leftrightarrow {}^2B_{1u}$, being Stokes shifted by about 0.9 eV (§9.2.1).

Transition-metal ions such as Ti^{3+} and Cr^{3+} occur in distorted octahedral symmetry sites in many potential laser gain media. Optical absorption and luminescence spectra in the $^2T \leftrightarrow {}^2E$ transition of $Ti^{3+}:Al_2O_3$ shown in Fig. 9.9 are strongly polarized. According to the energy level diagram of Ti^{3+} in D_{3d} sites, Fig. 5.9, there are two overlapping absorption bands centred on $10Dq = 19\,300\ cm^{-1}$ and separated by $2200\ cm^{-1}$. A single luminescence band in Fig. 9.9b peaks at 945–950 nm (10 580–10 525 cm^{-1}). Absorption measurements at $T = 4.2$ K reveal only one of the three zero-phonon lines implied in Fig. 5.9 as a consequence of the depopulation of the upper levels of the $^2T_2$ ground state. All three zero-phonon lines are observed in luminescence at wavelengths of 616.1 nm (A), 617.7 nm (B) and 620.2 nm (C).

In the MO-description of $Ti^{3+}:Al_2O_3$ delocalized LCAO wavefunctions are constructed from $3d$ orbitals on the central Ti^{3+} ion and s-, p_σ- and p_π-orbitals on nearest ligand ions, Eq. (8.70). In Fig. 8.7 the ξ-orbital involves the $3d$-$|yz\rangle$ orbital mixed with p_π-ligand orbitals of ions 2, 3, 5 and 6 (Fig. 8.1) by the $T_{1u}(\pi)$ and $T_{2u}(\pi)$ odd-parity distortions (Fig. 3.2). This octahedral basis set is easily transformed to an axial basis appropriate to the trigonal symmetry of Al_2O_3 or YAG or tetragonal symmetry of YAP. The matrix elements of the electric dipole moment operator $D = -e\mathbf{r} \cdot \mathbf{E}$ were calculated using the electronic MO-wavefunctions of ground and lowest excited states: the admixed MO-wavefunctions and the induced transition probabilities were separately evaluated for the X-, Y-; and Z-; components of the $T_{2u}(\pi)$, $T_{1u}(\pi)$ and $T_{1u}(\sigma)$ distortions. The relative transition probabilities induced by a $T_{2u}(\pi)$ distortion shown in Fig. 8.8 are compared with experiment below. Similar diagrams of intensities induced by $T_{1u}(\pi)$ and $T_{1u}(\sigma)$ displacements are given in Yamaga *et al.* (1991a).

In principle, such relative intensities apply only to the zero-phonon lines A, B and C. However, the relative magnitudes of the transition probabilities calculated with

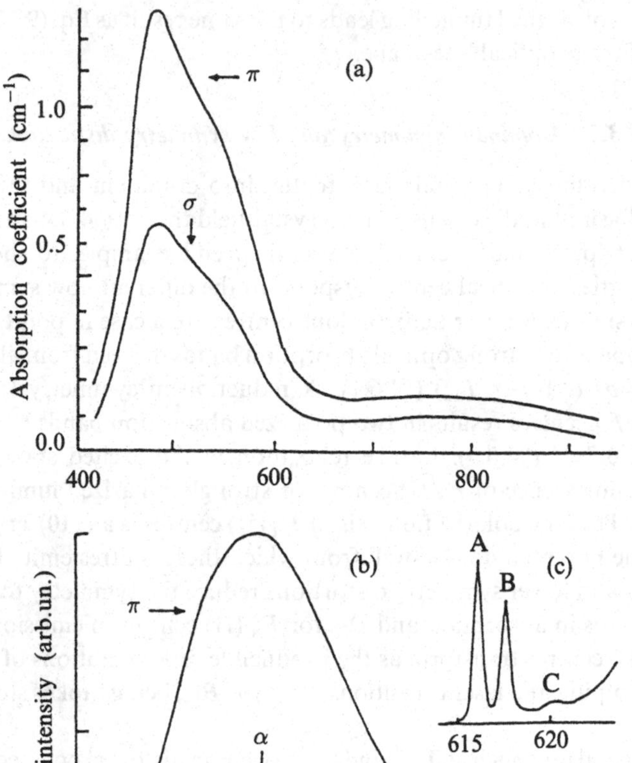

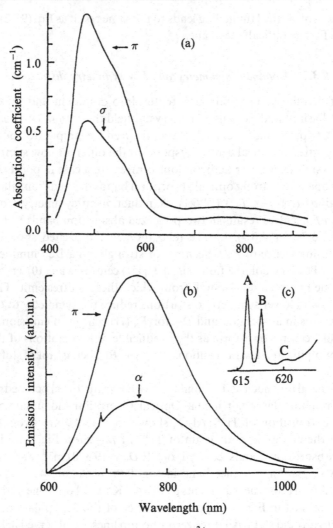

Figure 9.9. Polarized optical spectra of $Ti^{3+}:Al_2O_3$ (a) optical absorption, (b) luminescence and (c) zero-phonon lines at $T = 4$ K. In π-polarization $E \parallel c$ and in α-polarization $E \perp c$ [after Henderson *et al.* (1990a)].

vibronic wavefunctions are exactly the same as those in Fig. 8.8 except for a nuclear overlap reduction factor. The zero-phonon lines A, B and C have integrated intensities in the ratio $I_A : I_B : I_C = 1 : 0.8 : 0.1$ and polarized intensities $I_\pi / I_\sigma = 2.1$. For the broadband $I_\pi / I_\sigma \cong 2.2 : 1$. These results are consistent with the zero-phonon lines being induced by the static odd-parity distortion with the broadband receiving intensity induced by both static and dynamic odd-parity distortions. The polarization intensities induced by the T_{2u}-distortions for Ti^{3+} ions in Al_2O_3, shown in Fig. 8.8, indicate that zero-phonon line C is forbidden when the enabling interaction is the Z-component of the $T_{2u}(\pi)$ distortion, whereas it is of similar intensity to lines A and B when promoted by both $T^x_{2u}(\pi)$ and $T^y_{2u}(\pi)$ modes. Furthermore, lines A, B and C have finite probabilities when induced by $T_{1u}(\pi)$ and $T_{1u}(\sigma)$ distortions. Since the experimentally observed lines A and B are very intense and line C is very weak it is evident that the Z-

component of $T_{2u}(\pi)$ is dominant in exciting radiative decay. The calculated relative intensities of the A, B and C lines, Fig. 8.8, show that $I_A:I_B:I_C=1:1:0$ only for the $T^z_{1u}(\pi)$ displacement, close to the experimental result $I_A:I_B:I_C=1:0.8:0.1$. The small difference between theory and experiment can be explained by including second-order spin–orbit coupling. The polarization characteristics derived from the $T_{2u}(\pi)$ distortion, Fig. 8.8, show that I_π/I_σ, where π- and σ-components correspond to Z and Y (or X)-polarized light, is 2 : 1, which agrees well with the observations on both the zero-phonon lines and broadband. In consequence, the $T^z_{2u}(\pi)$ distortion is recognized as the primary agent in exciting radiative decay via both zero-phonon lines and broadband.

The molecular orbital model of the $(AlO_6)^{9-}$ complex was also used to derive the relative strengths of radiative transitions of Ti^{3+} : YAP (point group D_{2h}), in which Ti^{3+} substitutes for Al^{3+} on sites with C_i symmetry. In this material optical absorption (400–550 nm) and emission (540–800 nm) bands occur at shorter wavelengths than in $Ti^{3+}:Al_2O_3$, consequent on the reduced Ti^{3+}-ligand ion separation in YAP relative to Ti^{3+}-sapphire and are strongly polarized in the plane perpendicular to the distortion axis [Yamaga *et al.* (1991a)]. The tetragonal component of the crystal-field splits the $^2T_{2g}$ ground state of Ti^{3+} into $^2B_{2g}$ ($|0,\pm1/2\rangle$) and ^2E_g states separated by 90 cm^{-1}, measured by the energy separation of the 537.1 nm (weak) and 539.7 nm (intense) zero-phonon lines. The ^2E_g state is further split into $|\pm1,\pm1/2\rangle$ and $|\pm1,\pm1/2\rangle$ states by spin–orbit interaction, which also splits the excited ^2E_g state into $|\varepsilon\rangle$ and $|\theta\rangle$ levels. In the molecular orbital model of the luminescence, the zero-phonon line $|\varepsilon,\pm1/2\rangle \rightarrow |0,\pm1/2\rangle$ is allowed but the $|\varepsilon\rangle \rightarrow |\pm1,\pm1/2\rangle$ and $|\pm1,\pm1/2\rangle$ lines are forbidden by $T^{x,y}_{1u}(\pi)$ and $T^{x,y}_{2u}(\pi)$ odd-parity distortions. In contrast, $T^z_{1u}(\sigma)$, $T^z_{1u}(\pi)$ and $T^z_{2u}(\pi)$ distortions enable $|\theta,\pm1/2\rangle$, $|\varepsilon,\pm1/2\rangle \rightarrow |\pm1,\pm1/2\rangle$ and $|\pm1,\pm1/2\rangle$ by the electric dipole mechanism but forbid $|\theta,\pm1/2\rangle, |\varepsilon,\pm1/2\rangle \rightarrow |0,\pm1/2\rangle$ transitions. The observed polarizations of the zero-phonon lines are consistent with $|\varepsilon\rangle$ being the lowest lying excited state and the $|\varepsilon\rangle \rightarrow |0,\pm1/2\rangle$ and $|\pm1,\pm1/2\rangle$ transitions account for the 539.7 nm and 537.1 nm zero-phonon lines, respectively, being allowed by the X- and Y-components of $T_{1u}(\pi)$ or $T_{2u}(\pi)$ distortion. Only very inefficient laser action has been reported, because ESA of the pump radiation by Ti^{3+}–Ti^{4+} complexes reduces the pump efficiency and increases threshold power considerably [Wegner and Petermann (1989)]. Basun *et al.* (1996) suggest that stimulated emission from the ^2E state of Ti^{3+} is also limited by photo ionization from this state.

In the elpasolites Cr^{3+}-dopants occupy rigorously octahedral sites; their unpolarized $^4T_2 \rightarrow {}^4A_2$ bands are induced by coupling to odd-parity phonons [§5.4.1]. The B-sites in the garnets have trigonally-distorted octahedral symmetry at which the inversion centre is preserved despite the T_{2g} distortion. This even-parity distortion splits the optical spectra of Cr^{3+} impurities, without enhancing their intensities. Odd-parity vibrations raise the Laporte selection rule against the vibronic $^4A_2 \rightarrow {}^4T_2$, $^4T_1$ transitions. In colquiriite crystals there are even- and odd-parity distortions of MF_6-octahedra, both of which play an important role in the crystal-field spectra of laser active Cr^{3+} and Ce^{3+}, (see §10.3, §10.4). The polarized absorption/luminescence spectra of Cr^{3+}-doped ruby, alexandrite and emerald all reflect the particular combination of odd/even distortions of the $(CrO_6)^{9-}$ octahedra in such materials.

The isoelectronic $3d^2$ ions (V^{3+}, Cr^{4+}, Mn^{5+} and Fe^{6+}) substitute at four-fold coordinated sites in such molecular anions as $(PO_4)^{3-}$ $(SiO_4)^{4-}$ and $(VO_4)^{5-}$. In the fluoroapatites and fluorovanadates of Sr^{2+} and Ca^{2+}, e.g. $Ca_5(PO_4)_3F$, an even-parity extension along the triad axis reduces the symmetry at the P^{5+} site from T_d to C_{3v}. A further displacement of the P^{5+} ion to an off-centre position in the $(PO_4)^{5-}$ octahedron further lowers the point symmetry to C_s. This is also the point group at the Si^{4+} site in forsterite. These symmetry reductions from tetrahedral, T_d, vary in detail from one crystal to another. In the oxide garnets a tetragonal expansion along $\langle 100 \rangle$ axes reduces the C-site occupied by Cr^{4+} ions from T_d symmetry to D_{2d}. These minor symmetry-lowering distortions introduce further splittings of the $3d^2$-ion spectra and modify the optical selection rules (§5.4.2). The Tanabe–Sugano diagram, Fig. 4.8, identifies $^3A_2$, ^1E, $^3T_1$ and $^1A_2$ states of $3d^2$ in T_d symmetry, in order of increasing energy. The numbers and symmetries of $3d^2$ levels in distorted crystals are deduced by comparing the characters of the symmetry elements in the group character tables of the dominant and reduced symmetries.

In Cr-doped forsterite the Cr-impurity takes on several charge states and occupies a number of different symmetry sites. A similar problem arose in Cr : Ca : YAG, leading to problems in establishing the energy level structure of Cr^{4+} dopants in this host [Eilers *et al.* (1993, 1994)]. In §5.4.2 it was established that in tetrahedral (T_d) symmetry $^3A_2 \rightarrow {}^3T_1$ transitions are allowed in the electric dipole approximation, with $^3A_2 \rightarrow {}^3T_2$ being much weaker magnetic dipole transitions. In D_{2d} symmetry $^3B_1\,(^3A_2) \rightarrow {}^3A_2(^3T_1)$ is an allowed electric dipole transition, whereas $^3B_1\,(^3A_2) \rightarrow {}^3E(^3T_1, {}^3T_2)$ are partially-allowed electric dipole transitions. Finally, $^3B_1\,(^3A_2) \rightarrow {}^3B_2\,(^3T_2)$ is a magnetic dipole transition, as shown in Fig. 5.7. This diagram also shows the fine-structure levels that result from spin–orbit coupling. The only indicated transitions then involve the lowest energy levels resulting from the ground state ($^3B_1\,(^3T_2) = E + B_2$) and lowest excited state ($^3B_2\,(^3T_2) = B_1 + E$) in D_{2d} symmetry. The selection rules for polarized absorption and emission transitions involving all spin-orbitals are given in Table 9.4. Although the number of levels, their symmetry characteristics and the polarization selection rules can be determined from group theory, the splittings by low symmetry distortions are not determined. Hence the ordering of levels cannot be decided, except with guidance from experiment and/or calculation.

Table 9.4. *Selection rules for transitions of $3d^2$ ions in D_{2d} symmetry including the effects of spin–orbit coupling*

Final state	Initial state				
	A_1	A_2	B_1	B_2	E
A_1	—	MD(π)	—	ED(π)	ED(σ) MD(σ)
A_2	MD(π)	—	ED(π)	—	ED(σ) MD(σ)
B_1	—	ED(π)	—	MD(π)	ED(σ) MD(σ)
B_2	ED(π)	—	MD(π)	—	ED(σ) MD(σ)
E	ED(σ) MD(σ)	ED(σ) MD(σ)	ED(σ) MD(σ)	ED(σ) MD(σ)	ED(π) MD(π)

The energy level structures of the $3d^2$ ion in low symmetry sites can be calculated using the intermediate crystal-field approach (§4.4.3, 5.4.2) or ligand field theory (§8.1). In the former the crystal-field energies are expressed in Table 5.3. Once the various transitions have been identified in terms of the wavefunctions in Table 5.3 then values of Dq, C_2 and C_4 can be evaluated. The angular overlap model (AOM) (§8.4.4) has been used to analyse the absorption/luminescence spectra of Cr^{4+} : Mg_2SiO_4 and Ca_2GeO_4 [Hazenkamp *et al.* (1996)] and Cr^{4+} : YAG and YGG [Riley *et al.* (1999)]. In their ligand field calculation for Cr^{4+} : YAG, Riley *et al.* (1999) allowed the ligand field (e_σ, α), spin–orbit coupling and inter-electron repulsion parameters B and C to vary using a diagonalization of the full d^2 basis set on each iteration of a nonlinear least squares algorithm to fit the observed spectral features, with several notable results. The splitting of the $^3T_1$ state by the D_{2d} distortion ($\sim 6500\ \mathrm{cm}^{-1}$) was much larger than the $^3T_2$ splitting ($\sim 2500\ \mathrm{cm}^{-1}$). The splitting of the ^1E state was also significant (*ca* 1500 cm^{-1}). From the measured splittings of the $^3T_1$ and $^3T_2$ states, and using Table 5.3, the tetragonal crystal-field parameters in the intermediate crystal-field approximation, Eqs. (5.138), are estimated to be $C_2 = 460\ \mathrm{cm}^{-1}$ and $C_4 = 10\ \mathrm{cm}^{-1}$. The ordering of Cr^{4+} energy levels in YAG relative to the $^3B_1\,(^3A_2)$ ground state was determined using ligand field theory to be $^3B_2\,(^3T_2)$, $^3A_2\,(^3T_1)$, $^1A_1\,(^1E)$, $^3E\,(^3T_2)$, $^1B_1\,(^1E)$ and $^3E\,(^3T_1)$ assuming that $e_\sigma = 15\,586\ \mathrm{cm}^{-1}$, $e_\sigma/e_\pi = 3$, $B = 430\ \mathrm{cm}^{-1}$, $C = 3585\ \mathrm{cm}^{-1}$, $\zeta = 220\ \mathrm{cm}^{-1}$ and $\alpha = 4.66°$ in accord with detailed spectroscopic results for Cr^{4+} : YAG and YGG.

9.4 Laser efficiency and threshold

9.4.1 Strength of optical transitions

The starting points for this part of the discussion are the stimulated absorption and stimulated emission processes of the solid state gain medium. For amplification of the beam there must be population inversion, i.e. $N_b > (g_b/g_a)N_a$. The intensity of the beam is then given by

$$I(L) = I_0 \exp(\gamma(\nu)L - \Delta), \tag{9.14}$$

where Δ refers to losses by such other processes as absorption at the laser wavelength, spurious reflections, scattering and diffraction by cavity components and transmission by the output coupler. The small gain coefficient, $\gamma(\nu)$, is determined solely by the stimulated processes of absorption and emission, and is given by Eq. (1.22) as

$$\gamma(\nu) = \frac{\lambda^2 \Delta N}{8\pi n^2 \tau_R} g(\nu), \tag{9.15}$$

in which $g(\nu)$ is the shape function defined so that $\int g(\nu)\,d\nu = 1$, and the inverse lifetime, τ_R^{-1}, is equal to the Einstein coefficient against spontaneous decay, A_{ba} (§5.1.5). For an electric dipole process, and ignoring the local field correction factor in Eq. (5.65), A_{ba} takes the form

$$A_{ba} = \frac{1}{4\pi\varepsilon_0} \cdot \frac{4n\omega^3}{3\hbar c^3} \cdot \frac{1}{g_b} \sum |\langle a|\mu_D|b\rangle|^2, \tag{9.16}$$

where μ_D = the electric dipole operator $\sum_i er_i$ is summed over i electrons. Substituting from Eq. (9.16) into (9.15) gives

$$\gamma(\nu) = \frac{\pi}{3\hbar\varepsilon_0\lambda} \cdot \frac{1}{g_b}\left(\sum |\langle a|\mu_D|b\rangle|^2\right) g(\nu), \tag{9.17}$$

where the summation is over the individual states in levels a and b. For d–d transitions this matrix element is only of finite magnitude if the transition is induced by an odd-parity component of the crystal field. The gain coefficient can then be written to within a constant as

$$\gamma(\nu) \propto \frac{\pi}{3\hbar\varepsilon_0\lambda} \cdot \frac{1}{g_b}\left(\sum |\langle a|V_{\text{odd}}\mu_D^\alpha|b\rangle|^2\right) g(\nu), \tag{9.18}$$

where V_{odd} is the odd-parity field and μ_D^α is the appropriate form of the dipole moment for light of polarization, α. Applications of the Wigner–Eckart theorem to the evaluation of these matrix elements are given in standard texts [Griffith (1961), Sugano *et al.* (1970), Henderson and Imbusch (1989)]. The specific form of the opposite parity states admixed into the states involved in the $a \rightarrow b$ transition were assumed to be higher lying orbitals on the central ion as discussed in §5.3.3 and §5.3.4. However, as discussed by Yamaga *et al.* (1990a, 1991a,b) and in §8.6 other admixtures are at least as probable.

The strengths and polarizations of d–d and f–f transitions cannot be calculated within the point-ion model. Since the d- and f-orbitals in octahedral symmetry have even parity then d–d transitions and f–f transitions of the $3d^n$ and $4f^n$ configurations, respectively, are parity forbidden. Mechanisms must be introduced that break these parity selection rules involving the static and/or dynamic interactions of these configurations with their ligand neighbours. If these interactions result from odd-parity displacements of ligand ions then the parity selection rules are raised through the mixing of opposite-parity wavefunctions into the ground and excited states. For example, the d–d transitions of the $3d^n$-configuration ions may have admixtures of odd-parity wavefunctions associated with higher lying $4p$ orbitals [Sugano *et al.* (1970)], nearby $3s$ and $3p$ orbitals on the neighbouring ligand ions [Yamaga *et al.* (1989a), (1990a), (1991a)] or fully-occupied $3p$ orbitals on the central $3d$-ion. However, the consequences of each type of admixture are analogous, in that the matrix elements in Eqs. (9.16)–(9.18) are then finite. The effects of static and dynamic distortions can be identified spectroscopically. Intensity enhanced by odd-parity crystal-fields are evident in f–f transitions of $4f^n$ ions [Judd (1962), Ofelt (1962)]. Static perturbations of odd-parity determine the low temperature transition rate (and hence radiative lifetime). Odd-parity lattice vibrations introduce a temperature dependent enhancement of the transition rate (§6.4). A detailed analysis of the strengths of optical transitions induced by T_{1u} odd-parity displacements of the octahedrally-disposed ligands only in Cr^{3+} : Al_2O_3, was sufficient to explain the observed absorption and emission spectra [Tanabe and Sugano (1958)]. Odd-parity distortions of T_{2u} symmetry were needed to account for excited state absorptions from the ^2E and $^4T_2$ levels [Shinada *et al.* (1966), Klauminzer *et al.* (1966)].

9.4.2 *Quenching of luminescence and laser efficiency*

Various mechanisms within the gain medium subtract from laser efficiency and increase the threshold power, including absorption bands that compete with the laser transitions at pump and emission wavelengths and phonon-assisted nonradiative transitions that result in crystal heating. ESA can result in the absorption of pump and/or luminescence photons, and having done so cause the centre either to emit from an even higher excited state (upconverted luminescence) or to de-excite nonradiatively to the ground state. These processes reduce the luminescence efficiency, Eq. (7.27), and the slope efficiency of the laser also, Eq. (7.28). The luminescence efficiency, $\eta(T)$, is also reduced by thermal and concentration quenching, which have the effect of reducing the luminescence lifetime with increasing concentration of active ions. In consequence, there is an optimum concentration of active ions below which quenching of the lifetime can be compensated for by the increased absorption and emission intensities that accompany higher concentrations. For example, alexandrite laser rods usually contain up to 0.23 at.% Cr^{3+}, at which concentration there is no concentration quenching. In such materials the quantum yield at 300 K is $95 \pm 5\%$ and the observed fluorescence decaytime is essentially the radiative lifetime [Shand (1983)]. In consequence, the quantum efficiency can be as high as $\nu_p/\nu_1 \cong 70\%$ for Kr^+ pumping at 647 nm. This behaviour contrasts strongly with that of the $Co^{2+}:MgF_2$ laser, in which nonradiative transitions reduce the quantum yield and so degrade the laser performance that practical use is prevented above *ca* 80 K in CW operation and 225 K in pulsed mode. As noted in §7.3 for Pr^{3+} in $La_xPr_{1-x}Cl_3$, efficient laser performance can be obtained across the entire composition range $0 < x < 1$ as it can from Cr^{3+} in $LiSrAl_xCr_{1-x}F_6$, in which cases luminescence quenching is very weak, even at 300 K, and laser efficiency is hardly affected. In contrast, Nd^{3+} luminescence from $NdCl_3$ is undetectably weak [Pelletier-Allard and Pelletier (1987)].

Thermal and concentration quenching of luminescence both reduce laser efficiency and increase the pump power required to cross the laser threshold. Since these processes reduce the excited state lifetime they compete with radiative decay in returning a population to the terminal state. Since the pre-exponent τ_{NR}^{-1} is larger than the radiative decay rate τ_R^{-1} in Eq. (6.38), nonradiative processes can be dominant when the gap energy is relatively small. In oxides the mean phonon energy is large (300–500 cm^{-1}), the energy gap over which radiative decay predominates is reduced, relative to halide crystals. In $Co^{2+}:MgO$ phonon-assisted nonradiative decay at 300 K is so efficient that luminescence in the infrared is too weak to detect. In contrast, the $Co^{2+}:MgF_2$ laser is efficient enough at cryogenic temperatures to provide the basis of a commercial laser system. Ni^{2+} ions are the active component in the $Ni^{2+}:MgO$ and $Ni^{2+}:MgF_2$ lasers, which also operate at infrared wavelengths. MgO is an almost ideal laser host, but one that is very difficult to grow. Nevertheless, in operation at $T = 100$ K, 10 watts of average power has been achieved in CW operation for $Ni^{2+}:MgO$ [Moulton (1985)]. In comparing the properties of $Cr^{3+}:LiCAF$ and $Cr^{3+}:LiSAF$, the latter is usually taken as the more desirable gain medium, because of the larger tuning range of $Cr^{3+}:LiSAF$ than $Cr^{3+}:LiCAF$. However, the excited state decay dynamics of Cr^{3+} ions are different in the two compounds. In $Cr^{3+}:LiCAF$ multiphonon nonradiative decay is unimportant up to 500 K, whereas such processes are fairly efficient close to room temperature in $Cr^{3+}:LiSAF$. In consequence, the optothermal and optomechanical properties of Cr^{3+} : LiSAF laser rods are inferior to those of LiCAF under conditions of high average laser power. Large single crystals of this compound, Cr^{3+} : $LiSr_{0.8}Ca_{0.2}AlF_6$, can be grown by the Czochralski technique. Some improvement in the optothermal and thermomechanical properties of LiSAF was anticipated by substituting some 20% of Sr^{2+} ions by Ca^{2+} [Chai *et al.* (1992)].

9.4.3 Excited state absorption

ESA spectra that overlap the absorption and emission transitions of a laser active ion will reduce laser efficiency, especially when the excited state is long-lived. The ${}^4T_2 \rightarrow {}^4T_1$ ESA spectrum of Cr^{3+} : $BeAl_2O_4$ illustrated in Fig. 9.10 shows that the ESA loss at the peak of the ${}^4T_2 \rightarrow {}^4A_2$ emission is minimal, becoming more important in reducing laser efficiency at both longer and shorter wavelengths than the peak [Shand and Walling (1982), Shand and Jenssen (1983)]. Andrews *et al.* (1986a) concluded that ESA limits the tuning range of the Cr^{3+} : GSGG laser at long wavelength to 800 nm, as Petermann (1990) subsequently confirmed. The studies of numerous tunable $3d^3$ ion lasers suggest

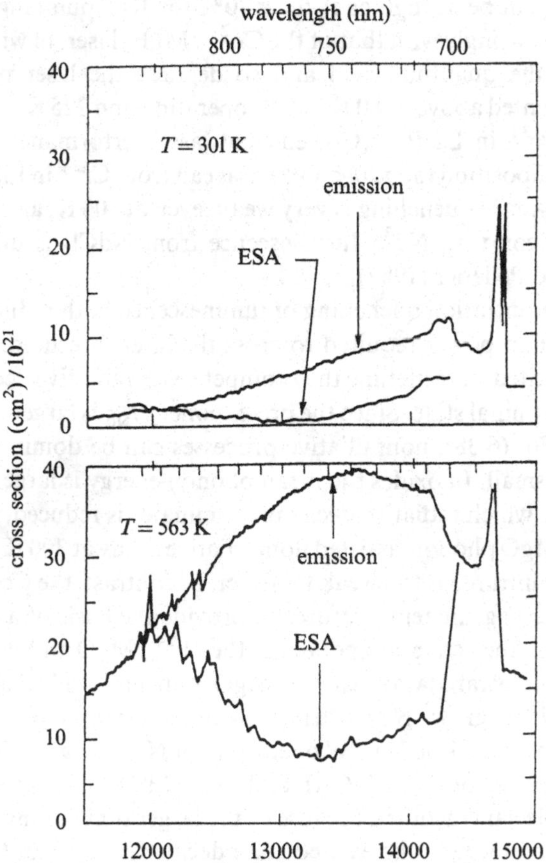

Figure 9.10. The luminescence and ESA cross sections of Cr^{3+} : $BeAl_2O_4$ measured at 301 K and 563 K [after Shand and Jenssen (1983)].

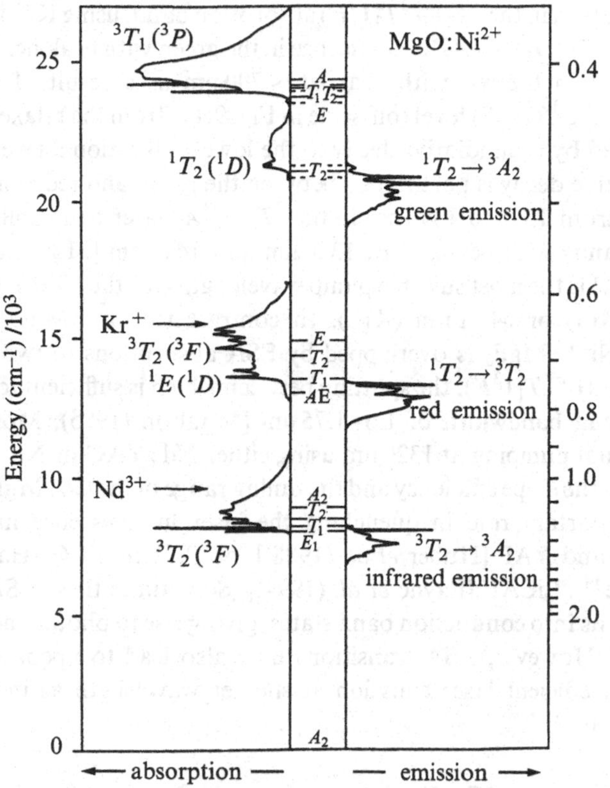

Figure 9.11. The low temperature absorption (LHS) and emission (RHS) spectra of Ni^{2+} in MgO, and the estimated positions of the zero-phonon spin–orbit components of the energy levels [after Moncorgé and Benyattou (1988)].

that ESA reduces the slope efficiency and increases the threshold power, but to different extents in all such lasers. To fully understand the optical bandshapes and polarization intensities requires that quite subtle effects of the electron–phonon interaction in the excited states must be considered [Caird *et al.* (1988), Lee *et al.* (1989)].

The state reached in an ESA transition may decay radiatively or nonradiatively to the ground or other lower lying state, or by some combination of such processes, and it is necessary to fully understand the competition between radiative and nonradiative processes in order to determine the optimal operating conditions of all solid state lasers. Spectroscopic studies of Co^{2+} and Ni^{2+}-doped MgF_2 and MgO show clearly the interplay of radiative and nonradiative transitions [Moncorgé and Benyattou (1988)]. For example, the low temperature (80 K) absorption and luminescence spectra of Ni^{2+} ions in MgO, Fig. 9.11, define several wavelength regimes for these decay channels. The 1.06 μm line from a Nd^{3+} laser directly populates the metastable $^3T_2$ level, which subsequently decays radiatively into the broad infrared luminescence band with peak at ~1.35 μm with radiative decay time of $\tau_R = 3.6$ ms. The small quantum defect between vibronic absorption and luminescence bands of *ca* 0.3 eV per absorbed photon is a minor source of crystal heating during laser operation. A more important source of heating is nonradiative decay from the vibrationally relaxed $^3T_2$ state to the ground state, a rather weak function of temperature in the range 4–80 K, but dominant at

300 K. Laser excitation in the ${}^3A_2 \rightarrow {}^3T_1\,({}^3F)$ absorption band, using Kr^+ ion laser lines at 647 nm leads to ${}^1T_2\,({}^1D) \rightarrow {}^3A_2$ luminescence in the green with peak near 500 nm and ${}^1T_2 \rightarrow {}^3T_2$ emission in the red with peak at $\sim$790 nm as a result of excited state upconversion into the ${}^1T_1\,({}^1G)$ level (off-scale in Fig. 9.11) from the relaxed excited ${}^3T_1$ $({}^3F)$ state, followed by nonradiative decay to the lowest vibrational level of ${}^1T_2\,({}^1D)$. Subsequent radiative decay is partitioned between the green and red emission bands. The strong ESA from 1350 to 1390 nm in the ${}^3T_2 \rightarrow {}^3A_2$ laser band splits the tuning curve into two bands with peaks at *ca* 1320 nm and 1410 nm [Moulton (1985)]. As indicated in Fig. 9.11, the most suitable pump wavelengths for the Ni : MgO laser are at 1062 nm (Nd : YAG) or 647.1 nm (Kr^+). In contrast to Ni^{2+} : MgO the infrared luminescence in $Ni^{2+} : MgF_2$ is overlapped by ESA absorptions to two higher lying orbital components of ${}^3T_1\,({}^3F)$, the splitting between which is sufficient to allow useful gain over the lasing bandwidth of 1.6–1.75 μm [Moulton (1985), Moncorgé *et al.* (1985)] with optimal pumping at 1320 nm using either Nd : YAG or Nd : YLF laser.

ESA limits both the slope efficiency and the tuning range of the $Co : MgF_2$ laser. ESA also plays an important role in quenching the laser luminescence in Ti^{3+} : YAG [Schepler (1986)] and YAP [Huber *et al.* (1988)], of Ce^{3+} in YAG [Hamilton *et al.* (1989)] and of Ce^{3+} : LiCAF [Payne *et al.* (1994)]. Sometimes these ESA transitions may excite electrons into conduction band states, giving rise to photoconductivity and defect formation. However, ESA transitions may also lead to upconverted photoluminescence and efficient laser emission at shorter wavelength as in $Er^{3+} : LiYF_4$ [Brede *et al.* (1993)].

9.5 Energy transfer processes

Luminescence life times indicate the presence of various excited state processes including nonradiative decay and site-to-site energy transfer. Early work on the Cr^{3+}-doped colquiriites, especially Cr^{3+} : LiCAF, showed that the single-exponential decay at 300 K with luminescence lifetime $\tau = 170 \pm 5$ μs is little changed by the Cr^{3+} concentration up to 5.5 mol. % Cr^{3+} [Payne *et al.* (1988a,b), Lee *et al.* (1989)]. Indeed, in the Cr^{3+} : LiSAF system efficient laser performance is achieved at even higher Cr^{3+} concentrations, up to 100% Cr^{3+} substitution on Al^{3+} sites. Such spectroscopic insensitivity to the Cr^{3+} concentration probably results from the fact that the neighbouring substitutional Al^{3+} (Cr^{3+}) sites do not share common F^- ions so that Cr^{3+}–F^-–Cr^{3+} super-exchange paths do not exist. The situation is different in ruby, in which radiative energy transfer from one excited Cr^{3+} ion to another ground state Cr^{3+} ion can be repeated many times before the emitted photon leaves the crystal. Such *radiative trapping* is signalled by the luminescence lifetime being longer in finely dispersed $Cr^{3+} : Al_2O_3$ powders than in $Cr^{3+} : Al_2O_3$ single crystals [Varsanyi *et al.* (1959)]. Since the emission cross section σ_{em} at wavelength λ is given by

$$\sigma_{em} = \left(\frac{\ln 2}{\pi}\right)^{1/2} \frac{\eta}{4\pi c n^2 \tau} \frac{\lambda^4}{\Delta\lambda}, \tag{9.19}$$

where η is the quantum efficiency and τ is the luminescence lifetime, it is evident that radiative energy transfer effectively reduces the emission cross section, and requires to be minimized in laser crystals. Fortunately, in most gain media radiative trapping only becomes significant at high dopant concentrations.

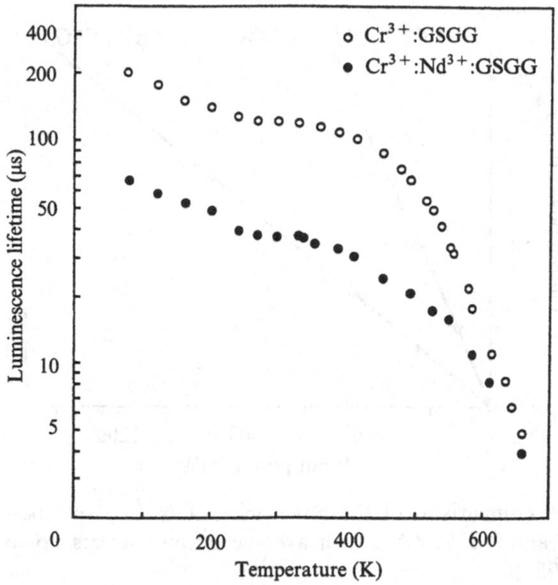

Figure 9.12. The temperature dependence of the luminescence lifetime of Cr^{3+} ions in Cr^{3+} : GSGG and Cr^{3+} : Nd^{3+} : GSGG. The luminescence was excited at 460 nm and detected at 760 nm through an interference filter [after Armagan and DiBartolo (1986)].

Energy transfer has been put to good use in a number of RE^{3+}-garnet laser systems to activate Nd^{3+}, Er^{3+}, Ho^{3+} or Tm^{3+} ions using Cr^{3+} sensitizers. The luminescence decay patterns of Cr^{3+} donors and Nd^{3+} acceptors co-doped YAG, Fig. 7.13, under white light excitation were discussed in §7.4.3. The best characterized material, however, is Cr^{3+} : Nd^3 : GSGG in which Cr^{3+} to Nd^{3+} energy transfer may approach 100% because of the excellent spectral overlap of the broad $^4T_2 \rightarrow {}^4A_2$ emission of Cr^{3+} ions and some of the lower lying absorption lines of Nd^{3+} (see Fig. 7.11). A comparison of the lifetime-temperature characteristics of Cr^{3+} in two GSGG crystals, one containing $0.8 \times 10^{20}\,cm^{-3}$ of Cr^{3+} and the other $2.0 \times 10^{20}\,cm^{-3}$ of both Cr^{3+} and Nd^{3+} ions, is given in Fig. 9.12 [Armagan and DiBartolo, (1986)]. At low temperature (< 250 K) the Cr^{3+} luminescence excited at 460 nm appears as an admixture of ^2E, $^4T_2 \rightarrow {}^4A_2$ transitions with non-exponential fluorescence decays. However, at higher temperatures vibronic interaction results in homogeneous emission bands and exponential decay patterns. These results show that Nd-Cr co-doping strongly quenches the Cr^{3+} luminescence. The measured decay times of the $^4T_2 \rightarrow {}^4A_2$ luminescence band at 300 K of 120 μs for Cr^{3+} : GSGG crystals and 38 μs for the co-doped Cr^{3+} : Nd^{3+} : GSGG samples represent graphically the influence of D–A energy transfer in the co-doped material. The temperature dependences of the decay times of the two materials in Fig. 9.12 converge above 550 K where nonradiative decay is dominant. The Nd^{3+} fluorescence lifetime (at short times) is little changed as a consequence of the Cr^{3+}–Nd^{3+} transfer with characteristic time $\tau = 290$ μs.

Since laser radiation at 460 nm excites the $^4A_2 \rightarrow {}^4T_1$ absorption band of Cr^{3+} ions exclusively, only these ions are excited and Nd^{3+} fluorescence results from the $Cr^{3+} \rightarrow Nd^{3+}$ energy transfer. When the concentrations of sensitizer and activator ions

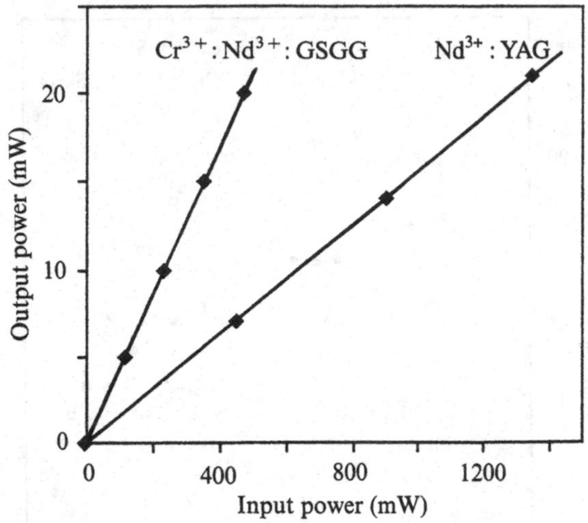

Figure 9.13. A comparison of the efficiencies of Nd^{3+} lasers based on Cr^{3+}: Nd^{3+}: GSGG and Nd^{3+}: YAG with average output powers up to 20 W [after Pruss *et al.* (1982)].

exceeds $\sim 10^{20}\,cm^{-3}$ the energy transfer efficiency exceeds 85%, this value varying with the actual Cr^{3+} and Nd^{3+} concentrations. The overlap integral between the ${}^4T_2 \rightarrow {}^4A_2$ luminescence of Cr^{3+} and the ${}^4I_{9/2} \rightarrow {}^4F_{3/2}$, ${}^4F_{5/2} + {}^2H_{9/2}$ and ${}^4F_{7/2} + {}^4S_{3/2}$ absorption transitions of Nd^{3+} (see Fig. 7.11) measures, to within a constant, the radiative energy transfer rate. Armagan and DiBartolo (1986) showed that the radiative transfer rate decreases slowly with increasing temperature, and did not exceed 5% of the $Cr^{3+} \rightarrow Nd^{3+}$ energy transfer at room temperature. The nonradiative energy transfer, in view of Eq. (7.5), may be written as

$$W_{DA}^{nr} = \tau^{-1} - \tau_0^{-1}, \tag{9.20}$$

where τ_0 and τ are the life times of Cr^{3+} luminescence in Cr^{3+}: GSGG and Cr^{3+}: Nd^{3+}: GSGG samples, respectively. The transfer rate by the nonradiative electric dipole mechanism varies from *ca* $10^4\,s^{-1}$ at 80 K to $5 \times 10^4\,s^{-1}$ at 600 K. The relative performances of Nd^{3+} lasers based on Cr^{3+}: Nd^{3+}: GSGG relative to Nd^{3+}: YAG are illustrated in Fig. 9.13: white light pumping of the Nd^{3+} laser action is more efficient for co-doped GSGG gain media than for Nd : YAG. Similar results have been reported for Cr^{3+}: Nd^{3+} co-doped GSAG and YSAG crystals [Duczynski *et al.* (1986a,b), Pruss *et al.* (1982), Shcherbakov (1986)]. Improved performance of rare-earth sensitized solid state lasers using Cr^{3+} ion co-doping with Nd^{3+}, Er^{3+}, Ho^{3+}, Tm^{3+} and Yb^{3+} has been studied in yttrium and gadolinium-based garnets. Cr^{3+} to Er^{3+} energy transfer in YSGG by the electric dipole–dipole interaction is characterized by a *D–A* micro-parameter of $\alpha_{DA}^6 = 2 \times 10^{-39}\,cm^6 s^{-1}$. Values of α_{DA}^6 for Cr^{3+} to Tm^{3+} and Cr^{3+} to Ho^{3+} transfer in YSAG are $4.2 \times 10^{-39}\,cm^6\,s^{-1}$ and $2.8 \times 10^{-40}\,cm^6\,s^{-1}$, respectively. The rather small value for Cr^{3+} to Ho^{3+} transfer is due to very poor spectral overlap between the Cr^{3+} emission and the ${}^5I_8 \rightarrow {}^5I_7$ transition on Ho^{3+} ions.

9.6 Empirical rules for transition-metal and rare-earth ions

9.6.1 Unit cell dimensions

The size of the unit cell is determined by the sizes of the constituent ions, whereas the structure of the unit cell is determined by the ratios of the cationic and anionic radii, it being assumed that the ions may be treated as hard spheres, that unlike ions must just touch along some particular crystal direction and that like ions may never touch. Alkali halide crystals with formula MX, where $M^+ = Li^+$, Na^+, K^+, Rb^+ and Cs^+ and $X^- = F^-$, Cl^-, Br^- and I^- are either face centred cubic (Figs. 2.4a and 4.4a) or simple cubic (Figs. 2.4b and 4.4b) depending upon the radius ratio r_x/r_m. When r_x/r_m is in the range $1.0 < r_x/r_m < 1.37$ the anions and cations are eight-fold coordinated with each other in the simple cubic CsCl- structure, O_h^1(Pm3m, #221). Crystals for which $1.37 < r_x/r_m < 2.41$ have six-fold coordinated anions and cations in a face-centred cubic (NaCl) structure. In the case of the caesium halides M^+ and X^- form the body-diagonal of the cube, along which each M^+ ion touches two nearest neighbour anions X^- along a particular $\langle 111 \rangle$-direction. In CsCl, where $r_x/r_m = 1.81/1.75 = 1.04$, the length of this body diagonal is $(1.81 + 3.50 + 1.81) \times 10^{-10}\,\text{m} = \sqrt{3}a$, and the cube edge length is $a = 0.41$ nm, precisely the value determined in X-ray diffraction measurements. In the rocksalt structure, anions and cations are in contact along $\langle 100 \rangle$ directions: for NaCl the cube edge has length $a = (0.98 + 3.62 + 0.98) \times 10^{-10}\,\text{m} = 0.558$ nm. Although empirical ionic radii have been used for anions and cations, the resulting calculated value for the lattice parameter, a, is equal to that measured by X-rays to within *ca* 0.01%. The reason that the rocksalt-structured alkali halides do not crystallise in the cubic Cs-chloride structure is that they would be unstable because the large uninegative halide ions would touch along $\langle 100 \rangle$ directions.

Extension to the alkaline earth difluorides MF_2, where M refers to the cations Mg^{2+}, Ca^{2+}, Zn^{2+}, etc. is straightforward. When r_+/r_- is in the range 0.7 to 1.0, the fluorite structure is stable, as it is for CaF_2, SrF_2 and BaF_2, where the radius ratios are 0.75, 0.85 and 0.99, respectively. This cubic structure is generated schematically from that of CsCl by removing alternate M^+ ions from each layer and replacing the remainder by dipositive cations (M^{2+}), each of which is coordinated to eight F^- ions arranged on cube corners. Every anion (F^-) is at the centre of a regular tetrahedron, the four corners of which are occupied M^{2+} ions. Thus the difluorides of metal ions with large ionic radii have the (8 : 4) fluorite coordination. In contrast, metal ions with small ionic radii, within the range $0.5 < r_+/r_- < 0.7$ have the rutile (TiO_2) cation/anion coordination (6 : 3), of which MgF_2 ($r_+/r_- = 0.5$) and ZnF_2 ($r_+/r_- = 0.56$) are typical. The Cs-halides, alkali halides and alkaline difluorides have important applications as optical crystals, the alkali halides being hosts for $Tl^0(1)$ and colour centre lasers, Ni^{2+} and Co^{2+} ions acting as efficient laser activators in MgO, MgF_2 and ZnF_2 and the Cs halides as scintillators. Some laser crystals have more complex structures. For example, in Al_2O_3, $r_+/r_- = 0.43$ is just greater than the upper stability limit for tetrahedral symmetry (0.225–0.414) and the lower stability limit for octahedral symmetry (0.414–0.732). Trigonally distorted coordination octahedra result. In contrast, Al^{3+} occupies both octahedral and tetrahedral sites in YAG, and exclusively the tetrahedral site in YSAG and GSAG.

Each of the cation sites in the colquiriites is sixfold-coordinated to F^- with r_+/r_- ratio of 0.59 (Li^+), 0.79 (Ca^{2+}), 0.43 (Al^{2+}), 0.96 (Sr^{2+}) and 1.08 (Ba^2). For all crystals

Li^+ and Al^{3+} are within the stability limit for octahedral coordination, i.e. $0.414 < r_+/r_- < 0.732$. However, the alkaline earth ions Ca^{2+}, Sr^{2+} and Ba^{2+} lie outside this range, resulting in distorted cation sites. The ionic displacements at cation sites vary from compound to compound and are accommodated elastically, except in $LiBaAlF_6$ where they are so large that the colquiriite structure is unstable. LiCAF and LiSAF isomorphs are completely soluble with each other in the solid state, forming solid solution compounds, $LiSr_xCa_{1-x}AlF_6$ in the range $0 < x < 1$, from which large single crystals can be grown at the congruent melting composition $LiSr_{0.8}Ca_{0.2}AlF_6$ [Chai *et al.* (1992)]. These alloys are compositionally disordered due to the random distribution of Ca^{2+}/Sr^{2+} ions on nearest neighbour dipositive ion sites. There are seven distinct configurations (m, n) of the second nearest neighbour cations in the colquiriite structure, Fig. 3.4, the total number of Sr^{2+} (m) and Ca^{2+} (n) positions in the second nearest neighbour sites being $m + n = 6$. These configurations introduce small changes to the ground state energies that are manifest in ESR spectra as zero-field splittings [Yamaga *et al.* (1999)]. ESR measurements of Cr^{3+} doped solid solutions, $LiSr_xCa_{1-x}AlF_6$, reveal small numbers (2–3) of discrete statistical configurations. Ordered configurations (0,6) and (6,0) are the only incumbents of pure LiCAF and LiSAF, respectively. Configurations with the largest probabilities; (2,4), (3,3) and (4,2) in $LiSr_{0.5}Ca_{0.5}AlF_6$ and (4,2), (5,1) and (6,0) in $LiSr_{0.8}Ca_{0.2}AlF_6$ were all identified by ESR. The measured zero-field splittings decreased with increasing Sr^{2+} content such that the $(CrF_6)^{3-}$ units achieve almost perfect octahedral symmetry with $x = 0.8$. The distortions of the octahedron that correspond to these crystal-field parameters have important consequences for the optical properties of these laser active materials.

The interaction of the electronic centres with the crystal-field was discussed in Chapter 4, emphasis being given to the utility of defining the dominant and lower symmetry components of the crystal field. Despite current computational efficiency it is not possible to predict accurately the variations of the crystal-field splittings from crystal to crystal. Even so, the point ion model of the energy levels of transition-metal and rare-earth ions facilitates an understanding of the spectroscopic trends, which have led to extensions of the formal theory to include finite ion sizes. Such complex developments in ligand field theory have been applied to only a limited number of laser materials (§8.6). It is now appropriate to examine the evidence of general spectral information for guidance in the design of novel optical materials, ignoring the subtleties introduced by low symmetry perturbations in order to recognize the implications of the empirical series that were developed to allow for the unreliability of the point-ion model (§4.4.8).

9.6.2 Spectrochemical series

The crystal-field splitting, $\Delta = 10Dq$, §4.4.8, is a convenient empirical parameter of the system, which can be factored into a product of ligand and metal ion functions, f and g, respectively, thus:

$$\Delta \cong f(\text{ligands}) \cdot g(\text{metal}), \tag{9.21}$$

so as to compare effects of the different ligands and the different central metal ions. For a given metal the contributions to Δ by the ligands increases from *ca* 0.7 to *ca* 1.0 along the *spectrochemical series*, $I^- < Br^- < Cl^- < S^{2-} < F^- < O^{2-}$, and is approximately independent of metal ion. A second series for metal ions, $Mn^{3+} < Ni^{2+} < Co^{2+} < V^{2+} < Fe^{3+} < Cr^{3+} < Co^{3+} < Ru^{3+} < Mo^{3+} < Rh^{3+} < Pd^{4+} < Ir^{3+} < Pt^{4+}$ is approximately independent of ligand, with g(metal) rising from $8 \times 10^3\,cm^{-1}$ for Mn^{3+} to $36.0 \times 10^3\,cm^{-1}$ for Pt^{4+}. For isoelectronic $3d^3$ ions g(metal) decreases along the series $V^{2+} < Cr^{3+} < Mn^{4+}$, as it does for $3d < 4d < 5d$ elements. The *spectrochemical series* of ligand and metal ions show that the highest values of Δ occur in fluorides or oxides, and in the higher charge states of a particular isoelectronic series.

9.6.3 *Nephelauxetic effect*

The interelectron repulsion parameters express the interactions of electrons on metal ions among themselves: for each LS-level they are represented by a sum of Slater integrals, $F^{(k)}$, or of reduced Slater integrals, F_k, or in terms of the Racah parameters A, B and C. Values of the Racah parameters of free transition-metal ions are listed in Table 4.3 and for free rare-earth ions in Table 4.10. The reduction of the interelectron repulsion parameters in a crystal relative to the free ion is called the nephelauxetic effect (§4.4.8). For rare-earth ions F_k-values are reduced by about 2%. Much larger nephelauxetic effects are reported for transition-metal ions, as is evident for the Cr^{3+} ion in several crystals quoted in Table 9.5. The nephelauxetic series for transition-metal ions determined by $1 - \beta = B/B_0$, factored into separate ligand and metal ion contributions,

$$1 - \beta = h(\text{lig}) \cdot k(\text{met}), \tag{9.22}$$

is not well defined with k(met) values $Mn^{2+} \sim V^{2+} < Ni^{2+} \sim Co^{2+} < Mo^{2+} < Cr^{3+} < V^{3+} < Fe^{3+} < Mn^{4+}$ and increasing h(lig) values is $F^- < H_2O < OH^- < O^{2-} < Cl^- < Br^- < S^{2-} < I^- < Se^{2-}$. The transition-metal ion series suggests that Fe^{2+} and Fe^{3+} are more covalent than Ni^{2+} and Cr^{3+}, respectively.

9.6.4 *Crystal-field stabilization energies*

The crystal-field experienced by the dopant ion reflects the symmetry of the environment. It is convenient to discuss the crystal-field in terms of three high symmetry sites occupied by optical centres in crystals. In *octahedral* symmetry, Fig. 4.4a, the central cation is surrounded by six anions each with charge $-Ze$ at a distance a from the origin along the orthogonal $\pm x$, $\pm y$, $\pm z$ axes; the crystal-field in *tetrahedral* symmetry, Fig. 4.4c, is due to point charges $-Ze$ at the alternate vertices of a cube; and in *cubic* symmetry, Fig. 4.4b, there are eight equivalent anions, $-Ze$, each located at the cor-

Table 9.5. *Slater and Racah parameters for some* Cr^{3+} *octahedral complexes (in* cm^{-1}*)*

	Free ion	Al_2O_3	K_2NaCrF_6	LiCAF
B	920	640	760	740
C	3680	3300	3020	3080
F_2	1446	1112	1192	1180
F_4	105	94.5	86.2	88

ners of a cube. The cubic cells that circumscribe these simple structural units have the same dimensions, $2a$. For a $3d$ electron at the centre of such an octahedron, tetrahedron or simple cube, the crystal-field strengths from Eqs. (4.50) are related by

$$Dq(\text{oct.}) = -\tfrac{9}{4}Dq(\text{tetrah.}) = -\tfrac{9}{8}Dq(\text{cube}). \tag{9.23}$$

The crystal-field expansion in tetrahedral symmetry contains a term in r^4, which is of the same functional form as in the crystal-field Hamiltonian in octahedral symmetry, which yields the energy $-\frac{9}{4}Dq$. There is also an odd-parity term that varies as r^3. There are no matrix elements of this odd parity field within the same (nl) configuration; it makes no contribution to the energy level splittings of the $3d$ electron. This term does have significant effect on the strength of the radiative transitions. The simple cubic cell may be considered as two tetrahedral arrangements which, taken together, yield a unit cell with inversion symmetry. These odd-parity components, of equal magnitude but opposite phase, cancel, whereas the two even-parity terms add together. Hence, the crystal-field splitting in the eight-fold coordinated unit cell is twice as large as that in the four-fold symmetry.

Transition-metal ions prefer to occupy sites with near octahedral symmetry in ionic crystals. In contrast, rare-earth ions prefer eight-fold coordinated sites, although six-fold coordination is not precluded. Frequently the symmetries are reduced from perfect octahedral, etc., by small displacements of the near neighbour ions as we have discussed in Chapters 2–4. The splittings of spectral terms of an optical centre by low symmetry distortions were treated in §4.4.7. The following general rules may be invoked in choosing crystal-fields suitable for transition metal ions:

- Coordination polyhedra involving F^- or O^{2-} ions are most suitable;
- Strong crystal-field sites (i.e. large Dq) with six-fold anionic coordination are preferred;
- Octahedral crystal-field sites with weak local distortions are beneficial;
- Laser active ions should be substituted at crystal-field sites with the same charge;
- Weak electron–phonon coupling to low energy vibrations minimizes nonradiative decay.

The strong crystal fields introduce larger splittings of energy levels, which can facilitate operation at shorter wavlengths and enhance the stability of metastable levels against nonradiative decay. Small splittings of the octahedral levels by even-parity distortions may extend the tuning ranges of vibronic lasers by removing degeneracies of the ground and excited levels. In laser hosts that are activated by rare-earth ions, such splittings are typically only of order 50–300 cm^{-1}. Since most rare-earth ion lasers are single frequency devices, the crystal-field interaction introduces small shifts in the luminescence wavelength at which the lasers operate. Since the even-parity distortions can also remove degeneracies of the RE^{3+} levels there may be emission into more sharp laser lines. At room temperature this fine structure is partially eliminated by temperature-dependent homogeneous broadening. Distortions of odd-parity do not change the energies of states. Rather do they admix opposite parity wavefunctions into the central $3d$ and $4f$ orbitals, resulting in shorter life times and polarization of absorption and emission transitions, in contrast to the characteristics of $3d^n$ and $4f^n$ ions in perfectly octahedral symmetry sites (§5.3, §8.6). Optical centres that occupy differently charged sites will require charge compensation by lattice defects or alternate valence states, which can lead to undesirable absorption bands, luminescence quenching and reduced concentrations of the desired optical species. For example, laser action associated with Cr^{3+} ions substituted for Zn^{2+} ions in $KZnF_3$ and $ZnWO_4$ occurs with efficiencies of only 14% and 5% respectively, as a consequence of parasitic absorptions and strong nonradiative processes.

Some transition-metal and rare-earth ions enter ionic crystals in more than one valence state. For example, Ti may enter Al_2O_3 as Ti^{3+} or Ti^{4+}. The simultaneous presence of both can result in poor laser efficiency since Ti^{3+}–Ti^{4+} pairs undergo infrared absorption transitions that overlap the laser emission band. The presence of Ti^{4+} ions reduces the maximum concentration of Ti^{3+} in the crystal. To minimize the Ti^{4+} content requires special treatments; growth or post-growth annealing in a reducing atmosphere stabilizes Ti^{3+}. Similarly, $Mn^{2+}/Mn^{3+}/Mn^{4+}$, Cr^{2+}/Cr^{3+}, Fe^{3+}/Fe^{2+}, Co^{3+}/Co^{2+} and Ni^{2+}/Ni^{+} may appear together in some crystals. The stability of the most common transition-metal ions against oxidation and reduction are tabulated in Table 9.6. The rare-earth ions may also appear as RE^{2+} or RE^{3+} at divalent sites or trivalent sites, respectively. For example, Tm^{2+} has been much studied in the alkaline earth fluorides, in which its optical properties reflect those expected of a $4f^{13}$ ion. Like Yb^{3+}, this $4f^{13}$ ion has two energy levels, $^2F_{7/2}$ and $^2F_{5/2}$, which are efficiently pumped in the broad $4f^{13} \rightarrow 4f^{12}5d$ absorption bands. Laser action has been achieved on the $^2F_{5/2} \rightarrow F_{7/2}$ emission transition in $Tm^{2+}:CaF_2$. In contrast, when present in YLF or YAG, Tm^{3+} is the preferred valence state which is the activator of an efficient laser transition in the infrared.

The strong preference of most transition-metal ions for octahedral sites is easily understood. Consider Ti^{3+} ($3d^1$) in octahedral symmetry: the crystal-field stabilization energy (CFSE) is given by the energy of one electron in a t_{2g} orbital, i.e., $-4Dq$ (oct)

Table 9.6 *Relative stability of some $3d^n$ configuration ions* [*Caird* (1986)]

Configuration	Ion	Against oxidation	Against reduction
$3d^0$	Ti^{4+}	high	moderate
$3d^1$	Ti^{3+}	low	high
	V^{4+}	moderate	low
$3d^2$	V^{3+}	high	moderate
	Cr^{4+}	low	high
$3d^3$	V^{2+}	low	high
	Cr^{3+}	high	high
	Mn^{4+}	high	low
$3d^4$	Cr^{2+}	low	high
	Mn^{3+}	low	low
$3d^5$	Mn^{2+}	high	high
	Fe^{3+}	high	low
$3d^6$	Fe^{2+}	moderate	high
	Co^{3+}	high	low
$3d^7$	Co^{2+}	high	high
	Ni^{3+}	high	low
$3d^8$	Ni^{2+}	high	moderate
	Cu^{3+}	high	low
$3d^9$	Cu^{2+}	high	low

Table 9.7 *Crystal-field stabilization energies of transition-metal configuration ions in octahedral and tetrahedral sites [Caird (1986)]*

Configuration	Octahedral stabilization energy kJ mol^{-1}	Tetrahedral stabilization energy kJ mol^{-1}	Excess stabilization energy kJmol^{-1}
d^1	87.6	58.7	28.9
d^2	160.5	106.8	53.6
d^3	220.5	67.0	158.0
d^4	135.8	40.2	95.5
d^5	0	0	0
d^6	49.9	33.1	16.8
d^7	93.0	62.0	31.0
d^8	122.3	36.0	86.3
d^9	90.5	26.8	63.7

relative to the centre of gravity of the $3d^1$ configuration. In tetrahedral symmetry the *e*-orbitals lie lowest and the crystal-field stabilization energy for one electron in an *e*-orbital is $-6Dq$ (tet). From Eq. (9.23), the excess crystal-field stabilization energy for Ti^{3+} in octahedral sites is $-[4-(8/3)]Dq(\text{oct}) = -(4/3)Dq(\text{oct})$. For $3d^3$ ions each electron occupies the t_{2g} orbital in octahedral symmetry, and the resultant crystal-field stabilization energy is $-4\,Dq \times 3 = -12Dq$. In tetrahedral symmetry, where e lies below t_2, only two electrons occupy the *e*-orbital, the third electron occupying the higher lying t_2 orbital. Hence the stabilization energy in tetrahedral sites is given by $-6Dq(\text{tet}) \times 2 + 4Dq(\text{tet}) \times 1 = -8Dq(\text{tet})$. The excess stabilization energy is then given by ~ 8.5 Dq(oct) in favour of the octahedral site; this, being $\cong 28.9$ kJ mol^{-1} is a substantial fraction of the binding energy per cation. Extending this calculation to all $3d^n$-configuration ions (Table 9.7) shows that Cr^{3+} has by far the largest crystal-field stabilization energy. Furthermore, no ion shows a preference for tetrahedral symmetry, although one distinguishes the $3d^5$ configurations which shows no preference for either structure. The excess CFSE for Ti^{3+} in octahedral sites should not be a problem in crystals lacking tetrahedral sites such as Al_2O_3. Even when octahedral and tetrahedral sites co-exist, as in the garnets, Ti^{3+} occupies the octahedral site to the exclusion of the tetrahedral site [Schepler (1986), Gao *et al.* (1993)].

Finally, the desirability of using hosts with strong crystal-fields is not always compatible with the energy level structure of the dopant ion, except for $3d^1$ and $3d^9$ configuration ions for which there is but one spin state. These configuration ions should demonstrate greater resistance to nonradiative decay in strong crystal-field, which is not necessarily the case for d^2, d^3 and d^8 configurations as a consequence of a level crossing between the two lowest energy excited states at some critical value of Dq. The ^2E–$^4T_2$ levels of the Cr^{3+} ion cross at $Dq/B \cong 2.3$. Below this value the emission occurs via the $^4T_2 \rightarrow {}^4A_2$ band. At higher crystal-field levels Cr^{3+} ions emit in the narrow, spin-forbidden R-lines. In general, crystal-field sites with $Dq/B > 2.30$ are to be avoided, given that the R-line emission leads to three-level laser action, with rather small emission cross sections. Similar situations obtain for $3d^2$ ions (V^{3+}, Cr^{4+}, Mn^{5+}) and $3d^8$ ions (Ni^{2+}) both of which feature $^1E \rightarrow {}^3T_2$ level crossings near $Dq/B \cong 1.5$–1.6. Above the level crossing (e.g. V^{3+} in Al_2O_3 or Mn^{5+} in Ca-fluoroapatite) the $^1E \rightarrow {}^3T_1$

emission is narrow-line, below this value the $^3T_2 \rightarrow {}^3T_1$ transition is broadband and spin-allowed. The $3d^4$, $3d^5$, $3d^6$ and $3d^7$ configurations all have low spin multiplets that become the electronic ground state in moderate crystal field. Consider, for example, Fig. 4.8 for the $3d^7$ ion in octahedral crystal fields, e.g. Co^{2+} in MgO, where $Dq/B = 0.9$. At this value the $^4T_2$ level is just lower in energy than ^2E. However, the ^2E-level decreases rapidly in energy with increasing Dq-value and becomes the lowest energy level at and above $Dq/B \cong 2.1$. The crystal-field strength should be kept below this value to avoid the fluorescence transition becoming spin-forbidden. Unfortunately, this leads to longer wavelength emission and stronger nonradiative decay. A cursory examination of the Tanabe–Sugano diagrams for $3d^4$, $3d^6$ configurations confirms their similarity to $3d^7$ in respect of the potential for the low spin level becoming the ground state into which excited levels emit.

9.6.5 The $\sigma_e\tau_R$ product rule

There are qualifications to these rules. Octahedral symmetry allows weaker absorption/emission transitions and longer excited state life times. These characteristics are good for energy storage and flashlamp pumping, but bad for single pass gain, $\gamma(\nu)$. In consequence, distorted octahedral symmetry is more desirable, giving rise to polarized transitions, shorter life times and larger single pass gain. In flashlamp pumping, the excited state lifetime should be as long as possible consistent with a high cross-section for stimulated emission. Since the emission cross section σ_e and radiative lifetime τ_R are related by

$$\sigma_e = \frac{\lambda^2}{8\pi n^2 \Delta\nu} A_e, \tag{9.24}$$

in which A_e is the spontaneous emission rate, τ_R^{-1}, the lifetime–cross-section product $\sigma_e\tau_R$ cannot exceed

$$\sigma_e\tau_R = \frac{\lambda^2}{8\pi n^2 \Delta\nu}, \tag{9.25}$$

where λ is the peak wavelength and $\Delta\nu$ is the FWHM bandwidth of the fluorescence. Typically, $\sigma_e \sim 2 \times 10^{-20}$ cm^2 and $\tau_R \geq 100$ µs. In optically anisotropic media, the total decay rate is given by

$$\frac{1}{\tau_R} = \frac{1}{3}A_\pi + \frac{2}{3}A_\sigma. \tag{9.26}$$

When $A_\pi \gg A_\sigma$ the $\sigma_e\tau_R$ product

$$\sigma_e\tau_R = 3 \cdot \frac{\lambda^2}{8\pi n^2 \Delta\nu} \tag{9.27}$$

is a factor of 3 larger than in the isotropic case. The alexandrite laser is an example of a polarization-enhanced, tunable solid state laser. In $Cr^{3+}:BeAl_2O_4$, $A_\pi \cong 10A_\sigma$, although in this case of intermediate crystal-field the benefits are reduced by population distribution between the almost degenerate ^2E and $^4T_2$ levels. Some typical values of $\sigma_e\tau_R$ product are 1.40×10^{-24} cm^2 for Ti-Al_2O_3 and 1.35×10^{-24} s cm^2 for Cr^{4+}: YAG.

The interplay between the emission cross section and radiative decay time is an important characteristic of a laser material because the laser threshold varies as $(\sigma_e\tau_R)^{-1}$. Apparently, the larger this product, the lower is the power required to exceed the threshold, although there are limits to what may be achieved, given the interrelationship between σ_e and τ_R expressed in Eq. (9.27). Nevertheless, the advisability of using anisotropic optical crystals is properly stated.

9.7 All-solid-state lasers

The scientific basis of diode-pumped, laser gain media differs little from that of conventional bulk lasers. However, despite many successful host-active ion combinations the number of really useful laser gain media is small, due to both the intrinsic constraints of the materials and the extrinsic constraints of the laser pumping scheme. Improvements in semiconductor laser diodes (LDs) are revolutionising the development of all-solid-state lasers (ASSLs). The first report of laser diode operation by Nathan *et al.* (1962) was soon followed by news of a LD-pumped Nd : YAG laser [Ross (1968)]. The many advances in LDs and solid state lasers since guarantee widescale application of ASSLs in science and technology [Fan and Byer (1988), Ferguson (1994)].

Flashlamp pumping of ruby, YAG, alexandrite and Ti-sapphire lasers places materials requirements of size, robustness and optothermal integrity on the pumped medium as well as broad (perhaps multiple) absorption bands and resistance to solarization by high energy photons. Similar materials constraints apply to gain media excited by high-power ion lasers. All may be mitigated to some extent by efficient sample cooling. However, LDs emit photons over a narrow wavelength range, that can be temperature-tuned to the absorption bands of selected gain media, and operate at low peak powers. Hence, broadband absorption by the gain medium is no longer a prerequisite of the ASSLs and the gain medium should feature strong absorption over a narrow band of frequencies that matches the gain profile of the LD. This points to an important advantage of ASSLs. The reduced thermal loading of the pumped medium, consequent upon tight wavelength coupling of LD to the absorbing laser rod, minimizes material size constraints at work in bulk lasers. In consequence crystal growers can focus attention on developing lower yield production techniques to achieve improved optical quality at higher concentrations of optical centres without deleterious consequences for the optical, opto-mechanical and physico-chemical properties. The optical centres should occupy low symmetry sites that provide large absorption and emission cross sections with long fluorescence life times, thereby reducing the threshold pump power. In such environments the electron–phonon coupling broadens spectra and facilitates the wavelength match between diode output and absorption band without recourse to electronic feedback controls.

Early LDs were fabricated from GaAs/AlGaAs epilayers. They operated over a wavelength range of 780 nm to 820 nm with wall-plug efficiencies up to 50%. Although individual LDs produce only 40–50 mW of CW power, this output is sufficient to cause damage to the cleaved mirrors of the diode. In consequence LDs must have large output areas to minimize possible radiation damage. The beam quality of LDs is rather poor, being far from diffraction-limited and of unstable output wavelength. Also LDs have large linewidth and narrow tunability range. Such disadvantageous properties are eliminated in all-solid-state lasers (ASSLs) pumped by LDs. Although the earliest GaAs-based LDs produced barely tens of mW of output power at *ca* 800 nm, they were well matched to the homogeneously broadened $^4I_{9/2} \rightarrow {}^4F_{5/2}$ absorptions of Nd^{3+}-doped solids. Nevertheless, greater CW powers are needed from LDs to take advantage of the excellent coupling efficiency afforded by the narrow laser emission that overlaps the sharp crystal-field spectra of Nd^{3+}-doped materials. Arrays comprising 100 or so LDs emitting from a surface area of *ca* 500 μm × 1 μm can deliver up to 4 W of CW power. Mounting several diode arrays on a 10 mm long bar can output up to 20 W. In further optoelectronic miniaturization, Jewel *et al.* (1991) packed *ca* 10^6 microlasers on a single 10 × 10 mm^2 semiconductor chip. Output from these *microlasers* is perpendicular to the epilayer surface, in contrast to the edge emission of LDs. Whether LD arrays or microlaser arrays are used to pump ASSLs it is probable that cylindrical coupling optics or multi-mode fibres will be used to deliver excitation radiation to lasers using the end-pumped configuration. The example of an ASSL shown in Fig. 1.10 combines laser gain and harmonic generation media in a resonant frequency-doubling cavity for mode-locked operation. A pair of 3 W LDs operating at 800 nm pump a mode-locked Nd : YLF laser, the 1047 nm output of which is spatially mode-matched to an external ring-cavity containing a LiB_3O_5 (LBO) frequency-doubling crystal. In such resonant pumping the infrared-to-green conversion efficiency can be as high as 60%. In an alternative ASSL, a β-BaB_2O_4 (BBO) crystal frequency quadruples to 262 nm resonantly with an efficiency of 11%, yielding some 42 mW of usable ultraviolet output [Ferguson (1994)]. Apparently developments of LD-pumped ASSLs can be as much concerned with the development of nonlinear optical crystals as with laser gain media [Lin (1990)].

The diversity of laser gain media and nonlinear optical crystals promises ASSLs with complete spectral coverage from the infrared region (2000–3000 nm) into the UV region (200–250 nm), although greater pump versatility is required than is achievable from GaAs/AlGaAs LDs. However, there are numerous dopant-crystal combinations that are best pumped near 1000 nm including Er^{3+} ($^4I_{15/2} \rightarrow {}^4I_{11/2}$ transition), Pr^{3+} ($^3H_4 \rightarrow {}^1G_4$ transition), Dy^{3+} ($^6H_{15/2} \rightarrow {}^6H_{5/2}$ transition) and Yb^{3+} ($^2F_{7/2} \rightarrow {}^2F_{5/2}$) as fixed wavelength lasers, and Cr^{4+} ($^3A_2 \rightarrow {}^3T_1$ transition) as wavelength-tunable lasers. Strained layer InGaAs–GaAs super-lattices can be fabricated into LD-arrays that tune from 970 nm to 1030 nm and provide up to 1 W of CW power. LDs based on InGaAsP quantum wells with outputs of up to 100 mW of coherent radiation tunable from 1450–1550 nm are suitable for pumping the $^4I_{15/2} \rightarrow {}^4I_{13/2}$ absorption of Er^{3+}-doped glass fibre and waveguide amplifiers. Red LDs operating at 650 nm to 670 nm and configured as single element LDs (50 mW), multi-element arrays (~ 750 mW) or multi-array bars (5 W) are convenient for pumping Cr^{3+}-doped gain media in the $^4A_2 \rightarrow {}^4T_2$ band.

Finally, the III–V nitrides (AlN, GaN, InN) are the active medium in LEDs spanning the entire visible spectrum from 400 to 610 nm; laser diode operation at 405–420 nm has been demonstrated using GaInN/GaN quantum well lasers grown on Al_2O_3 substrates. The science and technology of blue LDs based on III–V nitrides is advancing rapidly. Several Japanese and US companies have fabricated blue LDs that operate in pulsed mode. None have matched the LDs manufactured by Nakamura *et al.* (1995), (1996), who not only produced the first GaN-based blue laser, but also have demonstrated CW lasing at 300 K with operating life times exceeding 10^4 hours. Once perfected, this technology will offer the simplest, and cheapest, means of producing blue laser light to satisfy demands from research laboratories and for applications in colour scanners, high density data storage, large-area displays, reprographics, underwater communications and more.

9.8 Optical nonlinearities

The light fields generated by lasers are many orders of magnitude more intense than those available from conventional light sources. At electric field strengths of order 100 kV/m optical properties that are normally linear in the electric field, **E**, become nonlinear functions of the field and the 'optical constants' are no longer independent of the light intensity ($I \propto E^2$). Equations 3.2–3.4 show that the macroscopic polarization, **P**, optical susceptibility, ψ, and dielectric constant, ε, contain terms that vary nonlinearly with the electric field, and these lead to the several optical nonlinearities that find application in solid state lasers. As discussed in Chapter 3, the second-order susceptibility is a tensor property of rank 3 (Eq. (3.5)) conveniently represented by the Kleinman d-tensor, the d_{im} coefficients of which are only non-zero in crystals that lack a centre of inversion. The particular d_{im} that are non-zero are determined by the crystal class and point group. Given the long range symmetry of crystalline solids, many of the d_{im} are zero and others may be equal. In consequence, knowledge of the crystal structure indicates which of the d_{im} must be measured. The number of independent coefficients of tensors of rank 1, 2 and 3 are listed in Table 3.1 for numerous inorganic crystals: the measured values of d_{im} for some important nonlinear optical crystals are collected in Table 3.2. Only occasionally are all the independent d_{im} coefficients reported (see Table 3.2 and the more extensive collection of data in Nikogosyan (1997)).

Although α-quartz is now mainly of historical interest in nonlinear optics, and is only occasionally used in devices, the presence of the tetrahedrally bonded SiO_4^{4-} molecular ions in this material pointed to the physical origins of the optical nonlinearities. Further indications of the involvement of particular molecular ion groups came from work on the ABO_3 perovskites, to which applied electric fields were used to probe the mechanisms of second harmonic generation. At room temperature $KTaO_3$ is cubic (O_h^1, Pm3m, #221), and the Kleinman coefficients $d_{im} = 0$ as a consequence of the Ta^{5+} ions occupying perfectly octahedral sites. The external electric field displaces the positive Ta^{5+} ions relative to the O^{2-} anions so that the MO_6 octahedron is distorted, resulting in a finite electric dipole moment. The larger the applied field the larger is the dipole moment and the induced polarization. Fujii and Sakudo (1976) applied electric fields along the $\langle 100 \rangle$ and $\langle 111 \rangle$ axes, reducing the O_h symmetry to tetragonal (C_{4v}) and trigonal (C_{3v}), respectively. The field-induced odd-parity perturbation along a $\langle 100 \rangle$ direction , V_{odd},

$$V_{odd} = (4qe/R_0^3)\Delta z, \tag{9.28}$$

is linear in the shift of the central ion, Δz, relative to the six ligand O^{2-} ions. An approximate result for the second order, nonlinear susceptibility, $\chi^{(2)}$, derived from quantum mechanics (§5.6, Eq. (5.168)) is

$$\chi^{(2)} \cong \frac{|\langle a|eV_{odd}|b\rangle|^2[\langle a|eV_{odd}|a\rangle - \langle b|eV_{odd}|b\rangle]}{[(\omega_{ab} - \omega)(\omega_{ab} - 2\omega)]}, \tag{9.29}$$

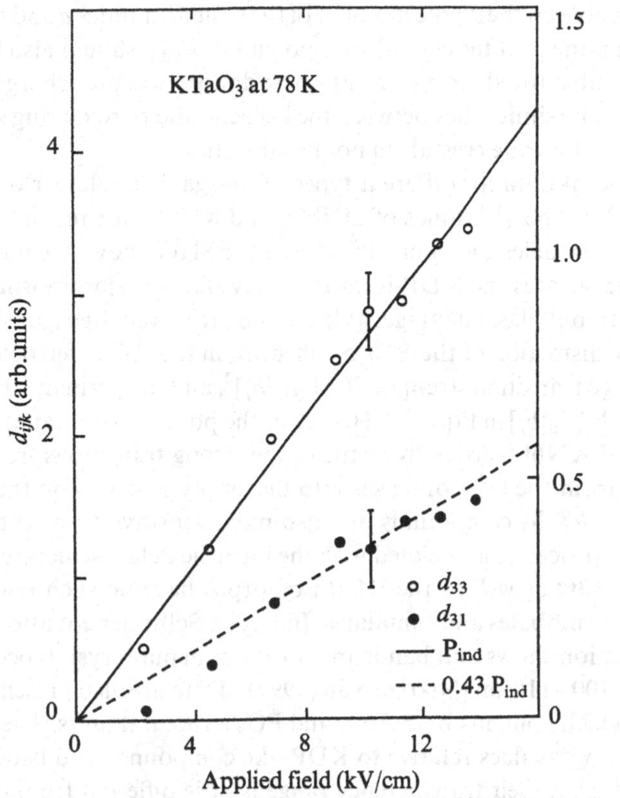

Figure 9.14. Electric field dependences of the Kleinman coefficients d_{31} and d_{33} for $KTaO_3$ at 78 K [after Fujii and Sakudo (1976)].

where the wavefunctions a and b are Hartree–Fock functions for occupied and unoccupied valence and conduction bands, respectively [Chen (1993)]. Equation (5.168) contains the product of an electronic transition moment and an electrostatic dipole moment. Experimental results for $KTaO_3$, shown in Fig. 9.14 for E-fields along [100] and [110] crystal directions, measure the $\chi^{(2)}_{333}$ and $\chi^{(2)}_{311}$ components, respectively. These results demonstrate that the electric field induces a $\chi^{(2)}_{ijk}$ proportional to the magnitude and direction of the displacement of the cation from the centre of the MO_6. In view of Eq. (9.29), it is evident that the induced electrostatic dipole moment is the dominant effect. Such induced nonlinear optical coefficients are an order of magnitude smaller than the intrinsic coefficients in the distorted perovskites $LiNbO_3$ and $KNbO_3$, which possess a *permanent* dipole moment associated with the intrinsic distortion of the NbO_6 octahedron. In the dihydrogen phosphates and arsenates there is similarly a built-in permanent dipole associated with MO_4-tetrahedral units. The quantum mechanical result, Eq. (9.29), illustrates how crystal-field engineering is used in designing new nonlinear optical materials. The first term in Eq. (9.29) should refer to an electric dipole transition at short wavelength so that frequency conversion is possible into the ultraviolet region. Furthermore, the denominator indicates that it is advantageous to work near to resonance with an allowed electronic transition because the squared transition moment $|\langle a|V_{\text{odd}}|b\rangle|^2$ is then large [Chen *et al.* (1990), Chen (1993)].

The difference between the dipole moment of the system in states a and b induced by the odd-parity component of the crystal-field potential, V_{odd}, should also be large. This is the case for the substituted benzene rings where donor–acceptor charge transfer bands are very strong. The similarities between the benzene and boroxal ring structures led to the applications of borate crystals in nonlinear optics.

A comparison of data for different types of inorganic nonlinear optical crystals is given in Table 3.2. The $\chi^{(2)}$ values of $LiNbO_3$ and $KNbO_3$ are much larger than other compounds. Nevertheless, neither can be used for SHG above 400 nm. At such short wavelengths the borates and KDP-isomorphs have far superior transmittance, and are more resistant to bulk laser damage, at least when relatively free from growth defects. The odd-parity distortion of the BO_6 octahedron in the ABO_3 ferroelectrics provides both for the large transition strength, $[\langle a|V_{\text{odd}}|b\rangle]^2$, and the permanent dipole moment $[\langle a|V_{\text{odd}}|a\rangle - \langle b|V_{\text{odd}}|b\rangle]$ in Eq. 9.29. However, the poor transmittance beyond 400 nm for $LiNbO_3$ and $KNbO_3$ arises by virtue of the strong transitions from filled valence bands derived from the O^{2-} ion levels into the empty $3d$-levels on the B^{5+} transition metal ions. The ABO_3 compounds are also more sensitive to radiation damage by photorefractive processes associated with the intrinsic defect structure of these materials. LBO and BBO as well as the KDP-isomorphs have no such low energy optical transitions as the niobates and tantalates. Indeed, a Sellmeier equation analysis of the indices of refraction shows that bandgap transitions in pure crystals occur at very short wavelengths ~ 100–110 nm [Nikogosyan (1997)]. Data are also presented in Table 3.2 for the KTP, which contains both TiO_6 and PO_4 structural units. These crystals have much improved $\chi^{(2)}$ values relative to KDP-like compounds and better birefringence also. Unfortunately, their transparency range is little different from the ABO_3 ferroelectric compounds.

9.9 Other considerations

The mechanical and thermomechanical properties are of particular importance for commercial devices. In the tensor form of Hooke's law the elastic stiffness, $\bar{e}$, and compliance, $\bar{s}$, are fourth rank tensors; 81 independent components are reduced to 36 by the symmetry relationships between stress and strain [Nye (1985)]. The number of independent components is further reduced by the symmetry of the material. An amorphous (isotropic) material has two independent elastic constants and a cubic crystal such as YAG will have three independent tensor coefficients. An optical material that absorbs light will experience heating as a consequence of phonon-induced nonradiative de-excitation and the quantum defect between absorption and emission, giving rise to a temperature difference between centre and edges of a sample. Thus there can be thermally-induced stresses which, in the presence of defects introduced by growth or post-growth preparative treatments, can cause failure by crack propagation [Cottrell (1964)]. A figure of merit for the fracture resistance of an absorbing slab of isotropic material convectively cooled on its major faces is:

$$R_{\text{T}} = \frac{K\kappa(1-\nu)}{\alpha E}, \tag{9.30}$$

where K is the fracture toughness, κ is the thermal conductivity, ν is Poisson's ratio, α is the coefficient of thermal expansion and E is Young's modulus. The fracture toughness,

K, measures a material's ability to resist crack propagation and is determined by the stress required to enlarge a crack of a given size. Typical values of K are 3 for Al_2O_3, 2.5 for YAG and 1.5 for $MgAl_2O_4$, in units of $MPa.m^{1/2}$. Obviously, care must be taken in producing laser rods to ensure that all surface flaws have been removed: the best surface condition may be achieved by chemical polishing [Marion (1985)].

The thermal conductivity, κ, in Eq. (9.30) is given by

$$\kappa = \lambda C_p \rho \tag{9.31}$$

where λ is the thermal diffusivity, C_p is the heat capacity and ρ is the density. Since K, α and λ are second rank tensors and E is of fourth rank, a proper determination of the directionally-dependent figure of merit R_T is quite tedious. Quite frequently, 'engineering values' are used, essentially isotropic measurements appropriate to fine-grained polycrystals. The reciprocal of the product, ρC_p, in Eqs. (9.30) and (9.31) measures the temperature rise associated with a given deposition of heat in the material. Woods *et al.* (1991) pointed out that $(\rho C_p)^{-1}$ is essentially insensitive to material for the six compounds LiCAF, YLF, CaF_2, $Be_2Al_2O_4$, YAG and GSGG within the range $0.32 < (\rho C_p)^{-1} < 0.39$. This material and structural insensitivity applies to an even wider range of crystals, including Al_2O_3, $BaTiO_3$, $SrTiO_3$, MgO, CaO, ZnO, $MgAl_2O_4$, MgF_2, SrF_2, BaF_2, LiF, $LiNbO_3$, and probably many others if the thermal data were available [Tropf *et al.* (1995)]. Apparently, the Debye model accounts for the value of $(\rho C_p)^{-1}$ and its constancy among insulating materials. Although the thermal conductivities κ (Eq. 9.31) are uncorrelated with general material type, the fracture toughness is material dependent, being significantly higher for oxides than for fluorides. Nevertheless, this does not exclude fluorides from applications as low and medium power laser gain media, because thermal lensing (reflected in dn/dT) is much more serious in oxides than in fluorides.

There are several radiation-matter interactions in laser crystals that result in crystal heating, with serious consequences for the operational characteristics of a solid state laser. All may be classified under a general heading of *nonradiative* transitions, resulting in radiant energy being downgraded to thermal energy. For example, solid state lasers are usually pumped in transitions higher in energy than the lasing transition. The excess energy, the quantum defect, excites lattice phonons, thereby increasing the temperature of the sample. Effective heat sinking of the laser crystal is needed to dissipate this additional thermal load. In addition, few optical absorption–emission cycles are 100% efficient so that fewer photons appear at the laser wavelength than are absorbed from the pump beam. This nonradiative loss returns the electronic centre involved in the absorption transition directly to the ground state without the emission of radiation. Excited state absorption and energy transfer processes, which may also be terminated by nonradiative relaxation, can contribute sample heating. In many instances such heating can result in thermal lensing and thermal shock, this last effect leading to rod failure by fracture. In designing a practical laser system some consideration must be given to the interplay of optothermal and thermomechanical properties, especially in high or ultra-high average power lasers.

In these situations there is an obvious advantage in cooling laser crystals to cryogenic temperatures. Significant increases in the average power abstracted from Ti-sapphire, Nd:YAG and Yb:YAG lasers has been obtained in operation at cryogenic temperatures, with consequent reduction in the appearance of thermally-induced aberrations [Schulz and Henion (1991), Brown (1997)]. These improvements occur simply because the nonradiative decay process is suppressed at low temperatures. Selectively pumping into transitions involving small quantum defect is an obvious advantage. In vibronic lasers such as Ti-sapphire and Cr^{3+} : LiSAF there is potential for pumping into broad absorption bands with monochromatic sources rather than with flashlamp systems. For Cr^{3+}-based lasers pumping into the low energy wing of the $^4A_2 \rightarrow {}^4T_2$ band has considerable potential for tunable lasers pumped by red (650–670 nm) laser diodes. Diode-pumping Nd^{3+}-based laser systems at 800 nm will obviously result in fewer materials problems than flashlamp pumping with white light, although there is a specific demand for such pulsed systems. Critical design of the sensitizer–activator energy level structure will also minimize the energy lost to thermal effects. Attention must be given to the surface condition of a laser rod or slab, since appropriate polishing will enhance the resistance to fracture by thermal shock by a factor of 3–5. Nevertheless, in seeking to design laser crystals it is essential to know something of the temperature dependences of such physical properties as the thermal conductivity and expansivity, Young's modulus and Poisson's ratio. Such knowledge may influence the choice of material for a particular laser application.

10

The crystal field engineered

The successful launch of the alexandrite laser by Allied Chemicals Inc. and the elucidation of essential design parameters [Walling *et al.* (1979), (1980)] spawned rapid growth of research into Cr^{3+}-based lasers [Caird and Payne (1991)]. Possible alternative gain media to alexandrite included the Cr^{3+}-doped garnets [Struve and Huber (1985)]. However, the development of Cr^{3+} : colquiriite lasers at Lawrence Livermore National Laboratory [Payne *et al.* (1988a), (1989a)] deflected attention away from the Cr^{3+} garnets, these mixed fluoride gain media being as efficient as alexandrite and almost as broadband as Ti-sapphire. Two distractions from the dominance of Cr^{3+} ion broadband tunable lasers were the inventions of the Ti-sapphire ($Ti^{3+}:Al_2O_3$) [Moulton (1982a,b)] and the Cr^{4+} : forsterite [Petricevic *et al.* (1988)] lasers. At present $Ti^{3+}:Al_2O_3$ and Nd^{3+} : YAG are the market leaders in solid state laser production against which new developments are assessed. Their pre-eminence derives in part from the quality and quantity of laser rods that can be produced at modest cost. Both materials have excellent photothermal and thermomechanical properties, and are robust components under laser operating conditions. However, despite much spectroscopic research Ti-sapphire is still the only usable Ti^{3+}-activated solid state laser. In contrast, operation of several $3d^2$-ion doped lasers have been reported giving broadband tunability at near-infrared wavelengths (1.0–1.7 μm) which have potential applications in optical communications, medical sciences and on remote sensing LIDAR platforms.

The approach taken here is to show how crystal-field engineering extended and/or improved laser performance of some traditional materials and identified new laser materials. The objectives of crystal-field engineering are attained by modifying the chemical make-up of optical centres, introducing even- and odd-parity distortions to cause spectral broadening and enhance polarized absorption/emission rates. The manner of the interactions of these distortions with the optical centre is analysed in terms of the relationships between families of compounds such as the garnets, gallogermanates, aluminoborates, colquiriites and so on. Examples are chosen that show not only the success of the approach, but also the potentially deleterious consequences of disorder, nonradiative relaxation and excited state absorption. The RE^{3+} rare-earth ions, Ce^{3+} and Yb^{3+} excepted, are dissimilar to the transition-metal ions since from Pr^{3+} ($4f^2$) to Tm^{3+} ($4f^{12}$) they absorb and emit in sharp lines. Many host crystals are of such low symmetry that all orbital degeneracies are lifted, resulting in the maximum possible number of lines. The splittings between these lines vary from host to host as the

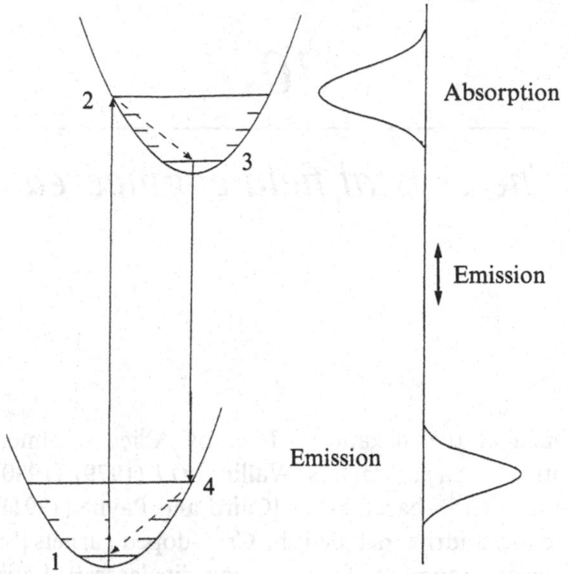

Figure 10.1. A configurational coordinate model of a vibronic laser.

strength and symmetry of even-parity distortions vary. Odd-parity distortions of the RE^{3+} ion symmetry enhance the oscillator strengths of laser transitions. In consequence, local site symmetry must be fully understood in respect both of transitions in and out of the ground state, and in excited state absorption and upconverted luminescence processes.

10.1 Tunable solid state lasers

Many lasers are based on vibronic transitions in various gain media, including organic dye-solvent combinations and colour centres, transition-metal ions and Ce^{3+} and Yb^{3+} ions in insulating crystals. Such media in a laser cavity are excited into stimulated emission at any wavelength within the shape function of the luminescence transition. Since these transitions are homogeneously broadened each centre is equally capable of contributing its energy to the required wavelength. Physically the electronic centres are coupled to the spectrum of vibrational excitations of the crystal environment and the ensuing bandwidths can be several times larger than the typical phonon energies. As described in Chapters 5 and 9, such *vibronic lasers* are four-level systems in which the transition energy in absorption and emission is partitioned between emitted photon and lattice phonons. However, when the laser is operated well above threshold, the broad spontaneous emission is suppressed and almost all the emission is concentrated at the wavelength of the selected mode. This spectral narrowing is a characteristic of lasers based on homogeneously broadened transitions.

A schematic illustration of the energy levels of a vibronic gain medium is shown in Fig. 10.1. The pump band is broad because the absorption transition samples the excited state configurational coordinate curve at level 2. The system then relaxes to the lowest vibrational state (level 3) within a few picoseconds. This *relaxed excited state* is the emitting state. Following the emission transition between levels 3 and 4, there is a further vibrational relaxation to the ground state in the normal configuration. The laser wavelength is determined by whichever of the vibrational levels is the terminus of the electronic transition. For this system the populations of levels 3 and 4 are inverted for any finite pumping rate, thus making it easier to obtain optical gain. From the gain coefficient per pass $\gamma(\nu)$, Eq. (1.22), it is straightforward to determine the gain cross section σ_p at the peak (λ_p) of a Gaussian-shaped band with full-width at the half-power points $\delta\lambda$ as

$$\sigma_p = \left(\frac{\eta}{8\pi n^2 \tau}\right)\left(\frac{c\lambda_p^4}{1.07\delta\lambda}\right), \tag{10.1}$$

in which n is the refractive index of the host, η is the quantum efficiency of luminescence with decay time τ, and c is the velocity of light. The gain coefficient at the peak is just $\gamma_p = \Delta N \sigma_p$, where the population inversion ΔN is effectively the population density in level 3. At the laser threshold ΔN is directly proportional to the pump beam intensity (I). Since the pump rate from the ground state NW_p is equal to the photon absorption rate $\alpha I/E$, where W_p is the pump rate, α is the absorption coefficient at the pump energy E, we have $\Delta N = NW_p\tau = \alpha(I/E)\tau$. Hence, the small gain coefficient at the emission peak is

$$\gamma_p = \left(\frac{\eta}{8\pi n^2}\right)\left(\frac{\lambda_p^4}{1.07\delta\lambda}\right)\left(\frac{\alpha I}{E}\right). \tag{10.2}$$

These are perfectly general equations for a four-level laser with gaussian bandshape, a reasonable approximation at high temperature. The vibronic bandshape is more accurately a Pekarian at low temperature, becoming more gaussian with increasing temperature as its peak shifts to longer wavelength (§5.2).

10.2 Colour centre lasers

Colour centres are electrons or holes trapped at vacancy or interstitial sites in ionic crystals [Henderson and Imbusch (1989), Spaeth *et al.* (1992)]. Since the neighbouring ions provide almost all the trapping potential, the energy levels of the trapped particle are strongly coupled to the phonon spectrum, resulting in rather wide, homogeneously broadened optical spectra. The present discussion has focused on the spectroscopic properties of electrons trapped at anion vacancies, i.e. *F*-type centres (§4.6, §9.2.1). *F*-centres in the alkali halides behave in absorption as localized optical centres with broad transitions that can be engineered to occur within the wavelength range 400–700 nm by varying the lattice constant (Fig. 4.11). Competing radiative and nonradiative decay from the excited state lead to very weak luminescence at 300 K. The broad $^2T_{1u} \rightarrow {}^2A_{1g}$ emission bands also have small emission cross sections ($\sigma < 10^{-19}$ cm^2) compared with the corresponding absorption transition ($\sigma \sim 10^{-17}$ cm^2) because of poor wavefunction overlap between the relaxed excited state and the ground state. The net result is that *F*-centres cause an optical loss rather than gain over their luminescence bands. The distribution of the *F*-electron charge outside the anion vacancy, implied by magnetic resonance spectra, provides a weak positive charge that can trap a second electron in the vacancy to form an F' centre.

The optical absorption spectra of the F' centres are broad and asymmetric, involving transitions to a continuum of states resonant with the conduction band and overlapping the F-bands on long and short wavelength sides. In consequence, the F' centre is thermally and optically unstable even below room temperature and acts mainly as a temporary repository for electrons in the formation of F-aggregate centres. The thermal ionization of F-centres leaves empty anion vacancies (F^+-centres), which, being mobile, readily participate in defect aggregation reactions. Anion vacancy centres (F^+, F and F^-) combine with one another and other defects in forming F-aggregate centres, several of which are laser-active. The aggregation processes are complex but can be controlled to yield reproducible concentrations of centres. F_A(II), F_B(II) and all F_2^+-type centres are four-level laser systems which have advantages relative to transition-metal and rare-earth ion lasers, given their large oscillator strengths that follow from allowed electric dipole transitions (§4.6). The optical spectra of F-aggregate centres involve allowed electric dipole transitions with large oscillator strengths ($f \sim 0.1-1.0$) and large absorption/emission cross sections ($\sigma \sim 10^{-17}\,\text{cm}^2$). F_A(I) centres are in some respects barely distinguishable from F-centres: their relaxed excited states are spatially diffuse giving rise to emission cross sections too small to sustain laser action. In contrast, F_A(II) centres undergo ionic relaxations in their excited states, Fig. 9.3, that result in emission from a symmetrical double-well potential with very large Stokes shift (1.4–1.5 eV), large oscillator strength ($f \sim 0.3$), gain bandwidth ($\sim 599-600\,\text{cm}^{-1}$) and peak emission cross section ($\sim 3 \times 10^{-16}\,\text{cm}^2$). Two polarized optical absorption bands overlap somewhat, enhancing the wavelength accessibility to a convenient pump laser. The emission is also polarized and short-lived ($\sim 10^{-7}$ s). However, the huge Stokes shift of F_A(II) (and F_B) centres limits laser output power since *ca* 70% of pump power is dissipated in the crystal. Thus the input and output powers of F_A(II) centre lasers are limited to 1–2 W and a few hundred mW, respectively. For optimal performance of F_A(II) centre lasers proper management is required of the reorientation of centres under polarized excitation in either F_{A1} or F_{A2} bands. Typical CW power tuning curves of F_A(II) and F_B(II) centre lasers are shown in Fig. 10.2 to illustrate the broad, overlapping tuning ranges provided by gain media comprising F_A (Li) centres in KCl and RbCl and F_B (Na) centres in KCl.

The family of alkali halides provides a net tuning range for F_2^+-centre lasers from 800 nm–2000 nm [Mollenauer (1985a), (1987)]. These lasers fade during operation by centre reorientation and migration, and are hardly practical devices for application outside the research laboratory. However, laser operation with $(F_2^+)^*$ and $(F_2^+)^{**}$ centres also offers high efficiency, small Stokes shift and low power threshold, typical of their progenitors, the F_2^+ centres. Furthermore, they are comparatively stable, their mild fading characteristic during CW laser operation decreasing rapidly with increasing pump wavelength within the absorption band. The data for the $(F_2^+)^{**}$ centres (later styled $F_2^+:O^{2-}$ centres) shown in Fig. 1.6 are quite typical of the genre. F_2^+ centres can also be stabilized by substitution of an alkali impurity ion in the nearest cation shell. Such centres are designated here as F_{2A}^+-centres, particular examples being F_{2A}^+(Li) centres in KCl, KBr, KI and RbI, F_{2A}^+(Na) centres in KCl [Schneider and Marrone (1979), Schneider and Pollock (1983), Schneider and Moss (1983)]. Their absorption and emission curves, Fig. 10.3, show that tunable laser emission is possible from *ca* 1.6–3.5 μm, suitable for applications to studies of the fundamental stretching vibrations of C–H bands. Using co-doped $Li^+:Na^+:KCl$ crystals, continuously

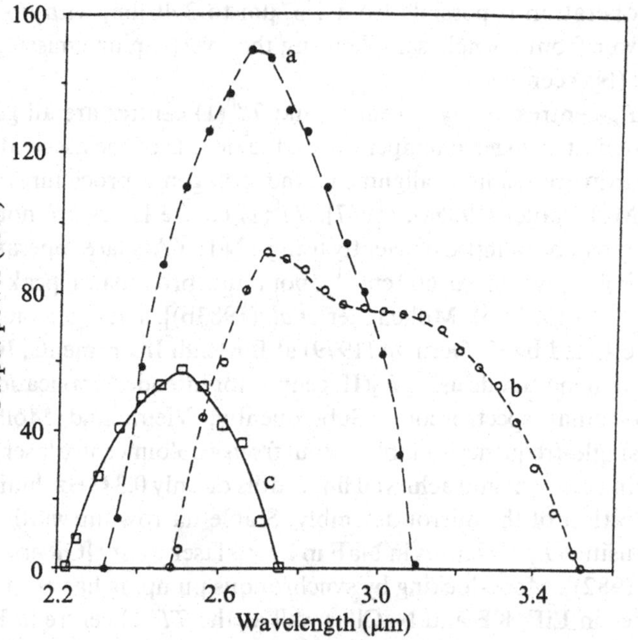

Figure 10.2. CW output power of F_A(II) and F_B(II) centre lasers as a function of wavelength (a) F_A(Li) centres in KCl pumped with 2W of power at 610 nm, (b) F_A(Li) centres in RbCl pumped with 2 W at 660 nm and (c) F_B(Na) centres in KCl pumped with 1.6 W at 595 nm [after German (1986)].

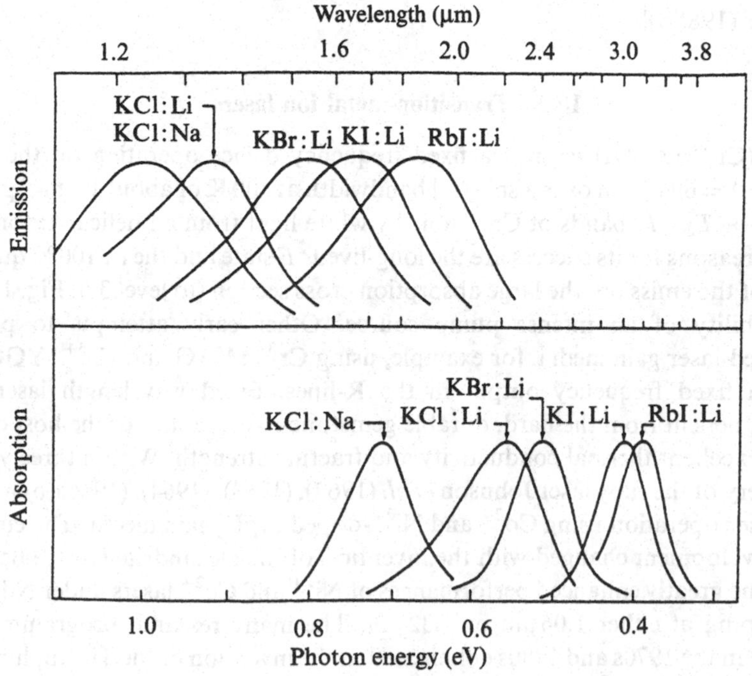

Figure 10.3. Normalized absorption and emission spectra of F_{2A}^+ centres in various hosts [after Mollenauer (1985a)].

tunable laser operation is possible from 1.67 μm to 2.46 μm, with up to 400 mW of stable CW power from a single sample using the overlapping emission bands of the F_2^+(Li) and F_2^+(Na) centres.

F_A(II) and F_B-centres, F_2^+-type centres and $Tl^0(1)$ centres are all gain media that must be operated at cryogenic temperatures: the details of the associated optical and mechanical hardware, sample alignment and cryogenic procedures are fulsomely described by Mollenauer (1985a), (1987). $Tl^0(1)$ centre lasers are not strongly host dependent: they can be pumped efficiently using a Nd : YAG laser operated at 1.062 μm to give a tuning range of *ca* 600 cm^{-1} about the broadband peak near 1500 nm [Gellermann *et al.* (1981a,b), Mollenauer *et al.* (1983b)]. The first commercial colour centre laser, designed by K. German (1979) at Burleigh Instruments, Inc. operated in stable single-frequency mode using F_A(II) centres for intended application in ultrahigh-resolution molecular spectroscopy. Subsequently, Vieira and Mollenauer (1985) achieved CW single-frequency tunable output from a colour centre laser using a grating as the sole tuning element and achieved linewidths of only 0.3 GHz, limited by the low-frequency vibration of the mirror assembly. Stable narrow linewidth operation was also achieved using $(F_2^+)^*$ centres in NaF in a ring laser cavity [Giberson *et al.* (1982), Trenec *et al.* (1982)]. Mode-locking by synchronous pumping has been achieved using F_2^+-type centres in LiF, KF and NaCl as well as the $Tl^0(1)$ centre in KCl with pulse widths in the range 4–10 ps [Mollenauer and Bloom (1979), Mollenauer *et al.* (1980), (1982)]. Such mode-locked pulses were shortened to a few tens of femtoseconds in the soliton laser, a synchronously pumped and mode-locked $Tl^0(1)$ centre laser coupled to a feedback cavity containing a single-mode polarization preserving optical fibre. Pulse compression and soliton formation in the fibre enable the colour centre laser to produce much shorter than normal optical pulses [Mollenauer and Stolen (1984), Mollenauer (1985b)].

10.3 Transition-metal ion lasers

The ruby (Cr^{3+} : Al_2O_3) laser is a fixed frequency device operating on the R-line emission at $\lambda = 694.3$ nm over a spectral bandwidth at 300 K of about 1.2 nm, pumped in the $^4A_2 \rightarrow {}^4T_2, {}^4T_1$ bands of Cr^{3+} ions by white light from an helical xenon flash-lamp. The reasons for its success are the long-lived ^2E state and the *ca* 100% quantum efficiency of the emission, the large absorption cross section (to level 3 in Fig. 1.3) and the availability of an intense pump source. Other early attempts to produce Cr^{3+} : doped laser gain media, for example, using Cr^{3+} : YAG and Cr^{3+} : YGG, also resulted in fixed frequency output in the R-lines. Fixed wavelength lasers, like Nd : YAG, benefit from the hard, durable gemstone-like qualities of the host crystals including excellent thermal conductivity and fracture strength. Within three years of the discovery of the ruby laser Johnson *et al.* (1962), (1963), (1964), (1966a,b) reported tunable laser operation using Co^{2+} and Ni^{2+}-doped MgF_2 gain media. The emphasis in laser development changed with the invention of the alexandrite (Cr^{3+} : $BeAl_2O_4$) laser and the greatly enhanced performances of Ni^{2+} and Co^{2+} lasers under Nd : YAG laser pumping at either 1.06 μm or 1.32 μm. The many research programmes that blossomed in the 1970s and 1980s culminated in the invention of the Ti-sapphire laser, Ce^{3+}-doped lasers, Cr^{3+} : colquiriite lasers and the forsterite laser, all broadband tunable.

The spectroscopic properties of $Ti^{3+}:Al_2O_3$ are determined by Ti^{3+} substitution for Al^{3+} at distorted $(AlO_6)^{9-}$ octahedra, the distortions from O_h to C_3 symmetry being conveniently envisaged in trigonal perspective, Fig. 3.1 (§3.3). Such distorted six-fold deployment of ligand ions around a central optically-active ion is not only the heart of the $Ti^{3+}:Al_2O_3$ gain medium but also of ruby, alexandrite, Cr^{3+}: garnets and Cr^{3+}: colquiriite laser crystals as well as several Ni^{2+} and Co^{2+}-doped materials. Indeed the distorted octahedral molecular ion is also the locus for optical interactions in the more complex structures of the rare-earth aluminoborates ($REAl_3(BO_3)_4$) and scandioborates ($RESc_3(BO_3)_4$), rare-earth titanyl niobates ($RETiNbO_6$) as well as certain SHG crystals ($LiNbO_3$, $KNbO_3$ and $KTiOPO_4$).

Usually the optical spectra of transition-metal ion complexes are described in terms of crystal-field theory which neglects both the finite spatial extent of ions and the wavefunction overlap of the optical centre and neighbouring ligands. However, given the power of modern computers, it is possible to carry out energy level calculations which include finite ion sizes [Woods *et al.* (1993), Aramburu *et al.* (1999)]. The spirit of such calculations was discussed in §8.6.4, with applications to Cr^{3+}-doped elpasolites (§8.6.1) and the $Tl^0(1)$ centre in the alkali halides. Consideration was also given to the $(CrF_6)^{3-}$ molecular ion in a perfect octahedron and under the action of a T_{2u} displacement characteristic of the Cr^{3+}: colquiriite laser crystals. This calculation emphasized the dominant role in parity mixing by odd-parity combinations of ligand orbitals rather than by metal $4p$ and $3p$ orbitals. Such mixing of s-, p_π- and p_σ-ligand wavefunctions into even-parity $3d$-wavefunctions was the basis of the molecular orbital model described in §8.6.3 for the case of Ti^{3+} ions in distorted octahedral environments. This model was applied by Yamaga *et al.* (1991a,b) to calculate the polarized relative intensities of d–d transitions induced by T_{1u} and T_{2u} distortions of crystals with trigonal, tetragonal and orthorhombic symmetries. The results of such calculations for Ti^{3+}-doped Al_2O_3, YAG and $YAlO_3$ were discussed in §9.3.2. As outlined below, Yamaga *et al.* (1990a,b) also applied this molecular orbital model to Cr^{3+}- activated laser gain media. The theory is readily adapted to other electron configurations, e.g., $3d^2$ (Cr^{4+}) and $3d^8$ (Ni^{2+}).

10.3.1 Alexandrite laser

The characteristics of most Cr^{3+}-based laser systems can be understood using the simplified Tanabe–Sugano diagram in Fig. 4.8, which shows the energy levels of Cr^{3+} in octahedral fields as a function of Dq/B with $C/B = 4.5$. The energy gap $E_0 = E(^4T_2) - E(^2E)$, which varies with crystal-field strength, determines the criteria for operation of $3d^3$ ion lasers at $T = 0$ and at finite temperatures (see Eqs. (5.134) and (5.135)). For ruby and Cr^{3+}: YAG E_0 is 2300 cm^{-1} and 1500 cm^{-1}, respectively. In these crystals at room temperature only the ^2E level is occupied and the resulting spin-forbidden $^2E \rightarrow {}^4A_2$ R-lines support fixed wavelength laser operation [Maiman (1960), Sevast'yanov *et al.* (1973)]. In contrast the $Cr^{3+}:BeAl_2O_4$ laser ($E_0 = 800\,cm^{-1}$), operates on the broad $^4T_2 \rightarrow {}^4A_2$ band at 300 K [Walling *et al.* (1979)]. Cr^{3+}: YGG ($E_0 = 600\,cm^{-1}$) and emerald ($E_0 = 400\,cm^{-1}$) are similar to alexandrite, the $^4T_2$ and ^2E levels being close enough together that a fraction of the excited ions (≅ 6% in alexandrite, 17% in YGG and 44% in emerald) can be maintained in the $^4T_2$ level at room temperature. Even though this fraction can be small, the spin-allowed $^4T_2 \rightarrow {}^4A_2$

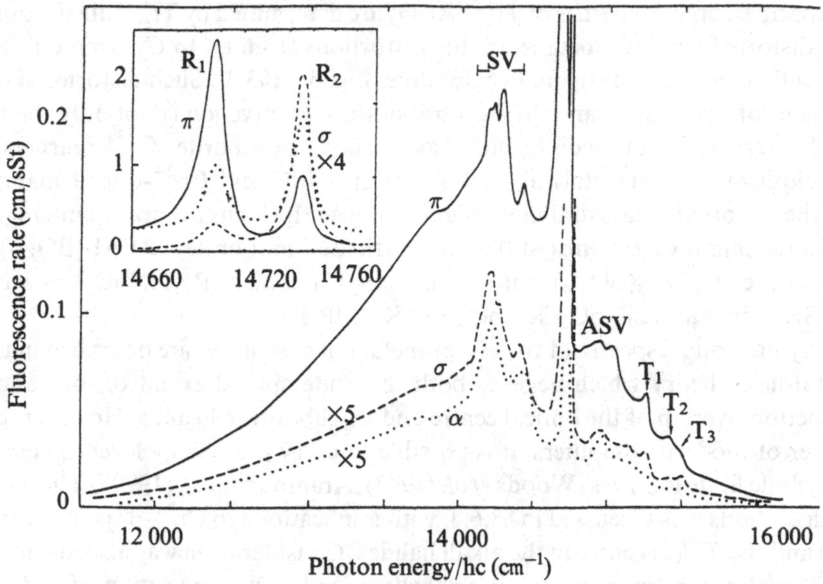

Figure 10.4. Polarized fluorescence Cr^{3+} : $BeAl_2O_4$ at 300 K showing Stokes (SV) and anti-Stokes (ASV) vibronic sidebands of the R-lines superposed on the $^4A_2 \rightarrow {}^4T_2$ broadband [after Walling *et al.* (1980)].

emission is dominant at room temperature because its transition probability is two or three orders of magnitude larger than that of the R-line. For Cr^{3+} in $KZnF_3$ $E_0 = -1500\ cm^{-1}$ and the vibronic $^4T_2 \rightarrow {}^4A_2$ luminescence band is the only emission at all temperatures [Dürr and Brauch (1986), Dürr *et al.* (1985)].

The crystal structure of $BeAl_2O_4$ in Fig. 3.3 shows that O^{2-} ions form two approximately close-packed hexagonal cells with Al^{3+} ions at their centres. The Cr^{3+} impurity ions substitute for Al^{3+} in these $(AlO_6)^{3-}$ cells. The combined density of Al^{3+} sites in $BeAl_2O_4$ is $3.509 \times 10^{22}\ cm^{-3}$. A typical laser rod measures 3–6 mm ϕ by 10 mm in length and contains 0.09–0.28 at.% Cr^{3+}. Although Cr^{3+} dopants substitute at both C_s and C_i sites, electron spin resonance studies show that $78 \pm 3\%$ of Cr^{3+} ions substitute on the C_s site in alexandrite [Forbes (1983)]. The site with slightly shorter Al–O bonds has inversion C_i symmetry. The octahedral component of the crystal field ($Dq/B = 2.85$, $\Delta E = 2400\ cm^{-1}$) is even stronger than that in ruby. The weaker octahedral field at the mirror, C_s site, characterized by $Dq/B = 2.47$ and $\Delta E = 800\ cm^{-1}$, provides a somewhat better fit for the Cr^{3+} ion. The R-lines and their vibronic sidebands are shown atop the short wavelength wing of the broad $^4T_2 \rightarrow {}^4A_2$ emission in Fig. 10.4. The R-line splitting of *ca* 40 cm^{-1} is larger than that in ruby (29 cm^{-1}), resulting from a stronger even-parity distortion of the $(CrO_6)^{9-}$ octahedron. Splittings of the $^4A_2 \rightarrow {}^4T_2$ transition are revealed in the polarized absorption spectra, Fig. 10.5 [Walling *et al.* (1980)]. In the molecular orbital model of the $(CrO_6)^{9-}$ octahedron, the $^4T_2$ state is split by orthorhombic crystal-field and spin–orbit coupling into $^4T_2$ (η_x), $^4T_2$ (η_y) and $^4T_2$ (η_z) levels described by wavefunctions that are linear combinations of the octahedral basis functions. The energies of these levels may be determined from the splittings in the polarized $^4A_2 \rightarrow {}^4T_2$ absorption spectra. The peaks in Fig. 10.5 occur at

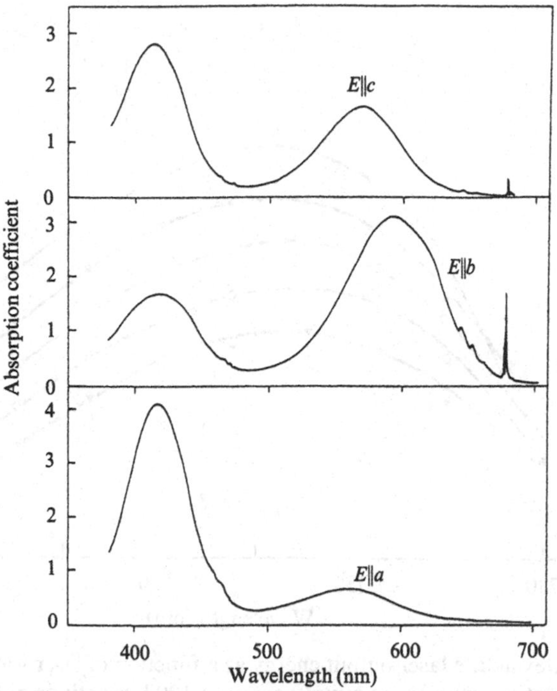

Figure 10.5. Polarized optical absorption spectra of an alexandrite crystal containing 2.2×10^{19} Cr^{3+} cm^{-3} [after Walling *et al.* (1980)].

563 nm (η_x), 573 nm (η_y), and 595 nm (η_z), the pseudo-trigonal splitting $E(\eta_x, \eta_y) - E(\eta_z) = 795\,cm^{-1}$ and the orthorhombic splitting $E(\eta_x) - E(\eta_y) = 310\,cm^{-1}$.

The combined mirror site plus inversion site absorption spectra shown in Fig. 10.5 can be resolved by selective excitation techniques to show that for the inversion site the $^4A_2 \rightarrow {}^4T_2, {}^4T_1$ absorption bands peak at 480 nm and 375 nm, respectively. There is a contribution of *ca* 50% of the total absorption at the centre of the trough between the $^4T_1$ and $^4T_2$ bands, indicating why the inversion site is unimportant in the laser emission band (Fig. 10.4). From Fig. 10.5, and knowing the concentration of Cr^{3+} ions on the mirror site, the peak cross sections for the $^4A_2 \rightarrow {}^4T_2, {}^4T_1$ transitions are determined as $1.36 \times 10^{-19}\,cm^2$ and $0.77 \times 10^{-19}\,cm^2$ measured at 573 nm and 417 nm, respectively. The emission cross section is determined from Fig. 10.4, given independent determinations of the quantum yield and the luminescence life time. Since the quantum yield determined by photoacoustic spectroscopy is $95 \pm 5\%$ [Shand (1983)], the radiative lifetime is essentially the fluorescence decay time. The measured decay time at 300 K is 260 μs for crystals containing $2 \times 10^{18}\,cm^{-3}$ up to $2 \times 10^{20}\,cm^{-3}$; there is no concentration quenching within the composition range used in laser applications. The effective stimulated emission cross section in the $^4T_2 \rightarrow {}^4A_2$ transition, after allowing for the effects of excited state absorption (Fig. 9.10) is of order $6\text{–}7 \times 10^{-21}\,cm^2$ in the wavelength range 700–800 nm [Shand and Jensen (1983)]. At 300 K the major component of excited state absorption comes from thermal occupancy of the $^4T_2$ levels, terminating in the higher lying $^4T_1$ level. In $Cr^{3+}:BeAl_2O_4$ the crystal field is close enough to the $^2E\text{–}^4T_2$ crossing to allow a significant population of the $^4T_2$ level at 300 K, which results in a temperature dependent gain spectrum.

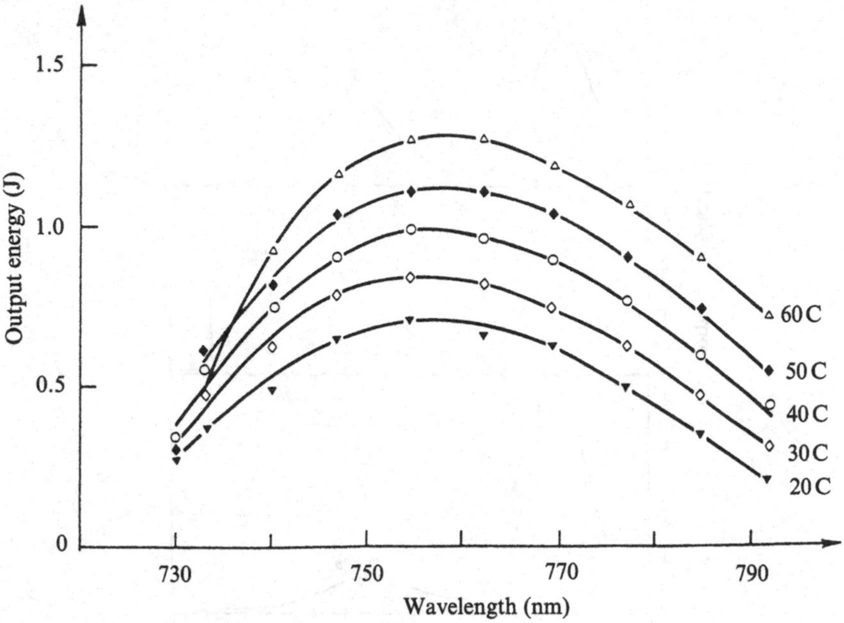

Figure 10.6. Alexandrite laser output energy as a function of laser wavelength for various temperatures with input pump energy of 100 J, repetition rate 4 Hz with R = 87% output mirror reflectivity [after Sam *et al.* (1980)].

The parity selection rule for *d–d* transitions is lifted by T_{1u} and T_{2u} distortions (Fig. 3.2) inducing covalent admixtures of ligand *s*- and *p*-orbitals into $^4A_2$ and $^4T_2$ states. The π- and σ-polarized probabilities for the $^4A_2 \rightarrow {}^4T_2$ (η_x, η_y, η_z) absorption transitions induced by *X*-, *Y*- and *Z*-components of the T_{1u} and T_{2u} distortions were calculated and presented diagrammatically in Yamaga *et al.* (1990a). The absorption coefficients at the band peaks, 1.64 cm^{-1} at 573 nm ($E \parallel c$), 0.63 cm^{-1} at 563 nm ($E \parallel a$) and 3.14 cm^{-1} at 595 nm ($E \parallel b$) show that the $^4A_2 \rightarrow {}^4T_2$ transition is induced by the T_{1u}^x distortion with a small contribution from T_{1u}^z. Luminescence from the lowest lying $^4T_2$ (η_z) level is strongly polarized with measured relative intensities $I(E \parallel b) : I(E \parallel c) : I(E \parallel a) = 1 : 0.1 : 0.06$, close to the calculated intensity ratio of 1 : 0 : 0 assuming that the dominant contribution to the polarized luminescence is induced by the T_{1u}^x mode.

The alexandrite laser was the most commercially developed tunable solid state laser until the advent of the Ti-sapphire laser. In addition to admirable optical qualities, the thermal conductivity ($\kappa = 0.23$ W/cm^2) and fracture stress ($\sigma_F \sim 8$–9×10^6 Pa) of alexandrite permit laser operation at very high average power. The effective gain cross section increases with temperature to well above 300 K because of the particular value of $\Delta E = 800$ cm^{-1}, and despite the decreased upper-state lifetime at 250–350°C it is still sufficiently long-lived to permit effective energy storage in Q-switched operation with 20–50 μs flashlamp pulses. Excited state absorption is significant for most Cr^{3+}-ion lasers, whether operating on the R-line or the vibronic band. In alexandrite the variety of ESA transitions from spin-doublet and spin-quartet levels results in absorption at both pump and laser wavelength. However, there is a fortuitous excited state absorption minimum at the peak laser wavelength (Fig. 9.10), and the rising ESA cross section on either side of the peak determines the long- and short-wavelength extremes of the tuning range. The tuning curve as a function of operating temperature is shown for flashlamp pumping in Fig. 10.6: the increase in output energy with increasing temperature occurs because the threshold is lowered by the increased gain cross section. Most modes of operation have been reported for the alexandrite laser. Q-switching under flashlamp pumping conditions can generate energies of 2J/pulse with pulse widths of 40 ns [Lai and Shand (1983)]. CW operation has been achieved by Hg or Xe arc pumping and by Kr^+- laser pumping at 647.1 nm. A quantum efficiency of 85% was calculated for a 4.6 mm long laser crystal pumped longitudinally by the 647.1 nm line in a near concentric laser cavity. Flashlamp pumping with passive mode-locking has been used to generate transform limit pulses with 38 ps duration at 750 nm: Lisityn *et al.* (1982) have reported pulses as short as 8 ps in the wavelength range 725–745 nm.

10.3.2 Cr^{3+} : colquiriite lasers

Colquiriite (LiCAF) and its isomorphs (LiSAF, $LiSr_xCa_{1-x}AlF_6$, LiSGaF and LiSCrAF) are ideal hosts for Cr^{3+} substituents on the tripositive cation sites. Payne and his colleagues at Lawrence Livermore National Laboratory (LLNL) first operated

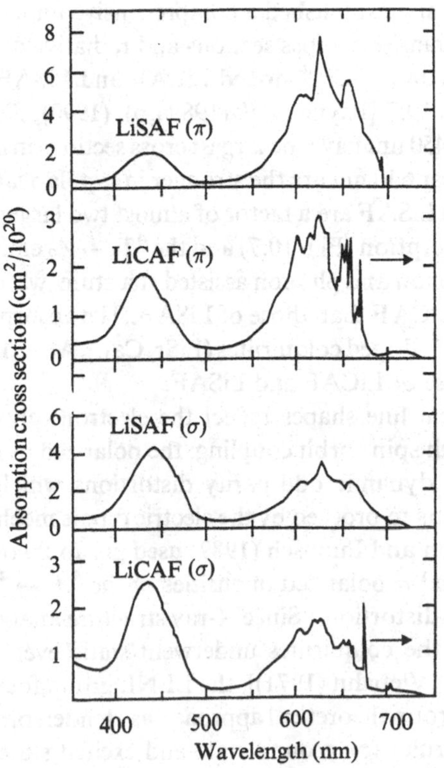

Figure 10.7. Polarized absorption spectra of Cr^{3+} : LiCAF and Cr^{3+} : LiSAF measured at $T = 20\,K$ for $E \parallel c(\pi)$ and $E \perp c(\sigma)$ polarization [after Payne *et al.* (1989a)].

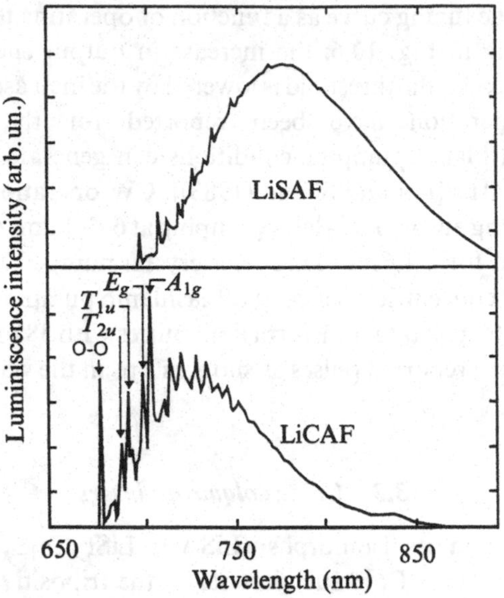

Figure 10.8. Photoluminescence spectra of Cr^{3+}:LiCAF and Cr^{3+}:LiSAF measured at $T=20$ K [after Payne *et al.* (1989a)].

Cr^{3+}: colquiriite lasers and established a comprehensive understanding of their large tuning ranges, strong transition cross sections and radiative decay processes. Typical optical absorption spectra for Cr^{3+}-doped LiCAF and LiSAF in π- and σ-polarizations are shown in Fig. 10.7 [Payne *et al.* (1989a,b), (1990), Smith *et al.* (1992)]. The $^4A_2 \rightarrow {}^4T_{1a}$ bands near 450 nm have the largest cross sections in σ-polarization whereas the $^4A_2 \rightarrow {}^4T_2$ bands near 640 nm are the stronger in π-polarization. Furthermore, the absorption strengths in LiSAF are a factor of almost two larger than those in LiCAF. Both the $^4A_2 \rightarrow {}^4T_2$ absorption (Fig. 10.7) and the $^4T_2 \rightarrow A_2$ emission (Fig. 10.8) bands show resolved zero-phonon and phonon assisted structure, which is much more prominent in the spectra of LiCAF than those of LiSAF. The absorption and luminescence spectra of the other Cr^{3+}: doped colquiriites ($LiSr_xCa_{1-x}AlF_6$, LiSGaF and LiSCrAF) are quite similar to those of LiCAF and LiSAF.

In general, the optical line shapes reflect the electron–phonon coupling to even parity modes acting with spin–orbit coupling: the polarized intensities are induced by coupling to static and dynamic odd-parity distortions which enable the normally forbidden *d–d* transitions to proceed by the electric dipole mechanism (§5.3.3, Sugano *et al.* (1970)). Henderson and Imbusch (1989) used group theoretical methods to calculate the relative π- and σ-polarized intensities of the $^4A_2 \rightarrow {}^4T_2$, $^4T_1$ bands of ruby induced by T_{1u} and T_{2u} distortions. Since X-ray structure analysis had shown that the $(MF_6)^{3-}$ octahedra in the colquiriites underwent static even-parity (T_{2g}) and odd-parity (T_{2u}) distortions [Viebahn (1971)], the LLNL group focused on the T_{2u} distortion in using the same group theoretical approach as Henderson and Imbusch (1989) to calculate the selection rules for ground state and excited state absorptions of Cr^{3+}: colquiriites. In contrast, Yamaga *et al.* [(1990a), (1991a)] used the molecular orbital model described in §8.6.3 to calculate the relative intensities of the spectra of Ti^{3+} and Cr^{3+} ions in distorted octahedral symmetries. Table 10.1 shows that I_π/I_σ, the theo-

Table 10.1. *The relative polarized intensities of* $^4A_2 \rightarrow {}^4T_2$ *transitions in trigonal symmetry induced by* $T^i_{1u}(\sigma)$, $T^i_{1u}(\pi)$ *and* $T_{2u}(\pi)$ *distortions, where* $i = X$, Y *and* Z *represent the orthogonal components of the odd-parity distortion*

Odd-parity component*	Relative $^4A_2 \rightarrow {}^4T_2$ polarized intensity		
	σ_X	σ_Y	π
$T^X_{1u}(\sigma)$	0	0	3
$T^Y_{1u}(\sigma)$	0	0	3
$T^Z_{1u}(\sigma)$	3	3	0
$T^X_{2u}(\pi)$	3	2	1
$T^Y_{2u}(\pi)$	2	3	1
$T^Z_{2u}(\pi)$	1	1	4

*The calculated intensities induced by $T^i_{1u}(\pi)$ and $T^i_{1u}(\sigma)$ are identical

retical polarized intensity ratios of the $^4A_2 \rightarrow {}^4T_2$ transitions, have the values 4 : 1 when induced by the $T^z_{2u}(\pi)$ distortion and 0 : 3 when induced by a $T^z_{1u}(\sigma)$ distortion. The calculated selection rules for the $^4A_2 \rightarrow {}^4T_1$ transitions are of opposite sense to those given in Table 10.1 for the $^4A_2 \rightarrow {}^4T_2$ transitions such that $I_\pi/I_\sigma = 0/3$ when induced by $T_{2u}(\pi)$ and 4/1 when induced by $T_{1u}(\sigma)$. The results shown in Figs. 10.7 and 10.8 for LiCAF and LiSAF are in general agreement with the theory. The larger cross sections of transitions in LiSAF than LiCAF arise out of the stronger odd parity T_{2u} distortion in LiSAF as measured by the counter-rotating displacement ϕ, which is only 4.3° in LiCAF compared with 7.2° in LiSAF. However, for a more complete understanding of the polarized transition intensities it is necessary to take into account the splittings of the excited states by the even-parity T_{2g} distortion.

ESR studies of the $LiSr_xCa_{1-x}AlF_6$ system, where $0 < x < 1$, have shown that the substitution of the Sr^{2+} ion for Ca^{2+} introduces a gradual change in the strength of the even-parity trigonal distortion of the $(CrF_6)^{3-}$ octahedron so that the *c*-axis compression in LiCAF becomes an extension in LiSAF [Yamaga *et al.* (1999)]. This T_{2g} distortion also splits the excited $^4T_1$ and $^4T_2$ states, by much more than the splitting of the $^4A_2$ ground state. In LiCAF this *c*-axis compression splits the $^4T_2$ state into $^4A_1(^4T_2)$ and $^4E(^4T_2)$ levels of which the former is lower, whereas in LiSAF $^4A_1(^4T_2)$ is above $^4E(^4T_2)$. The splitting and ordering of excited states has implications for the polarizations of the optical transitions of Cr^{3+}: colquiriites. Assuming from the ESR studies the presence of the excited $^4T_2$ state splittings, Yamaga *et al.* (1999) deduced that the $^4A_2 \rightarrow {}^4A_1(^4T_2)$ transition is π-polarized and induced by the *z*-component of T_{2u} alone. In contrast, the $^4A_2 \rightarrow E(^4T_2)$ transition is σ-polarized and induced by both $T_{2u}(z)$ and $T_{1u}(z)$ distortions. Theoretically $I_\pi/I_\sigma = 4/(1 + 3a)$, in which a is the relative σ-intensity induced by $T_{1u}(z)$. The experimental intensity ratios are 1.5 for LiCAF and 2.0 for LiSAF, and the corresponding values of a are 3/9 and 5/9, respectively, confirming that the difference between LiCAF and LiSAF is in the relative strengths of their $T_{2u}(z)$ distortions. In emission, where $I_\pi/I_\sigma = 1.5$ for LiCAF and 2.8 for LiSAF, it is the π-polarized transition from $^4A_1(^4T_2)$ that is dominant in both cases. Hence, in LiCAF $^4A_1(^4T_2)$ is lower than $^4E(^4T_2)$ both before and after vibronic relaxation, whereas in LiSAF $^4E(^4T_2)$ is lower than $^4A_1(^4T_2)$ in the unrelaxed excited state but *above* $^4E(^4T_2)$ after excited state relaxation. Similar considerations apply to the spectra of the

Table 10.2. *Polarized oscillator strengths* of the* $^4A_2 \rightarrow {}^4T_{1a,2}$ *absorption of* Cr^{3+}*: doped fluorite crystals deduced from measurements at* $T = 20$ *K [adapted from Payne et al.* (1989a)]

Transition	Oscillator strength ($\times 10^6$)	Host crystals				
		$Na_3Ga_2Li_3F_{12}$	$LiCaGaF_6$	$LiCaAlF_6$	$LiSrGaF_6$	$LiSrAlF_6$
$^4A_2 \rightarrow {}^4T_2$	$f(\pi)^*$	19	39	48	78	109
	$f(\sigma)^*$	10	26	28	34	53
$^4A_2 \rightarrow {}^4T_1$	$f(\pi)^*$	33	36	39	32	45
	$f(\sigma)^*$	33	45	51	65	98
$^4A_2 \rightarrow {}^4T_2$	$f_S(\pi)^*$	0	17	27	59	75
	$f_D(\pi)^*$	19	21	21	19	34
$^4A_2 \rightarrow {}^4T_1$	$f_S(\sigma)^*$	0	9	12	33	53
	$f_D(\sigma)^*$	33	36	39	32	45

*The magnitude of the $f_S(\pi)$ are calculated for a static T_{2u} distortion and assuming the results given in Table 10.1

$LiSr_xCa_{1-x}AlF_6$ solid solutions. The polarizations of the main absorption features of $LiSr_{0.8}Ca_{0.2}AlF_6$ are of opposite sense to those in LiCAF and LiSAF [Chai *et al.* (1992)] and the σ-component of the polarized $^4T_2$ band and the π-polarized component of the $^4T_1$ band are dominant. These polarizations are induced by $T_{1u}(z)$, $T_{2u}(x)$ and $T_{2u}(y)$ distortions [Yamaga *et al.* (1990a). The emergence of the $T_{2u}(x)$ and $T_{2u}(y)$ distortions arises out of the symmetry-lowering effects of disorder on the $(CrF_6)^{3-}$ octahedra in solid solution crystals.

The spectra in Figs. 10.7 and 10.8 were recorded in terms of the cross sections for absorption, σ_A, and emission, σ_E, which are related by

$$\int \sigma_A \, d\bar{\nu} = (g_a/g_e) \int \sigma_E \, d\bar{\nu}, \tag{10.3}$$

in which g_a and g_e are the spectroscopic weights of the starting states in absorption (a) and in emission (e). Using Eq. (5.41) the absorption cross section can be related to the transition oscillator strength (f) thus:

$$f = (9n/(n^2+2)^2)(mc^2/\pi e^2) \int \sigma \, d\bar{\nu}, \tag{10.4}$$

in which n, m, c and e have their usual significance. The oscillator strengths of the $^4A_2 \rightarrow {}^4T_{1,2}$ absorption bands of Cr^{3+}-doped fluoride crystals, of order 10^{-4}–10^{-5}, are typical of d–d transitions of $3d^n$ configuration ions in ionic crystals. The f-values of Cr^{3+} ions in centro-symmetric fluorides, such as the fluoride garnet, $Na_3Ga_2Li_3F_{12}$, and elpasolite, K_2NaScF_6, are some 3–5 times weaker than those of Cr^{3+}: colquiriites because in the absence of static crystal-field distortion of T_{1u} or T_{2u} symmetry the transitions are induced by coupling to odd-parity phonons alone. Polarized oscillator strengths calculated using Eq. (10.4) can be separated into their static (S) and dynamic (D) components using $f(\pi) = f_S(\pi) + f_D(\pi)$ and $f(\sigma) = f_S(\sigma) + f_D(\sigma)$. The results of such an analysis for $Na_3Ga_2Li_3F_{12}$ and four colquiriite-structured crystals are collected in Table 10.2. In the fluoride garnet the $(GaF_6)^{3-}$ units are perfectly octahedral and there are no contributions to the oscillator strengths from static distortions of the crystal-field: dynamic contributions induced by T_{1u} and T_{2u} phonons are significant

(19×10^{-6}) in both π- and σ-polarizations. Furthermore, that the strongly π-polarized $^4A_2 \rightarrow {}^4T_2$ emission varies from compound to compound is due mainly to the T_{2u}-intensity enabling distortion, except in the fluoride garnet, $Na_3Ga_2Li_3F_{12}$. The increasing oscillator strengths across the series $Na_3Ga_2Li_3F_{12} \rightarrow LiCaGaF_6 \rightarrow LiCaAlF_6 \rightarrow LiSrGaF_6 \rightarrow LiSrAlF_6$ follow from the increasing T_{2u} distortion as measured by the angle $\phi = 4.3°$. in LiCAF to 7.2 in LiSAF. The calculation predicts the opposite sense of polarization for the $^4A_2 \rightarrow {}^4T_{1a}$ absorption, mainly because there is no contribution from the static T_{2u} distortion in π-polarization (Table 10.1). Additionally, the dynamically induced contributions, $f_D(^4A_2 \rightarrow {}^4T_2)$ and $f_D(^4A_2 \rightarrow {}^4T_{1a})$ are constant, to within experimental error, having mean values of $(23 \pm 7) \times 10^{-6}$ and $(37 \pm 6) \times 10^{-6}$, respectively [Smith *et al.* (1992)]. Primarily, it is the volume of the CrF_6 octahedron in these compounds, approximately constant across the series, which determines the vibronic interaction. In contrast, the values of $f_S(\Gamma)$ induced by the static T_{2u} distortion are larger in LiSAF and LiSGaF, because in these compounds the T_{2u} distortions are largest.

The life times of the $^4T_2 \rightarrow {}^4A_2$ luminescence at low temperature (~20 K) in LiCAF and LiSAF are quite short, being 205 μs and 67 μs, respectively. In the temperature range 20–300 K the lifetime of the Cr^{3+} emission is almost independent of temperature for LiSAF and weakly dependent on temperature in LiCAF only above 300 K. The emission life times (τ) and emission cross sections (σ_E) are related by

$$\frac{1}{\tau} = \frac{8\pi c n^2}{\lambda^2} \int \sigma_E \, d\bar{\nu}, \tag{10.5}$$

which can be linked to the absorption spectrum and the oscillator strengths through Eqs. (10.3) and (10.4). At very low temperatures ($T < 20$ K) Eq. (10.5) should yield the temperature-independent radiative lifetime, τ_R, defined in Eqs. (1.18) and (5.67) and there related to the Einstein coefficient for spontaneous emission, A_{ba}. Calculating the Einstein coefficient from the spectral integration of the absorption and emission bands works quite well for the elpasolites and fluoride garnets, in which the CrF_6^{3-} cells are strictly octahedral. However, it is less accurate when Cr^{3+} ions occupy distorted octahedral sites since then the absorption integral is over all components of $^4A_2 \rightarrow {}^4T_2$, whereas the $^4T_2 \rightarrow {}^4A_2$ emission band involves only transitions from the lowest crystal-field component of the $^4T_2$ state. The measured low temperature (20 K) values of τ for Cr^{3+}: doped LiSAF, LiCAF and $Na_3Ga_2Li_3F_{12}$ are, in sequence, 67 μs, 205 μs and 530 μs, whereas the calculated values are 156 μs, 280 μs and 540 μs. The large trigonal splitting in the $^4T_2$ state is suggested to be the main source of disagreement in the case of Cr^{3+}: LiSAF [Payne *et al.* (1989a,b)].

The temperature dependence of the emission lifetime, Eq. (6.38), permits a determination of τ_R, τ_R^S, $\tau_R^{(\Gamma)}$, $\hbar\omega^{(\Gamma)}$, τ_{NR} and ΔE. Values of these parameters for $^4T_2 \rightarrow {}^4A_2$ emissions of the Cr^{3+} dopant in various mixed fluorides are presented in Table 10.3. In perfect octahedral symmetry at the Ga^{3+} and Sc^{3+} sites, respectively, in $Na_3Ga_2Li_3F_{12}$, and K_2NaScF_6, the $^4T_2 \rightarrow {}^4A_2$ emissions are vibronically induced and $(\tau_R^S)^{-1} = 0$. In consequence, the low temperature emission life times $\tau_R^{(\Gamma)}$ are long. For these eight compounds the enabling odd-parity phonons have mean energy, $\hbar\omega^{(\Gamma)} \cong 300$–$320\ cm^{-1}$, reasonably close to the energies of t_{1u} and t_{2u} phonons measured from the vibronic structure in the Cr^{3+} spectra in LiCAF, GFG and the elpasolites [Payne *et al.* (1989a), Dubicki *et al.* (1980)]. The materials dependences of the dynamically-induced decay rates, $(\tau_R^{(\Gamma)})^{-1}$, although small, are significant, being larger for the Sr-based than the Ca-based colquiriites. This materials dependency is rather stronger for $(\tau_R)^{-1}$, than

Table 10.3. *Fitting parameters for luminescence lifetime versus temperature dependence of* $^4T_2 \rightarrow {}^4A_2$ *luminescence, assuming Eq. (6.38) for various* Cr^{3+}*-doped fluorides*

Crystal	T_{2u} distortion	τ_R^+ (μs)	τ_R^{S+} (μs)	$\tau_R^{(\Gamma)+}$ (μs)	$\hbar\omega^{(\Gamma)}$ (cm^{-1})	τ_{NR} (10^{-14}s)	ΔE (cm^{-1})
$K_2NaScF_6^+$	0	1930	—	1930	300	8.3	7270
$Na_3Ga_2Li_3F_{12}$	0	530	—	490	310	—	—
LiCGaF	4°	210	350	440	300	—	—
LiCAF	4.3°	205	330	430	305	1.5	8900
LiSCAF	6.0°	80	105	310	315	14.0	4800
LiSAF	7.6	67	78	275	310	2.4	5175
LiSGaF	8.2	88	96	315	320	5.2	5520
LiSCrF	7.4	70	80	400	300	—	—

$^+$The measured radiative decay time is defined as τ_R: contributions to radiative decay induced by static T_{2u} distortions (τ_R^S) and by vibronic odd-parity distortions ($\tau_R^{(\Gamma)}$) are obtained by fitting to Eq. (6.38)

is the decay rate induced by the static T_{2u} distortion. However, the trends determined from the temperature dependences of the lifetime of Cr^{3+} : doped fluorides (Table 10.3) are in accord with those determined from an analysis of the oscillator strengths of the absorption spectra (Table 10.2). As τ_{NR} and ΔE in Table 10.3 show, thermal quenching of luminescence becomes important in the fluorides at a temperature that varies from host to host. Even so, nonradiative decay is not significant in the colquiriites at 300 K: that nonradiative decay is inefficient at room temperature follows from the small pre-exponent, $(\tau_{NR})^{-1}$, and large activation energy (ΔE) against nonradiative decay [Stalder *et al.* (1992)].

The luminescence spectra for Cr^{3+} : LiCAF and LiSAF [Payne *et al.* (1989a)] show resolvable zero-phonon lines and phonon-assisted sidebands at low temperature, Fig. 10.8: however, the structure in the Cr^{3+} : LiSAF emission is neither as intense nor as well resolved as it is in Cr^{3+} : LiCAF. Subsequent reports indicated that in LiSGaF and in LiSCAF the emission line shapes resemble that in LiSAF [Smith *et al.* (1992), Holliday *et al.* (1998)]. Indeed, the effect of disorder in the mixed colquiriite LiSCAF is to smear out the structure in the one-phonon sideband. The differences in the absorption and luminescence bandshapes are due to the different electron–phonon coupling strengths. According to the configurational coordinate model [§5.2.4, Henderson and Imbusch (1989)] the extent of the fine structure is determined by the offset in the equilibrium ionic positions of the ground and excited state, quantified for a totally symmetric, breathing mode (A_{1g}) by the Huang–Rhys factor, S, in Eq. (5.94). The values of S for LiCAF, LiSAF and LiSGaF are 4.2, 5.9 and 6.2 *et seq.*, implying that the relaxation of lattice around the Cr^{3+} ion is greater in LiSGaF and LiSAF than in LiCAF. In consequence, the fluorescence peak for Cr^{3+} : LiCAF is at shorter wavelength and is more structured than it is in Cr^{3+} : LiSAF (Fig. 10.8). The assignments of the one-phonon peaks in the emission band of Cr^{3+} : LiCAF were made by analogy with those in the perfectly octahedral CrF_6 of the elpasolites Cr^{3+} : K_2NaAlF_6 and K_2NaGaF_6 [Greenough and Paulusz (1979), Dubicki *et al.* (1980)]. The vibrations sampled by the Cr^{3+} ion will mimic the behaviour of the AlF_6 octahedron since the Cr^{3+} ions are substituents on the Al^{3+} sites. Accordingly, the 561 cm^{-1} peak is assigned to the A_{1g} breathing mode of the CrF_6 octahedra, whereas the E_g mode observed at 452 cm^{-1} corresponds to the coupled distortion of both F^- ions octahedrally disposed relative to the Cr^{3+} ion and the central Cr^{3+} ion in the plane perpendicular to the trigonal axis (Fig. 3.2). Coupling to T_{1u} and T_{2u} infrared modes gives rise to one-phonon peaks at 309 cm^{-1} and 217 cm^{-1}, respectively. The analogous peaks in K_2NaGaF_6 occur at 575 cm^{-1} (A_{1g}), 480 cm^{-1} (E_g), 330 cm^{-1} (T_{1u}) and 200 cm^{-1} (T_{2u}).

Careful examination of the Cr^{3+} : LiCAF and Cr^{3+} : LiSAF luminescence spectra show that there are four resolved zero-phonon lines in LiCAF. Of the three lines resolved in LiSAF, the dominant line may be an unresolved doublet [Payne *et al.* (1989a,b)]. The splittings of the zero-phonon line are caused by the interactions of the $^4T_2$ states with even-parity distortions and spin–orbit interaction. In the colquiriites and such other Cr^{3+} hosts as corundum and the garnets, the dominant T_{2g} distortion causes a first-order splitting of the $^4T_1$ and $^4T_2$ levels. In ruby and in alexandrite (Fig. 10.6) the splittings (*ca* 400–800 cm^{-1}) can be measured directly from $^4A_2 \rightarrow {}^4T_2$, $^4T_1$ absorption peaks measured in different polarizations [McClure (1962), Walling *et al.* (1980)]. That the splittings of the zero-phonon lines of the $^4A_2 \rightarrow {}^4T_2$ transition ($\cong$ 50–60 cm^{-1}) in Cr^{3+} : LiCAF and LiSAF are much reduced compared with both spin–orbit coupling (*ca* 180 cm^{-1}) and the crystal-field splitting (*ca* 220 cm^{-1}) is due to the quenching of spin–orbit and crystal-field operators by the dynamic Jahn–Teller effect [(§5.2.7) and Ham (1965)]. The dynamic Jahn–Teller effect also plays an important role in excited state absorption transitions in the Cr^{3+} : colquiriites and other fluoride hosts [Payne *et al.* (1989), Lee *et al.* (1989)].

The optical absorption and emission spectra of the Cr^{3+}:doped colquiriites, of which Figs. 10.7 and 10.8 are typical, hold out the promise of efficient laser operation being wavelength flexible in both pumping and output channels. This has been demonstrated in Cr^{3+}:doped LiCAF, LiSAF, LiSGaF and LiSCAF. Given the broadband spectra it is possible to excite tunable laser operation by flashlamp pumping, Kr^+-laser pumping at 647 nm and 676 nm, dye laser pumping, and semiconductor diode laser pumping at *ca* 670 nm. Smith *et al.* (1992) measured the slope efficiencies of a Cr^{3+} : LiSGaF laser from graphs of the laser output power measured as a function of power absorbed from the Kr^+ laser pump beam. Their data are displayed in Fig. 10.9 for two different output couplers ($T = 0.8\%$ and $T = 4.32\%$), giving slope efficiencies of 47.9% and 49.9%, respectively. Such similarity is consistent with very low passive losses in the laser rod, determined as $L = 0.05\%$ by applying Eq. (1.39), with an intrinsic slope efficiency of $\eta_o = 52\%$. The large difference between the threshold pump power as a function of output coupler, and indeed from run to run, is because of the sensitivity of P_{th}, Eq. (1.36), to the overlap of pump and laser modes. There is considerable similarity in the performance of the different colquiriite lasers: they have broad absorption and emission bands, although the emission band in Cr^{3+} : LiCAF (675–825 nm) is rather narrower than in LiSAF and LiSGaF (700–950 nm). They each have luminescence life times that makes them convenient for Q-switching at high repetition rates. The similarly large absorption and emission cross sections yield intrinsic laser efficiencies in excess of 50%. Apparently the Cr^{3+} : LiSGaF laser has better power handling capabilities, especially in CW operation, than LiCAF and LiSAF [Smith *et al.* (1992), Sorokina *et al.* (1996), Balembois *et al.* (1992), (1997)], primarily because of reduced thermal quenching of luminescence in Cr^{3+} : LiSGaF and the lower thermal expansion and anisotropy of thermal expansion of the LiSGaF compound.

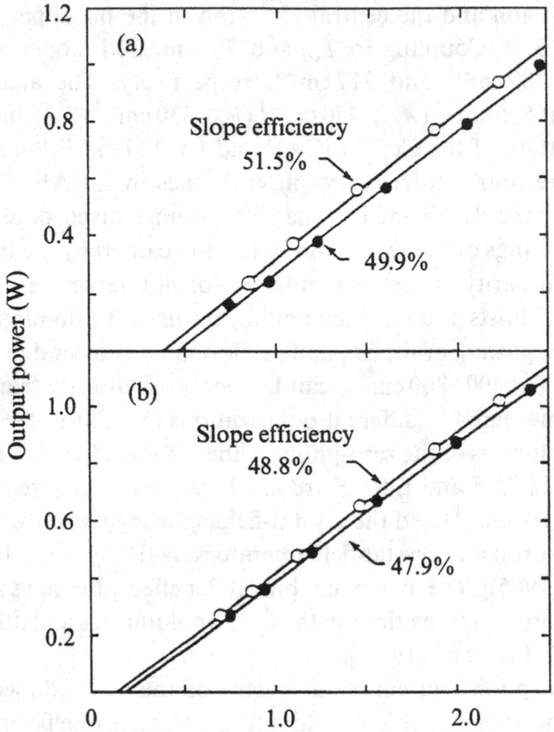

Figure 10.9. CW output power as a function of absorbed pump power for a Cr^{3+} : LiSGaF laser with output coupling transmission (a) 4.32% and (b) 0.80% [after Smith *et al.* (1992)].

10.3.3 Other Cr^{3+}-activated lasers

For many years the Nd^{3+} : YAG, Nd^{3+} : glass and Cr^{3+} : Al_2O_3 (ruby) lasers were the only commercial solid state lasers. Subsequently, the development of wavelength-agile lasers based on alexandrite and Ti-sapphire gave a dramatic boost to underlying research. Numerous other Cr^{3+}-oxide hosts were studied, including the garnets (GSGG, GSAG, YSGG, YSAG and LLGG), scalexandrite ($BeScAlO_4$), several borates ($SrBO_3$, $YAl_3(BO_3)_4$, $YSc_3(BO_3)_4$), the disordered gallogermanates and gallosilicates ($Ca_3Ga_2Ge_3O_{14}$, $La_3Ga_5GeO_{14}$ and $La_3Ga_5SiO_{14}$), a niobate $Ca_3Ga_{5.5}Nb_{0.5}O_{14}$ and zinc tungstate ($ZnWO_4$). Despite the obvious success of oxide hosts, there were attempts to develop fluorides also as practical laser hosts. Although fluorides have much lower fracture strength compared to oxides, they are mechanically robust enough for low and medium power applications. The development of Cr^{3+} : colquiriite lasers followed substantial research on such other hosts as MgF_2 and CaF_2 [Payne *et al.* (1987)], $KZnF_3$ [Brauch and Dürr (1984)], Cr^{3+} : $SrAlF_5$ [Lai and Jenssen (1984)] and Cr^{3+}-GFG [Payne *et al.* (1988a), Caird *et al.* (1988)]. In such crystals the Cr^{3+} ions are six-fold coordinated to nearby anions, as determined by considerations of the ionic radii of Cr^{3+} and host ions and the crystal-field stabilization energy [§9.6.4]. Among other reasons for the success of Cr^{3+} as the lasing

Table 10.4. *Properties of Cr^{3+} vibronic lasers*

Host	Free-running wavelength λ_o (nm)	Tuning range (nm)	Slope efficiency* η_o (%)	Reference
LiCAF	780	710–800	82	Smith *et al.* (1992)
Emerald	768	720–842	77	Lai (1987)
LiSAF	825	725–875	67	Smith *et al.* (1992)
LiSGaF	820	720–900	66	Smith *et al.* (1992)
Alexandrite	752	710–820	59	Walling *et al.* (1980)
$ScBO_3$	843	787–892	38	Lai *et al.* (1986)
Scalexandrite	792	—	37	Payne *et al.* (1989a)
GSGG	785	742–842	34	Struve & Huber (1985)
GFG	791	730–850	28	Caird *et al.* (1988)
$SrAlF_5$	932	825–1010	22	Caird *et al.* (1988)
GSAG	784	735–820	22	Drube *et al.* (1984)
$ZnWO_4$	1030	980–1090	20	Kolbe *et al.* (1985)
$KZnF_3$	820	785–865	18	Brauch & Dürr (1984), Dürr & Brauch (1986)
$La_3Ga_5SiO_{14}$	968	800–1150	10	Lai *et al.* (1986)
GGG	769	—	10	Struve & Huber (1985)
YSAG	767	730–815	5	Payne *et al.* (1990)
$La_3Ga_{5.5}Nb_{0.5}O_{14}$	1040	900–1250	5	Kaminskii *et al.* (1987)
$YAl_3(BO_4)_3$	745	700–870	5	Wang *et al.* (1997a,b)
CGGO	910	860–1210	<5	Kaminskii *et al.* (1987)
SGGO	840	900–1150	<5	Kaminskii *et al.* (1987)
$RbZnF_3$	830	790–860	<5	Dürr & Brauch (1986)
$V^{2+}:MgF_2$	1120	1070–1150	~ 0.1	Moulton (1985)
$V^{2+}:CsCaF_3$	1282	1240–1340	0.06	Brauch & Dürr (1984)

*Measured using a near-concentric cavity. The slope efficiencies are adjusted to take account of the quantum defect (λ_o/λ_p) for Kr^+ pumping at 647 nm

species in many crystals has been the stability of the Cr^{3+} state against oxidation and reduction, Table 9.6, and the resistance of the $^4T_2$ emitting state to nonradiative decay, resulting in the quantum efficiency of $^4T_2 \rightarrow {}^4A_2$ emission often being close to 100% at room temperature. Finally, Cr^{3+} ions preferentially occupy octahedral sites in crystals even though other site symmetries are available (as in the garnets, aluminoborates and gallogermanates).

Despite the many benefits of Cr^{3+}: doped laser materials there can be serious drawbacks. Much research is necessary to optimize the crystal growth process to eliminate parasitic scattering and absorption losses. Such problems are common to most laser crystals ahead of the successful implementation of quality control procedures. Many Cr^{3+}-doped garnets (GGG, GSGG, YSG, LLGG) experience defect formation during flashlamp pumping, which compete with Cr^{3+} as absorbers of pump radiation, albeit a lesser problem in the Al-based garnets GSAG and YSAG [Struve and Huber (1985), Drube *et al.* (1984), Barnes *et al.* (1986)]. Nonradiative decay resulting in low quantum yield at room temperature drives down the efficiency in $ZnWO_4$ and the Cr^{3+}: doped glasses, gallogermanates and gallosilicates [Andrews *et al.* (1981), Bergin *et al.* (1986), Rasheed *et al.* (1991), Macfarlane *et al.* (1996),

Grinberg *et al.* (1997) and references therein]. However, the more ubiquitous limitation of Cr^{3+} laser performance is excited state absorption.

Comparative data for a range of vibronic Cr^{3+} : lasers are presented in Table 10.4, the slope efficiencies have been adjusted for the quantum defect (λ_p/λ_L). The best performance is accorded to Cr^{3+} : LiCAF by virtue of the slope efficiency ($\eta_o = 82\%$) being very close to the quantum defect limit (83%) when pumped at 647 nm. LiSAF and LiSGaF are also excellent hosts combining good slope efficiencies with larger tuning ranges than LiCAF. The larger efficiency of the Cr^{3+} : LiCAF laser reflects in part the greater efforts made to reduce scattering losses in that material. However, of great importance are ESA transitions in the Cr^{3+}-doped colquiriites, fluoride perovskites, garnets, alexandrite and the Cr^{3+} : doped borates [Dürr *et al.* (1985), Caird *et al.* (1988), Moncorgé and Benyattou (1988), Andrews *et al.* (1986a), Shand and Jenssen (1983)]. In the case that ESA overlaps the laser band the effective emission cross section, σ_{eff}, is given by Eq. (7.24). The observed ESA transition varies host-by-host dependent on the ^2E or the $^4T_2$ excited state being the lower. In fluoride hosts (Table 10.4) $^4T_2$ is always the emissive level and the most probable ESA involves $^4T_2 \rightarrow {}^4T_1$ transitions. In LiCAF the $^4T_2 \rightarrow {}^4T_{1a}$ ESA peaks at *ca* 1000 nm with a peak cross section 0.17×10^{-20} cm^2, compared with the emission peak at 758 nm with cross section 1.3×10^{-20} cm^2. In consequence, the reduction of the emission cross section by ESA in Cr^{3+} : LiCAF is negligible.

Lee *et al.* (1989) discussed the ESA of Cr^{3+} : LiCAF in terms of the dynamic Jahn–Teller effect which couples E_g phonons to orbital triplets, which shifts and splits the $^4T_2 \rightarrow {}^4T_{1a}$ ESA band into two bands. According to Eq. (7.25), these bands are distinguishable from one another as $t_{2g} \rightarrow t_{2g}$ and $e_g \rightarrow e_g$ transitions and from the emission ($e_g \rightarrow t_{2g}$) band by their polarization characteristics (Table 10.5). Since the peak energy, $E_o = 7600$ cm^{-1}, predicted from the Tanabe–Sugano diagram and the observed $^4T_2 \rightarrow {}^4T_{1a}(2)$ transition occurs at 10 000 cm^{-1}, it is evident that $E_{JT} \cong 600$ cm^{-1}. This band is quite weak because there is no contribution to the oscillator strength from the static T_{1u} distortion, and is induced by odd-parity vibrations alone. In Cr^{3+} : LiCAF (and in Cr^{3+} : LiCGaF also) the $^4T_2 \rightarrow {}^4A_2$ emission is shifted by 30 nm to shorter wavelengths compared with this same emission in LiSAF and LiSGaF. The low oscillator strength, Table 10.2, minimal overlap of the $^4T_2 \rightarrow {}^4T_{1a}(2)$ ESA with the $^4T_2 \rightarrow {}^4A_2$ laser emission and weak thermal quenching of the fluorescence at 300 K accounts for the efficient operation of the Cr^{3+} : LiCAF laser. Greater overlap of the ESA with the $^4T_2 \rightarrow {}^4A_2$ emission in LiSAF, LiSGaF and GFG reduces the emission cross section by $\sim 30\%$.

The intensities of the ESA transitions and their splittings vary from host to host depending on the strength of Jahn–Teller and vibronic couplings to the $^4T_2$–$^4A_2$ transition. In consequence, laser performance is influenced because σ_{eff} (Eq. 7.24) is

Table 10.5. *Transition matrix elements for excited state* $^4T_2 \rightarrow {}^4T_1$ *transitions of* Cr^{3+} *ions induced by static* T_{2u} *distortions [adapted from Lee et al.* (1989)]

Transition	Transition strength $f_o^g(\Gamma, \alpha)$	
	$E \parallel c(\pi)$	$E \perp c(\sigma)$
$^4T_2 \rightarrow {}^4T_1(1)$	$f_S(\pi) = \frac{1}{6}\lvert\langle t_2 \Vert t_1 \Vert t_2\rangle\rvert^2$	$f_S(\sigma) = \frac{1}{24}\lvert\langle t_2 \Vert t_1 \Vert t_2\rangle\rvert^2 + \frac{1}{8}\lvert\langle t_2 \Vert r_2 \Vert t_2\rangle\rvert^2$
$^4T_2 \rightarrow {}^4T_1(2)$	$f_S(\pi) = \lvert\langle e \Vert a_2 \Vert e\rangle\rvert^2 = 0$ for *d*-orbitals	$f_S(\sigma) = 0$

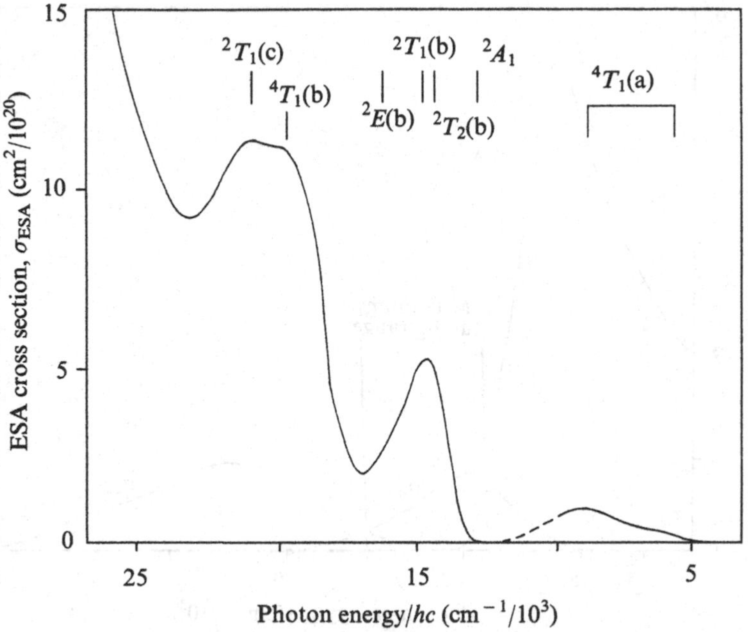

Figure 10.10. ESA transitions in Cr^{3+} : GSGG. The broken line between 12 000 and 9000 cm^{-1} is a double gaussian fit to the $^4T_2 \rightarrow {^4T_{1a}}$ band [after Petermann (1990)].

the appropriate emission cross section to insert in Eq (1.36). Although not clearcut in the case of Cr^{3+} : $KZnF_3$, ESA is responsible for reduced slope efficiency for most of the compounds listed in Table 10.4. In Cr^{3+} : $KZnF_3$ the requirement of charge compensation for Cr^{3+} ions substituting at Zn^{2+} sites may reduce the laser slope efficiency because some Cr^{3+} sites do not contribute to laser operation at all wavelengths in the gain bandwidth. A commercial Cr^{3+} : $KZnF_3$ was developed and marketed. In V^{2+}-doped crystals the laser emissions are shifted further into the near-infrared than are the Cr^{3+}-fluorides. Strong thermal quenching of fluorescence and ESA in V^{2+} : MgF_2 and $CsCaF_3$ militate against laser operation at 300 K. The Cr^{3+} : $SrAlF_6$ laser has almost the largest tuning range. However, the quantum efficiency, η_o, at room temperature is reduced by one or more of the four inequivalent Cr^{3+} sites in $SrAlF_5$ being non-radiative. Such phonon heating causes thermal lensing and laser pumping at 647 nm at 300 K is only possible with a 1–2% duty cycle.

Most Cr^{3+} : garnet lasers have been pumped using the 488 nm or 647 nm lines from an Ar^+ or Kr^+ laser, respectively [Struve *et al.* (1983), Drube *et al.* (1984)]. Analysis of the results for Kr^+ pumping in Table 10.4 suggests significant ESA at the pump wavelength in Cr^{3+}: GSGG. Indeed, Huber (1984) showed that with 647 nm Kr^+ pumping $\sigma_{ESA} \cong \sigma(^4A_2 \rightarrow {^4T_2})$ and for Ar^+ pumping at 488 nm $\sigma_{ESA} \cong 2.6\sigma$ $(^4A_2 \rightarrow {^4T_{1a}})$. Andrews *et al.* (1986a) observed ESA spectra for Cr^{3+} : GSGG due to $^4T_2 \rightarrow {^4T_{1a}}, {^4T_{1b}}$ peaked at 1100 nm and 520 nm, respectively, with intense charge transfer bands in the UV near 300 nm. Subsequently, Petermann (1990) reported the most complete ESA spectrum, Fig. 10.10, which he interpreted using the Jahn–Teller model developed by Caird *et al.* (1988) for the weak field Cr^{3+} : doped fluorides. For Cr^{3+} : GSGG, Jahn–Teller coupling to E_g phonons yields a broad peak at 1110 nm

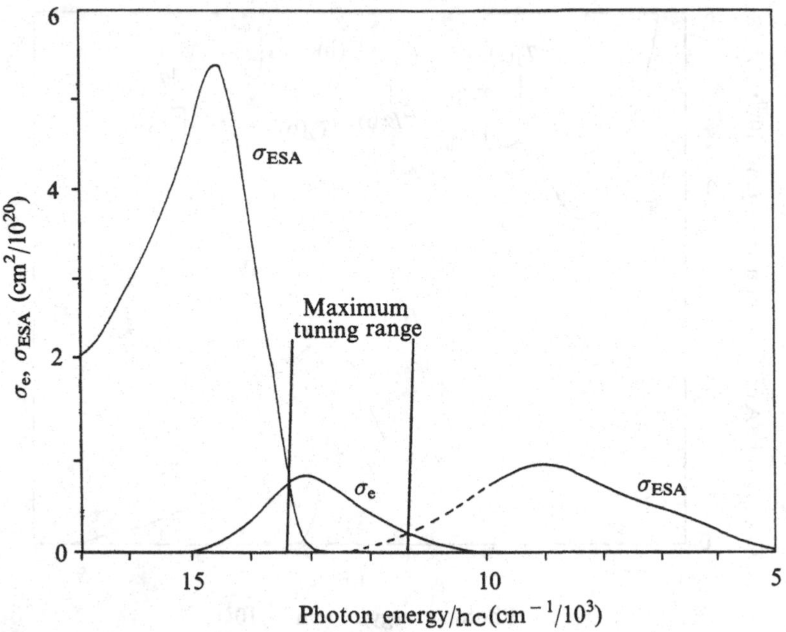

Figure 10.11. The cross sections for emission, σ_e, and ESA, σ_{ESA}, of Cr^{3+} : GSGG at 300 K [after Petermann (1990)].

with a shoulder on the long wavelength tail at 1500 nm, which Petermann deconvolved into overlapping gaussians peaked at $E_1 = 6500\,cm^{-1}$ and $E_2 = 8980\,cm^{-1}$ with FWHM values of $\Delta_1 = 1400\,cm^{-1}$ and $\Delta_2 = 2700\,cm^{-1}$, respectively. The value of E_o determined from the Tanabe–Sugano diagram for Cr^{3+} : GSGG is $6300\,cm^{-1}$, in excellent accord with experiment. Interpretations of the shorter wavelength components in Fig. 10.10 are more complex. However, given the admixed ^2E and $^4T_2$ wavefunctions, the $^4T_2 \rightarrow {}^4T_{1a}$ absorption may be dominant by virtue of the smaller ^2E component in the vibronic wavefunction at 300 K. The $^4T_2 \rightarrow {}^4T_{1b}$ band will be almost unaffected by the Jahn–Teller effect, appearing as a single band at 500 nm, and overlapping the weaker $^2E \rightarrow {}^2T_{1c}$ transition. A similarly located broadband was observed in the ESA spectra of V^{2+} : MgF_2 and $KMgF_3$ [Payne *et al.* (1988b)]. Narrow ESA bands near 660 nm and 490 nm in MgO, ruby and emerald were attributed to $^2E \rightarrow {}^2T_{2b}, {}^2T_{1b}, {}^2E_b$ and $^2E \rightarrow {}^2T_{1c}$, respectively [Fairbank *et al.* (1975)]. In Cr^{3+} : GSGG the lowest lying zero-phonon level of $^4T_2$ lies *ca* 30–60 cm^{-1} above ^2E, depending upon the particular Cr^{3+}-multisite [Monteil *et al.* (1990)], although the levels are probably inverted at 300 K as a consequence of thermal expansion [Henderson *et al.* (1988)]. Thus the ESA transitions in Cr^{3+} : GSGG near 650 nm and 500 nm (Fig. 10.11) probably originate on the vibronic ^2E and $^4T_2$ levels, and terminate on the spin doublets and quartets identified on the spectrum.

The rather disappointing laser results for the Cr^{3+} : garnets can now be interpreted. First, the $^4T_2 \rightarrow {}^4A_2$ transition strength is reduced by the admixing of ^2E wavefunction into the vibronic $\Psi'_T(r, Q)$ wavefunction and the consequential changes in the emission cross section σ_{em} in Eq. (7.24), and the Einstein coefficient, B_{ba}, for stimulated emission (Eqs. 5.63–5.65). Such an effect is greater for Cr^{3+} : garnets with stronger crystal-field sites. However, the slope efficiency of Cr^{3+} : garnet lasers is not determined by nonradiative decay, at least not at temperatures below 300–400 K [Armagan and DiBartolo (1986), Petermann (1990)]. The point is graphically made in studies of laser-pumped Cr^{3+} : GSGG (Fig. 1.7) and Cr^{3+} : LLGG (Fig. 9.2) lasers. The luminescence and ESA spectra for GSGG in Fig. 10.11 show the overlap on the short and long wavelength wings of the $^4T_2 \rightarrow {}^4A_2$ emission band. Since gain occurs when the emission cross section exceeds all losses, including the ESA, it is apparent that the tuning range will be reduced, in this case to *ca* 740 nm to 880 nm. In laser pumped-Cr^{3+} : GSGG laser tests Petermann (1990) used 488 nm and 647 nm pumping, measuring the laser output, P_o, and threshold, P_{th}, powers as a function of lasing wavelength. The data in Fig. 1.7 clearly reveal tuning range limitation by ESA overlap, at long wavelength by the $^4T_2 \rightarrow {}^4T_{1a}$ ESA and at shorter wavelength by absorption from the spin doublets ($^2T_{2b}$, $^2T_{1b}$ and ^2E_b). The different absorptions of the pump radiation by ESA are established by the greater output power and reduced threshold power for pumping at 647 nm. Using the threshold tuning curve in Fig. 1.7 with Eq. (1.36), having measured the pump efficiency, η_p, and the losses from the slope efficiency, Eq. 1.39, it is possible to calculate $\sigma_{em}(\lambda)$ as a function of wavelength. The excellent accord between the $\sigma_{em}(\lambda)$ versus λ curves calculated from the ESA spectra in Fig. 10.11 and the laser tuning curves at 488 nm and 647 nm confirm that the reduced efficiency and tuning range of the Cr^{3+} : GSGG laser are caused exclusively by ESA. Cr^{3+} : LLGG is perhaps the only genuine weak field host among the oxide garnets with energy gap $\Delta E = E(^4T_2) - E(^2E) \cong -1000\,\text{cm}^{-1}$. The shape of the luminescence spectrum of this material in Fig. 9.2 suggests that two Cr^{3+} centres exist in LLGG due to La/Lu disorder on dodecahedral sites and Lu/Ga disorder on octahedral sites. The resulting luminescence band which extends from 700 nm–1100 nm in Fig. 9.2 is inhomogeneously broadened just as in glasses. Estimates of the ESA from fundamental spectroscopic parameters suggest an even larger overlap of the $^4T_2 \rightarrow {}^4T_{1a}$ ESA with the long wavelength wing of the $^4T_2 \rightarrow {}^4A_2$ laser emission band, limiting laser action there. However, nonradiative decay by the weaker crystal-field sites in the disordered distribution quenches laser operation in this long wavelength region, as it does also in Cr^{3+}-doped glasses and gallogermanates [Henderson *et al.* (1992), Grinberg *et al.* (1997)]. Furthermore, the visible ESA near 650 nm is not expected to be host-sensitive, so that the restricted tuning range from 790–850 nm is not unexpected for much the same reasons as in the Cr^{3+} : GSGG laser.

Generally, the Cr^{3+} : garnet lasers have no advantages in tunable laser operation over Cr^{3+} : alexandrite and colquiriites lasers (Table 10.4). Nevertheless, Cr^{3+} : GSGG and Cr^{3+} : GSAG have received development for military and LIDAR applications, in which the high quantum yield, $\sim 100\%$ for concentrations up to $5 \times 10^{20}\,Cr^{3+}\,\text{cm}^{-3}$, is beneficial. Furthermore, Czochralski growth (§3.4.4) yields large single crystals of excellent optical quality [Zharikov (1991)]. Cr^{3+} : GSAG, in particular, has attractive growth characteristics in that Al_2O_3 is less aggressive than Ga_2O_3 in interacting with the Ir crucible, and also has a much lower vapour pressure. In consequence, the formation of Ir platelets within the GSAG boule is reduced relative to GSGG and a homogeneous, stoichiometric composition can be maintained over the whole length of the boule. Since $k_o(\text{Cr}) \cong 1$, Cr^{3+} doping is uniform throughout the crystal. In GSGG Cr^{3+}-multisites form during growth at both the congruent and stoichiometric compositions and since $k_o(\text{Cr}) \cong 3.0$ concentration gradients develop along the length of large boules. Because they are cubic the garnets emit unpolarized light and, although

Table 10.6. *Room temperature (300 K) absorption and luminescence characteristics of Ca-gallogermanate structured crystals doped with 0.1 at.% Cr^{3+}* [*after Kaminskii et al.* (1989)]

Crystal	Peak of absorption bands (μm)		Peak of luminescence bands (μm)	Δν (lumin.) (cm^{-1})	Decay time (μs)	
	$^4A_2\rightarrow{}^4T_2$	$^4T_2\rightarrow{}^4A_2$	$^4T_2\rightarrow{}^4A_2$		77 K	300 K
$Ca_3Ga_2Ge_4O_{14}$	0.445	0.625	0.910	~3645	~7	~1.8
$Sr_3Ga_2Ge_4O_{14}$	0.430	0.615	0.840	~4085	~6.2	~2.8
$La_3Ga_5SiO_{14}$	0.450	0.640	0.910	~2910	~12	~6
$La_3Ga_5GeO_{14}$	0.435	0.650	0.960	~2620	~14	~3.6
$La_3Ga_{5.5}Nb_{0.5}O_{14}$	0.460	0.660	1.030	~2825	~12	~1.3
$La_3Ga_{5.5}Ta_{0.5}O_{14}$	0.455	0.665	1.030	~2875	~16	~1.8

the inversion symmetry increases population storage time, the unwanted polarization component detracts from the gain cross section. Nevertheless, long storage times are beneficial in Q-switched, flashlamp pumped operation; this technology was actively developed in the 1980s. Unfortunately, broadband absorbing defects are generated in GSGG and other Ga-based oxide garnets during flashlamp pumping. GSGG and LLGG, as well as the isomorphic solid solution GSGG–GSAG–YSAG–YSGG are intrinsically disordered. In other Cr^{3+}-doped disordered solids, such as the inorganic glasses, gallogermanate and yttrium aluminate families of crystals, nonradiative processes can completely suppress large parts of the $^4T_2\rightarrow{}^4A_2$ emission band [Henderson *et al.* (1992), Grinberg *et al.* (1997)]. However, nonradiative decay is not an issue in Cr^{3+}: garnets.

Much interest was generated by the claimed record tuning ranges of Cr^{3+} ions in disordered crystals with the calcium gallogermanate structure [Kaminskii *et al.* (1989)]. These trigonal crystals, which have statistical distributions of host cations on the Ga^{3+} and Ge^{4+} sites in $Ca_3Ga_2Ge_4O_{14}$ (CGGO) and $Sr_3Ga_2Ge_4O_{14}$ (SGGO), provide weak and strong field sites for Cr^{3+} substituents on Ga^{3+} sites. Their spectroscopic properties are summarized in Table 10.6. Both the R-lines and broadband emission are inhomogeneously broadened by a distribution of low-symmetry crystal-field splittings and Huang–Rhys factors [Grinberg *et al.* (1995), Macfarlane *et al.* (1996)]. In CGGO and SGGO both trigonal and orthorhombic sites result from the disorder, whereas in LGGO and LGS only trigonal sites exist [Yamaga *et al.* (1998a)]. For the broadband $^4T_2\rightarrow{}^4A_2$ emission from Cr^{3+} ions in weak field sites the disorder causes the Huang–Rhys factor, S, to vary from site to site, resulting not only in the inhomogeneous broadening but also in rapid nonradiative decay. Sites with the largest values of S are also the sites shifted to lowest energy which, as temperature increases, are most likely to decay nonradiatively, thus (apparently) shifting the $^4T_2\rightarrow{}^4A_2$ band to shorter wavelengths, an effect also reported to occur in Cr^{3+} : doped glasses [Henderson *et al.* (1992), Yamaga *et al.* (1992)]. The nonradiative decay is signalled by rapid shortening of the luminescence lifetime from typically 12 μs at 4 K to *ca* 1 μs at 200 K in CGGO and LGS, with serious implications for laser operation. In their measurements of stimulated emission and laser operation Kaminskii *et al.* (1989) excited the Cr^{3+} : CGGO, SGGO, LGGO and LGS lasers with a pulsed ruby laser, observing tunable laser performance over some 75% of the emission bandwidth with efficiencies in the range 3–10%. Other workers have experienced difficulties in operating Cr^{3+}: lasers based on the gallogermanates in CW and synchronously-pumped mode-locked operation. The problem in these cases is linked to nonradiative decay from the excited $^4T_2$ state back to the ground state at 300 K. The associated local heating and strong thermal lensing precludes laser operation under all but the most reduced duty cycles.

The mixed borates with the formula $RX_3(BO_3)_4$ are of interest because of their potential in self-frequency doubling. These crystals, in which $R^{3+} = Y^{3+}$ or Gd^{3+} and $X^{3+} = Al^{3+}$ or Sc^{3+} are isostructural with $CaMg_3(CO_3)_4$, the uncommon mineral huntite [Mills (1962)]. This structure provides sites suitable for substitution by rare-earth ions (Nd^{3+}, Er^{3+} etc.) on the R^{3+} site and by transition-metal ions (Ti^{3+}, Cr^{3+}, Fe^{3+}) on the X^{3+} site. Large single crystals can be grown by the HTSG technique using $K_2Mo_3O_{10}$-based solvents at temperatures in the range 1250–1600 K. High dopant concentrations may be introduced on either R^{3+} or X^{3+} sites without serious problems associated with self-quenching of luminescence. Wang *et al.* (1995), (1996) grew large single crystals of Cr^{3+}-doped $YAl_3(BO_3)_4$ (YAB), $GdAl_3(BO_3)_4$ (GAB), $YSc_3(BO_3)_4$ (YSB) and $GdSc_3(BO_3)_4$ (GSB) and studied their optical properties, showing that the Cr^{3+}-doped aluminoborates (YAB and GAB) are strong field, much like alexandrite. Accordingly, the fluorescence spectra reveal the sharp R-lines and their sidebands at low temperatures and the mixed R-lines and $^4T_2 \rightarrow {}^4A_2$ broadband at 300 K. In contrast, Cr^{3+}: YSB and GSB emit into the broadband $^4T_2 \rightarrow {}^4A_2$ transitions at all temperatures. The emission life times confirm the aluminoborates as strong field hosts and the scandioborates as weak field hosts. Indeed, for Cr^{3+} in YAB and GAB, where $\Delta \cong 600\ cm^{-1}$, only the ^2E state maintains a significant low temperature population and τ_R is 3.5 ± 0.01 ms for YAB and 2.1 ± 0.01 ms for GAB at 14 K. At $T = 300$ K where both ^2E and $^4T_2$ are occupied the mixed $^2E, {}^4T_2$ luminescence has a shorter decay time of $170 \pm 10\ \mu s$ for YAB and $380 \pm 10\ \mu s$ for GAB. Ions in the weak field sites of YSB and GSB, where $\Delta E - 400\ cm^{-1}$ decay with luminescence life times at low temperature (14 K) of $228 \pm 10\ \mu s$ and $270 \pm 10\ \mu s$, respectively.

These double borates can be efficiently pumped in the $^4A_2 \rightarrow {}^4T_2, {}^4T_{1a}$ bands with the major Ar^+ resonance lines at 488 nm and 514 nm or with the Kr^+ laser lines at 647 nm and 676 nm, or by flashlamp pumping. At 300 K the luminescence output of Cr^{3+}: YAB and GAB are very similar to Cr^{3+}: $BeAl_2O_4$, and the long-lived ^2E state will act as an effective population store for laser operation on the $^4T_2 \rightarrow {}^4A_2$ emission. However, the potential tuning ranges are little different from alexandrite, and except for the absence of health hazards during growth of Be-based compounds, offer no real advantages. A major interest in new Cr^{3+}-based lasers is in the potential for pumping with visible-range diode lasers. The presently available single stripe diodes are able to deliver up to 1 W into the wavelength range 660–680 nm where Cr^{3+}: YSB and GSB have absorption coefficients *ca* 50% larger than Cr^{3+}: LiSAF with lower sensitivity to pump polarization. However, although the peak output wavelength of these Cr^{3+}-doped double borates is shifted 30–40 nm to longer wavelength than Cr^{3+}: LiSAF and LiSGaF, the overall emission bandwidth is somewhat restricted. Stimulated emission has been reported for Cr^{3+}: YAB: the single-pass gain measured by a pump-probe technique exceeded $1.4\ cm^{-1}$ over the wavelength range 740–860 nm, with a maximum value of $1.92\ cm^{-1}$ at 820 nm. The reported ESA at shorter, fixed wavelengths than the band peak, presumably due to $^4T_2 \rightarrow$ spin doublet transitions, may limit tunability at shorter wavelengths. The lack of inversion symmetry precludes Jahn–Teller coupling to E_g-modes, which would split, broaden and shift the $^4T_2 \rightarrow {}^4T_{1a}$ ESA into partial overlap with the laser emission. In the absence of the Jahn–Teller effect the $^4T_2 \rightarrow {}^4T_{1a}$ transition is expected to peak near 1350 nm, as discussed above for Cr^{3+} : LiCAF and Cr^{3+} : garnets. The dual-purpose potential of the Cr^{3+} : $REX_3(BO_3)_4$ compounds as gain media and frequency converter was demonstrated by SHG in the UV region. The effective nonlinear optical coefficient measured for Cr^{3+} : YAB for the fundamental wavelength of 750 nm was measured as $d_{eff} = 1.35$ pm V^{-1} with a phase-matching angle of $\theta = 40.8°$. These results promise better conversion efficiency than LBO, for which $d_{32} = 1.17$ pm V^{-1} measured under the same conditions. It was proposed that tunable UV laser operation of Cr^{3+} : YAB around 360 nm was possible using an intracavity self-frequency doubling scheme [Iwai *et al.* (1995)].

10.3.4 Ti^{3+}-activated lasers

The spectroscopic results discussed in §9.2.3 show that the Ti^{3+} : Al_2O_3 system is an almost ideal four-level laser medium. The broad, double peaked absorption band with peak cross section in π-polarization of $9.3 \pm 1.0 \times 10^{-20}$ cm^2 [Albers *et al.* (1986a), Sanchez *et al.* (1986)] at 490 nm allows for flexible pumping schemes. Emission over a broadband with half-power points at 680–850 nm and peak cross section of $3.2 \pm 0.3 \times 10^{-19}$ cm^2 at 780 nm in π-polarization provides for broadband tunable laser operation from *ca* 650 nm to beyond 1 μm. Furthermore, the simple energy level structure provided by the $3d^1$ configuration ion precludes ESA at pump or laser wavelengths. The parasitic absorption by Ti^{4+}–Ti^{3+} pairs across the emission band region (620–950 nm) in Fig. 9.9(a) can be eliminated by careful growth and/or post-growth annealing treatments [§3.5, Fahey *et al.* (1986), Kokta (1986)]. As Fig. 10.12a shows, the emission decay time is little changed at 300 K from the low temperature value, implying that nonradiative decay has little deleterious consequence for laser performance. The short luminescence lifetime results from the T_{2u} distortion, as do the large absorption and emission cross sections. It follows that good absorption of the pump radiation can be achieved at low doping levels. Typically the optimum concentration for good quality laser rods was recognized as 0.13 at.%, although satisfactory laser performance can be achieved at concentrations up to 0.41 at.% Ti^{3+} [McKinnie *et al.* (1997)]. Nevertheless, in terms of a figure of merit (FOM) defined as FOM $= \alpha(514)/\alpha(820)$, where the α refer to absorption coefficients at pump (514) and emission (820) wavelengths, the best performance is achieved at 0.15 at.% Ti^{3+} where FOM $= 220$.

Moulton (1982b) first operated a Ti : sapphire laser by longitudinally pumping at 500 nm using a pulsed dye laser, achieving laser oscillation over a continuous range from 660–986 nm with several sets of cavity mirrors. Efficient conversion of dye laser pump radiation to laser output was observed, the laser slope quantum efficiency being 62% with a high-transmission output coupler. He also observed CW-pumped laser operation. The output power–input power plot in Fig. 10.12b was determined by Ar^+-laser pumping (488 nm, 514 nm) a 7.5 mm long crystal containing 0.07 at.% Ti^{3+} in a near concentric cavity with output coupling of about 1%. The slope efficiency of 24% is almost certainly reduced by the infrared absorption in this as-grown crystal. The tuning curve in Fig. 10.12c was obtained using 5 W of total pump power in the 488 and 514 nm Ar^+ laser lines and a plain output coupler with 8% transmission. These laser operating

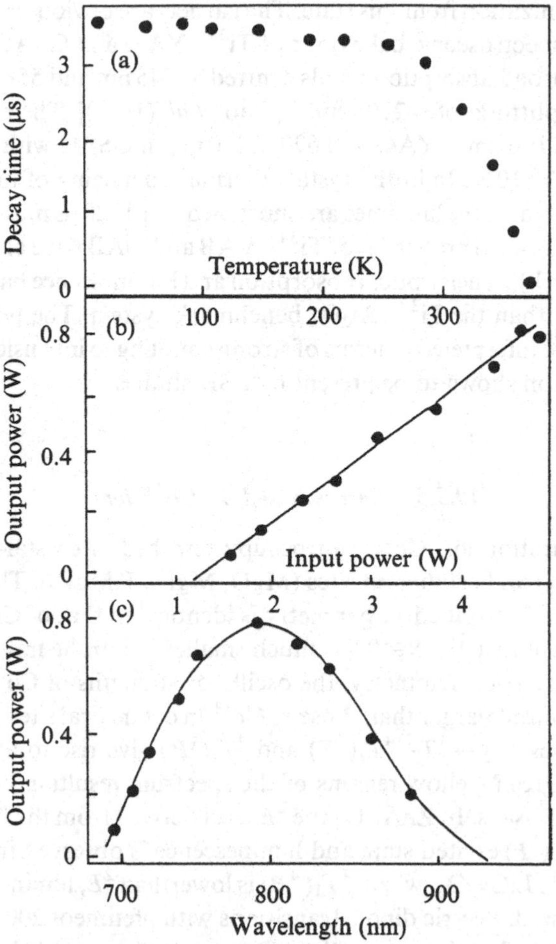

Figure 10.12. (a) The luminescence lifetime of $Ti^{3+}:Al_2O_3$ as a function of temperature, (b) the input–output power characteristic and (c) threshold power for an 0.13% Ti-sapphire laser using an 0.8% transmission output coupler.

data are comparable with other published results [Sanchez *et al.* (1986), Albers *et al.* (1986b)]. The rather short radiative lifetimeof 3.2 μs at 300 K is a distinct drawback in flashlamp pumping and in Q-switched operation.

The Ti-sapphire laser is an outstanding commercial success. There are no other successful Ti^{3+}-activated lasers despite research into Ti^{3+}-doped YAG [Bantien *et al.* (1987)], $YAlO_3$ [Schepler (1986), Wegner and Petermann (1989)], $BeAl_2O_4$ [Segawa *et al.* (1987), Alimpiev *et al.* (1986)], GSAG [Gao *et al.* (1993)], $MgAl_2O_4$ [Strek *et al.* (1987)], $ScBO_3$ [Aggarawal *et al.* (1987)], $YAl_3(BO_3)_4$ and $GdAl_3(BO_3)_4$ [Wang *et al.* (1997a,b)] and $CaGdAlO_4$ [Yamaga *et al.* (1994)]. The lasing of $Ti:BeAl_2O_4$ has been observed, but it offers no real advantage over $Ti:Al_2O_3$ either in terms of longer lifetime (for flashlamp pumping) or tuning range. Only very inefficient laser action has been reported for $Ti^{3+}:YAlO_3$ [Kvapil *et al.* (1988)], apparently because ESA of the pump radiation by Ti^{3+}–Ti^{4+} complexes reduces pump efficiency and increases threshold power appreciably [Wegner and Petermann (1989)]. Basun *et al.* (1996) suggest that stimulated emission from the ^2E state of Ti^{3+} is also limited by photo ionization from this state. The garnets are obvious candidates as hosts for Ti^{3+}ions. The spectroscopic behaviour of Ti^{3+} : YAG and GSAG are very similar with overlapping broad absorption bands centred at 545 nm and 558 nm, respectively, and Jahn–Teller splittings of $\sim 2700\,cm^{-1}$ [Gao *et al.* (1993)]. Their emission spectra stretch from 650–1100 nm in YAG and 690–1200 nm in GSAG with long decay times of $\tau_R = 50$ ms at $T = 10$ K. In both crystals thermal quenching of luminescence near room temperature where the life times are shortened to 10–20 μs mitigates against CW lasing. Finally, the optical properties of Ti^{3+} : YAB and GAB have also been examined [Wang *et al.* (1997a,b)]. Their optical absorption and luminescence bands are shifted to longer wavelength than the $Ti^{3+} : Al_2O_3$ benchmark system. The polarization of the optical spectra were interpreted in terms of strong coupling to intrinsic $T_{1u}^x(\pi)$ or $T_{2u}^z(\pi)$ odd-parity distortion shown to be present by ESR studies.

10.3.5 Lasers based on Co^{2+} ions

The $(3d)^7$ configuration ion Co^{2+} can occupy tetrahedral crystal-field sites (ZnO, $ZnAl_2O_4$, $LiGa_5O_8$) and octahedral sites (MgO, MgF_2, $KMgF_3$). The splitting of the Co^{2+} free ion states in tetrahedral symmetry is identical to that of Cr^{3+} in octahedral sites, Fig. 4.8, except that $Dq/B \cong 0.5$ is much smaller in tetrahedral sites. Since tetrahedral sites lack inversion symmetry, the oscillator strengths of Co^{2+} transitions are 2–3 orders of magnitude larger than those of Cr^{3+} in octahedral sites. The strong, spin-allowed absorptions $^4A_2 \rightarrow {}^4T_2$, $^4T_1(^4F)$ and $^4T_1(^4P)$ give rise to absorptions in the infrared, red and green–yellow regions of the spectrum resulting in the typical blue colouration of Co^{2+}-salts. In $ZnAl_2O_4$ the ^2E level derived from the free ion ^2G state is lower than the $^4T_1(^4P)$ excited state and luminescence is observed from the $^2E \rightarrow {}^4A_2$ transition. In Co^{2+}: $LiGa_5O_8$, where $^4T_1(^4P)$ is lowerthan ^2E, luminescence occurs via $^4T_1 - {}^4A_2$ spin-allowed, electric dipole transitions with lifetimeof 200 ns.

Co^{2+} ions in octahedral sites normally behave as weak crystal-field ions and the free ion ground term, ^4F, splits into $^4T_1$, $^4T_2$ and $^4A_2$ in order of increasing energy. The absorption bands of $Co^{2+} : MgF_2$ corresponding to $^4T_1 \rightarrow {}^4T_2$, $^4A_2$ transitions are centred at *ca* 1250 nm and 550 nm, respectively. The near-infrared luminescence

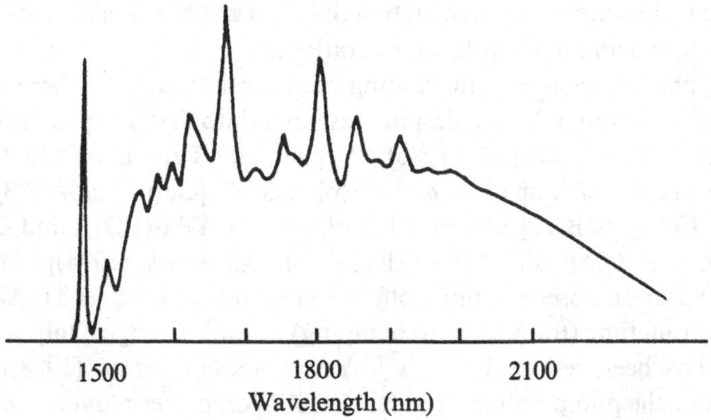

Figure 10.13. The $^4T_2 \rightarrow {}^4A_2$ luminescence spectrum of $Co^{2+} : MgF_2$ at 77 K.

spectrum due to the ${}^4T_2 \rightarrow {}^4A_2$ transitions, shown in Fig. 10.13, shows sharp zero-phonon lines and phonon-assisted structure atop a broad sideband across the range 1500–2300 nm. This band in $Co^{2+}:MgF_2$ and other fluorides has been the basis of tunable laser operation at cryogenic temperatures. In $Co^{2+}:MgO$ the nonradiative process is so efficient that luminescence is totally quenched above 77 K. In the laser-active crystals $Co^{2+}:MgF_2$ and $Co^{2+}:KZnF_3$ ESA in the ${}^4T_2 \rightarrow {}^4A_2({}^4F)$ and ${}^4T_2 \rightarrow {}^4T_1({}^4P)$ transitions are important in limiting efficient laser action by competition with ground state absorption for pump photons [Manna and Moncorgé (1990)].

The original Co^{2+}-lasers utilized flashlamp pumping of $Co^{2+}:MgF_2$, ZnF_2 and $KMgF_3$ crystals cooled in cryogenic liquids [Johnson *et al.* (1964), (1966b)]. CW operation of a $Co^{2+}:MgF_2$ laser pumped in the ${}^4T_1 \rightarrow {}^4T_2({}^4F)$ absorption band with the CW Nd : YAG laser line at 1.32 μm is possible even at 300 K, although the threshold power is slightly higher and slope efficiency reduced by nonradiative decay [Moulton and Mooradian (1979b), Welford and Moulton (1988)]. The overlapping CW power tuning curves in Fig. 10.14 were measured at 80 K using two different output couplers for excitation by *ca* 1.5 W of absorbed CW pump power in the 1.32 μm Nd : YAG laser line. $Co^{2+}:MgF_2$ lasing can also be achieved by pumping CW in the ${}^4T_1 \rightarrow {}^4T_1({}^4P)$ with the 514 nm line from the Ar^+ laser. Kunzel and Dürr (1981), (1982) have also reported CW operation of a $Co^{2+}:KZnF_3$ laser pumped with an Ar^+ laser at 80 K. Threshold power levels as low as 50 mW were achieved, with output up to 120 mW and tuning range of 1.65–2.07 μm. Both Q-switched and mode-locked operation has been obtained using astigmatically-compensated three-mirror cavities. In pulsed operating mode pumped with a Nd : $YAlO_3$ laser a $Co^{2+}:MgF_2$ laser has been tuned over the

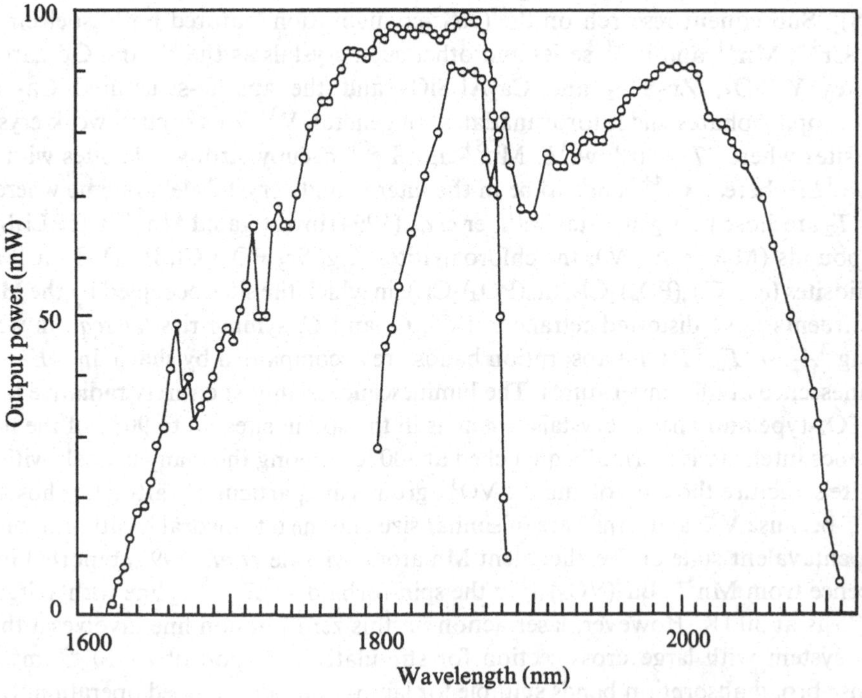

Figure 10.14. The CW output tuning curve for a $Co^{2+}:MgF_2$ laser pumped with 1.5 W of absorbed power at 1320 nm from a Nd : YAG laser measured at 80 K.

entire wavelength range (1.5–2.3 μm) of the $^4T_2 \rightarrow {}^4T_1(^4F)$ transition [Moulton (1985)]. The $Co^{2+}:MgF_2$ gain medium is the basis of a moderately successful commercial laser system [Welford and Moulton (1988)].

10.3.6 Lasers based on $(3d)^2$ and $(3d)^8$ configuration ions

The Tanabe–Sugano diagrams for $3d^2$ ions in *tetrahedral* symmetry sites and $3d^8$ ions in *octahedral* sites are similar, and resemblances are expected of the spectra of isoelectronic $(3d)^2$ ions Ti^{2+}, V^{3+}, Cr^{4+}, Mn^{5+} and the archetypal $(3d)^8$ ion Ni^{2+}, in which cases the free ion levels ^3F, ^1D, ^3P, ^1G and ^1S, in order of increasing energy, span some 50 000 cm^{-1}. Figure 4.8 shows that the ^3F free ion level splits into $^3A_2$, $^3T_2$ and $^3T_1$ in order of increasing energy. The crystal-field splitting of ^1D yields ^1E and $^1T_2$ levels and the strength of the crystal field determines whether ^1E or $^3T_2$ is the lowest-lying excited state. Emission from ions in strong crystal-field sites results in sharp zero-phonon lines and weak vibronic sidebands from the $^1E \rightarrow {}^3A_2$ transitions. Weak field $^3T_2 \rightarrow {}^3A_2$ transitions are broad, strong vibronic bands spanning several hundred nanometres. However, the electronic structure of $3d^2$ ions in octahedral sites is somewhat different, being characterized by broad spin-allowed absorptions due to $^3T_1(t_2^2) \rightarrow {}^3T_2(t_2e)$, $^3A_2(e^2)$ transitions. Although the energy gap $\Delta E = E(^3T_2) - E(^3T_1)$ is small, luminescence across it has been observed for $Ti^{2+}:MgCl_2$, in which coupling to low energy phonons is weak [Jacobsen *et al.* (1986)].

The $Cr^{4+}:Mg_2SiO_4$ (forsterite) gain medium permits tunable laser operation from 1130 nm to 1370 nm at 300 K [Petricevic *et al.* (1988)]. Tunable laser operation was reported over the wavelength range 1309–1596 nm in the Cr^{4+} : YAG [Angert *et al.* (1988)]. Subsequent research on the $(3d)^2$-configuration featured both isoelectronic V^{3+}, Cr^{4+}, Mn^{5+} and Fe^{6+} series and other such crystals as the Y- and Gd-garnets, silicates Y_2SiO_5, Zn_2SiO_4 and $Ca_2Al_2SiO_7$ and the apatite-structured Ca- and Sr-fluorophosphates and fluorovanadates. In general, V^{3+} ions occupy weak crystal-field sites where $^3T_2$ is below ^1E, Mn^{5+} and Fe^{6+} occupy strong field sites with $^3T_2$ above ^1E, whereas Cr^{4+} tends to be in the intermediate crystal-field regime where ^1E and $^3T_2$ are close to degenerate. Oetliker *et al.* (1994) investigated Mn^{5+} in the Li_3MO_4 compounds (M = P, As, V), the chloroapatites (e.g. $Sr_5(PO_4)_3Cl$, $Ba_5(VO_4)_3Cl$) and spodiosites (e.g. $Ca_4(PO_4)_2Cl_2$, $Sr_4(PO_4)_2Cl_2$), in which the sites occupied by the Mn^{5+} substituents have distorted tetrahedral C_2, C_3 and C_s symmetries, *et seq.*, in which strong $^3A_2 \rightarrow {}^3T_2, {}^3T_1(^3F)$ absorption bands are accompanied by sharp line $^1E \rightarrow {}^3A_2$ luminescence at all temperatures. The luminescence is almost entirely radiative in the Li_3MO_4-type and apatite crystals, whereas in the spodiosites up to 90% of the luminescence intensity is thermally quenched at 300 K. Among the many crystals with the apatite structure those involving the VO_4^{3-} groups are particularly attractive hosts for Mn^{5+} because V^{5+} and Mn^{5+} are of similar size and the tetrahedral location stabilizes the pentavalent state of the aliovalent Mn atom. Merkle *et al.* (1995) reported luminescence from $Mn^{5+}:Ba_5(VO_4)_3F$ in the spin-forbidden $^1E \rightarrow {}^3A_2$ line with a lifetime of 475 μs at 300 K. However, laser action on this zero-phonon line involves a three-level system with large cross section for stimulated emission of *ca* 10^{-19} cm^2 and intense broad absorption bands suitable for lamp- or diode-pumped operation. Laser operation at room temperature is of low efficiency (*ca* 0.5–1.5%) as a consequence of $^1E - {}^1T_1, {}^1T_2$ ESA at the pump wavelength.

The most active research has concerned the Cr^{4+} ion in weak crystal-field sites that support laser operation on the vibronic $^3T_2 \rightarrow {}^3A_2$ band. In the olivine-structured silicates and germanates the four-fold coordinated $(SiO_4)^{4-}$ or $(GeO_4)^{4-}$ tetrahedra are stretched along one of the bond axes (i.e. the *c*-axis) reducing the symmetry from T_d to C_{3v}. Further minor adjustments of ionic positions within the tetrahedra reduce the symmetry at the Si(Ge)-site to C_s [Deka *et al.* (1992), Merkle *et al.* (1992), Hazenkamp *et al.* (1996)]. The splittings of the low-lying energy levels of Cr^{4+} (and other isoelectronic ions) for this descent of symmetry $T_d \rightarrow C_{3v} \rightarrow C_s$ are determined from the character tables of these symmetry groups, as described in §5.4.2. Such considerations also apply to the isomorphs of the fluoroapatites $Ca_5(PO_4)_3F$ (CFAP), fluorovanadates ($Sr_5(VO_4)_3F$ (SVAP), spodiosites ($Ca_4(PO_4)_2Cl_2$) and mellilites $SrGdGa_3O_7$ (SGGM). In the Cr^{4+}-doped garnets the tetragonal extension along ⟨100⟩-axes reduces the T_d point group symmetry to D_{2d}, with the spectroscopic consequences discussed in §5.4.2 and §9.3.2. The selection rules of polarized optical transitions of $3d^2$ ions in D_{2d} (including spin–orbit coupling), given in Fig. 5.7 and Table 9.4, are easily extended to lower symmetry point groups (§2.5.5).

In both $Cr^{4+}:Mg_2SiO_4$ and Cr^{4+} : YAG there was initial confusion over the laser-active species. This confusion arose from Cr^{3+} and Cr^{4+} concentrations being sensitive to total Cr-content and to their variability with growth and/or thermal processing in oxidizing or reducing atmospheres. Cr impurities effect three different valence states in forsterite. There are two unwanted Cr^{3+} octahedral sites, one with C_i symmetry and the other with mirror-plane, C_s, symmetry [Hoffman *et al.* (1991), Jia *et al.* (1991)]. Laser active centres involve Cr^{4+} ions on the C_s-distorted Si^{4+}-sites [Petricevic *et al.* (1989), Verdun *et al.* (1988)]. Early laser experiments used crystals with low (0.04 at.%) Cr-content, yielding absorption coefficients of only 0.69–0.75 cm^{-1} at the pump wavelength of 1.064 μm. In the isomorphic crystal Ca_2GeO_4 the Cr^{4+} ion is more easily stabilized on the Ge^{4+} than on the Si^{4+} site in Mg_2SiO_4, and rather small Cr^{3+} levels are observed [Hazenkamp *et al.* (1996)]. The Cr^{4+} : garnets have greater potential for useful application given their excellent optothermal and thermomechanical properties. Early studies of the Czochralski-grown Nd^{3+} : GSGG containing trace amounts of Mg^{2+} and Ca^{2+} impurities led to parasitic absorption at 1.06 μm. A similar absorption in Cr^{3+} : YAG after an oxidizing anneal led to laser action at 1.35–1.45 μm. The nature of the absorbing/laser centre was spuriously identified with Cr^{3+}-oxygen vacancy complexes, Cr^{5+} and Cr^{2+} ions on octahedral B-sites. The nature of the emission in oxidized Cr^{3+} : YAG was identified as Cr^{4+} ions in tetrahedral B-sites by careful magneto- and piezo-optic spectroscopy [Eilers *et al.* (1994) and references therein]. To subdue the Cr^{3+} concentration on octahedral B-sites, co-doping with Cr and Ca is necessary for Czochralski growth in an oxidizing atmosphere to enhance the formation of Cr^{4+} on tetrahedral sites. In YAG and YGG the optimized incorporation of Cr^{4+} requires Ca : Cr ratios of 1 : 4–1 : 5, in crystals annealed in O_2 atmospheres above 1800 K subsequent to Czochralski growth in a 2% O_2 : 98% Ar growth atmosphere.

The $^3A_2 \rightarrow {}^3T_1$ transition is the only allowed electric dipole transition of Cr^{4+} ions in T_d symmetry. In D_{2d} symmetry appropriate to the C-sites in YAG and YGG this transition appears as two vibronically broadened bands $^3B_1(^3A_2) \rightarrow {}^3A_2(^3T_1)$ and $^3B_1(^3A_2) \rightarrow {}^3E(^3T_1)$, the latter occuring as a structured band with peak at *ca* 639 nm in YAG and 667 nm in YGG. The $^3B_1(^3A_2) \rightarrow {}^3A_2(^3T_1)$ absorption bands, shown in Fig. 10.15 for both Cr^{4+} : YGG and Cr^{4+} : YAG, confirm that there is a large splitting, $\sim 6500\ cm^{-1}$, of the tetrahedral $^3T_1$ state by the D_{2d} distortion. The broad-

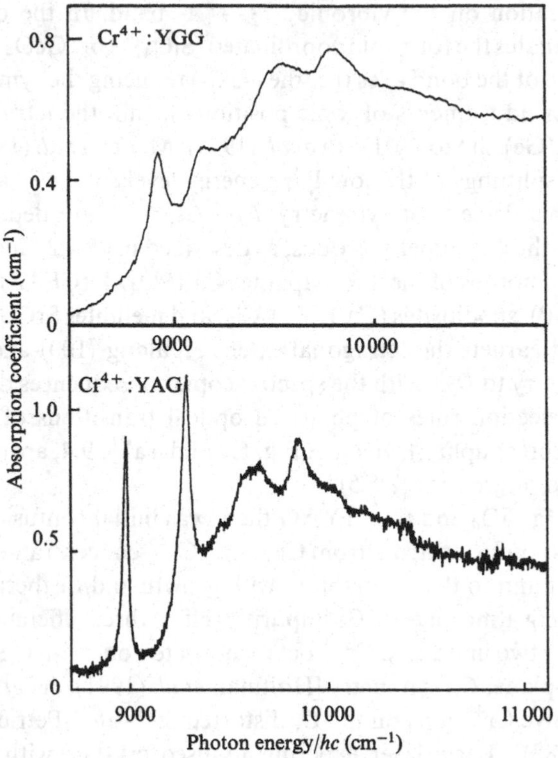

Figure 10.15. The near-infrared optical absorption bands of Cr^{4+} in (a) YGG and (b) YAG measured at 77 K and attributed to the ${}^3B_1({}^3A_2) \rightarrow {}^3A_2({}^3T_1)$ transition.

ening of structural features in the Cr^{4+} : YGG spectrum is caused by disorder induced by a random distribution of Ga^{3+} ions between the A and B sites in these crystals [Yamaga *et al.* (1988)]. The identification of this transition followed from polarized absorption and uniaxial stress spectroscopy on the zero- and one-phonon lines at 1114 nm (8977 cm^{-1}) and 1077 nm (9281 cm^{-1}) in Cr^{4+} : YAG : these absorption lines are shifted to 1123 nm (8919 cm^{-1}) and 1093 nm (9140 cm^{-1}) in Cr^{4+} : YGG. The reduced phonon energy in Cr^{4+} : YGG (240 cm^{-1}) relative to Cr^{4+} : YAG (304 cm^{-1}) implies coupling to an even-parity phonon involving the cooperative motions of Ga^{3+} ions and Al^{3+} ions, respectively.

The nature of the uncertainty in assigning the excited states of Cr^{4+} in YAG and YGG is evident from the spectra in Fig. 10.16; the energy differences between the strongest zero-phonon lines in absorption and emission exceed 1000 cm^{-1}. Tissue *et al.* (1990) identified the luminescence spectrum in Cr^{4+} : YAG with the ${}^1A_1({}^1E) \rightarrow {}^3B_1({}^3A_2)$ transition, a conclusion at variance both with the short luminescence life times of 30.6 μs and 20.0 μs, respectively, in YAG and YGG at $T = 15$ K, and the luminescence bandshapes. For a spin-forbidden transition the radiative decay time should be *ca* milliseconds rather than microseconds, and the zero-phonon line many times more intense than the vibronic sideband, as in the cases of ${}^1E \rightarrow {}^3A_2$ transition for Mn^{5+} in $Ba_{10}(VO_4)_6F_2$ and the ${}^2E \rightarrow {}^4A_2$ R-line in ruby. Conventional wisdom suggests that these emissions result from ${}^3B_2({}^3T_2) \rightarrow {}^3B_1({}^3A_2)$ transitions, the band-

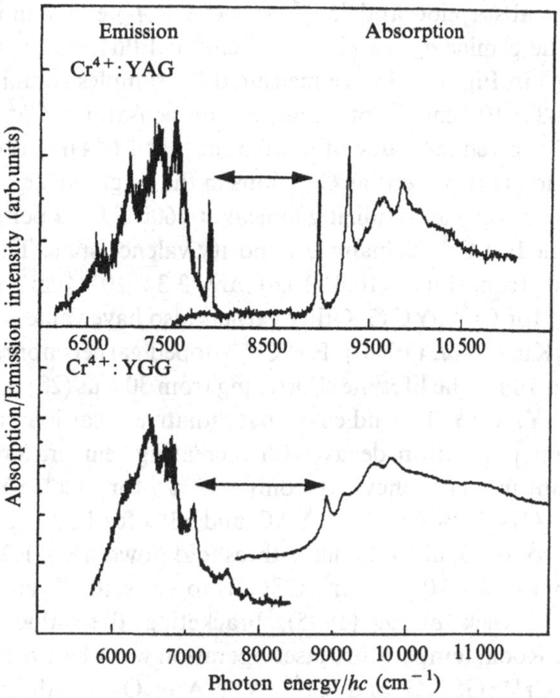

Figure 10.16. The near-infrared $^3B_1(^3A_2) \rightarrow {}^3A_2(^3T_1)$ absorption and $^3B_2(^3T_2) \rightarrow {}^3B_1(^3A_2)$ emission spectra of Cr^{4+}:YAG and YGG measured at 77 K and 20 K, respectively.

shapes reflecting coupling to vibrational modes with $S \approx 1-2$, as has been confirmed by polarized absorption/emission [Eilers *et al.* (1994)], and magnetic circular dichroism (MCD) and ligand field theory [Riley *et al.* (1999)].An absorption peak at 1280 nm (7813 cm^{-1}) is due to the zero-phonon line of $^3B_1(^3A_2) \rightarrow {}^3B_2(^3T_2)$ transition, in accord with the zero-phonon lines and band peak observed in the emission. With increasing temperature the structure shown in Figs. 10.15 and 10.16 gradually broadens: at 300 K structureless bands are observed for these and other garnets [Kück *et al.* (1995)]. The spectra in Fig. 10.16 show weak but finite Cr^{4+} absorption and emission in YAG and YGG in the wavelength region 1120–1330 nm [Eilers *et al.* (1994)]. In a hydrostatic pressure study, that beautifully illustrates the continuous tunability of the crystal field, Shen *et al.* (1997) identified the $^3T_2/^1E$ level crossing in Cr^{3+}:YAG, placing the $^3B_1(^3A_2) \rightarrow A_1(^1E)$ zero-phonon transition near 1044 nm. This is confirmed by the ligand field calculation of the A_1–B_1 splitting of the ^1E state, which also demonstrated that for all reasonable values of ligand field parameters (e_α, α) only $A_1(^1E)$ can become the emitting state. However, assignments of other weak spectra are more speculative. Nevertheless, the spectroscopic studies clearly identify that the ^1E level of Cr^{4+} in YAG and YGG is not lower in energy than either $^3B_2(^3T_2)$ or $^3A_2(^3T_1)$, so that Cr^{4+} ions occupy weak crystal-field sites in both hosts.

Insofar as Cr^{4+}-laser operation at $T = 300$ K is concerned it is evident that the structure in Figs. 10.15 and 10.16 has broadened to such an extent that structureless bands are observed (see Fig. 9.5). However, the cross sections for the

$^3B_1(^3A_2) \rightarrow {}^3A_2(^3T_1)$ absorption and $^3B_2(^3T_2) \rightarrow {}^3B_1(^3A_2)$ emission are rather large. Eilers *et al.* (1993) determine $\sigma_a = 5.17 \times 10^{-18}\,\text{cm}^2$ at 1.064 μm: the Cr^{4+} : YAG and YGG spectra shown in Fig. 10.15 were measured for samples containing 7×10^{18} Cr ions cm^{-3} and $2.8 \times 10^{19}\,\text{cm}^{-3}$ of charge compensating Ca^{2+} ions. Hence $N(Cr^{4+}) \cong 10^{17}\,\text{cm}^{-3}$, given the value of $\alpha = 0.5\,\text{cm}^{-1}$ at 1.064 μm, implying that only *ca* 1.5–2.0% of Cr ions are present as Cr^{4+} ions in the as-grown crystals. This can be increased to 3–4% or so by an oxidizing anneal at 1600 C for a period of 30–60 hrs. Nevertheless, tetrahedral Cr^{4+} remains a minority valence state. The emission cross section determined from Eqs. (10.4–10.6) are $3.3 \times 10^{-19}\,\text{cm}^2$ for Cr^{4+} : YAG and $4.3 \times 10^{-19}\,\text{cm}^2$ for Cr^{4+} : YGG. Other garnets also have values of σ_e in the range $(3–5) \times 10^{-19}\,\text{cm}^2$ [Kück *et al.* (1994)]. For Cr^{4+}-doped garnets nonradiative decay is ever present, even at 10 K. The lifetime shortening from 30.6 μs (20 μs) at 10 K to 4.1 μs (1.9 μs) at 300 K for YAG (YGG) indicates that radiative decay is a rapidly decreasing fraction of the total population decay with increasing temperature. At 300 K the luminescence quantum efficiency is only 12% for Cr^{4+} : YGG, 21% in $Cr^{4+} : Y_3Sc_{0.22}Al_{4.78}O_{12}$, 22% for Cr^{4+} : YAG and 33% for LAG. The $\sigma_e\tau_E$ figure of merit is inversely proportional to the laser threshold power (Eq. 1.36). This product varies at 300 K from $0.68 \times 10^{-24}\,\text{s cm}^2$ (GSGG) to $1.9 \times 10^{-24}\,\text{s cm}^2$ (YAG) for the garnets studied by Kück *et al.* (1995), bracketing the value for Ti-sapphire ($1.4 \times 10^{-24}\,\text{s cm}^2$). Room temperature laser operation was observed for Cr^{4+} : YAG, $Cr^{4+} : Lu_3Al_5O_{12}$, Cr^{4+} : GSGG and $Cr^{4+} : YSc_xAl_{5-x}O_{12}$ (with $x < 0.5$) in either pulsed or CW mode when pumped with the Nd : YAG 1.064 μm line. In view of their potential for wavelength agility in the wavelength range 1.2–1.55 μm, Cr^{4+}-activated lasers have potential applications as broadband sources in near-infrared spectroscopy, as ultrafast lasers for fibre communications test systems and in remote sensing and medicine. In this context Taylor and colleagues have developed compact, high repetition rate, all-solid-state, Kerr-lens mode-locked Cr^{4+} : YAG and $Cr^{4+} : Mg_2SiO_4$ lasers that are diode-pumped via Nd-YVO_4 or Yb-fibre lasers [Tong *et al.* (1997), Mellish *et al.* (1998)]. Ultrashort pulses as short as 43 fs tunable in Cr^{4+} : YAG from 1.505 to 1.550 μm have been produced with output powers of *ca* 300 mW at repetition rates up to 1 GHz.

Initial development of $Ni^{2+}(3d)^8$ ion lasers pre-dated that of the $(3d)^2$ ion lasers by more than two decades. The host crystals; MgF_2, MnF_2, ZnF_2, $KMgF_3$ and MgO, provide weak crystal-field sites in which the $^3T_2 \rightarrow {}^3A_2$ emission bands occur via magnetic-dipole transitions. The visible absorption spectra of Ni^{2+}-doped crystals by $^3A_2 \rightarrow {}^3T_2, {}^3T_1(^3F)$ and $^3T_1(^3P)$ transitions result in the typical green colouration of all but $Ni^{2+} : MnF_2$ crystals. The $3d$–$3d$ transitions among the host Mn^{2+} ions result in the reddish colour of this host. Mn^{2+} to Ni^{2+} energy transfer improves the luminescence efficiency of Ni^{2+} ions under broadband, lamp excitation. These crystals emit in the green ($^1E_2 \rightarrow {}^3A_2$), red ($^1E \rightarrow {}^3A_2$) and near-infrared ($^3T_2 \rightarrow {}^3A_2$) under appropriate excitation conditions (see §9.3). However, the weaker crystal fields in fluoride hosts determine that pumping of the laser emission is most efficient using the 1.32 μm Nd : YAG line. $Ni^{2+} : MgF_2$ crystals, conduction cooled to *ca* 80 K, produce ~2 W of output power for *ca* 10 W of CW pump power at 1.32 μm with a slope efficiency of 28%. The advantage of Ni^{2+} : MgO over Ni^{2+} : fluorides is that the $^3A_2 \rightarrow {}^3T_2$ absorption can be pumped with the 1.06 μm Nd : YAG output, a much more efficient pump than Nd : YAG at 1.32 μm. The CW output power versus input power plot in Fig. 10.17

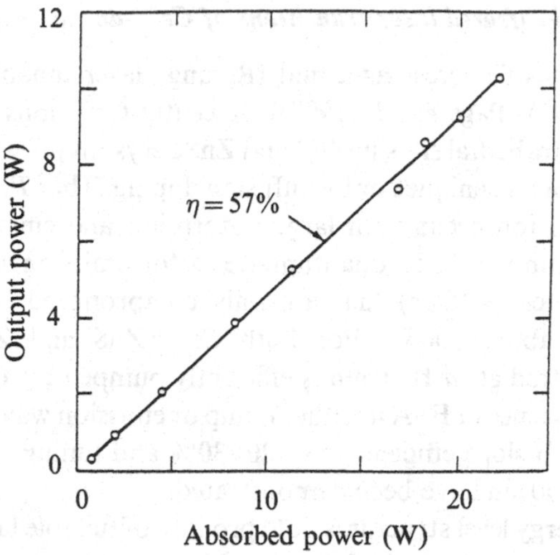

Figure 10.17. CW input-versus-output power of a Ni^{2+}: MgO laser at ≥80 K pumped at 1065 nm with a Nd^{3+}: YAG laser and operating at 1.318 μm [after Moulton (1985)].

shows that with a crystal cooled to ~80 K the Ni^{2+}: MgO laser produces 10 W of output at 1.318 μm with a slope efficiency of 57% [Moulton and Mooradian (1979b), Moulton (1985)]. The laser tuning curve shows operation limited to two wavelength ranges, one centred on the main vibronic peak at 1.32 μm and the other on a weaker peak at 1.41 μm in the fluorescence spectrum. This limited tuning range is the result of the overlap of the emission with excited state and ground state absorption [Moncorgé and Benyattou (1988), Koetke *et al.* (1993a,b)]. The Ni^{2+}-activated lasers developed to date have not operated at room temperature, although such had been expected [Iverson and Sibley (1979)]: Moncorgé and Benyattou (1988) have reported CW-operation of the Ni^{2+}: MgO laser at 240 K.

Searches for broadband tunability and room temperature operation led Koetke *et al.* (1991), (1993a) to study the Ni^{2+}-doped garnets (YAG, YGG, YSGG, GSGG, etc.) and perovskites ($LaGaO_3$ (LGO), $YAlO_3$ (YAP), etc.). In both families, since Ni^{2+} ions substitute for trivalent ions, the crystals were co-doped with Zr^{4+} or Ti^{4+} to maintain charge neutrality. In Ni : Zr co-doped YGG and GGG some 5% of Ni^{2+} occupy tetrahedral sites, whereas 50% of Ni^{2+} ions substitute at tetrahedral sites in the other garnets studied (YAG, YSAG, GSAG, YSGG and GSGG). Emission was not observed from tetrahedrally-coordinated Ni^{2+} in these garnets. There are no tetrahedral sites in YAP and LGO and the absorption/luminescence bands were exclusively due to Ni^{2+} in octahedral sites. Spectroscopic analysis showed that Ni^{2+} : Zr^{4+}-doped GGG and YAP have the best potential for laser operation given their insensitivity to nonradiative decay even at 300 K. The quantum efficiency is unity, and the emission cross section is *ca* twice that of Ni^{2+} : MgO. The failure to achieve CW laser action for Ni^{2+} : Zr^{4+} : GGG was attributed to ESA in the long wavelength wing of the emission and overlap with the GSA at short wavelengths.

10.3.7 *Mid-infrared laser transitions of Cr^{2+}-doped chalcogenides*

The $(3d)^4$ ion Cr^{2+} is of interest as a mid-IR range laser tunable from *ca* 2–4 μm [DeLoach *et al.* (1996), Page *et al.* (1997)]. Recently Cr^{2+} ions were shown to be accommodated on tetrahedral sites in ZnS and ZnSe crystals grown by Bridgman and vapour phase transport techniques or by diffusion doping. The $^5T_2 \rightarrow {}^5E$ transitions of this ^5D configuration ion occur with large absorption and emission cross sections ($\sim 10^{-18}\,cm^2$), high luminescence quantum yields for emission in the range 2000–3000 nm and life times ($\sim 10\,\mu s$) that are only compromized by phonon-induced nonradiative decay above 300 K. For both Cr^{2+} : ZnS and ZnSe the $^5T_2 \rightarrow {}^5E$ absorption band centred at *ca* 1800 nm is efficiently pumped by a Co^{2+} : MgF_2 laser. There is very little evidence of ESA at either pump or emission wavelengths. The lasers operate at 300 K, with slope efficiencies of 20–30% and tuning ranges of *ca* 650 nm about the peak of 2500 nm have been demonstrated.

Although their energy level structures hold promise of tunable laser action (Fig. 4.8, §9.2.3), there have been no reports of successful laser gain media associated with the optical transitions of materials activated by $3d^6$ configuration ions.

10.4 Tunable rare-earth ion lasers

Several RE^{n+}-host crystal combinations have attracted attention as potential tunable lasers. The first of these ions, Sm^{2+} $(4f^6)$, has a ^7F_J ground term split by spin–orbit coupling into $^7F_0$, $^7F_1$–$^7F_6$ manifolds spread over *ca* 5000 cm^{-1} in energy. In SrF_2 the next highest state is ^5D_J, also split by LS coupling so that $^5D_0$ is lowest, *ca* 15 000 cm^{-1} above the $^7F_0$ ground state. The usual $4f$–$4f$ transitions are sharp lines with long luminescence decay times. In the isomorphic CaF_2 the $4f^5\,5d^1$ configuration of Sm^{2+} is below $^5D_0$ and broad $4f$–$5d$ bands occur as allowed electric-dipole transitions with large oscillator strengths ($f \cong 10^{-2}$) and short radiative life times (~ 10–100 ms). The broad absorption bands of Sm^{2+} : CaF_2 are centred at *ca* 350 nm, 475 nm and 650 nm as a result of crystal-field and spin–orbit splittings of the $(4f)^5(5d)^1$ configuration. Excitation in these bands results in luminescence into the $^7F_1$ manifold characterized at low temperature ($T < 80$ K) by an intense zero-phonon line at 708.5 nm attendant upon a vibronically-broadened band stretching to 770 nm. The structure broadens above *ca* 100 K. Sorokin and Stevenson (1961) operated the Sm^{2+} : CaF_2 laser at $T = 20$ K on the zero-phonon line by flashlamp pumping. Subsequently, laser action on the vibronic band was reported at $T > 80$ K [Vagin *et al.* (1969)], the free-running laser wavelength shifting from 708.5 nm to the bandpeak at 745 nm. The Sm^{2+} : CaF_2 laser is a four-level system because the $^7F_1$ terminus level is essentially empty at low temperatures, being 263 cm^{-1} above the $^7F_0$ ground level.

The first vibronic laser to be operated at *room temperature* used the $^5I_7 \rightarrow {}^5I_8$ emission of Ho^{3+} : BaY_2F_8 at 2.171 μm. For better coupling to the white light continuum of the xenon pump lamp the crystal was co-doped with Er^{3+} and Tm^{3+} to take advantage of $^4I_{13/2}(Er^{3+}) \rightarrow {}^3H_4(Tm^{3+}) \rightarrow {}^5I_7(Ho^{3+})$ energy transfer pumping. The broad, structured emission band is a mixture of zero-phonon lines terminating on the 17 possible crystal-field components of the $^5I_8$ state of Ho^{3+} and weak vibronic sidebands [Johnson and Guggenheim (1971)]. A common host for Ho^{3+} has been YLF in which the $^5I_7 \rightarrow {}^5I_8$ transitions emit in the wavelength range 1880–2100 nm with lifetime of 12 ms at 300 K. However, the best performance of Ho^{3+} lasers, using either

Ho : Er : Tm : YLF [Erbil and Jenssen (1980)] or Ho : Er : YAG [Johnson *et al.* (1966b)] gain media, is at 77 K where thermal depopulation of the terminal levels and energy transfer dynamics are improved.

The electronic structures of $Ce^{3+}(4f^1)$ and $Yb^{3+}(4f^{13})$ are similar, there being one electron and one hole, respectively, in the 4*f*-shell. The low-lying $^2F_{5/2,7/2}$ levels have spin–orbit splittings of order 2000 cm^{-1} in Ce^{3+} and −10 000 cm^{-1} in Yb^{3+}, resulting in their having $^2F_{5/2}$ and $^2F_{7/2}$ ground states, respectively. In consequence, Yb^{3+}-doped crystals feature narrow absorption lines at 0.92–1.0 μm depending on host crystal, involving transitions between the crystal-field split levels of $^2F_{7/2}$ and $^7F_{5/2}$ states, and which can be pumped at *ca* 980 nm by a laser diode. The low-lying $^2F_{5/2,7/2}$ levels of Ce^{3+} are separated by ~30 000–40 000 cm^{-1} from the next highest configuration, $(5d)^1$. The $4f \leftrightarrow 5d$ transitions, being parity-allowed, have large oscillator strengths. Although the 4*f* orbitals are spatially-compact and well shielded from the crystalline electric field, the 5*d* orbitals are spatially-diffuse and poorly shielded from the lattice, so that $4f \leftrightarrow 5d$ spectra have large bandwidths. Because of the large energy gap between the lowest 5*d*-level and the 4*f*-levels, nonradiative decay is weak, and the quantum efficiency of 5*d*–4*f* emission is near unity.

The first operational Ce^{3+}-lasers used YLF and LaF_3 host media containing *ca* 1 at.% Ce^{3+}, pumped with KrF excimer lasers [Erlich *et al.* (1979), (1980)]. Crystal-field transitions from the lowest 5*d* level into the $^2F_{7/2}$, $^2F_{5/2}$ levels result in broad emission bands peaking at 325 nm in YLF and 290 nm in LaF_3. Assuming that the luminescence quantum efficiency is near unity led to estimations that the peak gain cross section was 8×10^{-18} cm^2 in Ce : YLF and 7×10^{-18} cm^2 in Ce : LaF_3 [Yang and DeLuca (1977)]. Low-symmetry distortions result in the $^2F_{5/2} \rightarrow 5d$ absorption being split into at most five bands. However, the wavelength ranges of both absorption and emission spectra are determined by the dominant crystal-field component, leading to a good deal of variation among the different materials investigated. Recently Ce^{3+} : doped LiCAF and LiSAF were shown to outperform YLF, $LuLiF_4$ and LaF_3 hosts when pumped with the fourth harmonic of Nd : YAG at 266 nm [Dubinskii *et al.* (1993), Marshall *et al.* (1994)]. The crystal structure of LiCAF in Fig. 3.4 shows each cation to be located in a trigonally-distorted octahedron of F^- ions. The ionic radii of Li^+ (0.078 nm), Ca^{2+} (0.106 nm) and Al^{3+} (0.057 nm), suggests that Ce^{3+} (0.118 nm) will substitute on Ca^{2+} sites, the replacement of Ca^{2+} by Ce^{3+} being charge compensated by Li^+ vacancies [Yamaga *et al.* (1998b)]. Ce^{3+} ions that occupy unperturbed Ca^{2+} ion sites are charge compensated by Li^+ vacancies remote from the unit cell. Two other Ce^{3+} sites involve near neighbour Li^+ vacancies. The *g*-values of the three centres measured by ESR provide a fairly accurate description of the three component Kramers doublets of the $^2F_{5/2}$ ground state that result from mixing the $|5/2, \pm 1/2\rangle$-, $|5/2, \pm 3/2\rangle$- and $|5/2, \pm 5/2\rangle$-functions by spin–orbit coupling and trigonal and orthorhombic crystal-field interactions. These Kramers doublets are labeled $|\pm g_i\rangle$ with $i = 1, 2$ and 3. The first excited level of $^2F_{5/2}$ is some 210 cm^{-1} above $|\pm g_1\rangle$ according to an analysis of the temperature dependence of the ESR linewidth. The degeneracy of the five ^2D free ion orbitals is lifted by the octahedral component of the crystal field into $^2T_{2g}$ and ^2E_g levels, the remaining orbital degeneracy being removed by the combined effects of spin–orbit interaction and the lower symmetry crystal field. The excited states comprising the five Kramers doublets $|\pm e_j\rangle$ with $j = 1–5$ are each broadened by the electron–phonon interaction. Their wavefunctions were calculated assuming the

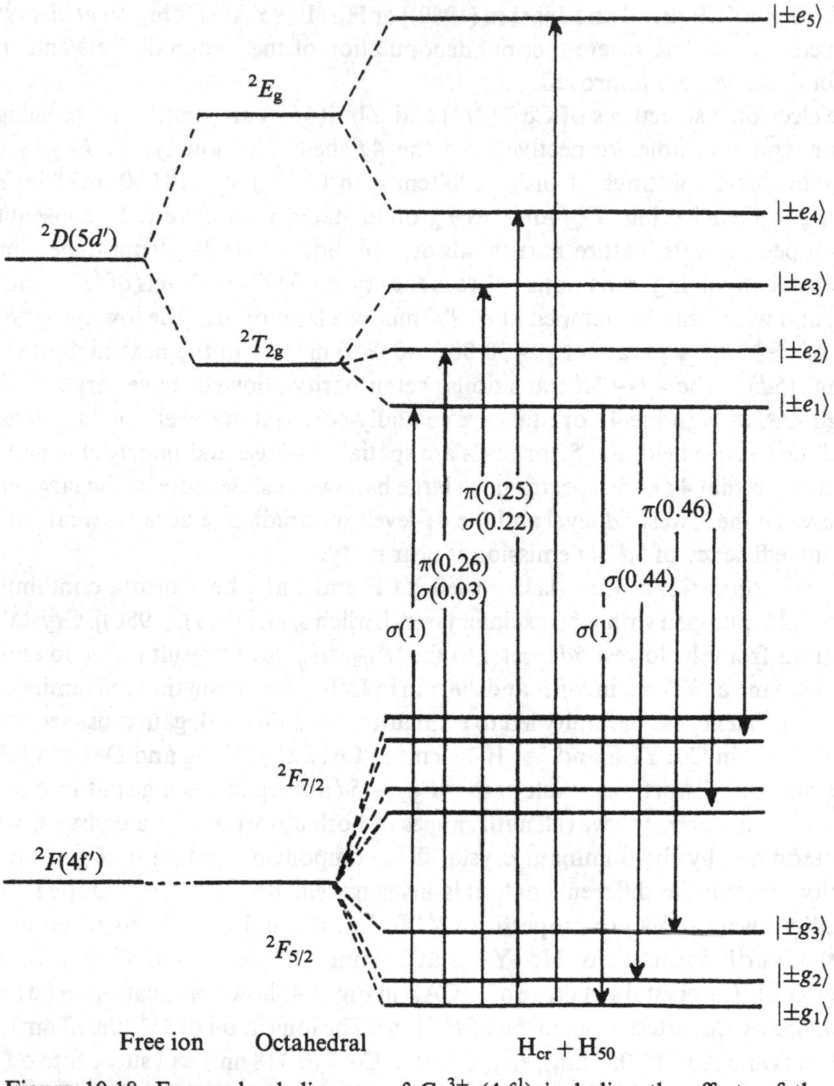

Figure 10.18. Energy level diagram of Ce^{3+} $(4f^1)$ including the effects of the dominant octahedral crystal field and trigonal field and spin–orbit perturbations [after Yamaga *et al.* (1998b)].

octahedral splitting to be much larger than the trigonal splitting, and included $|\tilde{l}, s_z\rangle$-eigenfunctions admixed to second order in spin–orbit coupling [Abragam and Bleaney (1970)]. The level structure for the excited $(5d^1)$ configuration in Fig. 10.18 also shows the polarized absorption and emission rates calculated from

$$A_{eg} = \text{const.} |\langle g_i|\mu^a|e_j\rangle|^2, \tag{10.6}$$

in which $\mu^a = ez$ or $e(x \pm iy)$ is the electric dipole operator for π- and σ-polarized transitions, respectively, between ground ($|g_i\rangle$) and excited ($|e_j\rangle$) states [Yamaga *et al.* (1998b)]. The selection rules are the usual ones for electric dipole transitions, i.e., $\Delta l = \pm 1$ for σ-polarization and $\Delta l_z = 0$ for π-polarization with $\Delta s_z = 0$.

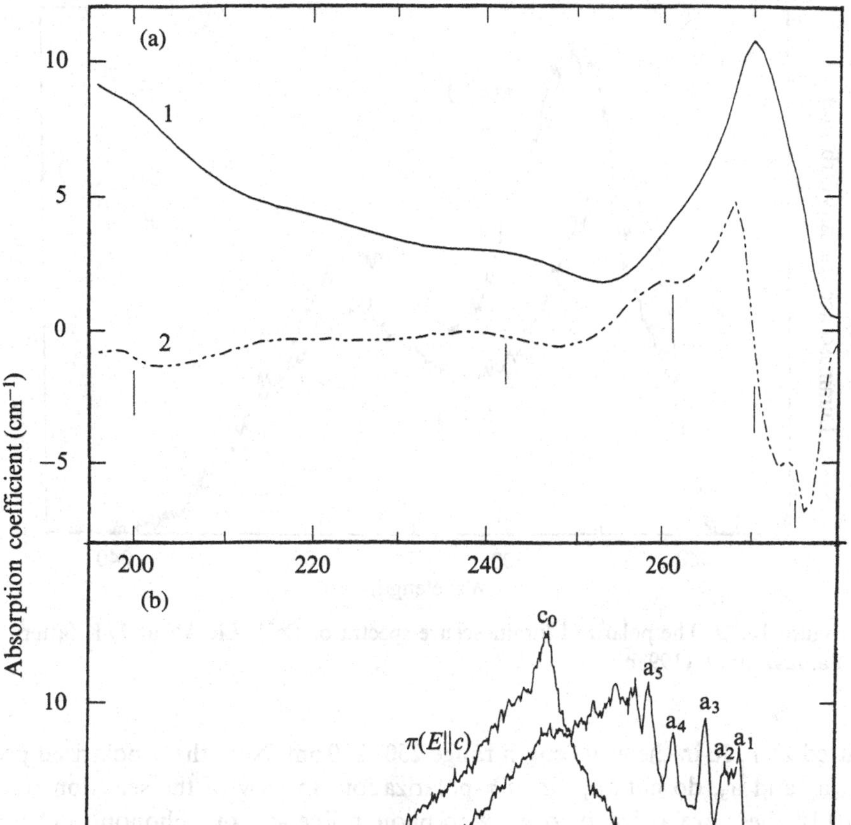

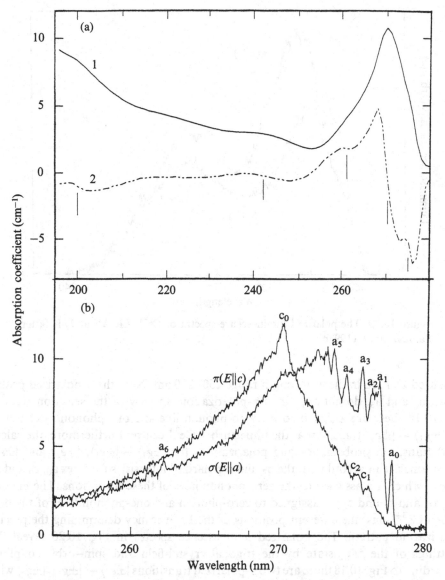

Figure 10.19. Optical absorption spectra of Ce^{3+} : LiCAF, measured at 17 K: (a) low resolution, first derivative presentation and (b) high resolution [after Yamaga *et al.* (1998b)].

The optical absorption spectrum of Ce^{3+} : LiCAF grown from high purity (4N) fluorides in a vertical Stockbarger–Bridgman furnace using a semiconductor-grade graphite crucible is shown in Fig. 10.19. The first derivative of the unpolarized low-resolution spectrum, Fig. 10.19a, shows broad peaks at 276 nm, 270 nm, 262 nm, 243 nm and 201 nm due to optical transitions from the ground level of $^2F_{5/2}$ ($|\pm g_1\rangle$) to the five ($|\pm e_j\rangle$) levels of the ^2D excited state of Ce^{3+}. They are indistinct because of vibronic broadening and the overlap of bands from the three different centres. Figure 10.19b shows the high resolution π- and σ-polarized absorption spectra

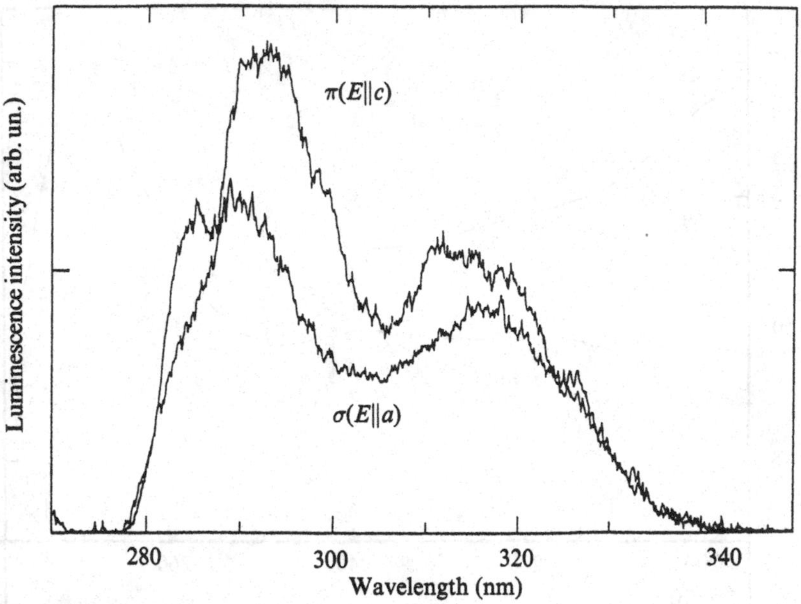

Figure 10.20. The polarized luminescence spectra of Ce^{3+}: LiCAF at 77 K [after Yamaga *et al.* (1998b)].

measured at 17 K in the wavelength range 250–280 nm. Note the σ-polarized peaks, a_0, a_1, a_2 and a_3, do not appear in π-polarization. In view of the selection rules in Fig. 10.18 these are assigned to the zero-phonon line and one-phonon sideband of the $|\pm g_1\rangle \rightarrow |\pm e_1\rangle$ transition at the unperturbed Ce^{3+} centre. Furthermore, the calculated transition probabilities and polarizations for $|\pm g_1\rangle \rightarrow |\pm e_1\rangle$, $|\pm e_2\rangle$ and $|\pm e_3\rangle$ transitions are in accord with the σ- and π-polarized intensities for lines a_0, c_0 and a_6 *et seq.*, which are assigned to the zero-phonon lines of these transitions. The lines a_4 and a_5, and c_1 and c_2 are assigned to zero-phonon and one-phonon lines of the perturbed Ce^{3+} ions, the different positions of the Li^+ vacancy determining the precise polarization pattern. The polarized luminescence spectra in Fig. 10.20 reveal the splittings of the $^2F_{5/2}$ state by the trigonal crystal-field and spin–orbit coupling. According to Fig. 10.18 there are two σ-polarized transitions $|\pm e_1\rangle \rightarrow |\pm g_1\rangle$, $|\pm g_2\rangle$ with relative intensities 1 and 0.44, respectively, corresponding to the broadband peaks at *ca* 285 nm and 289 nm in Fig. 10.20. The π-polarized peak at 292 nm is assigned to the $|\pm e_1\rangle \rightarrow |\pm g_3\rangle$ transition, the calculated intensity of which is 0.46 relative to the σ-polarized $|\pm e_1\rangle \rightarrow |\pm g_1\rangle$ transition. These properties indicate what might be expected of Ce^{3+} : LiCAF used as a tunable laser gain medium. The absorption coefficient at 270 nm in π- and σ-polarizations, respectively, of 12 cm^{-1} and 10.5 cm^{-1} corresponds to peak cross sections 7.3×10^{-18} cm^3 and 6.3×10^{-18} cm^2. The emission cross-sections at 290 nm are 9×10^{-18} cm^2 and 6×10^{-18} cm^2 in π- and σ-polarizations, respectively, and were obtained by numerical integration of the spectra. Such cross-sections correspond to allowed electric dipole transitions, as does the emission life time of 29 ns [Marshall *et al.* (1994)].

A major step forward was made by pumping Ce^{3+} : LiCAF and Ce^{3+} : LiSAF lasers at 266 nm with the pulsed fourth harmonic output from a Q-switched Nd : YAG laser

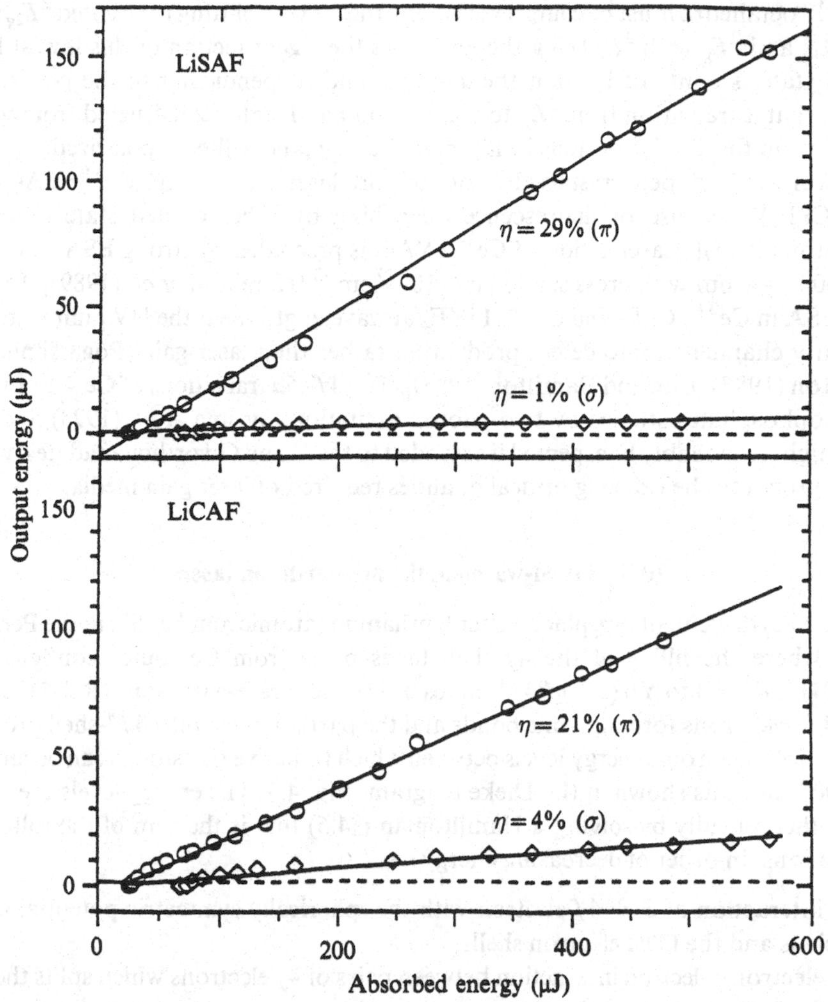

Figure 10.21. Laser output energies at 292 nm as a function of 266 nm pump energy for Ce^{3+}:LiSAF and Ce^{3+}:LiCAF lasers, using longitudinal pump polarization parallel (π) and perpendicular (σ) to the optical axis [after Marshall *et al.* (1994)].

[Dubinskii *et al.* (1993), Marshall *et al.* (1994)]. The slope efficiencies of Czochralski-grown crystals containing $\sim 0.6 \times 10^{18}\,cm^{-3}$ and $0.9 \times 10^{18}\,cm^{-3}$ of Ce^{3+} in LiCAF and LiSAF, respectively, are shown in Fig. 10.21 to be strongly sensitive to the pump polarization. The slope efficiencies were 21 to 29% for π-polarized pumping, and only 1 to 4% in σ-pumping. Given the rather weak luminescence anisotropy in Fig. 10.20, the polarization dependent laser operation was attributed to anisotropic ESA transitions, involving the promotion of an electron from $5d^1$ into the host conduction band, the anisotropy reflecting the layered LiCAF and LiSAF crystal. This is in accord with the phenomenological model of the electronic structure deduced from the ESR and polarization results. The *g*-values of Ce^{3+} in LiCAF show that the $|\pm g_1\rangle$ ground state wavefunctions contain a dominant $|5/2, \pm 5/2\rangle$ component, according to which the

$(CeF_6)^{3-}$ octahedra must be compressed along the *c*-axis, splitting the excited $^2T_{2g}$ state into $^2A_{1g}$ and ^2E_g with ^2E_g being the lower. As the eigenfunction of this lowest lying excited state is contained within the *aa*-plane and perpendicular to the *c*-axis, it is evident that a transition from ^2E_g to conduction band states constructed from wavefunctions on the Ca^{2+}/Sr^{2+} ions in layers in the *aa*-plane will be σ-polarized.

Several Ce^{3+} : doped crystals do not support laser action (e.g. Ce^{3+} : YAG and Ce^{3+} : CaF_2), because of fluorescence quenching by other excited state processes [Hamilton (1985)]. Laser action of Ce^{3+} : YAG is precluded by strong ESA extending from 600–800 nm with cross section $\sigma \sim 10^{-17}\,cm^{-2}$ [Hamilton *et al.* (1989)]. In contrast, ESA in Ce^{3+} : CaF_2 and Ce^{3+} : $LiYF_4$ at wavelengths near the UV pump bands is efficiently channelled into defect production rather than laser gain [Pogatshnik and Hamilton (1987), Lim and Hamilton (1988)]. The $4f$–$5d$ transitions of Ce^{3+} in Lu and Y orthophosphates studied by two-photon excitation [Sytsma *et al.* (1993)] indicate what might reasonably be expected if extended to the Y and Gd orthovanadates, which can be grown to the exacting optical qualities required of laser gain media.

10.5 Fixed-wavelength rare-earth ion lasers

The rare-earth elements are placed after lanthanum (atomic number 57) in the Periodic Table where the filling of the $4f$ shell takes place from Ce (outer configuration $5s^25p^64f^15d^16s^2$) to Yb ($5s^25p^64f^{13}5d^16s^2$). Trivalent rare-earth ions (RE^{3+}) lose all $5d$ and $6s$ electrons forming ionic bonds and the partially-occupied $4f^n$-shell provides the multiple electronic energy levels between which radiative transitions can occur over $\sim 40\,000\,cm^{-1}$, as shown in the Dieke diagram, Fig. 4.9. The energy levels are determined theoretically by solving a Hamiltonian (§4.5) that is the sum of the following interactions, in order of decreasing energy:

- the interaction of each $4f$ electron with the spherically symmetric potential of the nucleus and the filled electron shells;
- the electron–electron interaction between pairs of $4f$ electrons which splits the $4f^n$ configuration into the LS-terms;
- spin–orbit coupling splits LS-terms into $|LSJM_J\rangle$ J-multiplets, each with different J and M_J-values. Each J-multiplet is $(2J+1)$-fold degenerate with levels characterized by M_J. The number of levels into which the J-levels are split by the even parity crystal field, Tables 2.19, 2.20, can be calculated by group theory (§2.5.5 and §4.5).
- H_{CF} represents the interaction between the $4f^n$ electrons and the crystal field. Odd-parity components of the crystal field do not contribute to state energies but may induce electric dipole transitions between $4f^n$ states by admixing opposite-parity wavefunctions into the multiplets [Judd (1962), Ofelt (1962)].

There is some similarity between the term structures of the conjugate configurations $4f^n$ and $4f^{14-n}$. The $(14-n)$ electron system is treated as *n-positively charged* electrons, since, apart from a constant term, the mutual coulomb interactions between n-positive electrons and n-negatively-charged electrons are identical. In consequence, the LS-terms and their energy orderings are the same for $4f^n$ and $4f^{14-n}$ configurations (§4.2.2). However, since the spin–orbit energy changes sign when the electronic charge changes sign, the splitting of LS-terms into the J-fine structure levels is reversed

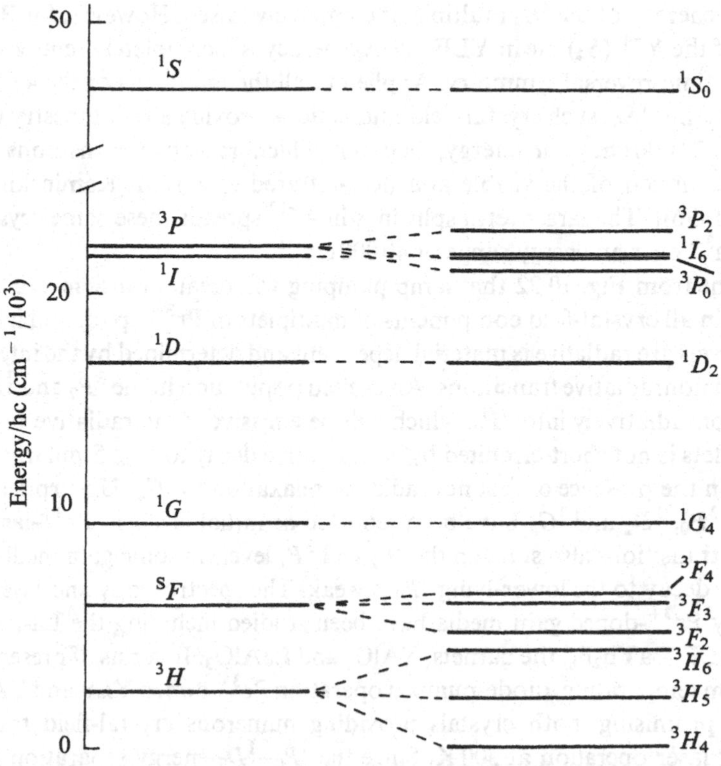

Figure 10.22. Term splitting by LS coupling for the $4f^2$ configuration of Pr^{3+}. Note the breakdown of the Landé interval rule for ^3H_J and ^3F_J multiplets caused by $J=4$ mixing. For the complementary $4f^{12}$ configuration of Tm^{3+} the effect of mixing is so large that $^3F_4$ lies lower than both $^3H_4$ and $^3H_5$.

for $4f^{14-n}$ ions relative to $4f^n$ ions (§4.2.3). For example, the ^4I_j term of Nd^{3+} splits into $^4I_{9/2}$, $^4I_{11/2}$, $^4I_{13/2}$ and $^4I_{15/2}$ levels in order of increasing energy. This ordering is reversed for Er^{3+}. The magnitude of the spin–orbit coupling parameter, ζ, increases by a factor of 3–4 across the rare-earth ions from $4f^1$ to $4f^{13}$. In consequence, term splittings are larger in Er^{3+} than in Nd^{3+}. Additionally, there is a greater likelihood that the Landé interval will break down at large ζ because of term–term mixing, and to lesser extent spin–spin, orbit–orbit and spin–other-orbit coupling.

10.5.1 *Spectroscopy and laser transitions of Pr^{3+} and Tm^{3+}*

The (LS)-terms for the $4f^2$ configuration of Pr^{3+} are shown in Fig. 10.22. Spin–orbit interaction, H_{so}, splits each (LS)-term into its J-levels and mixes states with the same J from different terms. As a result $^3H_4$ and $^3F_4$ are strongly mixed and the Landé interval rule (Eq. 4.26) breaks down. Some mixing of $^1I_6$ with $^3H_6$, and of $^1D_2$, $^3F_2$ and $^3P_2$ also occurs. The spin–orbit splitting in the $4f^{12}$-configuration ion Tm^{3+} is of opposite sign and a factor three or so larger than for Pr^{3+} so that the ordering of the J-multiplets from equivalent LS-terms is inverted. Furthermore, the mixing of the $J=4$ levels is so large that $^3F_4$ lies *below* $^3H_4$ and $^3H_5$ [Armagan *et al.* 1989]. In garnet crystals, where RE^{3+} ions occupy the dodecahedral (D_{2h}) A sites, reference to Table 2.19 shows that the nine-fold degeneracy of the $^3H_4$ multiplet is completely raised. However, for RE^{3+} ion occupancy of the Y^{3+} (S_4) site in YLF the degeneracy is incompletely removed due to the effects of time-reversal symmetry. Applied to all the multiplets of the $4f^2$ configuration, excepting $^1S_0$, such crystal-field interactions provide a rich tapestry of levels spanning *ca* 22 000 cm^{-1} in energy, between which radiative transitions can be observed over much of the visible and near-infrared spectrum [Kaminskii (1989), (1991) and (1996)]. The larger term splittings in $4f^{12}$ spreads these same crystal-field levels for Tm^{3+} over an energy range of 40 000 cm^{-1}.

It is evident from Fig. 10.22 that lamp pumping will create transient excited state populations in all crystal-field components of multiplets of Pr^{3+} up to, and including, $^3P_2$: which levels are radiative is material dependent and determined by the interplay of radiative and nonradiative transitions. An excited population in the $^3P_2$ and $^3P_1$ levels will decay nonradiatively into $^3P_0$, which will be emissive if the radiative process to other multiplets is not short-circuited by nonradiative decay to $^1D_2$. Similarly $^1D_2$ will emit except in the presence of fast nonradiative relaxation to $^1G_4$. Unsurprisingly, all three-levels, $^3P_0$, $^1D_2$ and $^1G_4$ have been reported as initial levels in Pr^{3+}-laser operation. Laser transitions also start on the $^3P_2$ and $^3P_1$ levels in some gain media, where nonradiative decay to the lower-lying $^3P_0$ is weak. The spectroscopy and laser operation of many Pr^{3+}-doped gain media have been studied including the La-trihalides, YLF and LuLF, $BaYb_2F_8$, the garnets, $YAlO_3$ and $LaAlO_3$. In terms of present trends towards room temperature, diode-pumped operation Pr^{3+}-doped YLF and $YAlO_3$ are particularly promising, both crystals providing numerous crystal-field transitions that support laser operation at 300 K. Since the $^3P_0$–$^1D_2$ energy separation is about 3700 cm^{-1} in YLF and $YAlO_3$, where the phonon energies are *ca* 350 cm^{-1} and 600 cm^{-1}, respectively, nonradiative decay across this energy gap will be slower in YLF, where it is a ten-phonon process, than in $YAlO_3$ which involves a six-phonon process. In consequence, $^3P_0$ is the starting level for room temperature laser operation in Pr^{3+} : YLF at *ca* 479 nm ($^3H_4$ final state), 538 nm ($^3H_5$), 607 nm ($^3H_6$), 640 nm ($^3F_2$), 720 nm ($^3F_4$) and 907 nm ($^1G_4$). Given the excellent optical properties of the YLF host, it is evident that the technology exists for commercial development of Pr^{3+} : YLF lasers, as appropriate applications catalyze significant demand.

The first room temperature laser action in Pr^{3+} : YLF operated at 479 nm on a transition from $^3P_0$ to the lowest crystal-field level of the $^3H_4$ ground manifold [Esterowitz *et al.* (1977)]. Subsequently, detailed spectroscopic studies guided the assignment of many laser transitions from the blue into the mid-infrared [Adam *et al.* (1985), Kaminskii *et al.* (1987)]. Most early studies used flashlamp or dye laser pumping, which can be inefficient. More recently, the sharp-line emissions of Pr^{3+} : YLF have been pumped using Ar^+ laser lines, and efficient visible laser operation was reported at six wavelengths [Sandrock *et al.* (1994)]. A further fourteen CW lasing transitions were identified by Sutherland *et al.* (1996) with room temperature pumping using the 476 nm line from an Ar^+ laser. Several of these lines permitted modest tuning ranges and adequate bandwidth to support sub-picosecond pulses. The laser output wavelengths of 604.4 nm, 607.3 nm, 609.2 nm and 613.0 nm originate on the $^3P_0$–$^3H_6$ transition, which Table 2.19 shows to have thirteen possible terminal levels. The absence of laser output on some crystal-field lines is not uncommon [Esterowitz *et al.* (1977), Jenssen *et al.* (1975)]. Other multiple-laser line components of inter-multiplet transitions starting on $^3P_0$ included $^3F_2$, $^3F_3$ and $^3F_4$. Since a sizeable population is maintained in $^3P_1$ and $^1I_6$ levels, laser operation has been excited on $^3P_1 \rightarrow {}^3F_{2,3,4}$ and $^1I_6 \rightarrow {}^3F_4$ transitions. In low temperature operation ($T \sim 110$ K) laser action on the $^3P_0 \rightarrow {}^1G_4$ transition can be the first stage in a cascade laser acting on the $^3P_0 \rightarrow {}^1G_4 \rightarrow {}^3H_5$ transitions at 907 nm and 1347 nm in Pr^{3+} : YLF and on three laser transitions $^3P_0 \rightarrow {}^1G_4 \rightarrow {}^3H_5, {}^3F_4$ at 915 nm, 1.335 µm and 3.608 µm in Pr^{3+} : $BaYb_2F_8$ [Kaminskii (1991)].

Pr^{3+}-doped $YAlO_3$ also has potential for commercial development combining excellent spectroscopic and laser properties with such other qualities as hardness, mechanical strength, excellent resistance to thermal shock, large thermal conductivity and broadband transparency (250 nm–4000 nm). Large single crystals (250 mm × 35 mm ϕ) can be grown by the Czochralski technique with excellent tolerance of high concentrations of RE^{3+}-dopants. In operation at 300 K the Pr^{3+} : $YAlO_3$ laser gives efficient performance on several inter-manifold transitions; $^3P_0 \rightarrow {}^3F_3$ at 720 nm, $^3P_0 \rightarrow {}^3F_4$ at 747 nm and $^3P_0 \rightarrow {}^1G_4$ at 931 nm using crystals containing 1 at.% Pr^{3+}. However, in crystals containing greater than 4 at.% Pr^{3+} very efficient cross-relaxation transitions $^3P_0 \rightarrow {}^1D_2$ and $^3H_4 \rightarrow {}^3H_6$ provide efficient excitation of the $^1D_2$ level and concomitant laser operation at 300 K using the $^1D_2 \rightarrow {}^3F_3$ transition at 996 nm [Kaminskii *et al.* (1991)].

The interest in Tm^{3+} ($4f^{12}$)-activated gain media arises from potential applications in medical technology, in eye-safe optics, in laser range finding and remote sensing and in optical communications. The $^3P_0$, $^1D_2$ and $^1G_4$ emitting levels of Tm^{3+} are much higher in energy than their counterparts in Pr^{3+}. Absorption measurements on Tm^{3+} : YAG place transitions from the $^3H_6$ ground multiplet to crystal-field levels of $^3H_4$ centred at *ca* 780 nm, of $^3F_3$ near 680 nm, $^3F_2$ near 660 nm, $^1G_4$ near 460 nm and $^1D_2$ near 355 nm [Armagan *et al.* (1989)]. In consequence, access to the energy level structure of the emitting levels by pumping into $^1G_4$, $^1D_2$ or $^3P_0$ levels using the strong Ar^+ lines at 514 nm and 488 nm is not efficient. Nevertheless, the 476 nm line from Ar^+ will excite some crystal-field components of $^3H_6 \rightarrow {}^1G_4$, the 647.1 nm Kr^+ line will access the $^3F_3$ and $^3F_2$ multiplets and an 800 nm diode laser will pump the $^3H_4$ level. Tm^{3+} : YAG laser operation was first reported at 77 K in pulsed mode at 1.8834 µm and 2.0132 µm, and CW at the latter wavelength [Johnson *et al.* (1965)]. Caird *et al.* (1975) have also reported room temperature emission at 2.3 µm in both Tm^{3+} : YAG and Tm^{3+} : $YAlO_3$. There are many schemes for improving the pumping of Tm^{3+} lasers using laser-diodes. The difficulty is that the 2 µm laser transitions ending on the $^3H_6$ ground multiplet operate on a quasi-three-level pumping cycle with a lower level that is thermally populated at room temperature. Furthermore, the effective cross section for stimulated emission in Ho^{3+}-doped materials is low, requiring that a considerable fraction of the Tm^{3+} ions be excited during laser operation. Consequently, there is a reduced absorption of pump light by the depleted ground state, resulting in a significant increase in the threshold and reduced slope efficiency.Laser performance can be improved by co-doping with Tm^{3+} and Cr^{3+}, since the overlap between the Cr^{3+} emission and the Tm^{3+} absorption sensitizes the laser transitions on the Tm^{3+} ion. Using a Cr^{3+} : Tm^{3+} : YAG gain medium the threshold for pulsed laser operation on the 2 µm transitions at 300 K can be reduced to tens of joules compared to the hundreds of joules of energy needed to start laser operation in Tm^{3+} : YAG. Tm^{3+}-laser operation has been reported for numerous hosts including $CaWO_4$, $Ca(NbO_3)_2$, $YAlO_3$ and YLF and their isomorphs, as well as the Y and Gd garnets.

10.5.2 Nd^{3+} and Er^{3+}-activated lasers

The Nd^{3+} : YAG laser is still the most used solid state laser, especially in high power applications. It represents a mature technology with many applications. There has been much research and development of the Nd^{3+} : YAG laser; to improve it, to upsize it, down-size it and even to replace it. In this last sense Nd^{3+} : YLF, Nd^{3+} : $YAlO_3$ and Nd^{3+} : YVO_4 have had most commercial success. Nd^{3+} : YLF lasers are having an impact on the traditional Nd^{3+} : YAG market, especially in medium power applications, in producing ultrashort mode-locked pulses and in ultrashort pulse pumping of optical parametric oscillators (OPOs) and amplifiers (OPAs). Compared to YAG, the integrated absorption over the pump band at *ca* 800 nm is *ca* 20–25% larger, and the emission cross section for the $^4F_{3/2} \rightarrow {}^4I_{11/2}$ multiplet is similarly large and into a larger bandwidth of 7–8 nm compared with 2–3 nm in YAG. Furthermore, the upper state lifetime is a factor of two longer. The advantages of using Nd^{3+} : $YAlO_3$ are not so self-evident, although larger polarized absorption and emission cross sections with shorter decay times are not to be neglected. A superior branching ratio for the $^4F_{3/2} \rightarrow {}^4I_{13/2}$, $^4I_{11/2}$ transitions suggests Nd^{3+} : $YAlO_3$ to be the material of choice for operation at 1.34 μm. Other garnets, such as GSGG and YSAG, have been considered usually co-doped to improve efficiency and reduce threshold by Cr^{3+} : Nd^{3+} energy transfer. So far their performance has not surpassed that of Cr^{3+} : Nd^{3+} : YAG. However, efficiency should be improved in lower symmetry (coordination) sites and by a closer fit of the Nd^{3+} ion on the substitutional site in the host crystal. Hence Nd^{3+} substituted for Lu^{3+} in $Lu_3Al_5O_{12}$ leads to improved laser efficiency over Nd^{3+} : YAG, in contrast to Nd^{3+} : GGG, which performs less well. There is also much interest in the laser-active properties of Nd^{3+} : orthovanadates and orthosilicates, fluorovanadates and fluoroapatites both for diode-pumping and operation at 1.34 μm.

The role of bonding is important: On the basis of electronegativity, performance is expected to deteriorate along the electronegativity series from F → O → Cl → Br, and an abundance of research into fluorides was mandatory. Since Nd^{3+} more closely matches the ionic radius of Gd^{3+} than Y^{3+} it is evident that $LiGdF_4$ will permit larger Nd^{3+} concentrations than $LiYF_4$, without reduced crystal quality. However, the primary improvement is in the absorption coefficient at pump wavelengths. The promise of the anisotropic fluorides as hosts for rare-earth dopants is well recognized. Several groups have focused on the LiF–KF–Y(Gd)F_3 phase system and identified stable phases; LiY(Gd)F_4, KYF_4, KY(Gd)$_3F_{10}$, LiKY(Gd)F_5 and $Li_{1+3x}KY_{10x}F_6$, which can be fashioned into high quality single crystals [Goryunov *et al.* (1992), Chai *et al.* (1993)]. Each crystal structure contains distorted Y(Gd)F_8 polyhedra. The original Russian work focused on hydrothermal growth [Kaminskii and Khaidukov (1992)]. Chai *et al.* (1993) and Pham *et al.* (1994) developed a modified weight-feedback ADC Czochralski technique for growth at the appropriate peritectic composition which permitted the growth of large single crystals boules, typically 40 mm ϕ by 100 mm long. They showed that the Nd^{3+} distribution coefficient of 0.75 in $LiGdF_4$ compared with 0.34 in $LiYF_4$ allowed more uniform doping of Gd-fluoride hosts at higher concentrations. The laser performance under Ti-sapphire pumping at *ca* 794 nm in π-polarization gave typical slope efficiencies of 60–70% for Nd : YLF and GLF from thresholds of only 10–15 mW. Furthermore, the higher doping levels of Nd^{3+} : GLF

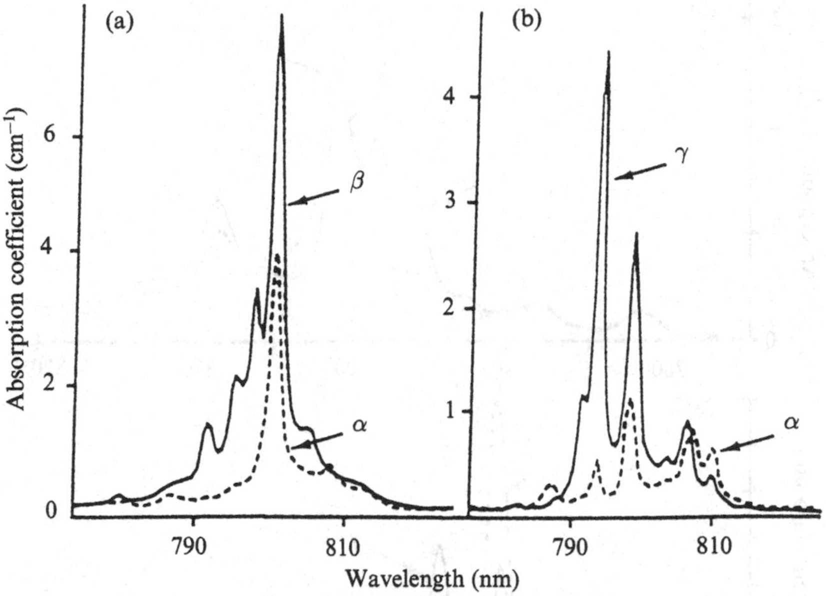

Figure 10.23. A comparison of the optical absorption spectrum corresponding to the $^4I_{9/2} \rightarrow {}^4F_{9/2} + {}^2H_{9/2}$ transitions for (a) $LiKYF_5$ and (b) $LiYF_4$ crystals containing 1 at.% Nd^{3+}, measured at *ca* 77 K.

facilitate shorter Nd^{3+} absorption lengths at 800 nm than for Nd^{3+} : YLF, which is much preferred for diode-pumping.

The most highly anisotropic of the fluoride gain media so far studied are the isomorphs $LiKYF_5$ (KLYF) and $LiKGdF_5$ (KLGF), which can be grown in cm^3 sizes containing up to 5 at.% Nd on the Y(Gd) site [Pham *et al.* (1994), Nicholls *et al.* (1997a)]. It is instructive to compare the optical properties of such materials with those of the established laser material Nd^{3+} : YLF [Krupke and Gruber (1964), Fan *et al.* (1986)]. Figure 10.23 shows the overlapping $^4I_{9/2} \rightarrow {}^4I_{9/2}$, $^2H_{9/2}$ absorption lines of KLYF and YLF crystals containing 1 at.% Nd measured at *ca* 77 K [Nicholls *et al.* (1998)]. Much of the fine structure observed at 4 K [Summers *et al.* (1994)] is masked by inhomogeneous broadening at the higher temperature. The absorption by Nd : KLYF is larger by a factor of almost two than Nd : YLF. The strongest absorption line in Nd : KLYF ($\alpha = 7.5\,\text{cm}^{-1}$) occurs at $\lambda = 800$ nm, near the optimal pump wavelength for AlGaAs laser diodes. In contrast, the strongest absorption line in Nd : YLF occurs near 792 nm ($\alpha = 4.5\,\text{cm}^{-1}$), at which wavelength AlGaAs LDs are less efficient. Polarization results from odd-parity distortions that are stronger in KLYF than in YLF, as indicated by the increased absorption. The tetragonal scheelite structure of YLF is uniaxial and there are two senses of polarization: in π-polarization $E \parallel c$ and in σ-polarization, $E \perp c$. In contrast, monoclinic KLYF is biaxial with three distinct principal axes. For β-polarization, used for the measurement in Fig. 10.24, $E \parallel b$, whereas α- and γ-polarizations refer to $E \perp b$ but parallel to one or other of two principal dielectric axes in the a–c plane, which are rotationally-displaced from the a- and c-axes. In KLYF they are 114° apart. There is advantage to pumping with γ-polarized light propagating along the b-axis into the primary absorption bandwidth of 7–8 nm centred at 800 nm (Fig. 10.24a). The emission so-excited but detected in β-polarization,

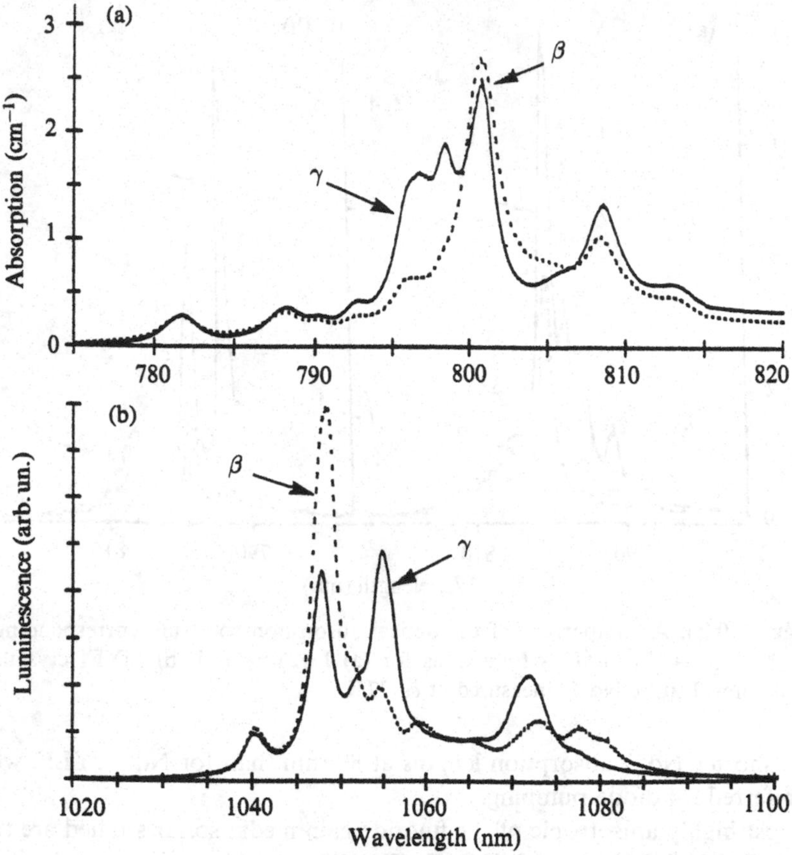

Figure 10.24. Polarized absorption and luminescence spectra of 3.4 at.% Nd^{3+} : KLYF measured at room temperature corresponding to (a) the $^4I_{9/2} \rightarrow (^2H_{9/2}, {}^4F_{5/2})$ absorption and (b) the $^4F_{3/2} \rightarrow {}^4I_{11/2}$ luminescence transitions [after Nicholls *et al.* (1998)].

Fig. 10.24b, is almost exclusively channelled into a single crystal-field component of the $^4F_{3/2} \rightarrow {}^4I_{11/2}$ transition at 1048 nm with a linewidth of 3–4 nm. Apparently, crystals of KLYF grown hydrothermally have a single substitutional site for occupancy by Nd^{3+} and other RE^{3+} ions. However, a different phase results from Czochralski growth, having identical chemical composition ($KLiYF_5$) but different unit cell composition $(KLiYF_5)_2$ and in which there are two slightly different YF_8 coordination polyhedra. In consequence, twice as many crystal-field transitions are observed in absorption and emission for Czochralski crystals than are observed in spectra of hydrothermally grown crystals. The spectroscopic evidence of dual-site occupancy is smeared out at 300 K by homogeneous broadening, but is resolvable on all inter-manifold transitions at low temperature. Weidner *et al.* (1994) showed that efficient energy transfer occurs between the two sites, which has an important role in the rich, multicoloured STEP and ETU luminescence in these crystals [Russell *et al.* (1997), Russell (1998)].

Kaminskii *et al.* (1991) reported flashlamp pumped Nd : KLYF laser operation at 1050 nm. Subsequently, Nicholls *et al.* (1997a) demonstrated room temperature Nd-KLYF and Nd-KLGF laser output at 1049 nm pumped at 799 nm with 200 ns pulses from a Cr : LiSAF laser. For crystals containing 3.4 at.% Nd the absorbed pump power

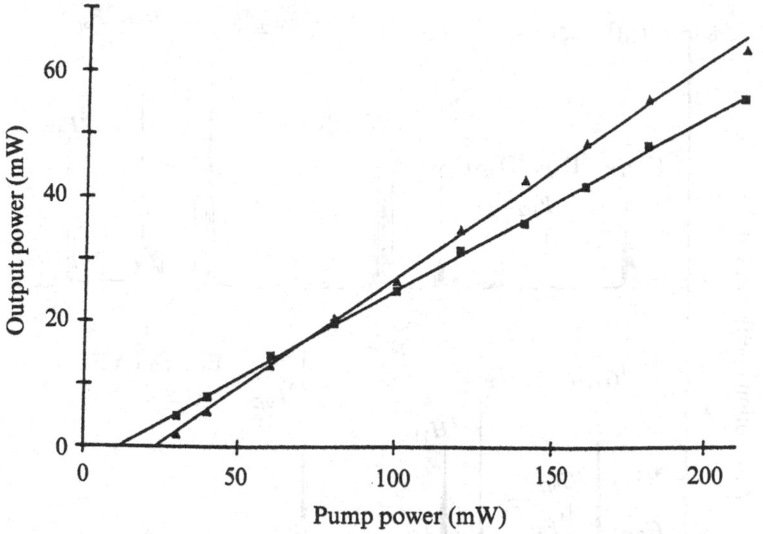

Figure 10.25. The CW input/output power of a Nd : KLYF laser with Ti: sapphire pumping at 800 nm and 300 K using 4% transmission (▲) and 10% transmission (■) output couplers [Nicholls *et al.* (1998)].

at threshold was only 120 μJ with a measured slope efficiency of 25% for a 5% output coupler. In CW-pumping at 800 nm with a Ti : sapphire laser, Fig. 10.25, the threshold was only 24 mW with a slope efficiency of 42% calculated relative to absorbed power and using a 10% output coupler [Nicholls *et al.* (1998)]. With output coupler transmission reduced to 4% the threshold power was 9 mW and the slope efficiency 34%. Even better pulsed and CW operation is expected with optimized crystal quality and pumping configurations.

A comparison of the $\lambda = 300$ nm–900 nm optical absorption spectra of Nd : KLYF and Er : KLYF is given in Fig. 10.26. For these Czochralski-grown crystals the dual Y-site feature of the crystal structures yields twice as many crystal-field transitions for each manifold of both dopants as is indicated in Table 2.19. As is usually the case, optical transitions from/into the $^4I_{15/2}$ ground state of Er^{3+} are weaker than those involving the $^4I_{9/2}$ ground state of Nd^{3+}. The $^4I_{9/2} \rightarrow {}^4F_{3/2}$ absorption on Nd^{3+} is 4–5 times stronger than the $^4I_{15/2} \rightarrow {}^4F_{3/2}$ absorption on Er^{3+}. In addition, the $^4F_{3/2} \rightarrow {}^4I_{15/2}$ emission of Er^{3+} is correspondingly weak. In terms of pumping with AlGaAs LDs at *ca* 800 nm, the rather weak $^4I_{15/2} \rightarrow {}^4I_{9/2}$ absorption of Er : KLYF, Fig. 10.26, is not conducive to infrared laser action pumped in this band. However, upconversion from $^4I_{9/2}$ results in efficient visible range laser operation in Er^{3+} : YLF [Brede *et al.* (1993), Heine *et al.* (1994)]: strong, green upconverted luminescence near to 550 nm has been reported for Er : KLYF pumped near 800 nm at both 15 K and 300 K [Nicholls *et al.* (1998), Smith *et al.* (1999)]. However, direct excitation of laser action in Er : KLYF may be possible using the $^4I_{15/2} \rightarrow {}^4F_{9/2}$ transitions near 650 nm, the $^4I_{15/2} \rightarrow {}^2H_{11/2}$ transitions near 515 nm and the $^4I_{15/2} \rightarrow {}^4F_{7/2}$ band near 485 nm. Nevertheless, in general the $^2H_{11/2}$, $^4F_{7/2}$ and $^4F_{3/2}$ levels decay radiatively only weakly or not at all. Rather do they decay nonradiatively to other nearby multiplets which then depopulate in luminescence transitions. The principal emitting levels are $^4S_{3/2}$ and $^4F_{9/2}$ in the visible region and $^4I_{11/2,13/2}$ in the infrared.

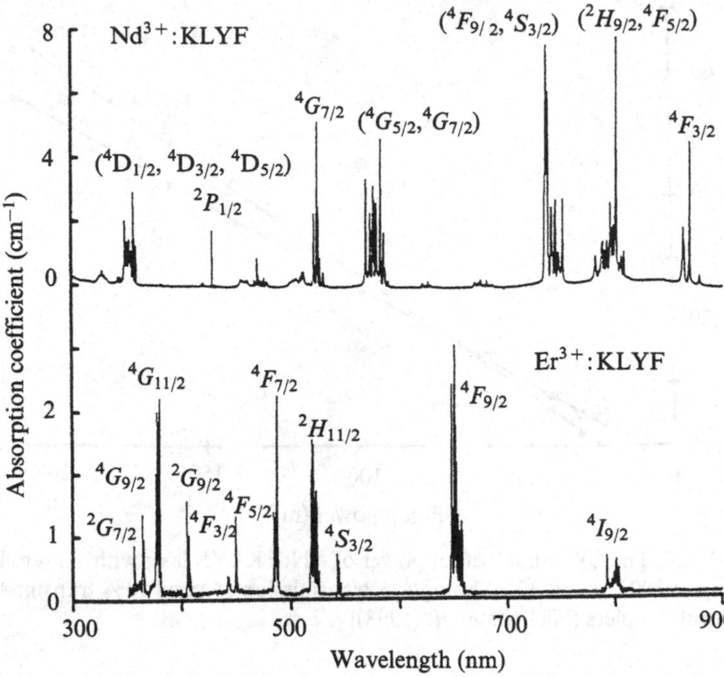

Figure 10.26. A comparison of the optical absorption spectra of 1 at.% Nd : KLYF and 1 at.% Er : KLYF measured at 77 K [after Nicholls *et al.* (1998)].

The infrared laser transitions $^4I_{11/2} \rightarrow {}^4I_{15/2}$, $^4I_{13/2}$ of Er^{3+} near 1.0 μm and 3.0 μm, respectively, and $^4I_{13/2} \rightarrow {}^4I_{15/2}$ near 1.5 μm have received particular emphasis for possible applications in micro-surgery, eye-safe coherent LIDAR, wind-shear detection and other short-range sensors. These transitions have been studied in $(Er_x Y_{1-x})_3 Al_5O_{12}$, where x can be varied between 0.01 and 1. For normal operation at 1.0 μm and 1.5 μm the Er^{3+} concentration is low with $x \sim 0.01$, at which composition laser operation at 3 μm is inefficient. Much more efficient laser operation is obtained at this wavelength, compared to 1% Er : YAG performance, by using long flashlamp pulses (170 μs) to excite $(Er_xY_{1-x})_3Al_5O_{12}$ crystals having $x = 0.33$ and $x = 0.50$. In these heavily doped oxides, Er^{3+}–Er^{3+} energy migration is important in depopulating the lower laser level and short-circuiting the quasi-three-level pumping from the long-lived $^4I_{13/2}$ level. In Er^{3+}–Er^{3+} cross-relaxation an ion in the $^4I_{13/2}$ level relaxes to $^4I_{15/2}$, simultaneously exciting a nearby ion in the $^4I_{13/2}$ level up to the $^4I_{9/2}$ level. Once in this level the ion relaxes nonradiatively into the upper laser level [Bass *et al.* (1986)]. Even so the lifetime of the $^4I_{11/2}$ level at 300 K varies somewhat with Er^{3+} content, reducing with increasing x from $\tau = 100$ μs at $x \cong 0.33$, to 75 μs in $Er_3Al_5O_{12}$. A value of $x = 0.33$ is optimal for diode-pumping at 800 nm. At this composition the large concentration of active ions compensates for their low oscillator strength without compromising radiative decay through concentration quenching [Moulton *et al.* (1988)]. Improved performance was reported also for $(Er_{0.5}Y_{0.5})AlO_3$ on the 1.663 μm, 1.677 μm, 1.706 μm and 1.72 μm lines of the $^4I_{11/2} \rightarrow {}^4I_{15/2}$ transition and the 2.7 μm line on the $^4I_{11/2} \rightarrow {}^4I_{13/2}$ transition [Weber and Lüthy (1986)]. Experiments with the $(Er_xY_{1-x})_3Sc_2Al_3O_{12}$ system reveal a longer fluorescence lifetimeof the $^4I_{11/2}$ level, $\tau \cong 1.4$ ms at $x < 0.3$, than for Er^{3+} ions on Y^{3+} sites in YAG.

10.5.3 *Other rare-earth ions*

The spectroscopic properties of the other rare-earth ions, Pm^{3+} ($4f^4$) through to Ho^{3+} ($4f^{10}$) as well as Yb^{3+} ($4f^{13}$) are well known: a comprehensive review of the multiplet structures of all RE^{3+} ions in LaF_3 is given by Carnall *et al.* (1988). The large spin–orbit coupling of Yb^{3+} causes the $^2F_{5/2}$ levels to be some 10 000 cm^{-1} above the $^2F_{7/2}$ ground levels, and the larger nuclear charge of Yb^{3+} moves the lowest-lying 5*d*-levels to much higher energies than in Ce^{3+}. In consequence, the crystal-field spectra of the $^2F_{5/2} \leftrightarrow {}^2F_{7/2}$ transitions occur near 900–1000 nm; they are the basis of CW laser operation in crystals and glasses when pumped by 970–980 nm LDs. Energy migration and cooperative upconversion make Yb^{3+}–Er^{3+} co-doping an effective means of pumping Er^{3+}-laser operation in the green. Pm^{3+} is strongly radioactive and is of no interest in laser applications. The complementary $4f^{10}$ configuration of Ho^{3+} has LS-terms ^5I_J, ^5F_J, ^5S_J, ^5G_J, ^3H_J, ^3K_J etc., spanning an energy range of order 40 000 cm^{-1}. Strong spin–orbit coupling, $\zeta = 2100\ cm^{-1}$, causes considerable mixing of multiplets from one term with those from another [Dieke and Pandey (1964), Caspers *et al.* (1970)]. There is much interest in Ho^{3+}-doped hosts as infrared laser gain media: the $^5I_7 \rightarrow {}^5I_8$ transition supports quasi-three-level laser operation at 2.0 μm and the $^5I_6 \rightarrow {}^5I_7$ transition is the basis of four-level lasing at 3.0 μm. However, laser operation on these Ho^{3+} transitions is not very efficient and it is appropriate to discuss Ho^{3+} in the context of sensitization by a variety of co-dopants. The remaining RE^{3+} ions (e.g. Sm^{3+}, Dy^{3+}, Eu^{3+} and Tb^{3+}) are usually found in such other optical applications as information storage materials, phosphors and scintillators.

10.6 Energy transfer and upconversion lasers

The primary purpose of co-doping with activator and sensitizer(s) was to improve the efficiency of flashlamp pumped laser action, especially for Er^{3+}, Tm^{3+} and Ho^{3+} ions which have fewer visible region absorption bands than Nd^{3+} to overlap with the output spectrum of conventional flashlamps. Sensitization can involve several co-dopants and different pumping pathways including energy transfer, cross-relaxation, energy migration and upconversion. The Cr^{3+} ion is an efficient sensitizing agent for various RE^{3+} acceptors. The $Cr^{3+}:Nd^{3+}$: garnets were among the earliest examples of laser operation by donor–acceptor transfer. The spectral overlap of the $^4T_2 \rightarrow {}^4A_2$ emission and the lower lying pump bands of Nd^{3+} is extensive at 300 K, as Fig. 7.15 shows, resulting in efficient broadband pumping of $Cr^{3+}:Nd^{3+}$: GSGG, GSAG, YSAG and YSGG lasers [Duczynski *et al.* (1986a,b), Shcherbakov (1986)]. Such an enhancement of broadband pumping of Nd^{3+} in $Cr^{3+}:Nd^{3+}$: YAG is less apparent than in $Cr^{3+}:Nd^{3+}$: GSGG because Cr^{3+} emits predominantly in the R-lines in YAG at 300 K. In Cr^{3+}-sensitized Tm^{3+}: YAG a fortuitous resonance of the R-line emission of Cr^{3+} with the $^3H_6 \rightarrow {}^3F_3$, $^3F_2$ absorption of Ho^{3+} permits very efficient transfer at 300 K and above. The rate of nonradiative Cr–Tm energy transfer in crystals containing 1 at.% Cr^{3+} and 5% Tm^{3+} is temperature dependent because of the thermal variation in the radiative decay probability of the Cr^{3+} ion, with quantum efficiency of near unity at 300 K [Armagan *et al.* (1989)]. At this temperature the Cr–Tm microscopic interaction parameter, $\alpha^6_{DA} \cong 8 \times 10^{-40}\ cm^6 s^{-1}$, and the characteristic energy transfer radius, $R_0 \cong 1.1$ nm, is less than one unit cell length. Apparently, excitation is transferred only between close $Cr^{3+}:Tm^{3+}$ pairs in YAG, as it is also in $Cr^{3+}:Nd^{3+}$:GSGG [Han *et al.* (1993)]. Site selective laser excitation studies in the latter host distinguish $Cr^{3+}:Nd^{3+}$ pairs, between which energy may be transferred, and isolated Nd^{3+} ions which must be directly excited by the pump radiation. Caird *et al.* (1975) have demonstrated room temperature lasing with $Cr^{3+}:Tm^{3+}$ co-doped YAG and $YAlO_3$ gain media, operating on crystal-field components of the ${}^3H_4 \rightarrow {}^3H_5$ transition.

For efficient sensitizer–activator transfer it is common practice to dope the gain medium with several sensitizer ions. In seeking lower threshold, higher efficiency, flashlamp pumped laser operation at 2–3 μm much attention was given to $Cr^{3+}:Tm^{3+}$, $Cr^{3+}:Ho^{3+}$ and $Cr^{3+}:Tm^{3+}:Ho^{3+}$ co-doped garnets. The superior growth characteristics and optomechanical properties led to co-doped YAG being particularly well studied. In this crystal the resonantly enhanced $Cr^{3+} \rightarrow Tm^{3+}$ energy transfer interaction leads to a quantum efficiency of unity in populating the ${}^3F_2$ and ${}^3F_3$ levels on the Tm^{3+} ions. Once populated, these levels decay very rapidly by nonradiative transitions into the ${}^3H_4$ level. Each Tm^{3+} ion in the excited ${}^3H_4$ level then cross-relaxes with a Tm^{3+} ion in the ${}^3H_6$ ground state to yield two Tm^{3+} ions in the ${}^3F_4$ emitting state. The ensuing ${}^3F_4 \rightarrow {}^3H_6$ transition is the basis of a Tm^{3+}:YAG laser operating at room temperature at 2.014 μm. In an alternative pumping scheme using $Cr^{3+}:Tm^{3+}:Ho^{3+}$:garnet the two Tm^{3+} ions in excited ${}^3F_4$ states transfer their energy to the ${}^5I_7$ levels on two Ho^{3+} ions, the upper levels in laser operation on the ${}^5I_7 \rightarrow {}^5I_8$ transition at 2.097 μm. Similar $Cr^{3+}:Tm^{3+}:Ho^{3+}$ co-doping schemes have been studied in YSGG and YSAG, in which Cr^{3+} ions occupy weak crystal-field sites and emit into broad ${}^4T_2 \rightarrow {}^4A_2$ bands [see for example Huber *et al.* (1988), Duczynski *et al.* (1986a,b)]. However, following a definitive study of flashlamp pumping of Tm^{3+} and Ho^{3+} lasers Quarles *et al.* (1990) commented on the need to judiciously choose both gain medium and concentration of the Cr^{3+} sensitizer. Apparently, in a YAG laser rod of 5 mm diameter only 8×10^{19} Cr^{3+} cm^{-3} is preferred as activator for Tm^{3+} in $Tm^{3+}:Ho^{3+}$ laser operation. In this strong crystal-field host resonant enhancement of the $Cr^{3+}:Tm^{3+}$ transfer via R-line emission is more efficient than ${}^4T_2 \rightarrow {}^4A_2$ emission in weak crystal-field hosts. For $Cr^{3+}:Tm^{3+}$:YAG flashlamp pumped at 300 K laser operation at 2.014 μm was observed with thresholds as low as 43 J, with output energies > 2 J and slope efficiencies of 4.5% [Storm *et al.* (1989)]. Furthermore, in room temperature operation of the $Cr^{3+}:Tm^{3+}:Ho^{3+}$:YAG laser on the 2.089 μm Ho^{3+} line output energies > 1.5 J were achieved with threshold energies as low as 38 J and a slope efficiency of 5.1%. Mode-locking of CW-pumped $Cr^{3+}:Tm^{3+}$:YAG and $Cr^{3+}:Tm^{3+}:Ho^{3+}$:YAG lasers at 300 K has been reported by Heine *et al.* (1992).

True CW operation at 1.864 μm has been reported for a Cr^{3+}-sensitized Tm^{3+}:YSGG laser at 300 K pumped with the 647.1 nm Kr-laser line: the quoted threshold power was less than 40 mW and the slope efficiency 0.8% [Duczynski *et al.* (1986a)]. They also reported CW laser wavelengths near 1.94 μm and 2.09 μm on the ${}^5I_7 \rightarrow {}^5I_8$ Ho^{3+} transition in YAG, YSAG and YSGG under Kr^+-laser pumping at 300 K. Typically the threshold powers were less than 25 mW and slope efficiencies *ca* 1.3%. Direct pumping of Ho^{3+} lasers via the Cr^{3+} sensitizer is not efficient given the small microscopic interaction parameter, $\alpha_{DA}^{(6)} = 2.8 \times 10^{-40}$ cm^6 s^{-1} (see also §7.4). When Tm^{3+} is added to mediate the Cr^{3+}–Ho^{3+} transfer in YSGG and YSAG it is usually at high concentration (8×10^{20} cm^{-3}) compared to both Cr^{3+} (2×10^{20} cm^{-3})

and Ho^{3+} ($5 \times 10^{19}\,cm^{-3}$). This facilitates the short-range Cr^{3+}–Tm^{3+} energy transfer and, after Tm^{3+}–Tm^{3+} energy migration, Tm^{3+}–Ho^{3+} energy transfer [Duczinski *et al.* (1986b)]. The low Ho^{3+} content is necessary to minimize resonant ${}^5I_8 \rightarrow {}^5I_7$ reabsorption losses. The Tm^{3+}–Tm^{3+} cross-relaxation downconverts the ${}^3F_4$ quantum of energy to the ${}^3H_4$ infrared region with a quantum efficiency of almost two: the overall Cr–Tm–Ho transfer efficiency is more than 50%. The Cr^{3+} ion is also used to sensitize Er^{3+} : YSGG and Er^{3+} : YAG lasers, where the Cr^{3+}–Er^{3+} transfer is complicated by Er^{3+}–Cr^{3+} back-transfer. Most of the energy is transferred to the Er^{3+} ion via the ${}^4I_{9/2}$ and ${}^4I_{11/2}$ levels, resulting in laser action at 3.0 μm and 1.5 μm on the ${}^4I_{11/2} \rightarrow {}^4I_{13/2}$, ${}^4I_{15/2}$ transitions, respectively.

In YLF-like crystals multiple RE^{3+} ions have been used to efficiently absorb light from a flashlamp or laser diode near 800 nm and transfer/downconvert energy into the infrared spectrum for laser operation. Such ions include Nd^{3+}, Er^{3+}, Tm^{3+}, Ho^{3+} and Yb^{3+}. Fluoride hosts are preferred because slower nonradiative decay compared to oxide hosts results in numerous long-lived emitting levels and large quantum efficiencies. Brenier *et al.* (1990) have studied the fluorescence dynamics of Er^{3+}–Tm^{3+}, Er^{3+}–Ho^{3+}, and Tm^{3+}–Ho^{3+} energy transfer in YLF after laser excitation at 800 nm. They find that in Er^{3+} : Tm^{3+} : YLF most of the Er^{3+} excitation in the ${}^4I_{15/2} \rightarrow {}^4I_{9/2}$ transition is transferred between the lowest excited states ${}^4I_{13/2}$ and ${}^3F_4$ on Er^{3+} and Tm^{3+}, respectively. Both Er^{3+}–Tm^{3+} transfer and Tm^{3+}–Er^{3+} back-transfer are so efficient that they induce a Boltzmann distribution between the populations of these states even at 77 K. As in the oxide hosts, in triply-doped Er^{3+} : Tm^{3+} : Ho^{3+} : YLF the incident radiation is absorbed by Er^{3+}, downconverted and transferred to Tm^{3+} ions, among which it migrates until it transfers to the laser active Ho^{3+} ions. This latter transfer mechanism was the basis of the first eye-safe laser that used $Er_xY_{1-x}LiF_4$ as the host, doped with 6.7 at.% Tm and 1.7 at.% Ho substituting for Y^{3+}, a composition established by Chicklis *et al.* (1971). Energy transfer was first used to improve Er^{3+} : glass laser operation at 1.5 μm using the ${}^2F_{7/2} \rightarrow {}^2F_{5/2}$ absorption of the Yb^{3+} sensitizer ions [Snitzer and Woodcock (1965)]. However, such pumping of Yb^{3+} can also lead to upconverted ${}^4S_{3/2} \rightarrow {}^4I_{15/2}$ green luminescence at 550 nm on the Er^{3+} activator, which degrades laser performance through enhanced re-absorption of the pump beam. Several RE^{3+} ions have been used as the activator in upconversion lasers, including Pr^{3+}, Nd^{3+}, Ho^{3+}, Er^{3+} and Tm^{3+}, in crystalline and glassy hosts operated in CW and pulsed mode, frequently below 300 K. Most upconversion lasers based on RE^{3+}-doped crystals operate at reduced temperature to minimize nonradiative decay, although in pulsed mode some upconversion lasers have been operated at 300 K. RE^{3+} : fibre lasers will operate CW at room temperature, low doping levels distributed over long fibre lengths minimize thermal problems inherent in heavily doped crystals.

Both crystalline and glassy gain media when doped with Pr^{3+} emit a rich spectrum of visible light when pumped in the strong ${}^3H_4 \rightarrow {}^3P_{0,1,2}$ (420–480 nm) and ${}^3H_4 \rightarrow {}^1D_2$ (590–600 nm) absorption channels. These luminescence spectra, which involve many crystal-field transitions of the ${}^3P_0 \rightarrow {}^3H_{6,5}$, ${}^3F_2$ and ${}^3P_1 \rightarrow {}^3H_5$ manifolds, are the basis of multi-line laser operation. These transitions can be pumped directly via the ${}^3P_2$, ${}^1I_6$ and ${}^3P_1$ levels using flashlamps or the UV-output from Ar^+ at 476 nm or by upconversion pumping from the ${}^1G_4$ state. This latter approach was taken by Smart *et al.* (1991) and others to pump Pr^{3+} : ZBLAN fibre lasers that operate at blue (491 nm), green (520 nm), orange (605 nm) and red (635 nm) wavelengths. The ZBLAN fluorozirconate glass fibre contained 56% ZrF_4 (by weight), 14% BaF_2, 6% LaF_3, 4% AlF_3 and 20% NaF : some 560 ppm of Pr^{3+} replaced La^{3+} in the 4–6 μm diameter core. A Ti-sapphire laser was tuned to pump the $^3H_4$–$^1G_4$ absorption at 1.1 μm while a second Ti-sapphire laser at 835 nm transferred population from $^1G_4$ to the thermally-coupled $^3P_1$, $^1I_6$ and $^3P_0$ levels. With a 10 m long fibre gain medium, laser operation was achieved in the red at 635 nm. Much shorter Pr^{3+} : ZBLAN fibres (1–2 m) were used for laser operation in the orange, green and blue wavelength regimes. The authors predicted much improved laser performance using fibres with lower background losses and reduced core diameters with judiciously chosen fibre length. Zhao and Poole (1994) reported improved room temperature laser performance of a CW Pr^{3+} : ZBLAN laser at 492 nm using a 1 m length of fibre containing 500 ppm of Pr^{3+} in a 3 μm diameter core. The total threshold launch power at 1017 nm plus 835 nm of only 105 mW resulted in a slope efficiency of 13% and output power greater than 9 mW. Laser operation in the blue (488 nm) and red (635 nm, 717 nm) using Nd^{3+} : Pr^{3+} co-doped ZBLAN fibres was pumped by a single source wavelength of 796 nm. A STEP scheme (§7.4.4) absorbs a 796 nm photon in exciting ground state Nd^{3+} ions ($^4I_{9/2}$) into the $^4F_{5/2}$ level: after nonradiative decay to $^4F_{3/2}$ a second absorbed 796 nm photon further excites the Nd^{3+} sensitizer into the $^2D_{3/2,\ 5/2}$ levels; these rapidly thermalize to the $^4G_{9/2,\ 11/2}$ levels in close resonance with the $^3P_1$ and $^1I_6$ levels on the Pr^{3+} ions. Resonant energy transfer Nd^{3+} ($^4G_{9/2,\ 11/2}$)–Pr^{3+} ($^3P_1$, $^1I_6$) results after nonradiative decay in a large population of excited Pr^{3+} ions in the emissive $^3P_0$ level. An absence of other visible Pr^{3+} emissions in the Pr^{3+} : ZBLAN fibre shows that STEP upconversion on the Nd^{3+} followed by resonant energy transfer to Pr^{3+} is responsible for the laser operating on the $^3P_0 \rightarrow {}^3H_4$ transition (488 nm) and $^3P_0 \rightarrow {}^3F_2$, $^3F_4$ transitions (635 nm and 717 nm), respectively [Goh *et al.* (1995)]. The significant advance of this energy transfer laser is operation in the blue (488 nm) pumped by a single wavelength (796 nm), at which cheap, reliable and robust laser diodes are available.

An unusual energy transfer process was used by Eichler *et al.* (1994) to flashlamp pump a Pr^{3+} : $LiYF_4$ laser at 607.2 nm, 639.5 nm and 720.9 nm. By doping the fused-quartz envelope of the flashlamp with Ce^{3+} ions the strong UV-component of the flashlamp was transferred via the Ce^{3+} ions to the Pr^{3+} pump bands near 420– 480 nm. At an excitation energy of 30 J the output energy in the $^3P_0$–$^3H_6$ channel was 4.7 mJ, 87 mJ in the $^3P_0$–$^3H_6$ channel at 639.5 nm and 30 mJ in the $^3P_0$–$^3F_4$ channel at 720.9 nm, all in operation at room temperature. Among other Pr^{3+}-doped crystalline gain media, Pr^{3+} : $LaCl_3$ operates with considerable efficiency via an AAP process excited at 677 nm and involving the $^3P_2 \rightarrow {}^3F_2$ emission at 644 nm [Koch *et al.* (1990)].

Upconverted luminescence leading to laser operation has been studied in various Nd^{3+}-doped crystals. The STEP and ETU processes (§7.4.4) were identified in Nd^{3+} : YLF [Fan and Byer (1986)] and shown to support CW laser oscillation in LaF_3 [Macfarlane *et al.* (1988)] and YLF [Lenth and Macfarlane (1990)] below 100 K. A laser that operates at 300 K on transitions involving sequential absorption of yellow photons to excite Nd^{3+} ions to the $^4I_{9/2}$ and $^4F_{3/2}$ states followed by emission in the violet and blue from the $^4D_{3/2}$ and $^2P_{3/2}$ states uses Nd^{3+} : ZBLAN fibres as the gain medium [Funk *et al.* (1994)]. However, in a number of cases, where measurements on Nd^{3+}-doped materials were made at 300 K, the crystal-field levels involved in upconversion have not been unambiguously determined [see Fan and Byer (1986), Stanley *et al.* (1993), Chuang and Verdun (1994)]. In contrast, site selective upconversion in

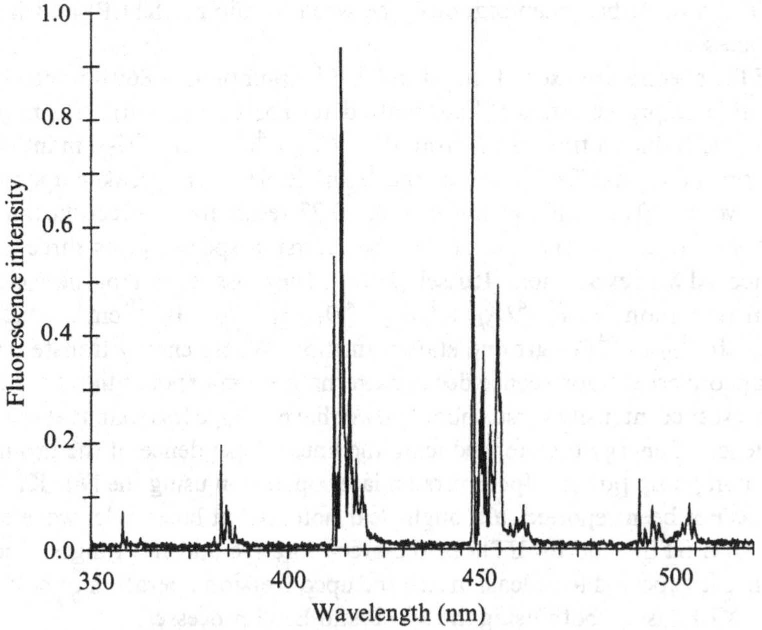

Figure 10.27. Typical unpolarized fluorescence spectrum in Nd^{3+} : KLYF excited at 585 nm [after Russell *et al.* (1997)].

Nd^{3+} : $KLiYF_5$ (Nd : KLYF) permits precise mechanisms to be determined. Despite the dual-site of KLYF, resulting in two slightly different crystal-field spectra, there are occasional degeneracies among the Stark levels on the two sites, the most important of which is the lowest crystal-field level of $^4F_{3/2}$ occurring at 11 510 cm^{-1} on both sites [Summers *et al.* (1994)]. Once in this metastable level Nd^{3+} ions on both sites can exchange energy very efficiently. In Nd : KLYF the Nd^{3+} ion population that shelves in the $^4F_{3/2}$ level after excitation at *ca* 580 nm in the $^4I_{9/2} \rightarrow {}^4G_{7/2} + {}^4G_{5/2}$ band decays along one of three paths: radiative decay to the ^4I_J fine structure level; emission at violet or blue wavelengths from $^4D_{3/2}$ and $^2P_{3/2}$ states following the absorption of a second 580 nm photon; energy transfer to another Nd^{3+} ion shelved in the $^4F_{3/2}$ state resulting in green and orange fluorescence from the ($^4E_{5/2}$, $^4G_{7/2}$) and $^4G_{7/2}$ fine structure levels into the crystal-field levels of $^4I_{9/2}$ [Russell *et al.* (1997)]. In dilute Nd^{3+}-doped crystals (0.2 at.% Nd^{3+}), energy transfer between sites is weak and STEP is dominant in exciting the upconversion luminescence (Fig. 10.27) on both sites by pumping at 585.3 nm. Since the Stark splittings within each fine structure level are different on each site, apart from accidental degeneracies, both ground and excited state absorption transitions occur at different wavelengths on the two sites. Hence the STEP upconversion fluorescence is site selective for absorption of both photons. The contributions of each site separately to the upconverted luminescence is resolved by exciting at 568.3 nm and 571.0 nm. ETU at Nd^{3+} ions in the $^4F_{3/2}$ levels on two different sites is dominant in crystals containing 3.4 at.% Nd^{3+}. Although the $^4I_{9/2} \rightarrow {}^4G_{7/2} + {}^4G_{5/2}$ resonance condition can be met on either site, the upconverted fluorescence spectrum contains lines originating on both sites. Site selectivity in the upconverted luminescence spectrum then occurs only during the second absorption stage. Site selectivity in

Nd : KLYF distinguishes unambiguously between intraionic (STEP) and interionic (ETU) processes.

Other ETU spectra are excited in Nd : KLYF by pumping at 869 nm into the $^4F_{3/2}$ manifold and comprise clusters of lines centred at *ca* 530 nm (green), 595 nm (orange) and 665 nm (red) due to transitions from the ($^4G_{5/2}+{}^4G_{7/2}$) and $^4G_{7/2}$ manifolds that terminate on the crystal-field levels of the ^4I_J multiplets. The weaker upconversion spectra between 350 nm and 460 nm in Fig. 10.27 result from three-photon upconversion. A rate equation analysis of the upconversion spectra gives three principal results in accord with experiment [Russell (1998)]. The cross section for the excited state absorption transition $^4F_{3/2}\rightarrow{}^4D_{3/2}+{}^4D_{5/2}+{}^4D_{1/2}$ is $1.96\times10^{-20}\,\mathrm{cm}^2$, a factor two larger than the $^4I_{9/2}\rightarrow{}^4F_{3/2}$ ground state transition. Where energy transfer rates are large the upconverted fluorescence decay patterns are non-exponential. Finally, plots of the fluorescence intensity versus pump power have a slope less than two (*ca* 1.83) as a consequence of energy transfer reducing the linear dependence of the ground state absorption on pump power. Upconversion laser operation using the Nd : KLYF gain medium has not been reported, although clear potential at blue/violet wavelengths is identified with STEP and by ETU in the green, orange and red ranges. Such laser performance is expected to at least match the upconversion operation of $Nd^{3+}:LaF_3$ and Nd^{3+} : YLF lasers, both using the STEP and ETU processes.

Green upconversion laser operation was reported for $Ho^{3+}:Ba(Yb_xY_{1-x})F_8$ pumped in the infrared by flashlamps to excite Yb^{3+} ions in the $^2F_{7/2}\rightarrow{}^2F_{5/2}$ transition [Johnson and Guggenheim (1971)]. Two successive energy transfer steps result in green emission from the $^5S_2\rightarrow{}^5I_8$ transition on Ho^{3+} at 551.5 nm. Yb^{3+} sensitized upconversion energy transfer is particularly attractive for Ho^{3+} and Er^{3+}, given the availability of high-power InGaAs laser diodes to directly pump the Yb^{3+} absorption bands near 970 nm. The original study realized Ho^{3+} upconversion laser operation at 77 K: to date there have been no reports of room temperature laser operation. Zhang *et al.* (1993) have shown that in YLF, KYF and BaY_2F_6 containing up to 20 at.% of the sensitizer ion (Yb^{3+}) and 0.4 at.% of the Ho^{3+} activator, the activator emission on the $^5S_2\rightarrow{}^5I_8$ transition near 550 nm is strongly temperature dependent. The concentration quenching of this green emission is caused by increasingly efficient $Ho^{3+}\rightarrow Yb^{3+}$ back-transfer at higher temperatures, $T\leq 300$ K. Careful choice of activator–sensitizer concentrations appear to be required to permit $Yb^{3+}\rightarrow Ho^{3+}$ transfer without efficient back transfer. The same Ho^{3+} transition supports laser operation between 540 nm and 553 nm at 300 K in Ho^{3+} : ZBLAN glass fibres pumped at 647 nm by a Kr^+ laser [Allain *et al.* (1990)].

Upconversion laser operation in the green has been reported also for several Er^{3+}-doped crystals using STEP (Er : YLF, Er : YAlO and Er : KYF) and ETU (Er : YLF and $Er:BaY_2F_8$) processes, mostly at low temperatures [Macfarlane (1994a,b)], in contrast to room temperature operation of STEP-excited Er^{3+} : ZBLAN fibre lasers pumped by 801 nm laser diodes [Whitley *et al.* (1991)]. This laser operated at 546 nm on the $^4S_{3/2}\rightarrow{}^4I_{15/2}$ transition of the Er^{3+} ion. Pulsed upconversion laser operation at room temperature has now been demonstrated at 551 nm using the $^4S_{3/2}\rightarrow{}^4I_{15/2}$ transition in Er : YLF pumped at 810 nm and at 561 nm using the $^2H_{9/2}\rightarrow{}^4I_{13/2}$ transition in Er : KYF pumped at 812 nm, both by sequential two-photon absorption [Brede *et al.* (1993)]. A CW Er^{3+} : YLF upconversion laser operated at 300 K also oscillated on the $^4S_{3/2}\rightarrow{}^4I_{15/2}$ transition. Output powers up to 40 mW of green emission at 551 nm were obtained using a near-concentric cavity with output coupling of 6.6% pumped by a Ti-sapphire laser tuned to 810 nm [Heine *et al.* (1994)].

Several Tm^{3+}-doped materials support upconversion laser operation at wavelengths in the blue. CW operation at 300 K was reported for a Tm : ZBLAN fibre laser pumped at 1.11 μm to give output at 480 nm in the $^1G_4 \rightarrow {}^3H_6$ transition on Tm^{3+} [Mackechnie *et al.* (1993)]. A Tm : ZBLAN fibre laser pumped at two Kr^+ wavelengths, 647.1 nm and 676.4 nm, operated on the $^1G_4 \rightarrow {}^3H_6$ (480 nm) and $^1D_2 \rightarrow {}^2F_4$ (455 nm) transitions, respectively, but at cryogenic temperatures [Allain *et al.* (1990)]. An energy transfer upconversion laser using a Tm^{3+} : Ba(YbY)F_8 gain medium pumped at 960 nm also operated at two laser wavelengths on the $^1D_2 \rightarrow {}^3H_6$ (455 nm), $^3H_5$ (510 nm) Tm^{3+} transitions at temperatures up to 200 K. An avalanche absorption pumped Tm : LiYF4 laser used 648 nm excitation to produce $^1G_4 \rightarrow {}^3H_6$ emission at 483 nm [Hebert *et al.* (1990)].

10.7 Glass fibre lasers

Although of superficial similarity, the optical spectra of glasses and crystals doped with transition-metal or rare-earth ions have significant differences. Perhaps of greatest importance is the disorder inherent in glassy solids. Glasses comprise random and continuous, three-dimensional networks that lack symmetry and periodicity. Usually they contain molecular-ion *network formers* (SiO_4, GeO_4, PO_3) and *network modifiers* (Na^+, Sr^{2+}, Al^{3+}), the latter being accommodated randomly adjacent to non-bridging anions (O^{2-}, F^-). Normally, RE^{3+} ions do not substitute at the molecular-ion network formers. Instead, they are situated near looser structures caused by the larger network modifiers (Na^+, Sr^{2+}), in which it is easy to accommodate ~1 at.% or more of RE^{3+} ions. The sharp lines characteristic of rare-earth ions in crystals are inhomogeneously broadened in glasses, where the optical line shape at low temperature is dominated by site-to-site broadening. Much of the research on laser-related properties of RE^{3+} activated silicate, phosphate, borate, germanate, tellurite and fluorophosphate glasses has required high resolution and site-selective laser spectroscopies to quantify the site-to-site variations in local fields, ion–ion and electron–vibrational interactions, and their effects on energy levels and relaxation phenomena [Weber (1981)]. However, homogeneous broadening is dominant at room temperature, where much of the fine structure splitting is smeared out by the combined effects of homogeneous and inhomogeneous processes (see §5.5.7 and §9.2.5). Site-to-site differences in radiative and nonradiative transition probabilities lead to non-exponential decay behaviours. However, the superposition of contributions from individual ions among the ensemble of local environments does not preclude laser operation in RE^{3+}-doped glasses. Rather does it confer on a glass laser a degree of tunability not normally associated with RE^{3+} : doped crystal lasers [Mears *et al.* (1986), Lëdig *et al.* (1990)]. The broadening of transition-metal ion spectra in glasses is more severe, dominating the shapes of sharp lines and broadbands [Andrews *et al.* (1981), Henderson *et al.* (1992)]. The sensitivity of transition-metal ions to local environment influences not only optical line shapes but also nonradiative decay rates such that the quantum efficiency can be reduced very appreciably. In Cr^{3+}-doped glasses the quantum efficiencies never exceed 25% and are frequently close to zero [Imbusch *et al.* (1990)]. The consequences are that RE^{3+} : doped glasses support laser operation whereas transition-metal ion doped glasses do not.

In view of the above comments, it may be surprising that Cr^{3+} co-dopants have been used as sensitizers in Nd^{3+} : glass and Er^{3+} : glass lasers. CW laser operation at 1.06 μm was reported for a Li-La-phosphate glass containing $10^{21}\,cm^{-3}Nd^{3+}$ ions and $10^{20}\,cm^{-3}\,Cr^{3+}$ ions [Härig *et al.* (1981)]. Direct pumping of Nd^{3+} ions at 300 K gave low threshold power (~ 1 mW) with slope efficiency of order 18%. Cross-pumping via the $^4T_2 \rightarrow {}^4A_2$ band of Cr^{3+}, for which the Cr^{3+}–Nd^{3+} transfer efficiency was *ca* 45%, resulted in a 1 mW threshold power at 647.1 nm with an 8% slope efficiency. $Cr^{3+}:Yb^{3+}:Er^{3+}$ energy transfer in fluorophosphate laser glasses that contain 80–85 mol.% of mixed (Mg, Ca, Sr, Al) fluorides, 15–20 mol.% of $Sr(PO_3)_2$ or $Ba(PO_3)_2$, $8 \times 10^{19}\,Cr^{3+}\,cm^{-3}$ and $0.0–1.2 \times 10^{21}\,Yb^{3+}\,cm^{-3}$ sensitizers and $5 \times 10^{19}\,Er^{3+}\,cm^{-3}$ activator ions is also efficient. The $^4T_2 \rightarrow {}^4A_2$ luminescence was completely quenched when the Yb^{3+} content was $\geq 6 \times 10^{20}\,cm^{-3}$ due to Cr^{3+} energy transfer to the $^2F_{5/2}$ state of Yb^{3+} ions [Lëdig *et al.* (1990)]. At this composition the Förster critical radius is $R_0 \cong 1$ nm, $D^* \rightarrow A$ energy transfer being characterized by $\alpha_{DA}^{(6)} = 2 \times 10^{-38}\,cm^6\,s^{-1}$. Once transferred from Cr^{3+} to Yb^{3+} the energy migrates among the Yb^{3+} ions before transferring to the $^4I_{11/2}$ state of Er^{3+} with 70–75% efficiency. In contrast to the Cr^{3+} luminescence, the $^2F_{5/2} \rightarrow {}^2F_{7/2}$ luminescence of the Yb^{3+} sensitizer was not fully quenched. These Er^{3+} : glass lasers have a tuning range of 1536–1596 nm and peak emission cross section $\sigma_0 = 6.2 \times 10^{-21}\,cm^2$ at $\lambda_0 = 1536$ nm under flashlamp pumping. Laser operation at 300 K required a threshold pump energy of 40 J: the slope efficiency was 0.4% using an output coupler of 13%. There have been numerous Pr^{3+}-, Er^{3+}- and Tm^{3+}-activated glass lasers operated at room temperature and below but none have been commercially developed, given the thrust towards miniaturization using LD pumping [Digonet (1993)].

Much impetus for fibre laser development followed the invention of the Er^{3+}-fibre amplifier [Mears *et al.* (1986)] and its deployment in optical communications. Succeeding generations of lightwave communications systems saw the transmission capacity of optical fibres increase from 10 Gbit-km/s in 1978, to a hundred Gbit-km/s in 1982 and *ca* one thousand Gbit-km/s by 1986, over a period in which the system wavelength changed from 870 nm to 1300 nm and then 1550 nm to take advantage of the reduced attenuation in the longer wavelength optical windows of silica glass and as transmitters changed from LEDs at 870 nm to LD chips operating at 1300 nm and then 1500 nm. The Er^{3+}-fibre amplifier, a length of Er^{3+}-doped silica fibre excited at 1550 nm by a laser diode, introduced into optical communication systems optical fibre transmission capacities enhanced by a further 100. Such enhancements are as important in optical communications via trans-oceanic and local area networks at the beginning of the new millennium as they will be to future coherent optics and soliton communications systems. Er^{3+} : doped fibre amplifiers require pump lasers that operate at wavelengths of 1470–1550 nm. CW and ultrashort pulse Er^{3+}-fibre lasers can be pumped at 980 nm in the $^4I_{15/2} \rightarrow {}^4I_{11/2}$ transition or at 1500 nm in the $^4I_{15/2} \rightarrow {}^4I_{13/2}$ transitions using strained layer InGaAs LDs and InGaAsP LDs, respectively. Both Er^{3+}-fibre amplifiers and lasers emit green light during operation due to parasitic upconversion luminescence. Spin-off developments from optical communications have included a wide range of fibre lasers for other wavelength regimes. Three competing technologies have focused on short wavelength lasers: semiconductor lasers based on the group II–VI sulphides and selenides and group III–V nitrides, nonlinear frequency conversion of lasers and upconversion lasers. Visible fibre (and crystal) lasers doped with Pr^{3+}, Nd^{3+}, Er^{3+} or Tm^{3+} have advantages of high brightness, good coherence and excellent beam quality over the other techniques.

As discussed in §10.6, Pr^{3+} ions in crystalline hosts can be directly pumped at 476 nm to give laser-active emission in the blue, green, orange, red and infrared spectral regions. This is also true of Pr^{3+}-doped fibres. A Pr^{3+}: doped fluoride fibre laser has been operated at 492 nm, 520 nm and 635 nm by upconversion processes [Smart *et al.* (1991)]. Nd^{3+}-doped fluoride fibre lasers operate at 381 nm and at visible wavelengths under upconversion pumping at 800 nm [Funk *et al.* (1994)]. Yb^{3+} ions have been used as co-dopants in $Pr^{3+}:Yb^{3+}$: ZBLAN fibre lasers, demonstrating simultaneous blue/green upconversion lasing at 490 nm and 520 nm pumped at 856 nm by a laser diode [Baney *et al.* (1996)]. This $Pr^{3+}:Yb^{3+}$: ZBLAN upconversion laser required 55 mW and 85 mW of threshold launch power at wavelengths of 520 nm and 490 nm, respectively, and yielded 1.4 mW total power for 350 mW incident power at 856 nm. The $Yb^{3+}:Pr^{3+}$: ZBLAN laser operates by energy transfer from the $^2F_{5/2}$ level of excited Yb^{3+} ions to the $^2G_4$ excited state of Pr^{3+}, which is further excited into the $^3P_0 + {}^1I_6$ states by absorption of a second 856 nm photon. The $^3P_0 \rightarrow {}^3H_4$ transition gives quasi-three-level laser oscillation at 490 nm, whereas $^3P_0 \rightarrow {}^3H_5$ transitions constitute a four-level scheme operating in the green (520 nm). There has been recent interest in Yb^{3+}-silica fibre as the gain medium in CW laser operation and broadband (975–1200 nm) fibre amplifiers [Pask *et al.* (1995), Paschotta *et al.* (1997)]. Among the attractive features of Yb^{3+}-doped fibres are broad gain bandwidth and high efficiency due to the absence of ESA and concentration quenching that typify other RE^{3+}-doped systems. Yb^{3+}-doped fibres find increasing applications as gain media in power amplifiers at special infrared wavelengths, as small-signal amplifiers in fibre sensors, in free-space lasers communications and in ultrashort pulse lasers and amplifiers.

There have been numerous improvements to the silica-based optical fibres that carry light from a transmitter. These fibres comprise a central core with a cladding that guides the light through the core by virtue of a slightly different index of refraction. As the optical signal passes through the core it broadens and weakens. The dispersion occurs because light of different frequencies moves through the core at different velocities. Attenuation is caused by defects and impurities in the glass that absorb and scatter the radiation. Improvements in fibre technology have reduced the core diameter and created *single-mode fibres*, in which light signals travel at almost uniform velocity, thereby reducing dispersion. Higher purity glasses have enhanced the transmission of fibres especially in the wavelength range 1.2–1.6 μm. The net result is that the electric field intensity in the core is 10^4–10^5 times greater than in bulk lasers and that consequently the threshold launch power is very much reduced. The Er^{3+}-fibre amplifier in optical communications detects, amplifies and re-emits signals at points along the fibre as did the repeater in telecommunications. However, there will be many fewer Er^{3+}-fibre amplifiers than repeaters for the reasons given above and because the fibre amplifier can handle much more data than could the electrical repeaters.

Optical amplifiers need not be fibres. Glass and crystalline slab waveguide amplifiers can be used. In either case, the waveguide has more or less the same dimensions as the core of the optical fibre, and must have a slightly larger index of refraction than the matrix in which it is fabricated. Generally, *planar waveguides* confine the radiation in one dimension and *channel waveguides* confine the light in two dimensions. In $LiNbO_3$

both types of guide can be produced by Ti in-diffusion [Brinkman *et al.* (1991)] or proton exchange techniques [Lallier *et al.* (1989), (1990)]. Liquid phase epitaxy has been used to fabricate planar waveguides in YAG or GGG-based garnets [Chartier *et al.* (1992)]. Buried channel waveguides are grown by liquid phase epitaxy and ion beam etching in combination [Katoh *et al.* (1992)]. Ion implantation followed by annealing has also been used to fabricate channel waveguides in numerous materials [Field *et al.* (1991)].

One means of producing a diode-pumped blue laser is to frequency double the $^4F_{3/2} \rightarrow {}^4I_{9/2}$, 946 nm line from a diode-pumped Nd : YAG laser. The terminus of this transition, the upper Stark level of the $^4I_{9/2}$ multiplet, is 857 cm^{-1} above the ground level. Since this level at 300 K will contain 0.7% of the total Nd population, it causes a significant absorption loss at the laser wavelength and consequently causes an increase in the threshold pump power. Using a waveguide system to confine both pump and laser radiation should give the requisite population inversion at modest excitation power. Hanna *et al.* (1993) used a planar waveguide with composition $(Nd_{0.01}Lu_{0.35}Y_{0.64})(Ga_{0.12}Al_{0.88})_5O_{12}$ grown epitaxially on a pure YAG substrate. A cladding layer of pure YAG was grown on top of the active layer, also by liquid phase epitaxy. At this composition the planar waveguide is lattice matched to substrate and cladding. The resulting difference between refractive indices of planar guide and substrate/overlayer was 1.4×10^{-2} at $\lambda = 633$ nm. Typically, losses of 0.1–0.2 dB/cm were

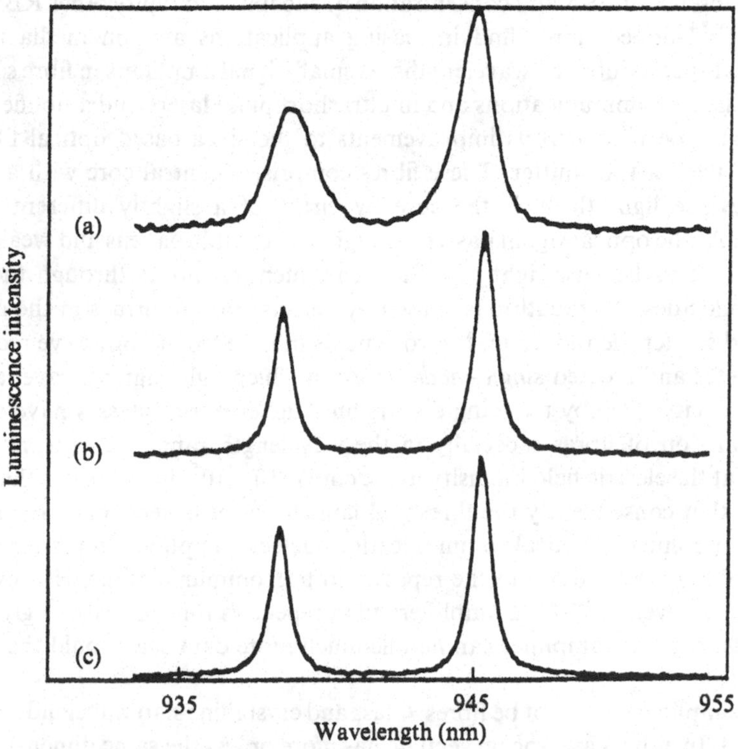

Figure 10.28. Fluorescence spectra at 930–955 nm of Nd^{3+} ions in (a) Ga : Lu : Nd : YAG and (b) Nd : YAG epitaxial waveguides compared with (c) bulk Nd : YAG [after Hanna *et al.* (1993)].

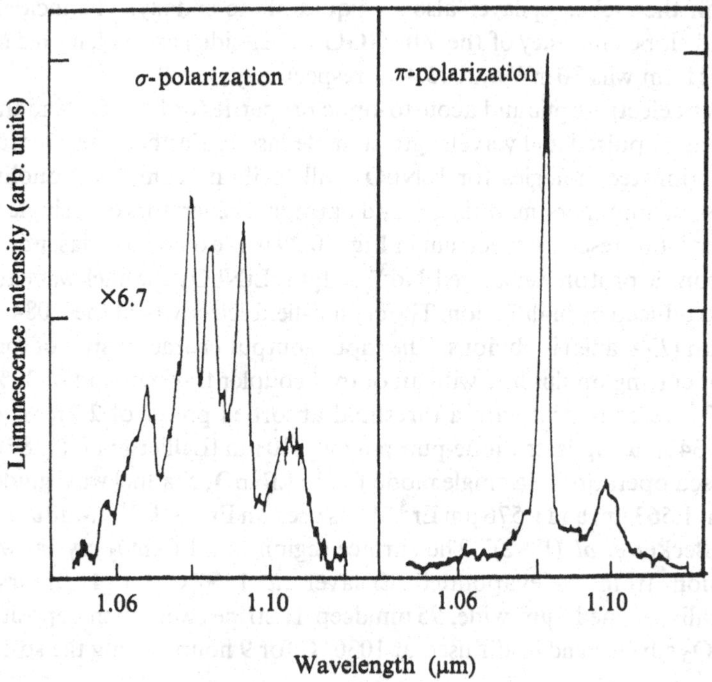

Figure 10.29. The polarized Nd^{3+} fluorescence excited by a laser diode tuned to 810 nm from a Nd^{3+}-doped MgO : $LiNbO_3$ proton exchanged channel waveguide [after Becker *et al.* (1992)].

obtained at a guide width of 3.8 μm. The fluorescence spectra of the Nd : Lu : Ga co-doped guide, Nd : YAG guide and bulk Nd : YAG crystal excited in the $^4I_{9/2} \rightarrow {}^4G_{5/2}$, $G_{7/2}$ band by a rhodamine G6 dye laser at 588 nm are shown in Fig. 10.28 to differ only in the greater linewidth of the co-doped guide induced by disorder on both cation sub-lattices. Within experimental error the fluorescence life times of these samples, $\tau = 240$ μs, are the same at 300 K. The absorption and emission cross sections of the Nd : Lu : Ga : YAG guide measured at 946 nm relative to bulk Nd : YAG, $\sigma_a = 2.3 \times 10^{-20}$ cm^2 and $\sigma_e = 3.1 \times 10^{-20}$ cm^2 are slightly different consequent on the vibronic interaction. Even without optimized waveguide length or cavity mirrors this planar waveguide laser operated at 946 nm with an absorbed threshold power at 588 nm of 1.2 mW and slope efficiency of *ca* 60%. With channel waveguides and optimized input and output mirrors the absorbed threshold power should reduce below 100 μW.

The prospects for low threshold power operation of waveguide lasers using the lower lying, crystal-field levels of the $^4I_{9/2}$ ground multiplet of Nd^{3+}, not usually laser active in bulk Nd : YAG lasers, are excellent as they are for other quasi-three-level systems, such as Er^{3+} : YAG at 1.6 μm, Ho^{3+} : YAG and Yb : YAG at 1.03 μm. Indeed, single-mode operation at 300 K has been reported for an Yb : GGG buried channel waveguide at 1.024 μm [Shimokozono *et al.* (1996)]. The waveguide with composition $Yb_{0.19}Nd_{0.11}Gd_{2.70}Ga_5O_{12}$ was grown on an $Y_{0.4}Gd_{2.6}Ga_5O_{12}$ substrate, the Nd^{3+} having been added for reasons of lattice matching and $Nd^{3+} \rightarrow Yb^{3+}$ energy transfer. A core waveguide ridge was formed in the planar guide by Ar^+ beam etching and subsequent soaking in hot phosphoric acid. An $Y_{0.4}Gd_{2.6}Ga_5O_{12}$ cladding overlayer was grown on the etched epilayer also by liquid phase epitaxy. The incident power threshold and slope efficiency of the Yb : GGG waveguide laser of length 5 mm when pumped at 941 nm was 80 mW and 13.4%, respectively.

The excellent electro-optic and acousto-optic properties of $LiNbO_3$ lend themselves to applications in pulsed and wavelength tunable lasers. Furthermore, mature waveguide production technologies for $LiNbO_3$ will facilitate complex monolithic integration of lasers, amplifiers, modulators and harmonic generators on a single chip. The polarized Nd^{3+}-fluorescence spectrum in Fig. 10.29 was excited by a laser diode tuned to 810 nm from a proton-exchanged Nd^{3+} : MgO : $LiNbO_3$ channel waveguide. The Nd^{3+} was introduced by in-diffusion. The crystal-field selectivity of the 1.084 μm line in π-polarization ($E \| c$-axis) is obvious. The input–output characteristics of the channel waveguide operating on this line with an output coupler transmission of 30%, shown in Fig. 10.30, are consistent with a threshold absorbed power of 2.7 mW and slope efficiency of 34% using laser diode-pumping at 810 nm [Lallier *et al.* (1989), (1990)]. CW and pulsed operation of a single mode Er^{3+} : $LiNbO_3$ channel waveguide laser on the 1.532 μm, 1.563 μm and 1.576 μm Er^{3+} lines used an Er^{3+}–Ti^{3+} in-diffused $LiNbO_3$ waveguide [Becker *et al.* (1992)]. The surface region of a $LiNbO_3$ crystal was doped by in-diffusion from an evaporated Er-layer at 1060°C for 41 hours. Photolithographically-defined 8 μm wide, 95 nm deep Ti-stripes were then deposited on the Er^{3+} : $LiNbO_3$ surface and in-diffused at 1030 °C for 9 hours. Using the so-fabricated

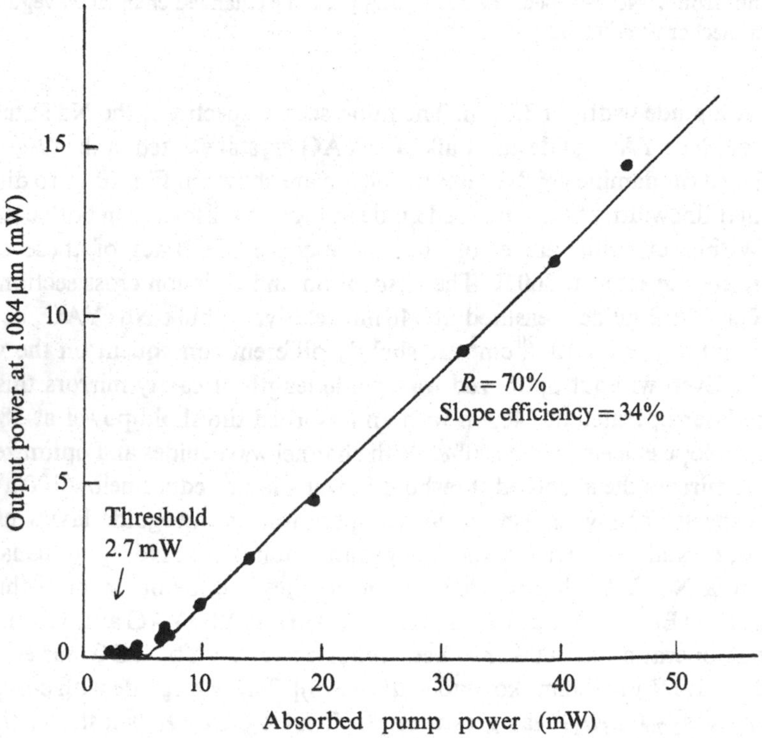

Figure 10.30. Input–output characteristics of a diode-pumped (810 nm) Nd : $LiNbO_3$ channel waveguide laser emitting at 1084 nm and operating with a 30% transmission output coupler [after Lallier *et al.* (1990)].

channel waveguides pumped in π-polarization at 1479 nm resulted in laser operation at 1576 nm and 13 mW threshold of coupled pump power. With coupled pump power exceeding 25 mW simultaneous laser operation occurred on both 1576 nm and 1563 nm lines. Using σ-polarized pumping the Er : $LiNbO_3$ channel waveguide operated only at 1563 nm with a slope efficiency of 3% and peak output power of 3 mW.

10.8 All-solid-state lasers (ASSLs)

For many years the pumping of solid state lasers was the domain of the flashlamp and the water-cooled Ar^+ laser. However, given the increasing availability of high power and choice of wavelength from semiconductor diode lasers the traditional technology may soon be eclipsed. At the conclusion of this monograph it is appropriate to review the progress made with all-solid-state lasers (ASSLs), which, using diode-pumping schemes, have the potential for compact and stable construction that make them attractive for mass production. Here we emphasize the spectroscopic features that are of major importance in the design of miniature solid state lasers.

10.8.1 Fixed wavelength LD-pumped solid state lasers

The first ASSL to be pumped by a LD was Nd : YAG. Although the ${}^4I_{9/2} \rightarrow {}^4F_{5/2}$ absorption of Nd : YAG is well-matched to pumping with GaAs/AlGaAs LDs it is not the optimal choice. Considerations of crystal growth restrict the concentration of active ion to 1 at.% in YAG and the resonance linewidth is only 1–2 nm. The emission linewidth of the LD is larger, typically 2–3 nm, and is not stable to better than ±2–3 nm. In consequence, there has been much effort to replace Nd : YAG with a gain medium more closely matched to the LD output characteristic. This point is illustrated by reference to Figures 10.23–10.25, the optical spectra and laser performance of Nd^{3+}: $LiKYF_6$, an unlikely commercial laser gain medium. Diode pumping at 800 nm with γ-polarization selects absorption into a homogeneously broadened band width of 7–8 nm. The sample used for the spectra in Fig. 10.24 contained 3.4 at.% Nd^{3+}, without deleterious effects on crystal quality or concentration quenching of luminescence [Nicholls *et al.* (1997a), (1998)]. This contrasts with results for Nd^{3+} : YAG and Nd^{3+} : YLF, in which the Nd^{3+} content of the host crystals cannot exceed more than 1 at.% Nd. In this crystal and its Gd isomorph it is the orientation of the Y(Gd)–F_8 polyhedra which dictate the maximum dopant concentration. However, note also the site selectivity of the odd-parity distortion in KLYF which feeds luminescence intensity almost exclusively into a single crystal-field line at 1048 nm with a linewidth of 3–4 nm when detected in β-polarization. The RE^{3+} site in KLYF has C_1 symmetry and is much more distorted than the RE^{3+} sites in YAG and YLF, thereby giving rise to larger absorption and emission cross sections. Both pulsed and CW laser operation have been reported at 1049 nm with very low threshold pump power and high efficiency. Without optimization of the gain medium (Nd^{3+} content, optical quality, fine polishing, etc.) a crystal containing 3.4 at.% Nd^{3+} pumped by a CW Ti-sapphire laser operating at 800 nm (to mimic diode laser pumping) supported CW laser action at 1049 nm with threshold of 24 mW and a slope efficiency of 42% relative to absorbed pump power. Under optimal conditions the efficiency is expected to be close to the quantum defect limit.

Various other Nd-doped materials have been assessed for suitability in LD-pumped ASSLs; each has slightly different optical characteristics that may influence choice for particular applications. Nd^{3+} : YLF has a long fluorescence lifetime of ~500 μs at 300 K and a gain bandwidth a factor of three larger than that of Nd : YAG. This gain medium is used in Q-switched and mode-locked ASSLs. Nd^{3+} : glasses have even larger gain bandwidths and can be used in somewhat inefficient short pulse lasers. Nd^{3+} : $YAlO_3$ lasers can be operated on six lines near 1060 nm, even at 300 K. The most efficient diode-pumped operation has been observed for Nd^{3+} : YVO_4 which has very strong and broad (*ca* 10 nm) absorption and emission lines. Both Nd-doped MgO : $LiNbO_3$ and $Y_3Al(BO_3)_4$ (YAB) have high gain and may find application in self-frequency doubling to generate visible radiation. Some Nd^{3+} : doped crystals find deployment using the ${}^4F_{3/2} \rightarrow {}^4I_{13/2}$ transitions near 1320 nm. In this context high efficiency, low threshold laser operation of Nd^{3+}-doped $Sr_{10}(PO_4)_6F_2$ and $Sr_{10}(VO_4)_6F_2$ at 1059 nm and 1328 nm have been demonstrated with performance that at least matches that of Nd^{3+} : YVO_4 lasers at both wavelengths [Zhang *et al.* (1994), Scott *et al.* (1994)]. Unfortunately the short upper state lifetime of Nd^{3+} in YVO_4 ($\tau = 90$ μs) makes this host less satisfactory in Q-switching. Better tolerance of higher Nd^{3+} content in the orthovanadates is afforded by the Gd-isomorph $GdVO_4$, given the better fit of Nd^{3+} on the Gd^{3+} site than on the Y^{3+} site in YVO_4. Two potential disadvantages of the fluoroapatites and fluorovanadates is that Nd^{3+} substitutes for the divalent alkaline earth ion (Sr^{2+}, Ca^{2+}) and this limits the possible doping concentration to < 1 at.%. Furthermore, SFAP and SVAP crystals usually contain Nd^{3+} multisites which permit concentration quenching of luminescence [Zhang *et al.* (1994)]. This latter problem has to some extent been overcome by careful control during growth of the fluoride content of crystals [Scott *et al.* (1997)]. In seeking new laser hosts it is necessary that the desired broadening of the pump band to effect absorption of the entire output of the diode array should be homogeneously induced, since inhomogeneous broadening by disorder and multisite formation can lead to surprising variations in slope efficiencies with output mirror transmission [Mermilliod *et al.* (1992), Barnes *et al.* (1990)].

It is possible to achieve high absorption of pump radiation by increasing the Nd^{3+} concentration using more suitable recipients for such dopants. Taken to a logical extreme, stoichiometric rare-earth compounds seem to have considerable promise. $LiNdPO_4$ has considerable potential despite major growth problems. The mixed borate $YAl_3(BO_3)_4$ (YAB) and its La- and Sc- isomorphs will also tolerate large RE^{3+} dopant concentrations with weak concentration quenching. Currently, $Nd_xY_{1-x}Al_3(BO_3)_4$ with $x = 0.001–0.004$ is available commercially for applications in diode-pumping with self-frequency doubling [Danielmeyer (1975), Huber (1980), Wang *et al.* (1991), Luo *et al.* (1989a,b)]. Such crystals are all grown by high temperature solution growth. There are two isomorphs of $LaSc_3(BO_3)_4$ (LSB), a high temperature phase (β-LSB) that can be grown by Czochralski pulling and a low temperature phase (α-LSB) grown by HTSG, both with up to 100% Nd^{3+} substitution on the La^{3+} site [Weyn *et al.* (1994)]. So far efficient diode-pumped laser operation has been reported only for the α-phase. The β-phase would be a preferred commercial option, given the shorter growth times, larger crystal yields and higher optical quality of Czochralski crystals. Both polymorphs offer higher absorption/emission cross sections than Nd : YAG by a factor of ~2 with homogeneously broadened pump bandwidth at 800 nm of *ca* 4 nm (FWHM).

Wide-ranging diversity of wavelength is achieved with diode-pumped ASSLs using RE^{3+}-doped gain media containing dopants other than Nd^{3+}. Pr^{3+}, Dy^{3+}, Er^{3+}, Ho^{3+} and Tm^{3+} ions all have absorption bands in the 780–820 nm region for pumping with first generation LDs. There is great interest in Er^{3+}: doped glasses and crystals for applications in optical communications and laser surgery using ultrashort pulse and high CW power regimes, respectively [Pollnau (1997)]. Yb^{3+} to Er^{3+} energy transfer pumping of Er^{3+}-laser operation has been demonstrated using $Yb^{3+}:Er^{3+}$ co-doped silicon yttrium oxysilicate ($SrY_4(SiO_4)_3O$), Kigre QE-7 and phosphate glass gain media in Q-switched [Hutchinson and Allik (1992)] and CW [Souria *et al.* (1994)] modes, respectively, pumped by 970-980 nm diodes. Blue-red upconversion lasing at 488 nm, 635 nm and 717 nm has been demonstrated in $Nd^{3+}:Pr^{3+}$:ZBLAN glass fibre using a single pump source LD at 796 nm, which is a significant operational advantage over the dual-wavelength pumped Pr^{3+}-doped and Tm^{3+}-doped blue lasers discussed in §10.5.3 and §10.6. The first diode-pumped Ho^{3+} ion laser used $Er^{3+}:Tm^{3+}$ sensitized $(Er_{0.6}Y_{0.4})Al_5O_{12}$ containing 3 at.% Tm^{3+} and 2 at.% Ho^{3+} [Esterowitz *et al.* (1986)]. This laser operated on the 2.1 μm $^5I_7 \rightarrow {}^5I_8$ transition on Ho^{3+}, with a slope efficiency of 17% and threshold of 3.4 mW at room temperature. Diode-pumped operation of Ho^{3+} lasers at 2210 nm using $Tm^{3+}:Ho^{3+}:YLiF_4$ and $LuLiF_4$ demonstrate the advantage of size-matching the active ions to the host substitutional site. In both cases the LDs matched the Tm^{3+} absorption band in both hosts at 794 nm, resulting in optical efficiencies 1.5 times greater for Tm^{3+}: Ho^{3+}: $LiLuF_4$ lasers than for Tm^{3+}: Ho^{3+}: $LiYF_4$ hosts. A compact, efficient and reliable LD-pumped Tm: YAG laser, delivering 115 W of CW power at 2.0 μm, has been developed by Honea *et al.* (1997). The pump source comprised a 23-module stack of 1 cm-long LD bars which produced up to 460 W of CW power at 804 nm!

There is considerable potential for high doping levels of RE^{3+}, other than Nd^{3+}, in laser hosts. The fully concentrated crystal $PrCl_3$ supports efficient laser operation on transitions out of the $^3P_0$ level: in the analogous compound $NdCl_3$ luminescence from Nd^{3+} ions is fully quenched even at low temperature, as discussed in §7.4. The alumino- and scandioborates of La, Y and Gd will accept high dopant levels of Er^{3+}, Pr^{3+}, Dy^{3+} and Tm^{3+}, in some cases as high as 100%, with only modest concentration quenching. Qi *et al.* (1996a,b) have speculated on the potential as laser gain media of the rare-earth titanyl niobates $RETiNbO_6$, where $RE^{3+} = Ce^{3+}$, Pr^{3+}, $Nd^{3+}, \ldots, Yb^{3+}$. In these crystals the density of active ions is very high ($\sim 8 \times 10^{21}\,cm^{-3}$), large absorption and emission cross sections ($\sim 5 \times 10^{-18}\,cm^2$) feature with large absorption linewidths (FWHM $\cong$ 4 nm) at 300 K. Strong absorption bands at *ca* 800 nm in the Nd^{3+} and Er^{3+} titanyl niobates suggest that these crystals may be amenable to LD pumping at this wavelength. $PrTiNbO_6$ absorbs strongly at 1020 nm in the $^3H_4 \rightarrow {}^1G_4$ transition, which matches the pump wavelength of 960–1040 nm LDs. Strong upconverted luminescence is observed at 300 K in $PrTiNbO_6$, $NdTiNbO_6$ and $ErTiNbO_6$.

10.8.2 Tunable solid state lasers pumped by laser diodes

Diode pumping of vibronic gain media is an alternative to RE^{3+}-doped laser gain media in applications that require broadband tunability. $Cr^{3+}:BeAl_2O_4$ (alexandrite) was the first tunable laser to be pumped by red LDs ($\lambda \sim$ 665–675 nm) [Scheps *et al.*

(1990)]. Crystals in which Cr^{3+} ions occupy weaker crystal field sites than in $BeAl_2O_4$ provide better overlap of the $^4A_2 \rightarrow {}^4T_2$ absorption band with the 630–670 nm wavelength range of current LD arrays. Among such lasers are Cr^{3+}: GSGG, Cr^{3+}: LiCAF, Cr^{3+}: LiSAF and Cr^{3+}: LiSGaF. The results of diode-pumped operation of Cr^{3+}: LiCAF, Cr^{3+} : LiSAF and Cr^{3+} : Nd^{3+}: GSGG have been reviewed by Scheps (1992), who reports low threshold and high slope efficiency when pumped with up to 1 W of diode power. Laser emission levels of almost 200 mW and slope efficiencies approaching 50% were demonstrated. Although the basic optical properties of Cr^{3+}-doped LiSAF and LiSGaF are very similar, Cr^{3+} : LiSGaF pumped in the same resonator configuration as Cr^{3+} : LiSAF demonstrates substantially higher output power in CW (900 mW) and mode-locked (200 mW) operation [Sorokina *et al.* (1996)]. In the Q-switched regime Cr^{3+} : LiSGaF produced four times more energy (12 μJ) at 10 kHz repetition rate than a Cr^{3+} : LiSAF laser under the same conditions [Balembois *et al.* (1997)]. In addition to CW and Q-switched ASSLs there is considerable demand for compact, stable, reliable and tunable ultrafast laser sources. The Cr^{3+} : colquiriites are especially well-suited as gain media in such diode-pumped lasers, given their broadband gain over a spectral range similar to Ti-sapphire but suitable for GaInP/AlGaInP laser diode-pumping. Various mode-locking systems have been used in the almost routine production of sub-50 fs pulses in diode-pumped, mode-locked Cr^{3+} : LiSAF and LiSGaF lasers [Rizvi *et al.* (1992), Dymott and Ferguson (1995), Sorokina *et al.* (1996)].

10.8.3 Optical nonlinearities and diode-pumped lasers

Nonlinear optical crystals have played an important role in laser development through extensions of the wavelength operating characteristics into the ultraviolet and infrared wavelength ranges not formerly accessible using conventional lasers. As discussed in §3.3, §5.6 and §9.8 optical nonlinearities result from the static odd-parity distortions that are present in certain crystals. Elements of the theory of nonlinear susceptibilities presented there imply that families of compounds should be considered for their potential as nonlinear optical media. Notable for their applicability were the ferroelectric compounds ABO_3 with distorted perovskite structure, the dihydrogen phosphates and arsenates, the titanyl phosphates and arsenates and the crystalline borates. Apparently no individual crystal has the appropriate combination of optical and physical properties to be the material of choice for the entire spectral range from 150 nm to 3000 nm and compromises must be made in choosing materials for particular applications. Crystals must have large second order d-coefficients in the spectral region of interest, Eq. (5.168), with relatively large birefringences to permit a wide range of phase-matching geometries for applications in harmonic generation and parametric oscillation. In general, nonlinear optical crystals should have wide transparency ranges from the infrared to the ultraviolet and large thresholds against photorefractive damage by high intensity laser beams. Finally, crystals should have excellent chemical and mechanical stability. Very few of the many different crystals studied find regular deployment outside the research laboratory. The theoretical background to the structure–property relationships of inorganic nonlinear optical crystals has been reviewed by Chen (1993) and their applications as tunable coherent sources are outlined by Lin (1990). Detailed materials considerations are reviewed by Bordui and Fejer (1993).

The outputs from conventional LDs, LD arrays and 1 cm bar-arrays are highly divergent and astigmatic. Malcolm and Ferguson (1991a) have described various cavity schemes that efficiently match these outputs to the gain mode volume using longitudinal end-pumped and transverse side-pumped (rod or slab) geometries. The evolution in diode-pumped ASSL cavity design has led to spectacular improvements in mode-locking performance and to novel all-solid-state ultrashort pulse lasers using active and passive mode-locking techniques. Combining these mode-locking techniques with Q-switching can yield exceptional peak powers useful for pumping tunable gain media and nonlinear optical crystals.

The non-availability of powerful blue-green LDs results in commercial Ti-sapphire lasers being pumped by bulky and expensive Ar^+ lasers, via the $^2T_2 \rightarrow {}^2E$ absorption band. Malcolm and Ferguson (1991b) developed an ASSL based on Ti-sapphire pumped at 523.5 nm by the frequency doubled output from a LD-pumped Nd : YLF laser simultaneously mode-locked and Q-switched to produce a high peak power envelope of mode-locked pulses. The Nd : YLF laser was pumped by two 1 W laser diodes and frequency doubled by non-critical phase matching in MgO : $LiNbO_3$ with 50% conversion efficiency. This ASSL had a threshold of 4.8 μJ absorbed energy and maximum output of 3 μJ for 15.6 μJ absorbed in the form of a 100 ns pulse of free-running linewidth 23 nm. Thus the peak power of this laser was 30 W. When a tuning element was used the linewidth was less than 1 nm, with tuning from 705–995 nm using two mirror sets.

The simplest technique for frequency doubling, i.e. single pass doubling, is too inefficient for CW and low average power mode-locked SSLs because of low peak intensities in the nonlinear crystal. Mode-locked and Q-switched diode-pumped lasers, such as the one described briefly for pumping Ti-sapphire, enable single pass doubling to be efficient into the visible and ultraviolet regions. However, of even greater efficiency are lasers that use the technique of resonant frequency doubling in an external ring enhancement cavity [Maker and Ferguson (1990a)]. Such a technique, illustrated in Fig. 1.8 and incorporating a LiB_3O_5 crystal as the nonlinear frequency converter, has been demonstrated to be 60% efficient, producing 600 mW of 523.5 nm output from 1 W of 1047 nm output from a diode-pumped Nd : YLF laser. Such sources provide a useful means of pumping optical parametric oscillators to achieve wide-ranging tunability and ultrashort pulses. A synchronously pumped all-solid-state OPO, illustrated in Fig. 10.31, was pumped by the frequency-doubled Nd : YLF laser described earlier (Fig. 1.8). The parametric medium was a LiB_3O_5 crystal cut at Brewster's angle. Careful adjustment of the cavity length results in a doubly resonant threshold of only 3 mW. By changing the crystal temperature it is possible to force singly resonant operation. The temperature tuning performance under non-critical phase matching is quite remarkable, Fig. 1.8, ranging from 650 nm to 2700 nm as the crystal temperature is changed from 120–200 °C [Maker and Ferguson (1990b)].

Major research on nonlinear optical conversion focuses on blue and green coherent light generation using diode lasers as pump sources. One approach is to use intracavity frequency doubling in a $LiIO_3$ crystal pumped by a Q-switched Cr-LiSAF laser, to give tunable (395 nm–435 nm), 230 ns pulses with 7 mW of average power at 407 nm [Balembois *et al.* (1992)]. It has also been possible to achieve CW frequency doubling of Cr : LiSAF output pumped by LD at 670 nm using a $KNbO_3$ crystal cut for non-critical phase matching. The second harmonic output was tunable from 427–443 nm with 13 mW of CW power for an absorbed pump power of 650 mW.

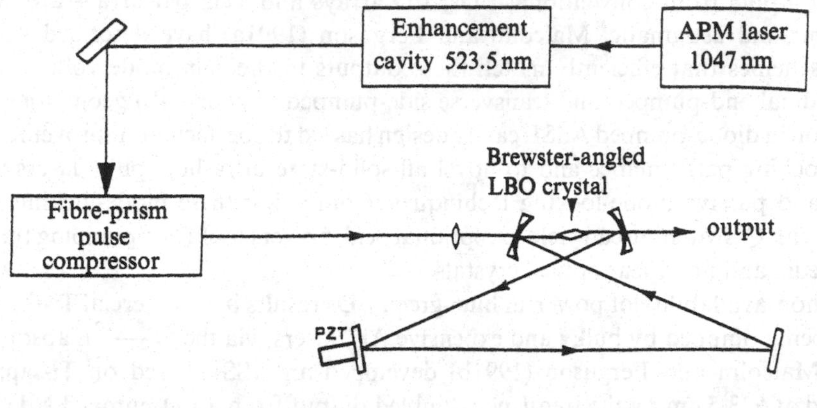

Figure 10.31. Schematic diagram of a synchronously pumped LBO optical parametric oscillator [after Ferguson (1994)].

An alternative technique for producing single frequency blue light is the direct frequency doubling of the LD output. Non-critical phase matching of ~ 800 nm LDs has been achieved using bulk crystals of $KNbO_3$ [Günter *et al.* (1979)] and $K_{3-y}Li_{2-x+y}Nb_{5+x}O_{15+2x}$(KLN) [Reid *et al.* (1992)]. However, $KNbO_3$ can easily be converted to multi-domain structure by simple mechanical shock and is not mechanically robust enough for real-life applications. KLN crystals have much superior physical and optical properties for SHG of laser diodes. Nevertheless, the complex phase relationships in the K_2O–Li_2O–Nb_2O_5 tertiary equilibrium make it difficult to determine exact growth conditions of crystals for frequency conversion of fundamental wavelengths near 800 nm. Nevertheless, Jiang *et al.* (1998) have grown crystals with compositions having different values of x and y satisfactory for NCPM for second harmonic generation in the wavelength range 1050 nm to 820 nm at 300 K. The visible range transmission cut-off occurs at 325 nm. The alkali boroxy-niobate family of compounds, XB_2NbO_6 with $X = K$, Rb, Cs,... have superior laser damage thresholds and UV cut-off wavelengths (275 nm) relative to $LiNbO_3$ and KLN and have potential applications in SHG and OPO devices [Nicholls *et al.* (1997b)]. There has also been much activity in frequency doubling LD radiation using periodically poled channel and planar waveguides in $LiNbO_3$, $LiTaO_3$, KTP and RTA [Myers *et al.* (1995), Yamamoto *et al.* (1993), Risk and Loicano (1996)]. The fundamental-to-second harmonic frequency conversion efficiency can be quite high (2–5%).

Finally, we comment on recent efforts to develop efficient laser systems for self-frequency-doubling in Nd-doped lasers. Kozlovsky *et al.* (1988) have reported preliminary studies of self-frequency doubling, 1060 nm to 530 nm, in Nd^{3+}: MgO : $LiNbO_3$. Nd : YAB has been studied both for the production of fundamental wavelengths of 1320 nm and 1060 nm and for the generation of their second harmonics at 660 nm and 530 nm, respectively using dye laser [Dorozhkhin *et al.* (1981)] and flashlamp pumping [Luo *et al.* (1989b)]. However, diode-pumped laser operation at 1064 nm and second harmonic generation at 532 nm using diode-pumping at 800 nm was subsequently reported [Lin (1990)]. A CW input power of 450 mW from the LD resulting in 1.5 mW of CW output at 1063 nm and 9 μW of green output at 532 nm.

10.8.4 Microchip lasers

Recently reviews of microchip lasers operating on their fundamental transition have been published by Zayhowski and Harrison (1996), by Zayhowski (1999) and by Sinclair (1999). The cavity in a microchip laser is typically $<1\,mm^3$ in volume, so short (<1 mm) that usually only a single longitudinal mode oscillates and in one polarization. The laser is inherently single frequency with very narrow linewidth. In dynamic applications short cavity life times and small mode volumes permit rapid changes in laser characteristics, leading to high rates of frequency modulation and ultrashort pulse generation. Although Nd : YAG was the first material to be used as the active medium in a microchip laser, various other host crystals have been used, including YLF, YVO_4, Y_2SiO_5 (YOS), SFAP, SFVAP and YAB and its La and Sc isomorphs. Rare-earth ion-activated microchip lasers include Yb : YAG at 1.05 μm, Er : Yb : glass for fibre-optic applications near 1.5 μm and Tm : Ho energy transfer devices operating at 2.0 μm and 3.0 μm for remote sensing and medical applications, respectively. A tunable Cr : LiSAF microchip laser pumped with a 670 nm diode has been used in the spectral region 0.8–1.0 μm. The materials used in microchip lasers must have high absorption coefficients at the pump wavelength to facilitate the use of thin crystals. This can be achieved by selection of materials with high intrinsic oscillator strength such as Nd : YVO_4 or Nd : $LiKYF_5$ or by using such stoichiometric rare-earth compounds as $LiNdPO_4$ or $Nd_xLa_{1-x}Sc(BO_3)_4$ (NLSB), where $0 < x < 1$. Although high RE^{3+} concentrations are tolerated in YVO_4, even higher concentrations may be substituted on the Gd^{3+} site in the $GdVO_4$ isomorph, as we commented in §10.8.1. For single mode operation it is helpful to have a narrow emission bandwidth: to obtain low pump thresholds and high slope efficiencies the material must exhibit a high $\sigma_{em}\tau_R$ product, Eq. (9.27). The optothermal properties of microchip laser materials are as important as spectroscopic properties. As the microchip cavity is optically pumped heat is deposited in the crystal. In materials such as Nd : YAG, where the refractive index increases with temperature, the resulting temperature profile results in a thermal waveguide, which is normally single mode. Microchip lasers, in consequence, usually operate on the fundamental transverse mode with extremely fine beam quality.

Many diode-pumped microchip lasers have been based on the resonance lines of Nd^{3+} that oscillate near 1.06 and 1.32 μm pumped with 805 nm laser diodes. By bonding a wafer of a piezoelectric material (e.g. $LiTaO_3$) to the gain medium frequency tunable CW lasers and high power, short pulse Q-switched microchip lasers can be produced that perform as well as or better than conventional ion pumped lasers. There are many potential applications of visible radiation at modest power and low cost that may be satisfied by mirochip lasers. These shorter operating wavelengths can be achieved by coherent upconversion through intracavity frequency doubling of diode-pumped solid state lasers, a technique first achieved by Baer (1986). Sinclair (1999) and his colleagues have used the frequency doubling route to achieve significant second harmonic outputs at red, green and blue wavelengths. The red laser doubled the 1342 nm line from Nd : YVO_4 in a non-critically phase matched LBO crystal, whereas the blue laser used the 946 nm transition of Nd : YAG doubled by a critically phase matched $KNbO_3$ crystal. Both suffer from materials problems which further research may resolve. The green laser produced an output of 158 mW from a Nd : YVO_4 /KTP sandwich pumped by a 1.2 W diode. The short cavity helps to keep losses small and to maintain single frequency operation. Nd : YVO_4 is now the material of choice for the $^4F_{3/2} \rightarrow {}^4I_{1/2}$ transition at 1.06 μm by virtue of its greater stimulated absorption and emission cross sections, pump transition bandwidth and $\sigma_{em}\tau_{em}$ product. However, it is possible that both Nd : YOS and Nd : SFAP outperform Nd : YVO_4 on the $^4F_{3/2} \rightarrow {}^4I_{9/2}, {}^4I_{13/2}$ transitions at 0.943 μm and 1.32 μm, respectively, because of better branching ratios to these transitions in these materials.

10.9 Concluding remarks

This monograph has presented theoretical and experimental phenomenology that underpins the spectroscopy of solid state laser materials, whether they be crystalline or non-crystalline. Crystal-field theory was developed in the early chapters and the static and vibrational symmetry of the environment of optical centres shown to be of paramount importance, even in doped glasses. When consideration is given to optothermal and optomechanical effects the concepts developed there are generally applicable to all-solid-state lasers, whether CW or pulsed, low powered or high powered, large or miniature. However, given the power of modern computers it is evident that molecular orbital theories that incorporate the finite sizes of constituent ions will be deployed increasingly in future, more realistic attempts to model laser-active centres in solids [Woods *et al.* (1993), Aramburu *et al.* (1999)]. In this context covalency is crucial, especially in low symmetry sites where directional bonding is significant. The illustrative examples discussed here included energy level assignments and polarized transition rates of Ti^{3+} and Cr^{3+} ions in distorted octahedral sites and of Cr^{4+} ions in distorted tetrahedral sites. Such techniques helped to resolve erroneous level assignments made in early work on Cr^{4+}-doped lasers. Nevertheless, simple crystal-field theory will continue to be a useful first approximation in interpreting the spectroscopic properties of novel laser materials.

In the commercial sense, lasers that provide massive power for materials processing in heavy industry will continue to feature among market leaders. They represent mature technologies that may not require much further development. Major development is more likely of miniature lasers, catalyzed by the emergence of low cost fibre lasers/amplifiers and laser diodes. Traditional CW and short pulse lasers based on Nd-YAG or Ar-ion laser pumping occupy whole laboratories, often requiring 50 kW or more of electrical power to produce 10–20 W of laser output. Such power consumption, requirements of heat exchange cooling, costs of upgrading and regular maintenance, limit their widespread application, except in the research laboratory.

Trends towards miniaturization, involving the pumping of laser gain media and nonlinear optical materials, were discussed in Chapters 1, 9 and 10. The fibre/waveguide/microchip geometry makes for the manufacture of rugged lasers that operate in extreme conditions as compact, stable, and power-efficient structures. These lasers have the potential for development over wavelengths from the near UV (250 nm) into the near-infrared (2500 nm). For example, a diode array at 980 nm used to excite an Yb^{3+} fibre amplifier plus LBO frequency doubler in a resonantly enhanced laser cavity can produce *ca* 10 W of laser output at 490 nm with a wall-plug efficiency of almost 50%. Such a laser would occupy a container with typical dimensions $20 \times 20 \times 10\,cm^3$. Similarly compact sources can be configured using a diode bar laser at 800 nm or 1064 nm to pump a miniature Nd-doped SSL that is then used to pump a tunable

Cr^{4+} laser operating around 1300 nm (Mg_2SiO_4) or 1500 nm (YAG), the second and third telecommunications windows, respectively. Alternative pumping schemes will permit tunable laser output around 700–1000 nm using Cr-colquiriite systems. The microchip laser approach is particularly suitable for low and medium power applications, operating either on the primary laser wavelengths or as nonlinear optical devices. Indeed, an UV microchip laser, containing fibre coupling to an 808 nm diode-pump laser and using proximity coupling to the IR microchip laser (Nd-YAG), Q-switch (Cr^{4+} : YAG) and nonlinear crystals (KTP and BBO), was fitted into a stainless steel can measuring only 25 mm long by 10 mm diameter. The IR to UV conversion efficiency of this device was over 6%, with more than 5 mW of 266 nm light generated at a pulse repetition rate of 10 kHz and peak powers of over 2 kW. The lifetime of the entire device was limited by the laser diode rather than by the microchip components [Zayhowski (1999)]. Although much miniaturization has proceeded with oxide-based materials there is considerable potential for devices based on fluoride hosts, especially glasses, where weak nonradiative decay of excited states of rare-earth ions allows efficient lasing at visible wavelengths pumped by laser diodes in the infrared. All these ASSLs are rugged and compact with good tolerance of the poor spectral characteristics of the pump diodes without loss of efficiency. Such miniature lasers are revolutionizing present day technology with real life and scientific applications in telecommunications, information technology, remote sensing, LIDAR, surgery and dermatology, spectroscopy and laboratory testing of fibre communications hardware. Given such wide-ranging applicability of ASSLs it is evident that the present level of feverish developmental activity will continue apace well past the forseeable future.

References

Abragam A. and Bleaney B. (1970), *Electron Paramagnetic Resonance of Transition Ions*, Oxford University Press, Oxford (Reprinted 1986, Dover, New York).

Adam J. L., Sibley W. A. and Gabbe D. R. (1985), *J. Lumin.*, **33**, 391.

Adams W. H. (1961), *J. Chem. Phys.*, **34**, 89.

Aegerter M. A. and Lüty F. (1971), *Phys. Status Solidi (b)*, **43**, 245.

Aggarawal R. L., Sanchez A., Fahey R. E., Stuppi M. M. and Strauss A. J. (1987), *CLEO Technical Digest*, **9**, 36.

Ahlers F., Lohse F., Spaeth J-M. and Mollenauer L.F. (1983), *Phys. Rev. B*, **28**, 1249.

Ahlers F., Lohse F., Hangleiter Th., Spaeth J-M. and Bartram R. H. (1984), *J. Phys. C: Sol. St. Phys.*, **17**, 4877.

Albers P., Jenssen H. P., Huber G. and Kokta M. (1986a), in *Tunable Solid State Lasers II* (Eds. Bugdor A. B., Esterowitz L. and DeShazer L. G.) Springer-Verlag, Berlin and Heidelberg, Springer Series in Optical Sciences, **52**, 208.

Albers P., Stark E. and Huber G. (1986b), *J. Opt. Soc. Am. B.*, **3**, 134.

Alig R. C. (1970), *Phys. Rev B*, **2**, 2108.

Alimpiev A. I., Bukin G. V., Matrosov V. N., Pestryakov E. V. and others (1986), *Sov. J. Quant. Elect.*, **16**, 579.

Allain J. Y., Monerie M. and Poignant H. (1990), *Electron. Lett.*, **26**, 166 and 261.

Andrews L. J., Lempicki A. and McCollum B. C. (1981), *J. Chem. Phys.*, **74**, 5526.

Andrews L. J., Hitelman S. M., Kokta M. and Gabbe D. (1986a), *J. Chem. Phys.*, **84**, 5229.

Andrews L. J., Lempicki A., McCollum B. C., Giunta C. J., Bartram R. H. and Dolan J. F. (1986b), *Phys. Rev. B.*, **34**, 2735.

Angert N. B., Borodin N. I., Garmash V. M., Zhitnyuk V. A., Okrimchuk A. G., Siyuchenko O. G. and Shestakov A. V. (1988), *Sov. J. Quant. Elect.*, **18**, 73.

Aramburu J. A., Moreno M., Doclo K., Daul C. and Barriuso M. T. (1999), *J. Chem. Phys.*, **110**, 1497.

Armagan G. and DiBartolo B. (1986), in *Tunable Solid State Lasers II* (Eds. Bugdor A.B., Esterowitz L. and DeShazer L.G.) Springer-Verlag, Heidelberg, Springer Series in Optical Sciences, **52**, 35.

Armagan G., DiBartolo B. and Buonchristiani A. M. (1989), *J. Lumin.*, **44**, 129 and 141.

Armstrong J. A., Bloembergen N., Ducuing J. and Pershan P. S. (1962), *Phys. Rev.*, **127**, 1918.

Austin B. J., Heine V. and Sham L. J. (1962), *Phys. Rev.*, **127**, 276.

Auzel F. (1966), *C. R. Acad. Sci. (Paris)*, **262**, 1016.

Auzel F. (1978), in *Luminescence in Inorganic Solids* (Ed. DiBartolo B.) Plenum Press, New York, p. 67.

Baer T. (1986), *J. Opt. Soc. Am.*, **B3**, 1175.

Balbashov A. M. and Egorov S. K. (1981), *J. Crys. Growth*, **52**, 498.

Balembois F., Georges P., Salin F., Roger G. and Brun A. (1992), *App. Phys. Lett.*, **61**, 2381.

Balembois F., Druon F., Falcoz F., Georges P. and Brun A. (1997), *Opt. Lett.*, **22**, 3387.

Ballhausen C. J. (1962), *Introduction to Ligand Field Theory*, McGraw-Hill, New York, San Francisco, Toronto and London.

Baney D. M., Rankin G. and Chang K. W. (1996), *App. Phys. Lett.*, **69**, 1662.

Bantien F., Albers P. and Huber G. (1987), *J. Lumin.*, **36**, 363.

Baraff G. A. and Schlüter M. (1986), *J. Phys. C: Solid St. Phys.*, **19**, 4383.

Bariuso M. T. (1999), *J. Chem. Phys.*, **110**, 1497.

Barnes N. P., Remelius D. K., Getterny D. J. and Kokta M. R. (1986), *Tunable Solid State Lasers II* (Ed. Bugdor A. B., Esterowitz L. and DeShazer L. G.), Springer-Verlag, Berlin and Heidelberg, Springer Series in Optical Sciences, **52**, 136.

Barnes N. P., Storm M. E., Cross P. L. and Skolaut M. W. (1990), *IEEE J. Quant. Elect.*, **26**, 558.

Barnett B. and Englman R. (1970), *J. Lumin.*, **3**, 55.

Barnett M. P. (1963), The evaluation of molecular integrals by the zeta function expansion. In *Methods in Computational Physics*, **2** (Eds. Alder B., Fernbach S. and Rotenberg M.) Academic Press, New York and London, p. 95–153.

Barnett M. P. and Coulson C. A. (1951), *Phil. Trans. Roy. Soc. (London)*, **A243**, 221.

Bartlett R. J. and Purvis G. D. (1978), *Int. J. Quantum Chemistry*, **14**, 561.

Bartram R. H. (1984), in Tunable Solid State Lasers (Eds. Hammerling P., Bugdor A.B. and Pinto A.) Springer-Verlag, Berlin and Heidelberg, Springer Series in Optical Sciences, **47**, 155.

Bartram R. H. (1990), *J. Phys. Chem. Solids*, **51**, 641.

Bartram R. H. and Henderson B. (1998), Unpublished.

Bartram R. H. and Stoneham A. M. (1975), *Sol. State Comm.*, **17**, 1593.

Bartram R. H. and Stoneham A. M. (1983), *Semicond. Insul.*, **5**, 297.

Bartram R. H. and Stoneham A. M. (1985), *J. Phys. C.*, **18**, L549.

Bartram R. H., Stoneham A. M. and Gash P. (1968), *Phys. Rev.*, **176**, 1014.

Bartram R. H., Charpie J. C., Andrews L. J. and Lempicki A. (1986a), *Phys. Rev. B.*, **34**, 2741.

Bartram R. H., Vassell M. O. and Zemon S. (1986b) *J. Appl. Phys.*, **60**, 4248.

Bartram R. H., Fockele M., Lohse F. and Spaeth J.-M. (1989), *J. Phys.: Condens. Matter*, **1**, 27.

Bass M. (Ed. in Chief) (1995), *Handbook of Optics* (2nd Edition) Volume **II** Part 4, McGraw-Hill, Inc., New York.

Bass M., Shi W. Q., Kurtz R., Kokta M. and Diegl H. (1986), in *Tunable Solid State Lasers II* (Eds. Bugdor A. B., Esterowitz L. and DeShazer L. G.) Springer-Verlag, Heidelberg and Berlin, *Springer Series in Optical Sciences*, **52**, 300.

Basun S. A., Danger T., Kaplyanskii A. A., McClure D. S., Petermann K. and Wong W. C. (1996), *Phys. Rev. B.*, **54**, 6141.

Bates D. R., Ledsham K. and Stewart A. L. (1953–54), *Phil. Trans. Roy. Soc. London*, Ser. **A 246**, 215.

Becker P., Brinkmann R., Dinand M., Sohler W. and Suche H. (1992), *App. Phys. Lett.*, **61**, 1257.

Bergin F., Donegan J. F., Glynn T. J. and Imbusch G. F. (1986), *J. Lumin.*, **34**, 307.

Bethe H. (1929), *Ann. Phys.*, **3**, 133.

Bjorken J. D. and Drell S. D. (1964), *Relativistic Quantum Mechanics*, McGraw-Hill, New York.

Blasse G. (1978), in *Energy Transfer Processes in Condensed Matter* (Ed. DiBartolo B.) Plenum Press, New York, p. 251.

Blasse G. (1984), in *Luminescence of Inorganic Solids* (Ed. DiBartolo B.) Plenum Press, New York.

Bloembergen N. (1965), *Nonlinear Optics*, Benjamin, New York.

Bloembergen N. and Shen Y. R. (1964), *Phys. Rev. A.*, **133**, 37.

Bogan L. D. and Fitchen D. B. (1970), *Phys. Rev.* **B1**, 4122.

Bordui P. F. and Fejer M. M. (1993), *Ann. Rev. Mat. Sci.*, **23**, 321.

Born M. and Oppenheimer J. R. (1927), *Ann. Physik*, **84**, 457.

Bosi L., Podini P. and Spinolo G. (1968), *Phys. Rev.*, **175**, 1133.

Botez D. and Scifres D., *Diode-Laser Arrays*, Cambridge Studies in Modern Optics, (Eds. P. L. Knight and A. Millar), (1994), Cambridge University Press, Cambridge, New York, Melbourne and Madrid.

Boys S. F. (1950), *Proc. Roy. Soc. (London)*, **A200**, 542.

Bradford J. N., Williams R. T. and Faust W. L. (1975), *Phys. Rev. Lett.*, **35**, 300.

Brandle C. D. and Barns R. L. (1973), *J. Cryst. Growth*, **20**, 1.

Brauch U. and Dürr U. (1984), *Opt. Comm.*, **49**, 61.

Brede R., Heumann E., Koetke J., Danger T., Huber G. and Chai B. H. T. (1993), *App. Phys. Lett.*, **63**, 729 and 2030.

Brenier A., Pedrini C. and Moncorgé R. (1990), *Opt. Quant. Elect.*, **22**, S153.

Bridgman P. W. (1925), *Proc. Am. Acad. Arts Sci.*, **60**, 303.

Brinkmann R., Sohler W. and Suche H. (1991), *Elect. Lett.*, **27**, 417.

Brown D. (1997), *IEEE J. Quant. Elect.*, **33**, 861.

Bruekner K. A. (1955), *Phys. Rev.*, **97**, 1353 and **100**, 36.

Buchenauer C. J. and Fitchen D. B. (1968), *Phys. Rev.*, **167**, 846.

Burns R. G. (1993), *Mineralogical Application of Crystal Field Theory*, 2nd Edition, Cambridge University Press, Cambridge.

Burns R. G. and Burns V. M. (1984), *Adv. Ceramics*, **10**, 46.

Burshtein A. L. (1972), *Soviet Physics JETP*, **31**, 882.

Burshtein A. L. (1985), *J. Lumin.*, **34**, 167.

Butcher P. N. and Cotter D. (1990), *The Elements of Nonlinear Optics*, Cambridge University Press, Cambridge.

Burt M. G. (1983), *J. Phys. C: Sol. St. Phys.*, **16**, 4137.

Byer R. L., Young J. F. and Feigelson R. S. (1970), *J. App. Phys.*, **41**, 2320.

Byrappa K. (1994) in *Handbook of Crystal Growth–Bulk Crystal Growth* (Ed. Hurle D. T. J.) Elsevier Science BV, North-Holland, Amsterdam, **2A**, 467.

Caird J. A. (1986), in *Tunable Solid State Lasers II* (Eds. Bugdor A. B., Esterowitz L. and DeShazer L. G.) Springer-Verlag, Berlin and Heidelberg, *Springer Series in Optical Sciences*, **52**, 20.

Caird J. A. and Payne S. A. (1991), *Handbook of Laser Science and Technology*, Supp. **1** Lasers (Ed. Weber M.J.) Boca Raton, Florida, 3.

Caird J. A., DeShazer L. G. and Nella J. (1975), *IEEE J. Quant. Elect.*, **11**, 874.

Caird J. A., Payne S. A., Staver P. R., Ramponi A. J., Chase L. L. and Krupke W. F. (1988), *IEEE J. Quant. Elect.*, **25**, 1077.

Callaway J. and March N. H. (1984), Density functional methods; theory and applications. In *Solid State Physics*, **38** (Eds. Ehrenreich H., Turnbull D. and Seitz F.), Academic Press, Orlando.

Car R. and Parrinello M. (1985), *Phys. Rev. Lett.*, **55**, 2471.

Carnall W. T., Goodman G. L., Rajnak K. and Rana R. S. (1988), *J. Chem. Phys.*, **90**, 3443 (with supplementary tables AIP/PAPS, JCPSA–90–3443–63) list energy levels of all RE3+ ions in LaF3.

Caspers H. H., Rast H. E. and Fry J. L. (1970), *J. Chem. Phys.*, **53**, 3208.

Chai B. H. T., Lefaucheur J-L., Stalder M. and Bass M. (1992), *Opt. Lett.*, **17**, 1584.

Chai B., Lefaucheur J., Pham A., Lutts G. and Nicholls J., (1993), *SPIE Proc.*, **1863**, 21 and 136.

Chartier I., Ferrand B., Pelenc D., Field S. J., Hanna D. C., Large A. C., Shepherd D. P. and Tropper A. C. (1992), *Opt. Lett.*, **17**, 810.

Chen C.T. (1993), *Laser Sci. Tech.*, **15**, 1.

Chen C. T., Wu Y. and Li R. (1990), *J. Cryst. Growth*, **99**, 790.

Chicklis E. P., Naiman C. S., Folweiler R. C., Gabbé D. R., Jenssen H. P. and Linz A. (1971), *App. Phys. Lett.*, **19**, 119.

Christiansen P. A., Lee Y. S., and Pitzer K. S. (1979), *J. Chem. Phys.*, **71**, 4445.

Chuang T. and Verdun H. R. (1994), *OSA Proc. Adv. Sol. State Lasers*, **20**, 77.

Cockayne B. (1977), *J. Cryst. Growth*, **42**, 413.

Cockayne B., Robertson D. S. and Bardsley W. (1964), *Brit. J. App. Phys.*, **15**, 1165.
Cockayne B., Chesswas M. and Gasson D. B. (1967), *J. Mater. Sci.*, **2**, 7.
Cockayne B., Plant J. G. and Clay R. A. (1981), *J. Cryst. Growth*, **54**, 407.
Cohen M. H. and Heine V. (1961), *Phys. Rev.*, **122**, 1821.
Condon E. U. (1926), *Phys. Rev.*, **28**, 1182 and (1928), *ibid*, **32**, 858.
Condon E. U. and Odabasi H. (1980), *Atomic Structure*, Cambridge University Press, Cambridge.
Condon E. U. and Shortley G. H. (1935), *The Theory of Atomic Spectra*, Cambridge University Press, Cambridge.
Cornwell J. F. (1984), *Group Theory in Physics*, Academic Press, New York and London.
Copley J. R. D., MacPherson R. W. and Timusk T. (1969), *Phys. Rev.*, **182**, 965.
Cottrell A. H. (1953), *Theoretical and Structural Metallurgy*, Edward Arnold, London.
Cottrell A. H. (1964), *The Mechanical Properties of Matter*, John Wiley, New York.
Czochralski J. (1917), *Zeit. Phys. Chem.*, **92**, 219.
Danielmeyer H. G. (1975), *Festkopferpröbleme*, **XV**, 253.
Davidson E. R. (1991), *MELD*, QCPE 23, Program No. 580.
Dawson R and Pooley D. (1969), *Phys. Stat. Sol.*, **35**, 95.
Dean P. J. and Choyke W. J. (1977), *Adv. Phys.*, **26**, 1.
Deka C., Chai B. H. T., Shimony Y., Zhang X. X., Munin E. and Bass M. (1992), *App. Phys. Lett.*, **61**, 2141.
DeLeo G. G., Kern R. C. and Bartram R. H. (1981), *Phys. Rev. B*, **24**, 2222.
DeLoach L. D., Payne S. A., Smith L. K., Kway W. L. and Krupke W. F. (1994), *J. Opt. Soc. Am. B*, **11**, 264.
DeLoach L. D., Page R. H., Wilke G. D., Payne S. A. and Krupke W. F. (1996), *IEEE J. Quant. Elect.*, **32**, 885.
Denner V. and Wagner M. (1984), *J. Phys. C: Sol. St. Phys.*, **17**, 153.
DeShazer L. G. (1994), *Laser Focus* (February issue).
De Vita A., Gillan M. J., Lin J. S., Payne M. C., Stich I. and Clarke L. J. (1992), *Phys. Rev. Lett.*, **68**, 3319.
De Yoreo J. J., Atherton L. J. and Roberts D. H. (1991), *J. Cryst. Growth*, **131**, 691.
Dexter D. L. (1953), *J. Chem. Phys.*, **21**, 836.
Dexter D. L., Klick C. C. and Russell G. A. (1955), *Phys. Rev.*, **100**, 63.
DiBartolo B. (1968), *Optical Interactions in Solids*, John Wiley, New York.
DiBartolo B. (1984), in *Energy Transfer Processes in Condensed Matter* (Ed. DiBartolo B.) Plenum Press, New York, p. 103.
Dick B. G. and Overhauser W. A. (1958), *Phys. Rev.*, **112**, 90.
Dieke G. H. (1968), *Spectra and Energy Levels of Rare Earth Ions in Crystals*, John Wiley, New York, London, Sydney and Toronto.
Dieke G. H. and Pandey B. (1964), *J. Chem. Phys.*, **41**, 1952.
Digonet M. J. F. (Ed.) (1993), *Rare Earth Doped Fibre Lasers and Amplifiers*, Marcel Dekker, New York.
Dirac P. A. M. (1930), *Proc. Cam. Philos. Soc.*, **26**, 376.
Dirac P. A. M. (1958), *The Principles of Quantum Mechanics*, 4th Ed., Oxford University Press, Oxford.
Dolan J. F., Kappers L. A. and Bartram R. H. (1986), *Phys. Rev. B*, **33**, 7339.
Dolan J. F., Rinzler A. G., Kappers L. A. and Bartram R. H. (1992), *J. Phys. Chem. Solids*, **53**, 905.
Donnelly C. J., Healy S. M., Glynn T. J., Imbusch G. F. and Morgan G. P. (1988), *J. Lumin.*, **42**, 117.
Donnerberg H. and Bartram R. H. (1996), *J. Phys.: Cond. Matter.*, **8**, 1687.
Dorozhkhin L. M., Kuratov I. I., Leonyuk N. I., Timuchenko T. I. and Shestakov A. V. (1981), *Sov. Tech. Phys. Lett.*, **7**, 555.
Drube J., Struve B. and Huber G. (1984), *Opt. Comm.*, **50**, 45.

Dubicki L., Ferguson J. and von Oosterhout B. (1980), *J. Phys. C.*, **13**, 2791.
Dubinskii M. A., Semashko V. V., Naumov A. K., Abdulsabirov R. Yu. and Korobleva S. L. (1993), *J. Mod. Opt.*, **40**, 1, *Laser Physics*, **3**, 216.
Duczynski E. W., Huber G. and Mizscherlich P. (1986a), in *Tunable Solid State Lasers II* (Eds. Bugdor A. B., Esterowitz L. and DeShazer L. G.) Springer-Verlag, Berlin and Heidelberg, Springer Series in Optical Sciences, **52**, 282.
Duczynski E. W., Huber G., Ostroumov V. G. and Shecherbakov I. A. (1986b), *App. Phys. Lett.*, **48**, 1562.
Dunning T. H. (1970), *J. Chem. Phys.*, **53**, 2823.
Dürr U. and Brauch U. (1986), in *Tunable Solid State Lasers II* (Eds. Bugdor A. B., Esterowitz L. and DeShazer L.G.) Springer-Verlag, Berlin and Heidelberg, Springer Series in Optical Sciences, **52**, 15.
Dürr U., Brauch U., Knierim W. and Schiller C. (1985), in *Tunable Solid State Lasers* (Eds. Hamerling P., Bugdor A. B., Pinto A.) Springer-Verlag, Heidelberg, Springer Series in Optical Sciences, **47**, 20.
Dymott M. J. P. and Ferguson A. P. (1995), *Opt. Lett.*, **20**, 1157.
Eckart C. (1930), *Rev. Mod. Phys.*, **2**, 305.
Eichler H. J., Liu B., Lu Z. and Kaminskii A. A. (1994), *App. Phys. B.*, **58**, 421.
Eilers H., Dennis W. M., Yen W. M., Kück S., Petermann K., Huber G. and Jia W. (1993), *IEEE J. Quant. Elect.*, **29**, 2508.
Eilers H., Hömerich U., Jacobsen S. M., Yen W. M., Hoffman K. R. and Jia W. (1994), *Phys. Rev. B.*, **49**, 15505.
Einstein A. (1917), *Physik. Zeits.*, **18**, 121.
Elwell D., Kway W. L. and Feigelson R. S. (1985), *J. Cryst. Growth*, **71**, 237.
Englman R. (1972), *The Jahn-Teller Effect in Molecules and Crystals*, John Wiley, London, New York, Sydney, Toronto.
Englman R. (1979), *Non-Radiative Decay of Ions and Molecules in Solids*, North Holland, Amsterdam.
Englman R. and Barnett B. (1970), *J. Lumin*, **3**, 37 and 55.
Englman R. and Jortner J. (1970), *Molec. Phys.*, **18**, 145.
Englman R. and Zgierski M. Z. (1985), *Chem. Phys.*, **94**, 187.
Erbil A. and Jenssen H. (1980), *App. Opt.*, **19**, 1729.
Eremenko V. V. and Petrov E. G. (1977), *Adv. Phys.*, **26**, 31.
Erlich D. J., Moulton P. F. and Osgood R. M. (1979), *Opt. Lett.*, **4**, 184 and (1980) *Opt. Lett.*, **5**, 339.
Ermler W. C., Ross R. B. and Christiansen P. A. (1988), *Adv. Quant. Chem.*, **19**, 139.
Esterowitz L., Allen R., Kruer M., Bartoli F., Goldberg L. S., Jenssen H. P., Linz A. and Nicolai V. O., (1977), *J. App. Phys.*, **48**, 650.
Esterowitz L., Allen R., Goldberg L., Weller J. F., Storm M. and Abella I. (1986), in *Tunable Solid State Laser II* (Eds. Bugdor A. B., Esterowitz L. and DeShazer L. G.) Springer-Verlag, Heidelberg and Berlin, Springer Series in Optical Sciences, **52**, 291.
Fahey R. E., Strauss A. J., Sanchez A. and Aggarwal R. L. (1986), in *Tunable Solid State Lasers II* (Eds. Bugdor A.B., Esterowitz L. and DeShazer L.G.) Springer-Verlag, Heidelberg and Berlin, Springer Series in Optical Sciences, **52**, 82.
Fairbank Jr W. M., Klauminzer G. K. and Schawlow A. L. (1975), *Phys. Rev. B.*, **11**, 60, and references therein.
Falcoz F., Balembois F., Georges P., Brun A. and Rytz D. (1995), *Opt. Lett.*, **20**, 1274.
Fan T . Y. and Byer R. L. (1986), *J. Opt. Soc. Amer. B.*, **3**, 1519.
Fan T. Y. and Byer R. L. (1988), *IEEE J. Quant. Elect.*, QE-**24**, 895.
Fan T. Y., Dixon G. D. and Byer R. L. (1986), *Opt. Lett.*, **11**, 204.
Farrell E. H., Fang J. H. and Newnham R. E. (1963), *Amer. Mineral.*, **48**, 9163.
Faulstich A., Baker H. J. and Hall D. R. (1996), *Opt. Lett.*, **21**, 594.
Feigelson R. S. (1986), *J. Cryst. Growth*, **79**, 669.

Fejer M. M., Nightingale J. L., Majel G. A. and Byer R. L. (1984), *Rev. Sci. Ins.*, **55**, 1791, see also Stone J. and Burrus C. A. (1979), *Fiber and Integ. Opt.*, **2**, 19.

Ferguson A. I. (1994), in *Nonlinear Spectroscopy of Solids* (NATO ASI Series B, Ed. DiBartolo B.) Plenum, New York, p. 225.

Ferguson J., Guggenheim H. J. and Wood D. L. (1971), *J. Chem. Phys.*, **54**, 504.

Feshbach H. (1962), *Ann. Phys.*, **19**, 287.

Field S. J., Hanna D. C., Large A. C., Shepherd D. P., Tropper A. C., Chandler P. J., Townsend P. D. and Zhang I. (1991), *Elect. Lett.*, **27**, 2375.

Fischer C. F. (1977), *The Hartree-Fock Method for Atoms*, John Wiley, New York, London, Sydney and Toronto.

Fisher A. J. (1991), *Rev. Solid State Sci.*, **5**, 107.

Fleischer M., Chao G. Y. and Francis C. A. (1981), *Amer. Mineralogist*, **66**, 868.

Fockele M., Ahlers F. J., Lohse F., Spaeth J.-M. and Bartram R. H. (1985), *J. Phys. C: Sol. St. Phys.*, **18**, 1963.

Foldy L. L. and Wouthuysen S. A. (1950), *Phys. Rev.*, **78**, 29.

Forbes C. E. (1983), *J. Chem. Phys.*, **79**, 2590.

Förster T. (1948), *Ann. Phys. (Leipzig)*, **2**, 55.

Fowler W. B. (Ed.) (1968a), *Physics of Color Centers*, Academic Press, New York and London provides a survey of colour centre physics in its pomp.

Fowler W. B. (1968b), in *Physics of Color Centers* (Ed. Fowler W. B.) Academic Press, New York.

Franck J. (1925), *Trans. Faraday Soc.*, **21**, 536.

Franken P. A., Hill A. E., Peters C. W. and Weinreich G. (1961), *Phys. Rev. Lett.*, **7**, 118.

Freeman A. J. and Watson R. E. (1965), in *Treatise on Magnetism* (Eds. Rado G. and Suhl H.) Academic Press, New York and London.

Frisch M. J. and thirty-four co-authors (1995), *Gaussian 94*, Gaussian Inc., Pittsburgh.

Frobenius G. and Schur I. (1906), *Berliner Belichte*, p. 186.

Fujii Y. and Sakudo T. (1976), *Phys. Rev. B.*, **13**, 1161.

Funk D. S., Carlson J. W. and Eden J. G. (1994), *Elect. Lett.*, **30**, 1859 and *Opt. Lett.*, **20**, 1474.

Gao Y., Yamaga M., Henderson B. and O'Donnell K. P. (1993), *Chem. Phys. Lett.*, **210**, 67.

Gächter B. F. and Koningstein J. A. (1974), *J. Chem. Phys.*, **60**, 2003.

Gasson D. B. and Cockayne B. (1970), *J. Mat. Sci.*, **5**, 100.

Gellermann W., Koch K. P. and Lüty F. (1981a), *Laser Focus*, (April), 71.

Gellermann W., Pollock C. and Lüty F. (1981b), *Opt. Comm.*, **39**, 391.

Gell-Mann M. and Brueckner K. A. (1957), *Phys. Rev.*, **106**, 364.

Gerloch M. and Slade R. C. (1973), *Ligand Field Analysis*, Cambridge University Press, Cambridge.

German K. (1979), *Opt. Lett.*, **4**, 68.

German K. (1986), *J. Opt. Soc. Am.*, **B3**, 149.

Geusic J. E., Marcos H. M. and Van Uitert L. B. (1964), *App. Phys. Lett.*, **4**, 182.

Giberson K. W., Cheng C., Dunning F. B. and Tittel F. K. (1982), *Appl. Opt.*, **21**, 172.

Giesecke, P., von der Osten W., Röder, U. (1972), *Phys. Stat. Solidi B*, **51**, 723.

Goh S. C., Pattie R., Byrne C. and Coulson D. (1995), *App. Phys. Lett.*, **67**, 768.

Goldstone J. (1957), *Proc. Roy. Soc. (London)*, **A239**, 267.

Goovaerts E., Andriessen J., Nistor S.V. and Schoemaker D. (1981), *Phys. Rev. B.*, **24**, 29.

Gordon J. P., Zeiger H. J. and Townes C. H. (1955), *Phys. Rev.*, **18**, 1264.

Goryunov A. V., Popov A. L., Khaidukov N. M. and Federov P. P. (1992), *Mat. Res. Bull.*, **27**, 213.

Gourary B. S. and Adrian F. J. (1957), *Phys. Rev.*, **105**, 1180.

Gourary B. S. and Fein A. E. (1962), *J. Appl. Phys. Suppl.*, **33**, 331.

Greenhough P. and Paulusz A. G. (1979), *J. Chem. Phys.*, **70**, 1967.

Griffith J. S. (1961), *The Theory of Transition-Metal Ions*, Cambridge University Press, Cambridge.

Grinberg M. (1993), *J. Lumin.*, **54**, 369.

Grinberg M., Macfarlane P. I., Henderson B. and Holliday K. (1995), *Phys. Rev. B.*, **52**, 3917.
Grinberg M., Jaskolski W., Macfarlane P. I., Henderson B. and Holliday K. (1997), *J. Lumin.*, **72–74**, 193.
Gryk T. J. and Bartram R. H. (1995), *J. Phys. Chem. Solids*, **56**, 863.
Guggenheim H. J. (1961), *J. App. Phys.*, **32**, 1337.
Günter P., Askbeck P. M. and Kurtz S. K. (1979), *App. Phys. Lett.*, **35**, 461.
Gurney R. W. and Mott N. F. (1939), *Trans. Faraday Soc.*, **35**, 69.
Gutsche E. (1982), *Phys. Stat. Sol. (b)*, **109**, 583.
Hahn T. (Ed.) (1995), *International Tables for Crystallography, Vol. A: Space Group Symmetry.* 4th Edn., Kluwer Academic Publishers, Dordecht, Boston and London.
Hall C. (1994), *Gemstones*, Dorling Kindersley, London.
Ham F. S. (1965) *Phys. Rev.*, **138A**, 1727 and (1968), *Phys. Rev.*, **166**, 307.
Ham F. S. (1972), in *Electron Paramagnetic Resonance* (Ed. Geschwind S.) Plenum Press, New York.
Ham F. S. and Grevsmühl U. (1973), *Phys. Rev B*, **8**, 2945.
Hamann D. R., Schlüter M. and Chiang C. (1979), *Phys. Rev. Lett.*, **43**, 1494.
Hamermesh M. (1962), *Group Theory*, John Wiley, New York.
Hamilton D. S., Seltzer P. M. and Yen W. M. (1977), *Phys. Rev.* **B 16**, 1858.
Hamilton D. S., Gayen S. K., Pogatschnik G. T., Chen R. D. and Miniscalco W. J. (1989) *Phys. Rev B*, **39**, 8807.
Hamilton D. W. (1985), in *Tunable Solid State Lasers* (Eds. Hammerling P., Bugdor A. B. and Pinto A.) Springer-Verlag, Berlin and Heidelberg, Springer Series in Optical Sciences, **47**, 80.
Han T. P. J., Scott M. A., Jaqué F., Gallagher H. G. and Henderson B. (1993), *Chem. Phys. Lett.*, **208**, 63.
Hanna D. C., Large A. C., Shepherd D. P., Tropper A. C., Chartier I., Ferrand B. and Pelenc D. (1993), *App. Phys. Lett.*, **63**, 7.
Hanna D. C., Large A. C., Shepherd D. P., Tropper A. C., Chartier I., Ferrand B. and Pelenc D. (1995), *Opt. Comm.*, **91**, 229.
Hansom F. and Poirier P. (1995), *J. Opt. Soc. Am.*, **B12**, 1311.
Harding J. H., Harker A. H., Keegstra P. B., Pandey R., Vail J. M. and Woodward C. (1985), *Physica*, **131B**, 151.
Härig T., Huber G. and Shcherbakov I. A. (1981), *J. Appl. Phys.*, **52**, 4450.
Hartree D. R. (1957), *The Calculation of Atomic Structures*, John Wiley, New York.
Hasegawa A. (1989) *Solitons in Optical Fibers*, Springer-Verlag, Berlin.
Hay P. J. and Wadt W. R. (1985), *J. Chem. Phys.*, **82**, 270 and 299.
Hazenkamp M. F. and Güdel H. U. (1996), *J. Lumin.*, **69**, 235.
Hazenkamp M. F., Güdel H. U., Atanosov M., Kesper U. and Reinen D. (1996), *Phys. Rev. B.*, **53**, 2367.
Hebert T., Wannemacher R., Lenth W. and Macfarlane R. M. (1990), *App. Phys. Lett.*, **57**, 1727.
Hebert T., Wannemacher R., Macfarlane R. M. and Lenth W. (1992), *Appl. Phys. Lett.*, **60**, 2592.
Hegarty J. (1976), *Ph.D. Thesis*, University College Galway (Unpublished).
Hehir J. P., Henry M. O., Larkin J. P. and Imbusch G. F. (1974), *J. Phys. C.: Sol. St. Phys.*, **7**, 2241, see also Henry M. O., Larkin J. P. and Imbusch G. F. (1975), *Proc. Roy. Irish Acad.*, **75**, 97.
Hehre W. J., Stewart R. F. and Pople J. A. (1969), *J. Chem. Phys.*, **51**, 2657.
Hehre W. J., Radom L., Schleyer P. v. R. and Pople J. A. (1986), *Ab Initio Molecular Orbital Theory*, John Wiley, New York.
Heine F., Heumann E., Hüber G. and Schepler K. L. (1992), *App. Phys. Lett.*, **60**, 1161.
Heine F., Heumann E., Koetke J., Danger T., Hüber G. and Chai B. H. T. (1994), *App. Phys. Lett.*, **65**, 353.
Heitler W. and London F. (1927), *Z. Physik*, **44**, 45.
Helmis G. (1965), *Ann. Phys. Leipz.*, **19**, 41.

Henderson B. and Imbusch G. F. (1989), *Optical Spectroscopy of Inorganic Crystals*, Oxford University Press, Oxford.

Henderson B., Marshall A., Yamaga M., O'Donnell K. P. and Cockayne B. (1988), *J. Phys. C: Sol. St. Phys.*, **21**, 6187.

Henderson B., Yamaga M. and O'Donnell K. P. (1990), *Opt. Quan. Elect.*, **22**, S167.

Henderson B., Yamaga M., Gao Y. and O'Donnell K. P. (1992), *Phys. Rev. B.*, **46**, 65.

Herman F. and Skillman S. (1963), *Atomic Structure Calculations*, Prentice-Hall, New Jersey.

Herman R. C., Wallis M. C. and Wallis R. F. (1956), *Phys. Rev.*, **103**, 87.

Hirai M. and Wakita S. (1983), *Semicond. Insul.*, **5**, 231.

Hoffman D. M., Lohse F., Paus H. J., Smith D. Y. and Spaeth J.-M. (1985), *J. Phys. C.*, **18**, 443.

Hoffman K. R., Casas-Conzalez J., Jacobsen S. M. and Yen W. M. (1991), *Phys. Rev. B.*, **44**, 12589.

Hoffman R. (1963), *J. Chem. Phys.*, **39**, 1397.

Hohenberg P. and Kohn W. (1964), *Phys. Rev.* **B136**, 864.

Holliday K., Russell D. L., Nicholls J. F. H., Henderson B., Yamaga M. and Yosida T. (1998), *App. Phys. Lett.*, **72**, 2232.

Holstein T., Lyo S. K., and Orbach R. (1981), *Laser Spectroscopy of Solids* (Eds. Yen W. M. and Seltzer P. M.) Springer-Verlag, Berlin, Heidelberg and New York, Springer Topics in Applied Physics, **49**, 39.

Hömmerich U. and Bray K. L. (1995), *Phys. Rev.* **B51**, 12133.

Honda S. and Tomura M. (1972), *J. Phys. Soc. Japan*, **33**, 1003.

Honea E. C., Beach R. J., Sutton S. B., Speth J. A., Mitchell S. C., Skidmore J. A., Emmanuel M. A. and Payne S. A. (1997), *IEEE J. Quant. Elect.*, **33**, 1592.

Huang K. (1981), *Sci. Sinica*, **24**, 27.

Huang K. and Rhys A. (1950), *Proc. Roy. Soc. London*, **A204**, 406.

Huber D. L. (1979), *Phys. Rev.* **B20**, 2037.

Huber D. L. (1981), *Laser Spectroscopy of Solids* (Eds. Yen W. M. and Seltzer P.M.) Springer-Verlag, Berlin, Heidelberg, and New York, Springer Topics in Applied Physics, **49**, 83.

Huber G. (1980), in *Current Topics in Materials Science* (Ed. Kaldis E.) North Holland, Amsterdam, **4**, 1.

Huber G. (1984), in private communication to Moulton (1985).

Huber G., Duczynski E. W. and Petermann K. (1988), *IEEE J. Quant. Elect.*, **QE-24**, 920.

Hückel E. (1931), *Z. Physik*, **70**, 204.

Hüfner S. (1978), *Optical Spectra of Transparent Rare Earth Compounds*, Academic Press, New York, San Francisco and London.

Hughes A. E. and Runciman W. A. (1965), *Proc. Phys. Soc. Lond.*, **86**, 615.

Hund F. (1927a), *Linienspectren und Periodisches System der Elemente*, Springer-Verlag, Berlin.

Hund F. (1927b), *Z. Physik*, **40**, 742.

Hunt W. J., Hay P. J. and Goddard (III) W. A. (1973), *J. Chem. Phys.*, **57**, 738.

Hurle D. T. J. (Editor) (1993–97) *Handbook of Crystal Growth*, Volumes **1–6**, North-Holland, Amsterdam, London, New York and Tokyo.

Hurle D. T. J. (1977), *J. Cryst. Growth*, **42**, 473.

Hurle D. T. J. and Cockayne B. (1994), *Handbook of Crystal Growth – Bulk Crystal Growth* (Ed. Hurle D. T. J.) North-Holland, Amsterdam and London, **2A**, p. 99.

Hurley A. C., Lennard-Jones J. E. and Pople J. A. (1953), *Proc. Roy. Soc. (London)*, **A220**, 446.

Hurley A. C. (1976), *Introduction to the Electron Theory of Small Molecules*, Academic Press, London, New York and San Francisco.

Hutchinson J. A. and Allik T. H. (1992), *App. Phys. Lett.*, **60**, 1424.

Huzinaga S. (1965), *J. Chem. Phys.*, **42**, 1293.

Imbusch G. F., Glynn T. J. and Morgan G. P. (1990), *J. Lumin.*, **45**, 63.

Inokuti M. and Hirayama F. (1965), *J. Chem. Phys.*, **43**, 1978.

Iverson M. and Sibley W. A. (1979), *J. Lumin.*, **20**, 311 (but see also Iverson M., Windscheif J. C. and Sibley W. A. (1979), *App. Phys. Lett.*, **36**, 183.)

Ivey H. V. (1947), *Phys. Rev.*, **72**, 341.
Iwai M., Mori Y., Sasaki T., Nakai S., Sarukura N., Liu Z. and Segawa Y. (1995), *Jap. J. Appl. Phys.*, **34**, 2338.
Jacobsen S. M., Smith W. E., Reber C. and Güdel H. U. (1986), *J. Chem. Phys.*, **84**, 5205.
Jahn H. A. and Teller E. (1937), *Proc. Roy. Soc.*, **A161**, 220.
Jenssen H. P., Linz A., Leavitt R. P., Morrison C. A. and Wortman D. E. (1975), *Phys. Rev. B.*, **11**, 92.
Jewell J. L., Harbison J. P., Scherer A., Lee Y. H. and Florez L. T. (1991), *IEEE J. Quant. Elect.*, **27**, 1332.
Jia W., Lizhu L., Tissue B. M. and Yen W. M. (1991) *J. Cryst. Growth*, **109**, 329.
Jiang Q., Han T. P. J. and Gallagher H. G. (1998), *Journ. Mat. Sci: Mat. in Elect.*, **9**, 193.
Johnson K. H. (1973), *Adv. Quant. Chem.*, **7**, 143.
Johnson L. F. and Guggenheim H. J. (1971), *App. Phys. Lett.*, **19**, 44.
Johnson L. F., Boyd G. D., Nassau K. and Soden R. R. (1962), *Phys. Rev.*, **126**, 1406.
Johnson L. F., Dietz R. E. and Guggenheim H. G. (1963), *Phys. Rev. Lett.*, **11**, 318.
Johnson L. F., Dietz R. E. and Guggenheim H. G. (1964), *App. Phys. Lett.*, **5**, 2.
Johnson L. F., Geusic J. E. and Van Uitert L. G. (1965), *Appl. Phys. Lett.*, **7**, 127.
Johnson L. F., Dietz R. E., Guggenheim H. J. and Thomas R. A. (1966a), *Phys. Rev.*, **149**, 179.
Johnson L. F., Geusic J. E. and Van Uitert L. B. (1966b), *Appl. Phys. Lett.*, **8**, 200.
Johnson L. F., Guggenheim H. J. and Bahnck D. (1983), *Opt. Lett.*, **8**, 371.
Joosen W., Goovaerts E. and Schoemaker D. (1985), *Phys. Rev.* **B32**, 6748.
Jørgensen C. K. (1971), *Modern Aspects of Ligand Field Theory*, North Holland, Amsterdam and London.
Jørgensen C. K. (1987), *Absorption Spectra and Chemical Bonding in Complexes*, Pergamon Press, Oxford.
Jortner J. (1979), *Phil. Mag.*, **B40**, 317.
Judd B. R. (1962), *Phys. Rev.*, **127**, 750
Judd B. R. (1963), *Operator Techniques in Atomic Spectroscopy*, McGraw Hill, New York, San Francisco, Toronto and London.
Kahn L. R., Hay P. J. and Cowan R. D. (1978), *J. Chem. Phys.*, **68**, 2386.
Kaminskii A. A. (1989), *Laser Crystals*, Springer Series in Optical Sciences, 2nd Edition.
Kaminskii A. A. (1991), *Phys. Stat. Sol. (a)*, **125**, K109 and references therein.
Kaminskii A. A. (1996), *Crystalline Lasers: Physical Processes and Operating Schemes*, CRC Press, Boca Raton, New York, London and Tokyo.
Kaminskii A. A. and Khaidukov N. M. (1992), *Phys. Stat. Sol.*, **129**, K65.
Kaminskii A. A., Kurbanov K. and Uvarova T. V. (1987), *Inorg. Mat.*, **23**, 940.
Kaminskii A. A., Butashin A. V., Demidovich A. A., Koptev V. G., Mill B. V. and Shkadarevich A. P. (1989), *Phys. Stat. Sol. (a)*, **112**, 197.
Kaminskii A. A., Petrosyan A. G., Markosyan A. A. and Shironyan G. O. (1991), *Phys. Stat. Sol. (a)*, **125**, 353.
Kaplyanskii A. A. (1959), *Optics Spectrosc. USSR*, **6**, 267, and (1964), *ibid*, **16**, 329 and 557.
Katoh Y., Sugimoto N. and Shibukawa A. (1992), *Jap. J. App. Phys.*, **13**, 3888.
Kayanuma Y. (1982), *J. Phys. Soc. Japan*, **51**, 3526.
Kayanuma Y. and Nasu K. (1978), *Solid St. Comm.*, **27**, 1371.
Keil T. H. (1965), *Phys. Rev.*, **140**, A601.
Kern R. C., DeLeo G. G. and Bartram R. H. (1981), *Phys. Rev.* **B24**, 2211.
Klauminzer G. K., Scott P. L. and Moos H. W. (1966), *Phys. Rev.*, **142**, 248.
Knochenmüss R., Reber C., Rajaskharan M. V. and Güdel H. U. (1986), *J. Chem. Phys.*, **85**, 4280.
Koch M. E., Kueny A. W. and Case W. E. (1990), *App. Phys. Lett.*, **56**, 1083.
Kodama N., Yamaga M. and Henderson B. (1998), *J. App. Phys.*, **84**, 5820.

Koechner W. (1976), *Solid-State Laser Engineering*, Springer Series in Opt. Sci. **1**, Springer-Verlag, Heidelberg and Berlin.
Koetke J., Petermann K. and Huber G. (1991), *J. Lumin.*, **48&49**, 564.
Koetke J., Petermann K. and Huber G. (1993a), *J. Lumin.*, **60&61**, 197.
Koetke J., Kück K., Petermann K., Huber G., Cerullo G., Denailev M., Magni V., Qian L. F. and Svelto O. (1993b), *Opt. Comm.*, **101**, 195.
Kohli M. and Huang-Liu N. L. (1974), *Phys. Rev.* **B9**, 1008.
Kohn W. (1957), in *Solid State Physics* (Eds. Seitz F. and Turnbull D.) Academic Press, New York and London, **5**, 258.
Kohn W. and Sham L. J. (1965), *Phys. Rev.*, **A140**, 1133.
Kokta M. R. (1985), in *Tunable Solid State Lasers* (Eds. Hammerling P., Bugdor A. B. and Pinto A.), Springer-Verlag, Berlin and Heidelberg, Springer Series in Optical Sciences, **47**, 105.
Kokta M. R. (1986), in *Tunable Solid State Lasers II* (Eds. Bugdor A. B., Esterowitz L. and DeShazer L. G.), Springer-Verlag, Berlin and Heidelberg, Springer Series in Optical Sciences, **52**, 89.
Kolbe W., Petermann K. and Huber G. (1985), *IEEE J. Quant. Elect.*, **21**, 1596.
Koster G. F. (1958), *Phys. Rev.*, **109**, 227–31.
Koster G. F., Dimmock J. O., Wheeler R. G. and Statz H. (1963), *Properties of the Thirty-Two Point Groups*, M.I.T. Press, Cambridge, MA.
Kozlovsky Y. C., Nabors C. D. and Byer R. L. (1988), *IEEE J. Quant. Elect.*, **24**, 913.
Kramers H. A. (1930), *Kominkl. Ned. Akad. Watenschap*, Proc. **33**, 959.
Krasatskuy N. J. (1983), *J. Appl. Phys.*, **54**, 1261.
Krumhansl J. A. and Schwartz N. (1953), *Phys. Rev.*, **89**, 1154.
Krupke W. F. (1966), *Phys. Rev.*, **145**, 325.
Krupke W. F. and Gruber J. B. (1964), *J. Chem. Phys.*, **41**, 1225.
Kubo R. (1952), *Phys. Rev.*, **86**, 929.
Kubo R. and Toyozawa Y. (1955), *Prog. Theor. Phys.*, **13**, 160.
Kueny A. W., Case W. E. and Koch M. E. (1989), *J. Opt. Soc. Am. B.*, **6**, 639.
Kück S., Petermann K., Pohlmann U., Schönhoff U. and Huber G. (1994), *Appl. Phys. B.*, **58**, 153.
Kück S., Petermann K., Pohlmann U. and Huber G. (1995), *Phys. Rev.* **B51**, 17323.
Kunz A. B. and Klein D. L. (1978), *Phys. Rev.* **B17**, 4614.
Kunz A. B. and Vail J. M. (1978), *Phys. Rev.* **B38**, 1058.
Kunzel W. and Dürr U. (1981), *Opt. Comm.*, **36**, 383.
Kunzel W. and Dürr U. (1982), *App. Phys. B.*, **28**, 233.
Kusunoki M. (1979), *Phys. Rev.* **B20**, 2512.
Kvapil J., Koselja M., Perner B., Skoda V., Kubelko J., Hamal K. and Kubacek V. (1988), *Czech. J. Phys.* **B38**, 237.
Kyropoulos S. (1926), *Zeit. Anorg. Chem.*, **154**, 308.
Lai S. T. (1987), *J. Opt. Soc. Am.*, **B4**, 1286.
Lai S. T. and Jenssen H. P. (1984), *IEEE and OSA Topical Meeting on Tunable Solid State Lasers*, Arlington, USA (1985) Technical Digest, FA71.
Lai S. T. and Shand M. L. (1983), *J. App. Phys.*, **54**, 5642.
Lai S. T., Chai B. H. T., Long M., Shinn M. D., Caird J. A., Marion J. E. and Staver P. R. (1986), in *Tunable Solid State Lasers II* (Eds. Bugdor A. B., Esterowitz L. and DeShazer L. G.) Springer-Verlag, Berlin and Heidelberg, Springer Series in Optical Sciences, **52**, 145.
Lai S. T., Chai B. H. T., Long M. and Shinn M. D. (1988), *IEEE J. Quant. Electron.*, **9**, 24.
Lallier E., Pocholle J. P., Papuchon M., Grezes-Besset C., Pelletier E., de Micheli M., Li M. J., He Q. and Ostrowsky D. B. (1989), *Elect. Lett.*, **25**, 1492.
Lallier E., Pocholle J. P., Papuchon M., Grezes-Besset C., Pelletier E., de Micheli M., Li M. J., He Q. and Ostrowsky D. B. (1990), *Elect. Lett.*, **26**, 927.
Landau L. (1932), *Phys. Z. Sowjetunion*, **1**, 88.

Lang D. V. (1982), *Ann. Rev. Mater. Sci.*, **12**, 377.

Lang D. V. and Henry C. H. (1975), *Phys. Rev. Lett.*, **35**, 1525.

Laudise R. A. (1970), *The Growth of Single Crystals*, Prentice-Hall, Englewood Cliffs, N.J.

Laudise R. A. (1991), *Am. Ass. Cryst. Growth*, **21**, 6.

Laudise R. A., Sunder W. A., Belt R. F. and Gashurov G. (1990), *J. Cryst. Growth*, **102**, 427.

Lax M. (1952), *J. Chem. Phys.*, **20**, 1752.

Lax M. (1974), *Symmetry Principles in Solid State and Molecular Physics*, John Wiley, New York, London, Sydney and Toronto.

Lëdig M., Heumann E., Ehrt D. and Seeber W. (1990), *Opt. and Quant. Elect.*, **22**, S107.

Lee H. W. H., Payne S. A. and Chase L. L. (1989), *Phys. Rev.* **B39**, 8907.

Lenth W. and Macfarlane R. M. (1990), *J. Lumin.*, **45**, 346.

Lenth W., Silversmith A. J. and Macfarlane R. M. (1988), *Advances in Laser Science III* (Eds. Tam A. C., Gole J. L. and Stwalley W. C.), *AIP Conference Proc.*, **172**, 8.

Leslie M. (1982), *Program CASCADE: Description of data sets for use in crystal defect calculations*, SERC Daresbury Rep., DL/SCI/TN31T.

Leung C. H. and Song K. S. (1980), *Sol. St. Commun.*, **33**, 907.

Li Y., Duncan I. and Morrow T. (1991), *J. Lumin.*, **50**, 69 and 333, (1992), *J. Lumin.*, **52**, 275.

Liehr A. D. (1963), *J. Phys. Chem.*, **67**, 389.

Lim K.-S. and Hamilton D. S. (1988), *J. Lumin.*, **40&41**, 319.

Lin J. T. (1990), *Opt. Quant. Elect.*, **22**, S283.

Lin S. H. and Bersohn R. (1968), *J. Chem. Phys.*, **48**, 2732.

Lisityn V. N., Matrosov N., Orekhova V. P., Pestryakov E. V., Sevast'yanov B. K., Trunov V. I., Zenin V. N. and Yu L. (1982), *Sov. J. Quant. Elect.*, **12**, 368.

Litfin G., Beigang R. and Welling H. (1977), *App. Phys. Lett.*, **31**, 381.

Longuet-Higgins H. C., Öpik U., Price M. H. L. and Sack R. A. (1958), *Proc. Roy. Soc. (London)*, **A244**, 1.

Loudon R. (1966), *Adv. Phys*, **17**, 243.

Loudon R. (1983), *The Quantum Theory of Light* (2nd Edn.) Oxford University Press, Oxford.

Löwdin P. O. (1955), *Phys. Rev.*, **97**, 1474 and 1509.

Löwdin P. O. (1956), *Ad. Phy.*, **5**, 96.

Luo Z. D., Jiang A., Huang Y. and Uiu M. (1989a), *Chin. Phys. Lett.*, **6**, 440.

Luo Z. D., Lin J. T., Jiang A. D., Huang Y. C. and Qui M. W. (1989b), *SPIE Proc.*, **1104**, 132.

Lutts G. B., Denisov A. L., Zharikov E. V., Zagumennyi A. I., Kozlikin S. N., Lavrishchev S. V. and Samoylova S. A. (1990), *Opt. Quant. Elect.*, **22**, S269.

Lüty F. (1968), in *Physics of Color Centers* (Ed. W.B. Fowler) Academic Press, New York and London, Chapter 3, 181.

Macfarlane P. I., Holliday K., Nicholls J. F. H. and Henderson B. (1995), *J. Phys. Cond. Matt.*, **7**, 9643, and references therein.

Macfarlane P. I., Henderson B., Holliday K. and Grinberg M. (1996), *J. Phys. Cond. Matt.*, **8**, 3933.

Macfarlane R. M. (1994a), in *Nonlinear Spectroscopy of Solids: Advances and Applications* (Ed. DiBartolo B.) Plenum Press, New York and London, p. 151.

Macfarlane R. M. (1994b), *J. de Physique IV, Supp. de J. de Physique III*, **4**, C4-289.

Macfarlane R. M., Wong J. Y. and Sturge M. D. (1968), *Phys. Rev.*, **166**, 250.

Macfarlane R. M., Tong F., Silversmith A. J. and Lenth W. (1988), *Appl. Phys. Lett.*, **52**, 1300.

Mackechnie C. J., Barnes W. L., Hanna D. C. and Townsend J. E. (1993), *Elect. Lett.*, **29**, 52.

McClure D. S. (1959), in *Solid State Physics* (Ed. Seitz F. and Turnbull D.) Academic Press, New York and London, **9**, 400.

McClure D. S. (1962), *J. Chem. Phys.*, **36**, 2757.

McConnell D. (1973), in *Apatite: Its Crystal Chemistry, Mineralogy, Utilization and Geological Occurrences*, Springer-Verlag, New York.

McKinnie I. T., Oien A. L., Warrington D. M., Tonga P. N., Gloster L. A. W. and King T. A. (1997), *IEEE J. Quant. Elect.*, **33**, 1221.

Maiman T. H. (1960), *Nature*, **187**, 493.

Maker G. T. and Ferguson A. I. (1990a), *Opt. Commun.*, **76**, 371.

Maker G. T. and Ferguson A. I. (1990b), *App. Phys. Lett.*, **56**, 1614.

Malcolm G. P. and Ferguson A. I. (1991a), *Cont. Phys.*, **32**, 305.

Malcolm G. P. and Ferguson A. I. (1991b), *Opt. Commun.*, **82**, 299.

Malcolm G. P. and Ferguson A. I. (1992), *Opt. Quant. Electron.*, **24**, 705.

Manna H. and Moncorgé R. (1990), *Opt. Quant. Electron.*, **22**, S219.

Manneback C. (1951), *Physica (Utrecht)*, **17**, 1001.

Maradudin, A. A., Montrol, E. W., Weiss, G. H. and Ipatova, I. P. (1971), *Theory of Lattice Dynamics in the Harmonic Approximation*, Academic Press, New York.

Marion J. (1985), *Appl. Phys. Lett.*, **47**, 694.

Markham J. J. (1966), *F-centers in Alkali Halides*, Academic Press, New York and London.

Marshall A., Henderson B., O'Donnell K. P., Yamaga M. and Cockayne B. (1990), *Appl. Phys. A.*, **50**, 565.

Marshall C. D., Speth J. A., Payne S. A., Krupke W. F., Quarles G. J., Castillo V. and Chai B. H. T. (1994), *J. Opt. Soc. Am. B.*, **11**, 2054.

Mazelsky R., Ohlmann R. C. and Steinbrugge K. B. (1968), *App. Opt.*, **7**, 905, see also (1968), *J. Electrochem. Soc.: Sol. State Sci.*, **115**, 69.

Mears R. J., Reekie L., Poole S. B. and Payne D. N. (1986), *Elect. Lett.*, **22**, 159.

Mears R. J., Reekie L., Jauncey I. M. and Payne D. N. (1987), *Elect. Lett.*, **23**, 1026.

Mehler F. G. (1866), *J. Math.*, **66**, 161.

Melcher C. L. and Schweitzer J. S. (1992), *Nuclear Instruments and Methods in Physics Research*, **A314**, 212.

Melius C. F. and Goddard (III) W. A. (1974), *Phys. Rev.*, **A 10**, 1528.

Mellish R., Chernikov S. V., French P. M. W. and Taylor J. R. (1998), *Elect. Lett.*, **34**, 552.

Merkle L. D., Allik T. H. and Chai B. H. T. (1992), *Opt. Mater.*, **1**, 91.

Merkle L. D., Guyot Y. and Chai B. H. T. (1995), *J. App. Phys.*, **77**, 474

Mermilliod N., Romero R., Chartier I., Garapon C. and Moncorgé R. (1992), *IEEE J. Quant. Elect.*, **28**, 1179.

Mills A.D. (1962) *Inorg. Chem.*, **1**, 960.

Moffitt W. and Liehr A. D. (1957), *Phys. Rev.*, **106**, 1195.

Moffitt W. and Thorson W. (1957), *Phys. Rev.*, **108**, 1251.

Mollenauer L. F. (1979), *Phys. Rev. Lett.*, **43**, 1524.

Mollenauer L. F. (1980), *Opt. Lett.*, **5**, 188.

Mollenauer L. F. (1981), *Opt. Lett.*, **6**, 342.

Mollenauer L. F. (1985a), in *The Laser Handbook* (Eds. Stitch M. and Bass M.) North Holland, Amsterdam, Chapter 3.

Mollenauer L. F. (1985b), *Phil. Trans. Roy. Soc.*, **A315**, 437.

Mollenauer L. F. (1987), in *Tunable Lasers* (Eds. Mollenauer L. F. and White J. C.) Springer-Verlag, Berlin, Heidelberg and New York, Springer Topics in Applied Physics, **59**, 225 and (1992) Second Edn. p. 225.

Mollenauer L. F. and Bloom D. M. (1979), *Opt. Lett.*, **4**, 247.

Mollenauer L. F. and Olson D. H. (1974), *J. Appl. Phys.*, **24**, 386.

Mollenauer L. F. and Stolen R. H. (1984), *Opt. Lett.*, **9**, 13.

Mollenauer L. F., Bloom D. M. and Guggenheim H. (1978), *App. Phys. Lett.*, **33**, 506.

Mollenauer L. F., Stolen R. H. and Gordon J. P. (1980), *Phys. Rev. Lett.*, **45**, 1095.

Mollenauer L. F., Vieira N. D. and Szeto L. (1982), *Opt. Lett.*, **7**, 414.

Mollenauer L. F., Stolen R. H., Gordon J. P. and Tomlinson W. J. (1983a), *Opt. Lett.*, **8**, 289.

Mollenauer L. F., Vieira N. D. and Szeto L. (1983b), *Phys. Rev.* **B27**, 5332.

Møller C. and Plesset M. S. (1934), *Phys. Rev.*, **46**, 618.

Monberg E. (1994) in *Handbook of Crystal Growth* (Ed. D. T. J. Hurle), North Holland, Amsterdam and London, **2A**, 53.

Moncorgé R. and Benyattou T. (1988), *Phys. Rev. B.*, **37**, 9177 and 9186.

Moncorgé R., Auzel F. and Bretau J. M. (1985), *Phil. Mag.*, **B51**, 489.
Moncorgé R., Cormier G., Simkin D. J. and Capabianco J. A. (1991), *IEEE J. Quant. Elect.*, **27**, 1144.
Monteil A., Garapon C. and Boulon G. (1988), *Proc. Electrochem. Soc.*, **88**, 72.
Monteil A., Nie W., Madej C. and Boulon G. (1990), *Opt. Quant. Electron.*, **22**, S247.
Moore C. E. (1952), *Atomic Energy Levels*, vol. II, NBS Circular No. **467**, U.S. Government Printing Office, Washington, D.C.
Moos H. W. (1970), *J. Lumin.*, **1**, 2, 106.
Morrison C. A. (1992), *Crystal Fields for Transition-Metal Ions in Laser Host Materials*, Springer-Verlag, Berlin and Heidelberg.
Mott N. F. (1938), *Proc. Roy. Soc. (Lond.)*, **A167**, 384.
Mott N. F. and Littleton M. J. (1938), *Trans. Faraday Soc.*, **34**, 485.
Moulton P. F. (1982a), *Opt. News*, November/December issue, p. 9.
Moulton P. F. (1982b), *IEEE J. Quant. Elect.*, **QE18**, 1185.
Moulton P. F. (1985) in *Laser Handbook* (Ed. Bass M. and Stitch M. L.) Elsevier, 203–850.
Moulton P. F. (1986a), *J. Opt. Soc. Am.*, **B3**, 125.
Moulton P. F. (1986b), *J. Opt. Soc. Am.*, **B.3**, 134 and references therein.
Moulton P. F. and Mooradian A. (1979a), *Appl. Phys. Lett.*, **35**, 127.
Moulton P. F. and Mooradian A. (1979b), *App. Phys. Lett.*, **35**, 838.
Moulton P. F., Manni J. G. and Rines G. A. (1988), *IEEE J. Quant. Elect.*, **24**, 960.
Mulliken R. S. (1928), *Phys. Rev.*, **32**, 186.
Mulliken R. S., Rieke C. A., Orloff D. and Orloff H. (1949), *J. Chem. Phys.*, **17**, 1248.
Muto K. and Awazu K., (1969), *Jap. J. App. Phys.*, **8**, 1360.
Myers L. E., Eckardt R. C., Fejer M. M. and Byer R. L. (1995), *Proc. SPIE*, **2379**, PD8.
Nakamura S. (1996), *Mat. Res. Soc. Symp. Proc.*, **395**, 879, see also Nakamura S., Senoh M., Iwasa N. and Nagahama S. (1995), *App. Phys. Lett.*, **67**, 1864.
Nassau K. (1961), *J. App. Phys.*, **32**, 1820.
Nassau K. (1971), in *Applied Solid State Science* (Ed. Wolfe R.) Academic Press, New York, Volume 2, p. 173.
Nassau K. (1983), *The Physics and Chemistry of Colour*, John Wiley, New York.
Nassau K. and Broyer A. M. (1964), *J. App. Phys.*, **33**, 3064.
Nasu K. and Kayanuma Y. (1978), *J. Phys. Soc. Japan*, **45**, 1341.
Nathan M. I., Dumke W. P., Burns G., Dill F. H. and Lasher G. J. (1962), *App. Phys. Lett.*, **1**, 92.
Neilsen J. W. and Monchamp R. R. (1970), in *Phase Diagrams* (Ed. Alper A. M.) Academic Press, New York, **3**, 1.
Nelson E. D., Wong J. Y. and Schawlow A. L. (1967), *Phys. Rev.*, **156**, 298.
Nesbet R. K. (1965), *Adv. Chem. Phys.*, **9**, 321.
Nicholls J. F. H., Zhang X. X., Bass M., Chai B. H. T. and Henderson B. (1997a), *Opt. Comm.*, **137**, 281 (but see also (1993) *SPIE Proceedings*, **1863**, 3).
Nicholls J. F. H., Henderson B. and Chai B. H. T. (1997b), *Optical Materials*, **8**, 215.
Nicholls J. F. H., Murray T. A., Russell D. L., Armstrong D., Chai B. H. T., Holliday K. and Henderson B. (1998), *Pure & Appl. Opt.*, **7**, 8.
Nie W., Boulon G. and Monteil A. (1989), *J. Phys. (Paris)*, **50**, 3309.
Nielson C. W. and Koster G. F. (1963), *Spectroscopic Coefficients for the* p^n, d^n and f^n Configurations, M.I.T. Press, Cambridge, MA.
Nijboer B. R. A. and De Wette F. W. (1957), *Physica*, **23**, 309.
Nikogosyan D. N. (1997), *Properties of Optical and Laser-Related Materials – A Handbook*, John Wiley, Chichester, New York.
Norgett M. J. (1974), *Harwell Report*, AERE-R 7650.
Nye J. F. (1985), *Physical Properties of Crystals* (2nd Edn) Oxford University Press, Oxford.
Ober M. H., Sorokin E., Sorokina I., Krausz F., Wintner E. and Shcherbakov I. A. (1992), *Opt. Lett.*, **17**, 1364.
O'Brien M. C. M. (1972), *J. Phys. C: Sol. St. Phys.*, **5**, 2045.

O'Donnell K. P., Marshall A., Yamaga M., Henderson B, and Cockayne B. (1989), *J. Lumin.*, **42**, 365.
Oetliker U., Riley M. J., May P. S. and Güdel H. U. (1992), *J. Lumin.*, **53**, 553.
Oetliker U., Herren M., Güdel H. U., Kesper U., Albrecht C. and Reinen D. (1994). *J. Chem. Phys.*, **100**, 8656.
Ofelt G. S. (1962), *J. Chem. Phys.*, **37**, 511.
Ohlmann R. C., Steinbruegge K. B. and Maselsky R. (1968), *J. App. Optics*, **7**, 905.
Orbach R. (1967), in *Optical Properties of Ions in Crystals* (Ed. Crosswhite H. M. and Moos H.W.) Interscience, New York, p. 307 and p. 445.
O'Rourke R. C. (1953), *Phys. Rev.*, **91**, 265.
Owen J. and Thornley J. H. M. (1966), *Rep. Prog. Phys.*, **29**, 675.
Page R. H., Schaffers K. I., DeLoach L. D., Wilke G. D., Patel F. D., Tassano J. B., Payne S. A., Krupke W. F., Chen K.-T. and Burger A. (1997), *IEEE J. Quant. Elect.*, **33**, 609.
Paschotta R., Nilsson J., Tropper A. C. and Hanna D. C. (1997), *IEEE J. Quant. Elect.*, **33**, 1049.
Pask H. M., Carman R. J., Hanna D. C., Tropper A. C., Mackechnie C. J., Barber P. R. and Dawes J. M. (1995), *IEEE Select Topics J. Quant. Elect.*, **1**, 2.
Pässler R. (1974), *Czech. J. Phys.*, **B24**, 322 and (1982) *Czech. J. Phys.*, **B32**, 246.
Pastor R. C., Robinson M. and Akutagawa W. M. (1975), *Mater. Res. Bull.*, **10**, 501.
Pauling L. (1939), *The Nature of the Chemical Bond*, Cornell University Press, Ithica.
Payne S. A., Chase L. L. and Krupke W. F. (1987), *J. Chem. Phys.*, **86**, 3455.
Payne S. A., Chase L. L., Newkirk H. W., Smith L. K. and Krupke W. F. (1988a), *IEEE J. Quant. Elect.*, **QE-24**, 2243.
Payne S. A., Chase L. L. and Wilke G. D. (1988b), *Phys. Rev B*, **37**, 998.
Payne S. A., Chase L. L. and Wilke G. D. (1989a), *J. Lumin.*, **44**, 167.
Payne S. A., Chase L. L., Smith L. K., Kway W. L. and Newkirk H. W. (1989b), *J. App. Phys.*, **66**, 1051.
Payne S. A., Chase L. L., Atherton L. J., Caird J. A., Kway W. L., Shinn M. D., Hughes R. S. and Smith L. K. (1990), *SPIE*, **1223**, 84.
Payne S. A., Smith L. K., DeLoache L. D., Kway W. L., Tassano J. P. and Krupke F. (1994), *IEEE, J. Quant. Elect.*, **30**, 170.
Peale R. E., Summers P. L., Weidner H., Chai B. H. T. and Morrison C. A. (1995), *J. App. Phys.*, **77**, 270, but see also Hong P., Zhang X. X., Peale R. E., Weidner H., Bass M. and Chai B. H. T. (1995), *J. App. Phys.*, **77**, 294.
Pekar S. I. (1954), *Untersuchungen über die Elektronentheorie der Kristalle*, Akademie-Verlag, Berlin.
Pelaez-Millas D., Faulstich A., Baker H. J. and Hall D. R. (1997), *SPIE Proc.*, **3092**, 25.
Pelletier-Allard N. and Pelletier R. (1987), *Phys. Rev.* **B36**, 4425.
Petermann K. (1990), *Opt. Quant. Electron.*, **22**, S199.
Petricevic V., Gayen S. K. and Alfano R. R. (1988), *App. Opt.*, **27**, 4162, (1988), *App. Phys. Lett.*, **52**, 1040 (with Yamagishi K., Anzai H. and Tamaguchi Y.) and 53, 2590 and references therein.
Petricevic V., Gayen S. K. and Alfano R. R. (1989), *App. Opt.*, **28**, 1610 and *Opt. Lett.*, **14**, 612.
Pfann W. G. (1966), *Zone Melting*, (2nd Edn.) John Wiley, New York.
Pham A., Lefaucheur J., Nicholls J., Lutts G. and Chai B. H. T. (1994), *OSA Proceedings of Advanced Solid State Laser Conference*, **15**, 178.
Phillips J. C. and Kleinman L. (1959), *Phys. Rev.*, **116**, 287.
Pinto J. F., Stratton L. W. and Pollock C. R. (1985) *Opt. Lett.*, **10**, 384.
Pinto J. F., Georgio E. and Pollock C. R. (1986), in *Tunable Solid State Lasers II* (Eds. Bugdor A. B., Esterowitz L. and DeShazer L. G.) Springer-Verlag, Berlin and Heidelberg, Springer Series in Optical Sciences, **52**, 261.
Pogatshnik G. T. and Hamilton D. S. (1987), *Phys. Rev.* **B36**, 8251.
Pollnau M. (1997), *IEEE J. Quant. Elect.*, **33**, 1982.

Pople J. A. and Beveridge D. L. (1970), *Approximate Molecular Orbital Theory*, McGraw-Hill, New York.

Pople J. A. and Nesbet R. K. (1954), *J. Chem. Phys.*, **22**, 571.

Pople J. A., Head-Gordon M. and Raghavachari K. (1987), *J. Chem. Phys.*, **87**, 5968.

Pruss D., Huber G., Beimowski A., Laptev V. V., Shcherbakov I. A. and Zharikov E. V. (1982), *App. Phys. B*, **28**, 355.

Pryce M. H. L. (1966), Interaction of Lattice Vibrations with Electrons at Point Defects, in *Phonons* (Ed. Stevenson R. W. H.) Plenum Press, New York, pp 403–448.

Qi X., Illingworth R., Gallagher H. G., Han T. P. J. and Henderson B. (1996a), *J. Cryst. Growth*, **160**, 111.

Qi X., Han T. P. J., Gallagher H. G., Henderson B., Illingworth R. and Ruddock I. (1996b), *J. Phys. Cond. Matter*, **8**, 4837.

Qi X., Han T. P. J., Gallagher H. G. and Henderson B. (1997), *Opt. Commun.*, **140**, 65.

Quarles G. J., Rosenbaum A., Abella I. D., Marquadt C. L. and Esterowitz L. (1990), *Opt. Quant. Electron.*, **22**, S141.

Racah G. (1942), *Phys. Rev.*, **62**, 438.

Racah G. (1949), *Phys. Rev.*, **76**, 1352.

Rand S. C. (1986), in *Tunable Solid State Lasers II* (Eds. Bugdor A. B., Esterowitz L. and DeShazer L. G.) Springer-Verlag, Berlin and Heidelberg, Springer Series in Optical Sciences, **52**, 276.

Rand S. C. and DeShazer L. G. (1985), *Opt. Lett.*, **10**, 481.

Rasheed F., O'Donnell K. P., Henderson B. and Hollis D. B. (1991), *J. Phys.: Cond. Matt.*, **3**, 1915 and **3**, 3825.

Reid J. J. E., Ouwerkerk M. and Beckers L. J. A. M. (1992), *Philips J. Res.*, **46**, 199.

Rhodes W. (1971), *Chem. Phys. Lett.*, **11**, 179.

Rhodes W. (1977), *Chem. Phys.*, **22**, 95.

Ridley B. K. (1978), *J. Phys. C: Sol. St. Phys.*, **11**, 2323.

Riley M. J., Krause E. R., Manson N. B. and Henderson B. (1999), *Phys. Rev.* **B59**, 1850.

Rinzler A. G., Dolan J. F., Kappers L. A., Hamilton D. and Bartram R. H. (1993), *J. Phys. Chem. Solids*, **54**, 89.

Riseberg L. A. and Moos H. W. (1968), *Phys. Rev.*, **174**, 429.

Riseberg L. A. and Weber M. J. (1975), in *Progress in Optics* (Ed. Wolf. E.) North Holland, Amsterdam.

Risk W. P. and Loicano G. M. (1996), *App. Phys. Lett.*, **69**, 311.

Rizvi N. H., French P. M. W. and Taylor J. R. (1992), *Opt. Lett.*, **17**, 1605.

Roothaan C. C. J. (1951), *Rev. Mod. Phys.*, **23**, 69.

Roothaan C. C. J. (1960), *Rev. Mod. Phys.*, **32**, 179.

Roothaan C. C. J. and Bagus P. S. (1963), Atomic self-consistent field calculations by the expansion method, in *Methods in Computational Physics*, 2 (Eds. Alder B., Fernbach S. and Rotenberg M.) Academic Press, New York and London, pp. 95–153.

Rose M. E. (1957), *Elementary Theory of Angular Momentum*, John Wiley, New York and London.

Rosenberger F. E. (1979), *Fundamentals of Crystal Growth I*, Springer Series in Solid-State Sciences, Vol. 5, Springer-Verlag, Berlin, Heidelberg, New York.

Ross M. (1968), *Proc. IEEE*, **56**, 2288.

Rotenberg M., Bivins R., Metropolis N. and Wooten J. K. (1959), *The 3-j and 6-j Symbols*, M.I.T. Press, Cambridge, MA.

Russell D. L. (1998), *Ph.D. Thesis*, University of Strathclyde (unpublished).

Russell D. L., Henderson B., Chai B. H. T., Nicholls J. F. H. and Holliday K. (1997), *Opt. Comm.*, **34**, 398.

Sam C. L., Walling J. C., Jenssen H. P., Morris R. C. and O'Dell E. W. (1980), *Proc. Soc. Phot. Opt. Inst. Eng.*, **247**, 130.

Sanchez A., Fahey R. E., Strauss A. and Aggrawal R. L. (1986), in *Tunable Solid-State Lasers II* (Eds. Bugdor A. B., Esterowitz L. and DeShazer L. G.) Springer-Verlag, Berlin and Heidelberg, Springer Series in Optical Sciences, **52**, 202.

Sandrock T., Danger T., Heumann E., Huber G. and Chai B. H. T. (1994), *App. Phys. B*, **58**, 149.

Sangster M. J. L. and Dixon M. (1976), *Adv. Phys.*, **25**, 247.

Schäffer C. E. and Jorgenson C. K. (1965), *Mol. Phys.*, **9**, 401.

Schawlow A. L. and Devlin G. E. (1961) *Phys. Rev. Lett.*, **6**, 96.

Schawlow A. L. and Townes C. H. (1958), *Phys. Rev.*, **112**, 1940.

Schawlow A. L., Piksis A. H. and Sugano S. (1961), *Phys. Rev.*, **122**, 1469.

Scheil E. (1942), *Zeit. Mettalkunde*, **34**, 70.

Schepler K. L. (1986), in *Tunable Solid State Lasers II* (Eds. Bugdor A. B., Estrowitz L., DeShazer L. G.), Springer-Verlag, Berlin and Heidelberg, Springer Series in Optical Sciences, **52**, 235.

Scheps R. (1992), *Opt. Mat.*, **1**, 1 and references therein.

Scheps R., Gately B. M., Myers J. F., Krasinski J. S., Heller D. F. (1990), *App. Phys. Lett.*, **56**, 2288.

Schneider I. and Marrone M. J. (1979), *Opt. Lett.*, **4**, 390.

Schneider I. and Moss S. C. (1983), *Opt. Lett.*, **8**, 7.

Schneider I. and Pollock C. R. (1983), *J. App. Phys.*, **54**, 6193.

Schröder, U. (1966), *Solid State Comm.*, **4**, 347.

Schulz P. A. and Henion S. R. (1991), *IEEE J. Quant. Elect.*, **27**, 1039.

Scott M. A., Gallagher H. G., Botheroyd I., Han T. P. J., Ferguson A. I. and Henderson B. (1994), *Proc. IEEE Conference Digest*, pp. 57–8.

Scott M. A., Gallagher H. G., Han T. P. J. and Henderson B. (1995), *Radiation Effects and Defects in Solids*, **136**, 47 and (1997), *J. Cryst. Growth*, **172**, 190.

Segawa Y., Sugimoto A., Kim P. H., Namba S., Yamagishi K., Anzai Y. and Yamagichi Y. (1987), *Jap. J. App. Phys.*, **26**, L291.

Seitz F. (1939), *Trans. Faraday Soc.*, **35**, 74.

Sennaroglu A. and Pollock C. R. (1991), *J. Lumin.*, **47**, 217.

Sevast'yanov B. K., Bagdasarov Kh. S., Pasternak L. B., Volkov S. Yu and Orekhova V. P. (1973), *JETP Lett.*, **17**, 47.

Shand M. L. (1983), *J. App. Phys.*, **54**, 2602.

Shand M. L. and Jenssen H. P. (1983), *IEEE J. Quant. Elect.*, **QE-19**, 480.

Shand M. L. and Walling J. C. (1982), *IEEE J. Quant. Elect.*, **QE-18**, 1152.

Shcherbakov I. A. (1986), in *Tunable Solid State Lasers II* (Eds. Bugdor A. B., Estrowitz L., DeShazer L. G.), Springer-Verlag, Berlin and Heidelberg, Springer Series in Optical Sciences, **52**, 293.

Shen Y. R. (1984), *The Principles of Nonlinear Optics*, John Wiley, Interscience, New York.

Shen Y. R., Hömmerich U. and Bray K. L. (1997), *Phys. Rev.* **B56**, R473.

Shimokozono M., Sugimoto N., Tate A., Katoh Y., Tanno M., Fukuda S. and Ryuoh T. (1996), *App. Phys. Lett.*, **68**, 2177.

Shinada M., Sugano S. and Kushida T. (1966), *J. Phys. Soc.* (*Jap*), **21**, 1342, first identified the role of odd parity distortions in promoting transition in $Cr^{3+}:Al_2O_3$.

Siegman A. (1986), *Lasers*, University Science Books, California.

Silversmith A. J., Lenth W. and Macfarlane R. M. (1987), *App. Phys. Lett.*, **51**, 1977.

Simánek E. and Sroubek Z. (1972), Covalent effects in EPR spectra-hyperfine interactions, in *Electron Paramagnetic Resonance* (Ed. Geschwind S.), Plenum Press, New York and London, pp. 535–74.

Simpson J. H. (1949), *Proc. Roy. Soc.* (*London*), **A197**, 269.

Sinanoglu O. (1964), *Adv. Chem. Phys.*, **6**, 315.

Sinclair B. D. (1999), *Opt. Mater.*, **11**, 217.

Slater J. C. (1929), *Phys. Rev.*, **34**, 1293.

Slater J. C. (1930), *Phys. Rev.*, **36**, 57.
Slater J. C. (1951), *Phys. Rev.*, **81**, 385.
Slater J. C. (1974), *The Self-Consistent Field for Molecules and Solids: Quantum Theory of Molecules and Solids, IV*, McGraw-Hill, New York.
Smart R. G., Hanna D. C., Tropper A. C., Davey S. T., Carter S. F. and Szebesta D. (1991), *Electron. Lett.*, **27**, 1307.
Smith A., Silversmith A. J., Manson N. B. and Henderson B. (2000), *Opt. Comm.* (in press).
Smith L. K., Payne S. A., Kway W. L., Chase L. L. and Chai B. H. T. (1992), *IEEE J. Quant. Elect.*, **28**, 2612.
Snitzer E. and Woodcock R. (1965), *App. Phys. Lett.*, **6**, 45.
Sobelman I. I. (1972), *Introduction to the Theory of Atomic Spectra*, Pergamon Press, Oxford, New York, Toronto, Sydney and Braunschweig.
Sobelman I. I. (1979), *Atomic Spectra and Radiative Transitions*, Springer-Verlag, Berlin, Heidelberg and New York.
Sorokin P. P. and Stevenson M. J. (1961), *IBM Journ. Res. Develop.*, **5**, 56.
Sorokina I. T., Sorokin E., Winter E., Cassanho A., Jenssen H. P., Noginov M. A. and Szipöcs R. (1996), *Opt. Lett.*, **21**, 204 and 1165.
Souria J. C., Romero R., Borel C., Wyon C., Li C. and Moncorgé R. (1994), *App. Phys. Lett.*, **64**, 1189.
Spaeth J.-M., Niklas J. R. and Bartram R. H. (1992), *Structural Analysis of Point Defects in Solids*, Springer-Verlag, Berlin and Heidelberg.
Stalder M., Bass M. and Chai B. H. T. (1992), *J. Opt. Soc. Am.*, **B9**, 2271.
Stanley A. T., Harris E. A., Searle T. M. and Parker J. M. (1993), *J. Non-Cryst. Sol.*, **161**, 235.
Steinbrugge K. B., Henningsen T., Hopkins R. H., Mazelsky R., Melamed N. T., Reidel E. P. and Roland G. W. (1972), *App. Opt.*, **11**, 999.
Stevens K. W. H. (1952), *Proc. Phys. Soc.*, **A65**, 209.
Stiles L. F., Fontana M. P. and Fitchen D. B. (1970), *Phys. Rev.* **B2**, 2077.
Stockbarger D. C. (1963), *Rev. Sci. Instrum.*, **7**, 133.
Stoneham A. M. (1969), *Rev. Mod. Phys.*, **41**, 82.
Stoneham A. M. (1985), *Theory of Defects in Solids*, 2nd Edition, Oxford University Press.
Stoneham A. M. and Bartram R. H. (1978), *Sol. St. Electron.*, **21**, 1325.
Storm M. E., Gettemy M. E., Barnes N. P., Cross P. L. and Kokta M. R. (1989), *App. Opt.*, **28**, 408.
Strek W., Deren P. and Jezowska-Trzebiatowska B. (1988), *J. Lumin.*, **40 & 41**, 421, but see also (1987), *J. de Physique*, **48**, 455.
Struck C. W. and Fonger W. H. (1975), *J. Lumin.*, **10**, 1, but see also Fonger W. H. and Struck C. W. (1978), *J. Chem. Phys.*, **69**, 4171.
Struck C. W. and Herzfeld F. (1966), *J. Chem. Phys.*, **44**, 464.
Struve B. and Huber G. (1985), *Appl. Phys. B.*, **36**, 195, first identified the effects of disorder in Cr^{3+} doped garnets.
Struve B., Huber G., Laptev V. V., Shcherbakov I. A. and Zharikov E. V. (1983), *Appl. Phys. B*, **30**, 117.
Sturge M. D. (1967), The Jahn-Teller Effect in Solids, in *Solid State Physics* (Eds. Seitz F., Turnbull D. and Ehrenreich H.) Academic Press, New York and London, **20**, pp. 91–211.
Sturge M. D. (1973), *Phys. Rev.* **B8**, 6.
Sugano S. and Tanabe Y. (1958), *J. Phys. Soc. Jap.*, **13**, 880.
Sugano S., Tanabe Y. and Kamimura H. (1970), *Multiplets of Transition-Metal Ions in Crystals*, Academic Press, New York and London.
Sumi H. (1980), *J. Phys. Soc. Jap.*, **49**, 1701.
Sumi H. (1982a), *Sol. St. Commun.*, **43**, 73.
Sumi H. (1982b), *J. Phys. Soc. Jap.*, **51**, 1745.
Summers P. L., Weidner H., Peale R. E. and Chai B. H. T. (1994), *J. App. Phys.*, **75**, 2194.
Sutherland J. M., French P. M. W., Taylor J. R. and Chai B. H. T. (1996), *Opt. Lett.*, **21**, 797.

Sverev G. M. and Shestakov A. V. (1989), *OSA Proceedings on Tunable Solid State Lasers* (Eds. Shand M. L. and Jensen H. P.), **5**, 66.
Swank R. L. and Brown F. C. (1963), *Phys. Rev.*, **130**, 34.
Sytsma J., Piehler D., Edelstein N. M., Boatner L. A. and Abraham M. M. (1993), *Phys. Rev.* **B 47**, 14 786.
Szabo A. and Ostlund N. S. (1989), *Modern Quantum Chemistry*, 1st ed. revised, McGraw-Hill, New York.
Tanabe Y. and Sugano S. (1954), *J. Phys. Soc. Jap.*, **9**, 753 and 766.
Tang D. Y. and Route R. K. (1988), *J. Cryst. Growth*, **91**, 81.
Taylor J. R. (Ed.), *Optical Solitons – Theory and Experiment*, Cambridge Studies in Modern Optics (Eds. Knight P. L. and Miller A., Cambridge University Press, 1994) Cambridge, New York, Melbourne and Madrid.
Teal G. K. and Buehler E. (1952), *Phys. Rev.*, **87**, 190.
Teal G. K. and Little J. B. (1950), *Phys. Rev.*, **78**, 647.
Tiller W. A. and Yen C. T. (1991), *J. Cryst. Growth*, **109**, 120.
Tinkham M. (1964), *Group Theory and Quantum Mechanics*, McGraw Hill, New York, San Francisco, Toronto and London.
Tissue B. M., Lu L. and Yen W. M. (1990), *J. Lumin.*, **45**, 20.
Tolksdorff W. (1994), in *Handbook of Crystal Growth–Bulk Crystal Growth* (Ed. Hurle D. T. J.), North-Holland, Amsterdam and London, **2A**, 563.
Tong Y. P., French P. M. W., Taylor J. R. and Fujimoto J. O. (1997), *Opt. Comm.*, **136**, 235.
Townsend P. D., Chandler P. J. and Zhang L., *Optical Effects of Ion Implantation* [Cambridge Studies in Modern Optics, (Eds. Knight P. L. and Miller A.). Cambridge University Press, (1992)] Cambridge, New York, Melbourne and Madrid.
Trenec G. P., Nacher P. J. and Leduc M. (1982), *Opt. Comm.*, **43**, 37.
Tropf W. J., Thomas M. and Harris T. J. (1995), in *Handbook of Optics*, Vol. II (Principal Ed. Bass M.) McGraw-Hill, New York, Chapter 33.
Tucker A. W., Birnbaum M., Fincher C. L. and Erler J. W. (1977), *J. App. Phys.*, **48**, 4907.
Vagin Yu S., Marchenko V. M. and Prokhorov A. M. (1969), *Sov. Phys. JETP*, **28**, 902.
Vail J. M. (1990), *J. Phys. Chem. Solids*, **51**, 589.
Van Puymbroeck W., Andriessen J. A., Nistor S. V. and Schoemaker D. (1981), *Phys. Rev.* **B24**, 29.
Van Vleck J. H. (1935), *J. Phys. Chem.*, **3**, 803 and 807, (1937), *J. Phys. Chem.*, **41**, 67.
Van Vleck J. H. and Sherman A. (1935), *Rev. Mod. Phys.*, **7**, 167.
Varsanyi F., Wood D. L. and Schawlow A. L. (1959), *Phys. Rev. Lett.*, **3**, 544.
Verdun H. P., Thomas L. M., Andrauskas D. M., McCollum T. and Pinto A. (1988), *App. Phys. Lett.*, **53**, 2593.
Viebahn V. W. (1971), *Z. Anorg. Allg. Chem.*, **386**, 335.
Vieira N. D. and Mollenauer L. F. (1985), *IEEE J. Quant. Elect.*, **21**, 195.
Wadt W. R. and Hay P. J. (1985), *J. Chem. Phys.*, **82**, 284.
Wagner M. (1982), *J. Phys. C: Sol. St. Phys.*, **15**, 5077.
Wahl A. C. and Das G. (1977), The multiconfiguration self-consistent field method. in *Methods of Electronic Structure Theory* (Ed. Schaefer (III) H. F.), Plenum Press, New York, p. 51.
Wakita S., Suzuki Y. and Hirai M. (1981), *J. Phys. Soc. Japan*, **50**, 2781.
Walling J. C. (1987), in *Tunable Lasers* (Eds. Mollenauer L. F. and White J. C.) Springer-Verlag, Berlin, Heidelberg and New York, Springer Topics in Applied Physics, **59**, 331.
Walling J. C., Jensen H. P., Morris R. C., O'Dell E. W. and Peterson O. G. (1979), *Opt. Lett.*, **4**, 182.
Walling J. C., Petersen O. G., Jensen H. P., Morris R. C. and O'Dell W. (1980), *IEEE J. Quant. Elec.*, **16**, 1302.
Walling J. C., Heller D. F., Samelson H., Harter D. J., Peta J. P. and Morris J. C. (1985), *IEEE, J. Quant. Elect.*, **QE-21**, 1568.

Wandt D., Gellermann W., Lüty F. and Welling H. (1986), in *Tunable Solid State Lasers II* (Eds. Bugdor A. B., Esterowitz L. and DeShazer L. B.) Springer-Verlag, Heidelberg, Springer Series in Optical Sciences, **52**, 252.

Wannemacher R. and Meltzer R. S. (1989), *J. Lumin.*, **43**, 251.

Wang G., He M. and Luo Z. (1991), *Mat. Res. Bull.*, **26**, 1085.

Wang G., Gallagher H. G., Han T. P. J. and Henderson B. (1995), *App. Phys. Lett.*, **67**, 3906.

Wang G., Gallagher H. G., Han T. P. J. and Henderson B. (1996), *J. Cryst. Growth*, **163**, 272.

Wang G., Gallagher H. G., Han T. P. J. and Henderson B. (1997a), *J. Cryst. Growth*, **181**, 48.

Wang G., Gallagher H. G., Han T. P. J., Henderson B., Yamaga M. and Yosida T. (1997b), *J. Phys.: Cond. Mater.*, **9**, 1649.

Wang W., Zou Q., Geng Z. and Feng D. (1986), *J. Cryst. Growth*, **79**, 706.

Watts R. K. (1975), in *Optical Properties of Ions in Solids* (Ed. DiBartolo B.) Plenum Press, New York, p. 307.

Weber H. P. and Lüthy W. (1986), in *Tunable Solid State Lasers II* (Eds. Bugdor A. B., Esterowitz L., DeShazer L. G.) Springer-Verlag, Heidelberg and Berlin, Springer Series in Optical Sciences, **52**, 30.

Weber M. J. (1971), *Phys. Rev.* **B4**, 2932.

Weber M. J. (1981), in *Laser Spectroscopy of Solids* (Eds. Yen W. M. and Selzer P.M.), Springer Topics in Applied Physics, **49**, 189.

Weber W. and Dick B. G. (1969), *Phys. Status Solidi*, **36**, 723.

Weeks J. D. and Rice S. A. (1968), *J. Chem. Phys.*, **49**, 2741.

Weeks J. D., Tully J. C. and Kimmerling L. C. (1975), *Phys. Rev B*, **12**, 3286.

Wegner T. and Petermann K. (1989), *App. Phys. B.*, **49**, 275.

Weidner H., Summers P. L., Peale R. E. and Chai B. H. T. (1994), *Mat. Res. Soc. Symp. Proc.*, **329**, 135.

Weissbluth M. (1978), *Atoms and Molecules*, Academic Press, New York and London.

Weissman Y. and Jortner J. (1978), *Phil. Mag B*, **37**, 24.

Welford D. and Moulton P. F. (1988), *Opt. Lett.*, **13**, 975.

Weyl H. (1931), *The Theory of Groups and Quantum Mechanics*, Princeton University Press, Princeton.

Weyn J. P., Jensen T. and Huber G. (1994), *IEEE J. Quant. Elect.*, **30**, 913.

Whitley T. J., Millar C. A., Wyatt R., Brierley M. C. and Szebesta D. (1991), *Elect. Lett.*, **27**, 1785.

Wigner E. P. (1931), *Gruppentheorie*, Friedrich Vieweg und Sohn, Braunschweig.

Wigner E. P. (1934), *Phys. Rev.*, **46**, 1002.

Wigner E. P. (1959), *Group Theory*, Academic Press, New York and London.

Williams R. T., Song K. S., Faust W. L. and Leung C. H. (1986), *Phys. Rev.* **B33**, 7232.

Wilson B. A., Hegarty J. and Yen W. M. (1978), *Phys. Rev. Lett.*, **41**, 268.

Wilson B. A., Hegarty J., Imbusch G. F. and Yen W. M. (1979), *Phys. Rev.* **B19**, 4238.

Winkler G. (1981), in *Magnetic Garnets* (Tracts in Pure and Applied Physics), Vieweg, Braunschweig, p. 23.

Wojtowicz A. J., Kazmierczak M., Lempicki A. and Bartram R. H. (1989), *J. Opt. Soc. Am. B*, **6**, 1106.

Wojtowicz A. J., Grinberg M. and Lempicki A. (1991), *J. Lumin.*, **50**, 231.

Wolfsberg M. and Helmholz L. (1952), *J. Chem. Phys.*, **20**, 837.

Woods A. M., Sinkovits R. S., Charpie J. C., Huang W. L., Bartram R. H. and Rossi A. R. (1993), *J. Phys. Chem. Solids*, **54**, 543.

Woods A. M., Sinkovits R. S. and Bartram R. H. (1994), *J. Phys. Chem. Solids*, **55**, 91.

Woods B. W., Payne S. A., Marion J. E., Hughes R. S. and Davis L. E. (1991), *J. Opt. Soc. Amer. B.*, **8**, 970.

Wybourne B. G. (1965), *Spectroscopic Properties of Rare Earths*, John Wiley, New York, London and Sydney.

Yamaga M., Marshall A., O'Donnell K. P., Henderson B. and Miyazaki Y. (1988), *J. Lumin.*, **39**, 335.

Yamaga M., Henderson B., Marshall A., O'Donnell K. P. and Cockayne B. (1989a), *J. Lumin.*, **43**, 139.

Yamaga M., Henderson B. and O'Donnell K. P. (1989b), *J. Phys.: Cond. Matter*, **1**, 9175.

Yamaga M., Henderson B. and O'Donnell K. P. (1990a), *J. Lumin.*, **46**, 397 and (with A. Marshall) **47**, 65

Yamaga M., Henderson B., O'Donnell K. P., Trager-Cowan C. and Marshall A. (1990b), *App. Phys. B.*, **50**, 425.

Yamaga M. Gao Y., Rasheed F., O'Donnell K. P., Henderson B. and Cockayne B. (1990c), *App. Phys. B.*, **51**, 329.

Yamaga M., Henderson B. and O'Donnell K. P. (1991a), *Appl. Phys. B.*, **52**, 122, and (with Rasheed F. F., Gao Yue and Cockayne B.) **52**, 225.

Yamaga M., Henderson B., O'Donnell K. P. and Gao Yue (1991b), *Phys. Rev.* **B44**, 652.

Yamaga M., Henderson B. and O'Donnell K. P. (1992), *Phys. Rev.* **B46**, 3273.

Yamaga M., Kodama N. and Naitoh Y. (1994). *Proc. XIII Int. Conf. on Defects in Insulating Materials* (Eds. Kanert O. and Spaeth J. M.) Singapore: World Scientific, p. 173.

Yamaga M., Macfarlane P. I., Henderson B., Holliday K., Takeuchi H., Yosida T. and Fukui M. (1998a), *J. Phys. Cond. Matt.*, **9**, 569.

Yamaga M., Lee D., Henderson B., Han T. P. J., Gallagher H. G. and Yosida M. (1998b), *J. Phys. Cond. Matt.*, **10**, 1.

Yamaga M., Henderson, B., Holliday K., Yosida T., Fukui M. and Kindo K. (1999) *J. Phys. Cond. Matt.*, **11**, 10499

Yamamoto K., Mizuuchi K., Kitaoka Y. and Kato M. (1993), *App. Phys. Lett.*, **62**, 2599.

Yang K. H. and DeLuca J. A. (1977), *App. Phys. Lett.*, **31**, 594.

Yang Y., von der Osten W. and Lüty F. (1985), *Phys. Rev.* **B32**, 2724.

Yangyang Ji, Shuqing Z. and Yijing H. (1991), *J. Cryst. Growth*, **112**, 283.

Yariv A, (1975), *Quantum Electronics*, John Wiley, New York.

Yen W. A. and Selzer P. M. (1981), *Laser Spectroscopy of Solids*, Springer Topics in Applied Physics, **49**, 141.

Yokota M. and Tanimoto O. (1967), *J. Phys. Soc.* (*Jap*), **22**, 779.

Zayhowski J. J. (1999), *Opt. Mater.*, **11**, 255.

Zayhowski J. J. and Harrison J. (1996), *Handbook of Photonics*, Ch. 8 (Ed. Gupta M. C.) CRC Press, Boca Raton.

Zener C. (1932), *Proc. R. Soc. London*, **A137**, 693.

Zener C. (1933), *Proc. R. Soc. London*, **A140**, 660.

Zhang X. X., Hong P., Bass M. and Chai B. H. T. (1993), *App. Phys. Lett.*, **63**, 2606.

Zhang X. X., Loutts G. B., Bass M. and Chai B. H. T. (1994), *App. Phys. Lett.*, **64**, 10, and **64**, 3205 (with Hong P. and Lefaucheur J.).

Zhao Y. and Poole S. (1994), *Elect. Lett.*, **30**, 967.

Zharikov E. V. (1986), in *Tunable Solid-State Lasers II* (Eds. Bugdor A. B., Esterowitz L. and DeShazer L. G.) Springer-Verlag, Berlin and Heidelberg, Springer Series in Optical Sciences, **52**, 64.

Zharikov E. V. (1991), *SPIE Proceedings*, **1839**, 46.

Zharikov E. V., Zagumennyi A. I., Lutts G. B., Smirnov V. A., Sorokina I. T. and Shcherbakov I. A. (1991), *Laser Physics* (English translation), **1**, 216.

Zhou F., De La Rue R. M., Ironside C. N., Han T. P. J., Henderson B. and Ferguson A. I. (1992), *Elect. Lett.*, **28**, 2041.

Ziegler T., Rauk A. and Baerends E. J. (1976), *Chem. Phys.*, **16**, 209.

Ziegler T., Rauk A. and Baerends E. J. (1977), *Theoret. Chim. Acta.*, **43**, 261.

Index